THE USE AND SIGNIFICANCE
OF PESTICIDES IN THE ENVIRONMENT

THE USE AND SIGNIFICANCE OF PESTICIDES IN THE ENVIRONMENT

F. L. McEWEN

G. R. STEPHENSON

University of Guelph
Guelph, Ontario, Canada

A WILEY-INTERSCIENCE PUBLICATION

JOHN WILEY & SONS, New York • Chichester • Brisbane • Toronto

Library of Congress Cataloging in Publication Data:

McEwen, Freeman Lester, 1926–
  The use and significance of pesticides in the
environment.

  "A Wiley-Interscience publication."
  Includes bibliographical references and index.
  1. Pesticides. 2. Pesticides—Environmental
aspects. I. Stephenson, Gerald Robert, 1942–
joint author. II. Title.
SB951.M38   363   78-23368
ISBN 0-471-03903-9

Printed in the United States of America

10 9 8 7 6 5 4 3

To the students of 34-303 "Pesticides in the Environment." Their search for an objective appraisal of this field made teaching the course a pleasure and dramatized the need for a text that placed pesticides in perspective.

The subject of pesticides in the environment is a large and controversial one. Its magnitude derives from the extent of its two major components. First, the number of different chemicals being used runs in the hundreds, and each of these is available in many formulations. Each chemical has its own set of chemical and biological properties, and in terms of environmental significance these may be modified by the formulation dispensed. The second, even larger and less well understood component is the environment. Included in this are all the biological, physical, and chemical components that make up our surroundings and the interactions among these dictated by principles that we can grasp only in a superficial way.

The controversial aspect of the subject also derives from two main components. In the first instance we do not know if or in what way pesticides may interfere with normal life processes and how effects might be translated into evolutionary patterns or species displacement over long periods of time. In the second place there is much argument in both the scientific and lay community about the necessity of using pesticides as widely as now practiced. This is especially true in North America, where most of the pesticide usage is in the production of food and fiber and where our land resource is more than adequate to meet our current needs. Fueling this latter controversy has been a lack of informative dialogue. The points of view that have received greatest press coverage have been those taking extreme positions of advocacy based on selected references and incidents, and little attention has been directed toward informing the public through a rational assessment of benefits and risks. Those on each side of the controversy have been equally to blame. In addition it is difficult for the majority of people in North America to appreciate the need for pesticides to protect food and fiber. Since the dawning of the modern age of pesticides in the late 1940s, food and fiber have been protected so well that products showing extensive damage from pests have not appeared in our supermarkets. People under the age of 50 have never done their grocery shopping in an era when products infested with insects or rotting from plant diseases found their way into most shopping baskets.

The failure of the scientific community to inform the public properly on the pesticides issue reflects, in part, the often too narrow disciplinary training provided within the universities. Many biologists have spoken out sharply against pesticides while knowing little about the products they were condemning. Similarly pesticides experts have extolled the wonders of their products for specific purposes without regard for or knowledge of their effects in a biological community.

This situation must not continue. Pesticides are an important and necessary part of the modern technology that ensures our food supply. Students in biology, agriculture, and forestry must develop a broader background concerning these new tools and carry this to the workplace to ensure that benefits are multiplied and risks reduced.

This book attempts to provide the student with a broad assessment of pesticides. It embraces a course we have offered to senior undergraduate students for several years with more than 200 students enrolled in each of the past 3 years. We believe an understanding of pesticides must include the reasons for their use in terms of perceived and demonstrable benefits, the nature of the pesticides involved and thus their potential for good or harm, the penetration of pesticides into the environment and residence there, the effects of such pesticides on target and nontarget organisms, and the legislative controls on pesticide use.

We recognize that it is not easy to cover such a broad series of topics in one text. We have tried, however, because our experience attests to the difficulty of teaching this material when no comprehensive text is available to the student and he must rely on numerous references, each covering a different aspect of the subject. This book does not attempt to provide the depth of coverage included in the many excellent reviews on specific aspects of the subject or to include all references. Instead we have tried to present an overview and selected references to demonstrate the more important points. We hope we have selected such references wisely and that the student is provided an objective text from which he will benefit whether his interest lies in the development of pest management programs, in the management of fisheries and wildlife, in environmental quality, or in the more general pursuit of knowledge about the world in which he lives and the factors that may operate to his benefit or detriment.

F. L. MCEWEN
G. R. STEPHENSON

Guelph, Ontario, Canada
January 1979

ACKNOWLEDGMENTS

We are indebted to R. Frank, C. R. Harris, J. R. W. Miles and D. Pimentel for supplying manuscripts and permitting us to use data not yet published. L. V. Edgington, Jr., provided valuable suggestions for the chapter on fungicides. D. Cornwell, H. Daniecki, R. Kostenuk, and M. Mitchell did the typing on the initial copies, and S. Helson typed the final draft. D. J. Hamilton prepared the illustrations and figures. We thank all of these people for their excellent assistance.

The senior author did much of the writing while on sabbatical at the University of Hawaii. He thanks W. C. Mitchell and the other members of the department of entomology for use of library and other facilities and for making his stay at that institution so pleasant. Finally we thank the University of Guelph and our colleagues in the department of environmental biology for advice and encouragement and our families for support and understanding during the long period when this book took precedence over normal family activities.

F. L. McE.
G. R. S.

CONTENTS

CHAPTER

THE USE AND SIGNIFICANCE
OF PESTICIDES IN THE ENVIRONMENT

1

INTRODUCTION

Since the beginning of history, mankind has had to compete with more than 1 million other species for the earth's goods essential to survival. A review of history leaves no question but that the struggle has been harsh, and from time to time catastrophic episodes of famine, pestilence, and war and the vicissitudes of nature have made his trek into modern civilization a torturous route. Along the way mankind learned to cultivate crops, raise livestock, and build shelters. In this way he was able to sustain life in areas where nature was too harsh for an unplanned existence, man's potential for survival improving as he devloped technology to cope with his surroundings.

Man's competitors adapted quickly to the human habitat. Many insects, rodents, and birds found his crops to their liking, and his dwellings provided shelters not only for man but also for a host of his competitors. As man continued to modify his surroundings, the competition became more intense, and organized patterns of "pest control" evolved. At first these efforts were minimal: scarecrows to discourage birds, hand removal, various types of traps for vertebrate and invertebrate pests, and a relentless use of the hoe and the cultivator to remove weeds from cultivated crops.

To a large extent it was with this arsenal that man approached the twentieth century. There were, however, some signs of change. Arsenic, known from ancient times as a potent poison, was found toxic to the Colorado potato beetle, an insect that had changed from a relatively uncommon species on wild Solanaceae in the mountain foothills to a destructive pest of the Irish potato, a main food item for North Americans. Whereas Paris Green was the first, other arsenicals would

1

soon be found and used for insect control on a wide variety of crops. This development was followed by the use of oils and salt for weed control and copper and mercury compounds for control of certain plant diseases. Two insecticides of plant origin, rotenone, from the roots of a South American plant, and pyrethrum, from the flowers of a chrysanthemum in Asia, were included in the nineteenth century insecticides and were useful for insect control in both plants and animals. Sulfur was used also, sometimes for disease control and sometimes for insect or mite control on plants and for mite and tick control on animals.

When viewed from the modern era, developements in pesticides* were slow, and by 1939, only about 30 were registered for use in the United States. Application techniques were limited largely to small sprayers and dusters, and applying pesticides was time consuming and the acreage of crops treated relatively small.

Under these circumstances the environmental impact of pesticides was small and restricted to the immediate area of application. Farm animals were often poisoned with the arsenicals, but these poisonings were confined largely to stock of the farmer applying the pesticide, and this emphasized to the grower the need for care in handling such materials. It would be learned later that the arsenicals were persistent and that their accumulation in some soils would lead to later problems. Since their reputed heavy use was, however, confined to orchards, the acreage involved was not great and was, from an environmental point of view, of little consequence.

World War II placed a strain on western civilization of unprecedented proportion. Able-bodied men went to war, and those left behind were given the imposing task of providing food and maintaining the factories to fuel the war. In addition, naval blockades and shipping priorities restricted the availability of imported pesticides, and in both North America and Europe pest control became more difficult. In Germany research to find replacement for pesticides no longer available was under way, a research effort that would lead to the development of the organophosphate insecticides. It was in this climate that 1,1,1,-trichloro-2-2-bis(parachlorophenyl)ethane (DDT) was discovered, an insecticide that would dominate insect control programs for 25 years and make pesticide use a national issue in many western countries.

DDT was the first of the "new" insecticides. It had been synthesized as part of a doctoral study by Zeidler in 1874, but its insecticidal

* The word "pesticides" is used in the generic sense and includes insecticides, fungicides, herbicides, and other categories of compounds used to kill pests.

properties were not recognized until 1939, when Müller, searching for something to control clothes' moths, repeated Zeidler's synthesis. This discovery of DDT and its effectiveness against a wide variety of arthropods earned for Müller the Nobel Prize in 1948. In a paper entitled, "A Century of DDT," Dr. R. L. Metcalf (1973) reviews the history of this compound and its impact on the world. The coincidence of its discovery and the outbreak of World War II meant that its first major use was in the war effort, where, in 1943, it eliminated an outbreak of typhus in Naples and saved the allied army from a disease that had been a major factor in the outcome of many military campaigns. The effectiveness of DDT against a wide range of insects led to its use against insect vectors of important world diseases such as plague, yellow fever, and, most importantly, malaria. In Sardinia its effectiveness against malaria was so outstanding that hopes for eradication of the disease were justified, and a global eradication was proposed by the World Health Assembly in 1955. The early success of this endeavor was outstanding. Residual treatment of homes in malarious climates resulted in the eradication of malaria in 37 countries by 1972 and a drastic reduction in cases in an additional 80 countries, bringing a heretofore undreamed of malaria relief to 1.5 billion people. In India alone the incidence of malaria decreased from 100 million cases annually in 1933–1935 to 150,000 cases in 1966, and death from this cause was reduced from $\frac{3}{4}$ million to 1500.

The effectiveness of DDT in agriculture was no less spectacular. During the early 1950s its low cost, coupled with the outstanding insect control it provided, made it the material of choice for most insect problems. In addition, its extensive use on military personnel and in homes had proved that it was of low hazard to human beings. The results were outstanding increases in yield and, in an unprecedented manner, the adoption of pesticides for insect control. In the United States alone production of DDT increased to more than 100 million lb annually, a production equal to that of all pesticides prior to World War II.

Developments were occurring also in other areas. Pesticide application by air had been carried out to a limited extent before 1939, but the availability of surplus aircraft after the war led to a revolution in application techniques. Aircraft were fitted to deliver either dust or spray formulations, and although the percentage of pesticide landing on the target was less than with good ground equipment, the aircraft had the advantage of speed and were caplable of treating a vast acreage in a short period of time. In addition, their operation was not restricted by muddy fields, and the damage caused by moving ground sprayers through agricultural plantings was eliminated. The airplane opened a new territory for pest control—the forest. Thus forest pests like the gyspy moth, the tussock moth, the hemlock looper, and the spruce budworm, against which man had been comparatively helpless, now became amenable to treatment. Using the airplane, man was now able to effect control. Thus the relatively small acreage treated annually before 1939 expanded to huge tracts with the availability of inexpensive insecticides and a rapid method for their distribution.

DDT was not alone. The chemist now had a nucleus around which to build, and in quick succession a number of compounds structurally related to DDT appeared. In addition the cyclodienes were synthesized, and these did for soil insects what DDT had done for foliar pests. During the next 20 years these insecticides were joined by the organophosphates, the carbamates, and, more recently, new compounds of various design attempting to mimic naturally occurring insect hormones but used in a way to disrupt their normal life processes. As a result hundreds of compounds in thousands of formulations are available and are used by all of us to protect our possessions from insect damage.

The rapid developments in insecticides were paralleled by new discoveries of compounds effective for plant disease and weed control. In the latter case the search was escalated by rising labor costs that made hand-weeding too expensive, and the pioneering discoveries of 2,4-D and atrazine provided selective herbicides that contributed immeasurably to the mechanization of crop production and the efficiency of labor inputs. As a result herbicides are the most extensively used pesticides in agriculture, their use in North America exceeding that of insecticides and fungicides combined. The new pesticide technology ushered in a wide range of compounds for other pest control problems. Products for rodent control, bird control, fish control, nematode control, and a host of other needs are available with formulations designed for practically any conceivable need.

The development of the modern pesticides, while undeniably a triumph in science and technology, has not been without its problems. DDT had been used only a few years when insects resistant to the compound developed. In addition it was learned rather early that nature, in its tremendous diversity, had provided a wide variety of predator and prey and that the balance in this relationship could be altered drastically by a broad spectrum insecticide like DDT. If all predators and prey were equally exposed and equally sensitive to DDT, this might not have happened, but this was not the case. Despite its broad toxicity to arthropods, selectivity was apparent. For example,

DDT killed predaceous beetles much more readily than it killed the aphids, scales, and mites upon which these fed. The result was the emergence of "new pest species" in situations where these had not been pests before. To counteract this, new pesticides were developed that could be added to DDT to kill those species DDT permitted to survive, but the results were use of more pesticides, developement of more resistant species to more compounds, and spiraling costs. It is easy to see how the "treadmill" developed but less easy to see how to get it stopped.

If the interplay of DDT and related pesticides had been confined to the predator-prey relationships of treated areas, its undesirable effects might never have received international prominence. It did not. The chemical nature of DDT is such that it persists in nature, its lipophilic properties resulting in its movement to living tissues. It accumulates in fat, and it was soon descovered that, where DDT had been used on animal feeds, residues occurred in tissues of the animals consuming such feeds and in products such as milk or eggs that they produced. Such residues could be quite high, and despite the pesticide's having a moderately low toxicity, it became necessary to establish maximum permitted levels to protect the health of the consumer. The problem was critical in milk, since this may be a major food source, expecially for the very young, the ill, or the aged, extra safeguards being needed because of possible extreme sensitivities in these categories of people. While these matters were being resolved, a much larger issue was developing: environmental contamination with DDT and the significance of such contamination to the natural inhabitants.

The issue was not new. It had been raised in a speculative way by Wigglesworth (1945) and had surfaced repeatedly when observers noted bird or fish mortality associated with large and small spray programs. Numerous reports concerning these incidents appeared in the literature, but they tended to be regarded as local and did not cause great concern. This attitude was shaken a bit by data from New Brunswick indicating massive fish kills associated with DDT spraying for spruce budworm control (Kerswill and Elson, 1955) and by other reports of widespread fish mortality. The clarion call for action was the publication of *Silent Spring* (Carson, 1962), a highly readable book emphasizing dramatically some instances of side effects from pesticide use and the ominous portents around us. Reports of reproductive failure in lake trout (Burdick et al., 1964) a catastrophe in wildlife in Clear Lake (Hunt and Bischoff, 1960) and the residues of DDT in a wide variety of wildlife (Keith and Hunt, 1966) left little doubt that widespread use of DDT had meant widespread contamination of the environment, the results of which could not be assessed. Improved analytical techniques demonstrated DDT, or its breakdown product DDE (1,1,-dichloro-2-2 bis (parachloropenyl)ethane), in measurable quantities in the air, oceans, and land.

DDT was not the only pesticide found widely in nontarget organisms. Dieldrin, heptachlor epoxide, and BHC (1,2,3,4,5,6-hexachlorocyclohexane) were detected in many wildlife tissues, and other pesticides such as endrin and endosulfan were being found causally associated with fish kills. In addition, several species of raptorial birds were noted to be declining in population, studies by Ratcliffe (1969), in England, pointing to pesticides as the responsible agents. In North America attention was focused on the Peregrine falcon, and a conference held in 1965 left little doubt that this bird was disappearing owing to poor breeding success brought about by exposure to pesticides and especially to DDT (Hickey, 1969).

More recently it has become clear that several pesticides have affected wildlife adversely. Others have caused a significant health problem among workers. Increasingly there is concern that exposure to some pesticides may cause human cancer or teratogenic or mutagenic effects. As a result DDT and several other compounds have been greatly restricted in use or banned in many western countries. The demand for effective pest control continues, however, and replacements for those pesticides that have been restricted are not without their problems. Indeed there are instances where a return to DDT has been argued (Devlin, 1974). Although it was banned in the United States in 1972, it was reinstated for use against the tussock moth in Oregon, Idaho, and Washington in 1974 (Gibney, 1974).

Numerous attempts are being made to curtail the use of all pesticides. Various alternatives are suggested, including a return to less rigid standards in food sanitation. The preponderance of evidence indicates that better pest surveillance, more discreet methods of application, and a better appreciation and use of natural methods of control can reduce significantly the amount of pesticides needed and change the type of pesticide being applied so that less environmental damage will result. There is equally good evidence that for the forseeable future pesticides will continue to play a major role in crop production. Whereas in agriculture and forestry it has been possible to replace some of those pesticides that pose a hazard to nontarget species, this is not yet true in some areas of public health. No suitable alternative to DDT for malaria control has been found, and this being the case, DDT will continue to be used. This use as a residual spray in human dwellings has, however, a minimum of deleterious environmental effects and demonstrates that even broad spectrum, persistent, environmentally hazardous pesticides can be used safely if used properly.

The public alarm raised about pesticides in the 1960s has been translated into legislative controls during the 1970s. The results have been more rigid testing of pesticides prior to their registration and attempts to restrict pesticides to competent people, knowledgeable about pesticides and about their safe and effective use. Concern about environmental contamination has been an important concept behind this legislation, and it is hoped that the experience of the past will ensure wiser planning in the future. Regardless of our best efforts it is likely that new pesticides will bring new problems, some with far-reaching environmental implications in terms both of their potential for benefit and their propensity for damage. These problems will be compounded by residues of pesticides and a host of other contaminants already present. It is our responsibility to ensure that the potential benefits from pesticides are maximized. This will occur only if we understand the environmental implications of their use and use them with such discretion that undesirable effects are avoided. It is to this end that this book is dedicated.

2

THE PESTICIDES DILEMMA

To use or not to use—that is the question. The question of whether or not pesticides should be used in a particular situation must concern us all. Recent years have seen phenomenal increases in the use of pesticides in attempts to increase food supply, protect our health and welfare, preserve our forests, and improve recreational opportunities. It is true that pesticides have contributed positively to each of these goals. It is equally true that, in some instances, while the intended function of the pesticide has been realized, undesirable side effects have accrued to despoil the environment, jeopardize human health, or, in the longer term consideration, intensify the problem the pesticide was supposed to solve. Although experience has taught us something, there is little evidence that we have really developed a philosophy on pesticides that will ensure the kind of discretionary use the nature of these materials requires. Why?

Perhaps the reason why this is so involves the nature of the issues that surround pesticides. It is a basic fact that pesticides are toxic, and while the scope and degree of toxicity vary widely among compounds, there are none that can be considered harmless. It is not surprising that the general public is therefore aroused when instances of careless or unnecessary use are brought to their attention. The public is not informed, and scientists have made little progress toward improving this situation. Thus the issue becomes one of emotion with good judgment replaced by political expediency. We must seek a forum where the benefits and risks can be assessed and decisions made that reflect the best advice available. Balancing benefits versus risks will not be easy. Each of us will have his own set of values, values rooted in our ancestry and modified or confirmed by personal experience. Thus the decisions may, at times, be difficult, since they involve many issues viewed from many differing points of view. While this complicates the picture and contributes to the dilemma, the solution will be found in dealing with the issues and balancing benefits and risks.

What are the considerations involved?

2.1 ECONOMIC CONSIDERATIONS

Advocates of pesticides point with pride to the high quality, variety, and volume of food produced with the aid of pesticides. Concurrent with the postwar pesticide era, yield of potatoes doubled. Although new varieties played a role in this increase, credit must go in a major way to control of the potato leafhopper, potato flea beetle, Colorado potato beetle, and the fungus disease late blight. In the United States control of the northern corn rootworm on corn and the development of effective herbicides revolutionized corn production with respect to both profit and volume. Reduced need for cultivation made it possible for much larger and efficient production units with greatly reduced manpower. Seed treatment chemicals to control insects and diseases in cereal seedlings, beans, and a host of other large-seeded vegetables reduced seed requirements per acre and provided greatly improved crop stands. Pest control in cotton, tobacco, citrus, and deciduous fruits contributed to increased production and decreased cost per unit. In agriculture, generally, United States estimates indicate a $1.00 to $10.00 return/$1.00 invested in pesticides, but the return differs greatly in different situations. Pimentel (1973) arrived at an overall dollar return of $2.82. Using a somewhat different system of computation, Headley (1968) arrived at a figure of about $4.00. Similar claims can be made for the forest industry (Marshall, 1975).

Economic benefits as a function of the World Health Organization's efforts to eradicate malaria can also be cited as phenomenal when one considers the number of ill days converted to working days through this program.

Weighed against the positive economic effects are some facts that make the generalizations of economic benefits less dramatic. Nova Scotia orchardists did not have a significant problem with spider mites until DDT was used for codling moth control; the red-banded leafroller was not a problem to apple growers in eastern North America until modern pesticides were used; fall panicum was not a problem in corn until the herbicide atrazine became used widely, and modern pesticides, as used during the past 25 years, have created as many problems as they have solved in citrus and cotton. In the United States the pesticide portion of the cost of production in agricultural crops rose from 1% in 1955 to 4.6% in 1968 (Neumeyer et al., 1969).

2.2   HEALTH CONSIDERATIONS

Pesticides in relation to human health is a two-edged issue. Most, if not all, pesticides are toxic, but the degree to which they are toxic is well known only in a few species of animals. We extrapolate from findings on mice, rats, and a variety of other test animals and use these extrapolations as a guide to human toxicity. Yet, despite many similarities, there are differences in metabolic processes between man and common test animals, and we can conjecture only on how significant these changes may be in terms of pesticide toxicity (see Chapter 7). If direct effects were our only concern, the matter would be much more simple, but such is not the case. Modern technology has exposed man to a plethora of new chemicals in the diet he consumes, the air he breathes, and the water he drinks. Effects of these new chemicals on a long-term-exposure basis can, at best, be speculative. But recent discoveries concerning chronic effects of mercury on those consuming it and on their offspring, the role of diethylstilbestrol (DES) in predisposing female progeny to some types of vaginal cancer, and the teratogenic effects of thalidamide and some dioxins give reason for concern and dictate that all new products be screened carefully and that exposure to new chemicals be accepted only where it can be shown clearly that these contribute to man's welfare.

The health record of pesticides is relatively good when viewed in the context of the total population. Although medical statistics on pesticide-related illnesses are sketchy at best, the number of cases reported indicates that in the general population this is not a major problem. Illnesses among people occupationally exposed in pesticide formulation, in application, and in working in treated crops are more common. Pesticides are involved also in suicides and accidental poisoning, especially in children, but this represents carelessness and cannot be singled out in terms of a health hazard.

Perhaps most importantly, some pesticides are now present in the human body as persistent deposits in fatty tissue. DDT, dieldrin, and heptachlor epoxide are commonly found in man in many regions of the globe and are secreted in mother's milk at levels much higher than we would wish. Although there is no evidence that exposure levels are harmful, we have not had such exposure for sufficient time to establish that they are not. Thus an uneasiness is not only understandable but also justified. Such uneasiness does not, however, suggest alarm. There is no evidence that any of the pesticides now in use are carcinogenic or teratogenic at exposure levels likely to be encountered in our food or environment or that they cause any adverse effects save for allergenic reactions. Such reactions are likely in a small percentage of the population and may take a variety of expressions. For example, the senior author becomes asthmatic in the presence of dichlorvos and develops a headache following prolonged exposure to DDT. The fungicide difolitan is likely to cause a rash on tomato harvesters working in recently treated plantings, and some dinitro herbicides and fungicides may also cause skin problems.

Whereas the possible negative aspects of pesticide usage and human health are of concern, especially in the developed nations, their role in improving world health is one of the outstanding chapters of preventive medicine. If one examines the developments of the past four decades and pinpoints those that have contributed most significantly to the health and welfare of mankind, included among the top 10 would be the discovery of antibiotics, the role of the triazine herbicides in increasing corn production, and the role of DDT in reducing insect-transmitted human diseases. North Americans may fail to appreciate the significance of this latter discovery, since malaria is now of minor significance and yellow fever is something we associate with events past, most notably the building of the Panama Canal. To the developing nations in Africa, Asia, and South America the picture looks quite different. Millions are alive today who would not be if it were not for DDT, and tens of millions enjoy good health rather than play host to malaria parasites. Contrary to the hopes expressed in 1955 DDT has not eliminated malaria from the face of the globe, nor is such a development likely. DDT and a succession of other pesticides have, however, reduced the incidence of infection with many insect-transmitted parasites to a tolerable level.

In North America pesticides have been used quite extensively for mosquito control in areas where encephalitides are known to occur. This use has been questioned, since the incidence of disease is usually low and its etiology so poorly known that the efficacy of preventive treatments is more presumed than demonstrated. Again the issue is seldom clear, and the benefits versus risks are difficult to quantify.

2.3   AESTHETIC CONSIDERATIONS

Decisions on pesticide use for economic or health reasons may be difficult, but in these cases the benefits and risks have a degree of tangibility lacking where the only justification for use is aesthetic. To some a weed-free lawn or a beautiful golf course is essential, while to others these represent whims of an affluent society. Thus the argument about what pesticides may be used to maintain such surroundings will be quite different depending on one's point of view.

To some extent aesthetic considerations can be given economic parameters. The cost of replacing a 50 year old elm tree may be excessive simply from the standpoint of its cost for removal. Provided the tree serves a desirable aesthetic function for shade or beauty, it may be more economical to employ expensive pesticide injections to prevent Dutch elm disease, or kill the bark beetles that transmit it than to permit the tree to die and have it removed. Shade trees defoliated for successive years by the gypsy moth are unlikely to survive. Given the relatively low cost of pesticide to control this insect, it makes better sense economically to apply control than to permit the tree to die.

2.4   POLITICAL CONSIDERATIONS

Although pesticides are an important part of our production system, their attributes are such that they evoke both positive and negative reactions in the general population. It is therefore understandable that they become political issues. Sir Winston Churchill recognized the role DDT played in stopping an outbreak of typhus among allied troops in Naples and in a speech (September 1944) referred to it as "that miraculous DDT powder." Twenty years later Rachel Carson dubbed that same DDT the "elixir of death."

The political sphere for pesticides breaks clearly into two camps, but the degree of activity in each camp is modified by the particular issue under debate. In general, agriculture, forestry, and the pesticide industry support continued and, in some cases, increased use of pesticides. They point to the return on the pesticide dollar and the contribution of pesticides in preserving our forests and providing, especially in North America, an abundance of high quality, nutritious foods in endless variety. Opposed to pesticides are organizations such as the Environmental Defense Fund, The Sierra Club, Pollution Probe, and a host of other groups who argue that America can produce an adequate food supply without pesticides (at least without the amount now used), that pesticides (especially DDT and other persistent, non-biodegradable chemicals) have damaged our wildlife, and that their continued use is "ecologically unsound." The argument against pesticides is usually punctuated with a number of statements about health risks. Invariably the possibility of carcinogenic and teratogenic effects is included. Depending on the issue involved, a voice may be heard from the medical field, but this voice is usually muted, for hard facts on pesticide risks to human health are lacking and, in North America, the need for extensive pesticide use as a health prophylactic is not established. A different situation prevails worldwide, the WHO supporting an extensive pesticide program in its battle against world disease.

Thus politics enters the field at local, state, national, and international levels. The general citizenry, unable to distinguish among pesticides, believes that what has made headlines about DDT and the defoliants used in Vietnam applies generally to pesticides and may become quite upset by county workers' using a herbicide for weed control on the local highway or treating mosquito breeding sites to reduce a local mosquito biting problem.

These same citizens may, however, demand of their public servants that the weeds be removed and mosquitoes eliminated. In addition there are always some local "experts" with a "biological method" to control everything. Add to this the requirement that minimum expenditure be involved, and the politician's dilemma is compounded.

2.5 ENVIRONMENTAL CONSIDERATIONS

Webster's dictionary defines environment as "the complex of climatic, edaphic and biotic factors that act upon an organism or an ecological community and ultimately determine its form and survival." Obviously pesticides have invaded every corner of the globe (more than 2 billion lb used in 1975), and since some are toxic to a wide range of organisms, one would not expect their effects to be negligible.

Although pesticides have been used for many years, our concern for their environmental effects is of recent origin. There are two main reasons for this. First, the number of pesticides used was rather limited and, second, the amount of pesticide used was not great prior to 1945. Despite these qualifications, however, the nature of some of the pesticides used during that period (e.g., the arsenicals, fluorides, and mercury) was such that we should have been concerned. The new pesticides developed during the last four decades present an entirely different picture. Not only has the number of compounds increased dramatically, but also the spectrum for their use has become almost unlimited. Herbicides have been developed with selectivity for practically all types of crops and with persistence that varies from almost immediate inactivation on soil contact to several years of effective vegetation control. Other pesticides may control foliage diseases, fruit diseases, algal growths, nematodes, and insects. Some of these may be selective in action while others kill a wide variety of plant or animal species. Add to these the piscicides, avicides, and rodenticides, and the list becomes large indeed. Found among the arsenal of chemical pesticides now available is a product designed for almost any pest problem. Little wonder that our environment has been bombarded with pesticides and that some of these have had effects other than those intended.

Most of our environmental concern has centered around aquatic ecosystems and, to a lesser extent, the forests. Certainly most of the documentation on effects of pesticides on nontarget organisms has come from these communities. This may be misleading, since most of the ecologists electing to study such effects have selected these areas for investigation. By contrast most workers in cropping systems have been concerned primarily with determining efficacy on target species and, with few exceptions, have seriously considered nontarget effects only in recent years.

Most of the environmental problems associated with pesticides have centered around DDT and the related persistent chlorinated hydrocarbon insecticides. In a number of instances these chemicals-when used in aquatic and/or forested areas, resulted in fish kills, and concentrations developed in some predatory birds large enough to affect their reproduction and nesting success. The degree to which this occurred is still poorly known, for, unfortunately, much of the data are conflicting. Before 1970 many of the chemical analyses reporting the level of DDT and its metabolites were not valid, for they failed to distinguish other environmental contaminants, particularly polychlorinated biphenyls (Sherman, 1973).

In recent years the use of DDT and the related compounds has been phased out in North America and throughout much of the world. Its use in vector control programs is likely to continue, for effective and practical substitutes have not been found. Gross, long-term environmental effects have not been identified with other pesticides except mercury, and with mercury the source of major contamination has not arisen from its use for pest control (Lepple, 1973).

The problem arises in defining the ideal environment and determining what represents first a significant and second a detrimental effect. The use of a pesticide such as fenitrothion on 5 million acres of forest will kill many insects, a few birds, and, possibly, a number of fish. Populations of all but the target species, the spruce budworm, rebound in a few months to "normal" levels (Varty, 1975). Because of the large acreage involved it is probable that in 1975 millions of birds were killed when almost 15 million acres of forest in Quebec and New Brunswick were treated for control of the spruce budworm. Data from the previous year indicate that, while populations of some species were adversely affected, others were not. Pretreatment and posttreatment counts showed differences in some cases, but on total, these differences were less than one bird/acre (Buckner and Sarrazin, 1975). One wonders how significant this is and whether saving the forest by use of pesticides may not be preferable environmentally to permitting the forest to be killed and its inhabitants subjected to all the environmental upheavals associated with such an event.

Early work in Nova Scotia (Lord, 1949) brought to our attention the role played by beneficial insects in the agricultural environment. There is no doubt but that modern pesticides have altered the distribution and ratios of various pest and beneficial species in modern cropping systems. Experts disagree sharply on the significance of such changes. Whereas some entomologists insist that an apple orchard is best as a "faunal desert," others extol the merit of maintaining parasites and predators to reduce problems from pest species. The preponderance of evidence supports the latter view. While ecologists espouse the importance of species diversity as a prerequisite for community stability, they provide few guidelines in terms of species numbers to define what constitutes a stable community and what relevance this has, in fact, to a country like Canada or the United States, where only a small percentage of the land area is under annual cultivation.

Thus the environmental concerns about pesticides are neither resolved nor well defined. There is no question but that indiscriminate use of pesticides has the potential to upset temporarily many ecological patterns. Whether this is significant in the long term remains to be seen. It should be remembered that nature is not static and that the "balance of nature" is a shifting one, the result of countless influences and an endless struggle among the inhabitants of any particular community. Pesticides add elements to the struggle and, at least temporarily, shift the balance. Whether this results in a better or worse environment cannot be determined until we learn what better or worse means. The concept that any change is bad can be embraced only if we assume that evolution has, in 1979, finally reached an optimum stage for all. We have difficulty accepting such a conclusion.

2.6 PSYCHOLOGICAL CONSIDERATIONS

To many, pesticide usage has a psychological effect, an effect that may be positive or negative. Some view pesticides with alarm as chemicals foreign to "nature" and the body system and therefore to be avoided. To support this feeling they go to great lengths to purchase food at "nature food" stores and often accept products damaged by insects or rodents or blemished or rotted by fungi in preference to the more attractive products offered in the supermarket. To these people some insects (fragments or whole), rodent hairs, or fecal pellets in food products are preferable to any trace of pesticide.

Others react quite differently. A green worm in a salad, sap beetles in a can of tomatoes, or a tiny maggot in an apple evokes a feeling of disgust that makes the food unusable. Others may be terrified of a mouse, shudder at the thought of cockroaches, or become hysterical when confronted with a bat. Just as there are fastidious housekeepers, there are equally fastidious groundsmen to whom a weedy lawn causes physical illness and mental distress. To these people pesticides provide assurance. Whether treatment is needed or not is not the issue. That treatment has been made provides a feeling of satisfaction and, just like the medical placebo, serves a function and fulfills a need.

2.7 MORAL CONSIDERATIONS

It is no secret that we live in a hungry world. The degree to which starvation persists varies widely from one region to another, but it is reasonably certain that at least one-third of the world's population goes to bed hungry. The United Nations through its Food and Agriculture Organization (FAO) strives to relieve the suffering from hunger, and major centers for research to improve crop production have been established through FAO and/or philanthrophic foundations. These efforts to increase food production through new varieties and improved production practices in older ones have been productive, the result having been an increased food supply. Despite this "green revolution" increased food production has barely kept pace with increased population. The "stork" has continued to "outrun the plow."

In this competition pesticides have played a large role. The stork has had the benefit of products like DDT that have greatly lessened mortality and morbidity from diseases whose causal agents are transmitted by insects. Drastic reductions in the incidence of malaria, typhus, bubonic plague, and yellow fever (to cite a few) have reduced childhood mortality and extended life expectancy, especially in the densely populated countries of Asia, Africa, and South America. Add to this the giant strides made in the medical profession and the improvements (although in some cases rather modest) in health delivery systems and the ingredients for exponential growth rate at an increased annual increment become apparent.

Pesticides have played an equal role in improving the productivity of the plow. Many instances can be cited to document the level of yield increase as a result of weed control, insect and disease control, or bird and rodent control in selected crops (Chapter 4). It is a fact that many of our most productive food crops are "biological weaklings," unable to compete in the natural environment without extensive supplements of nutrients and moisture and protection from pests. Most of the crops that provide the backbone of our food supply have been selected for high yields, attractive appearance, and nutritive value rather than for survival under adverse conditions. It is not surprising then that the "green revolution" introduced varieties whose maximum yield potential depended heavily on pesticides for disease, insect, and weed control. There is nothing new or unexpected in this. The world's most important food plant—rice, for example—is subject to attack by 70 species of insects alone, 20 of these major pests in some of the rice-producing areas of the world (Barr et al., 1975). Small wonder then that in the breeding selection programs, pest control would be included as part of production practice and the varieties selected unable to meet full potential in the absence of such support. It has been suggested that our crop selection programs should pay much greater attention to inherent susceptibilities or resistances to various pests. Whether or not this would be productive is a moot point. The appetites of many pests are sufficiently broad that selection within a species has produced resistance in relatively few instances. The success rate has been much better with plant pathogens than with insects, but outstanding success stories can be told for each. The danger of not following such a selection procedure was demonstrated dramatically in 1970, when large plantings of field corn in the United States and Canada were devastated by southern corn leaf blight. This occurred because a susceptibility to the disease was inherent in a major breeding line, a line that ensured in its hybrids high productivity and excellent quality. All too frequently the gene pool from which we select our crop varieties is becoming increasingly narrow, and we may run the risk of disaster, especially from plant pathogens whose ability to develop new strains has been amply demonstrated (National Academy of Science [NAS], 1975).

Thus the moral issues regarding pesticides are viewed differently, depending on one's own convictions. Their role in improving world health has added an increment to our human population, a population even now greater than we can feed. Our use of pesticides to increase food supply is argued by some as a further factor increasing world population. Our dependence on pesticides may, however, lead us to a false sense of security through reliance on cultivars that in the absence of pesticides are vulnerable to utter destruction. World populations must be curtailed, but few would suggest that the withdrawal of pesticides from the health field or removal of their contribution toward feeding people represents an acceptable course.

Another side of the moral issue involves the possible deleterious effects of pesticides on future generations. While there is heated debate on issues such as the use of defoliants in Vietnam and continued use of DDT; heptachlor; chlordecone; 2,4,5-T; and a variety of other pesticides, such debate has generated more heat than light. Certainly it is morally unacceptable to introduce now into our environment chemicals that will have medically significant effects on our progeny. At this time it would appear that we have not developed the technology to make such assessments (Chapter 19).

2.8   SAFETY CONSIDERATIONS

The safety aspects of pesticides to human beings have been discussed briefly under Health Considerations and are dealt with in detail in later chapters. Two other aspects should be discussed, highway safety and fire. Vegetation maintenance on highways is important, and the removal of the growth from highway intersections and road markers aids significantly in driver safety. A similar situation prevails for the railways, where visibility at road intersections is critical, especially at intersections not protected by traffic bars or light-bell signals. Perhaps of greater concern are the danger from fires and the risk of damage to property or loss of life that may result. A weedy roadside or railway bed, dried from autumn heat and maturity, is ignited readily from the sparks of engine wheels or a cigarette butt tossed carelessly from a passing vehicle. There is no question but that weed control in such situations is needed. The argument is by what means this should be achieved. Mowing is certainly possible, but the monetary cost is much greater than control with herbicides. In addition to railways and highways, right-of-way brush control on transmission and power lines has safety considerations in terms of fires and service disruption. Again the job of brush control could be accomplished by means other than herbicides if labor were plentiful and cost was of no concern.

In a somewhat different vein is fire following severe outbreaks of defoliating pests in the forest. Forest fires have always been with us, and the conditions leading to their eruption may be many and the ultimate cause questionable. Defoliation by insect pests and tree mortality that results is one factor that may be important. Perhaps more than any other insect, the spruce budworm is a good candidate in this regard. This insect prefers to work in a mixed stand of spruce and fir that is approaching maturity. Depending on the degree of defoliation and the vigor of the tree, the tree may survive attack for several years. In general nearly mature trees will be killed by 3 successive years of attack if the attack is sufficient to remove most of the current season's growth. Since this insect attacks vast acreages of timber, the potential for tree mortality over thousands of square miles is evident and the development of conditions favorable to forest fires apparent.

3

THE ROLE AND USE OF
PESTICIDES IN HEALTH

Insects play an important role in the transmission of human diseases.* While vectors are found in four orders of insects and among the ticks and mites, the Diptera are the most important order with vectors included in at least eight families (Mattingly et al., 1973). The role played by insects may be quite different. In diseases such as malaria the insect is an essential host of the pathogen for several of its developmental stages, while for other diseases, for example, bacillary dysentery and "pink eye," insects such as the housefly are vectors because they become contaminated with the causal bacteria and carry these to new hosts. Between these two extremes are many variations in complexity of relationship.

Early discoveries of the role of insects in disease transmission evoked attempts at control by control of the vector. These attempts included sanitation for diseases such as plague and typhus, screening of windows and bed netting to prevent biting by mosquitoes, and draining mosquito-infested swamps to reduce breeding areas. These methods realized a degree of success. Pesticides entered the picture around the turn of the twentieth century with the use of kerosene oil as a larvicide to control the mosquito vector of yellow fever during the building of the Panama Canal. But it was not until the emergence of the synthetic insecticides after World War II that pesticides were used as a major factor. No attempt is made to deal with all insect-transmitted

* For the sake of simplicity we refer to these as insect-transmitted diseases, but the reader should realize that the disease is not transmitted, only the pathogen.

diseases in this discussion. A few are selected to indicate the nature of the insect-disease interaction, the role played by pesticides in disrupting this relationship, and the value of this to society.

3.1 MALARIA

Of all diseases that have plagued man in his attempt to inhabit the globe, perhaps none has been more devastating than malaria. Although thought of normally as a tropical or near tropical problem, this disease has occurred in many parts of the temperate world as well. The United States, Canada, Russia, and the Scandinavian countries have all been host to this disease at some time in their history, and though the disease declined in incidence during the twentieth century, more than 5% of the world's population (200 million people) were annually infected during the 1950s.

The disease in man is caused by one or more of four species of protozoa of the genus *Plasmodium*. The severity of the disease varies greatly, depending on the species of pathogen involved, the death rate in human beings varying from low to as high as 10%. The disease is characterized by recurring sequences of chills, fever, and excessive perspiration, each sequence lasting a few to several hours and recurrence depending, in large measure, on the pathogen involved.

The malaria pathogen(s) has a complicated life cycle, some stages of which develop in man and others in mosquitoes. Both are essential to its survival.

Malaria is transmitted only by mosquitoes (rare exceptions involving direct blood contamination) and only by members of the genus *Anopheles*. There are more than 200 species in this genus, and 63 are considered major vectors (Russell, 1959). Small wonder that the disease is widespread, since few areas of the world can claim freedom from one or other of these vector species. In some areas of the world the number of vector species is low (e.g., United States, Canada, Japan), while in other areas (e.g., India, Thailand) many species occur.

Although there may be minor variations, depending on temperature, strain of malaria involved, and mosquito species, there is an extended period of perhaps 10 to 14 days between the time the mosquito bites a malarious person and the time at which the mosquito can transmit infective parasites to a new victim. This is an important consideration, since short-lived species are less likely to be good vectors than those with an extended life span.

The introduction of DDT to control malaria-carrying mosquitoes was one of the most significant developments of this century in terms of saving human lives. In 1955 it was estimated that 200 million cases of malaria occurred each year with an annual death rate of 2.5 million (Pampana and Russell, 1955). DDT was to be a turning point in this struggle, but as with so many other discoveries, the hope for a panacea had to be satisfied with a palliative.

At first it seemed as though everything was right for an insecticide like DDT to remove malaria as a major cause of human suffering. The insecticide was toxic to mosquitoes, and most of the mosquitoes important to the transmission of the disease spent most of their time in or near human dwellings. DDT was persistent, and once a surface had been treated, mosquitoes touching that surface during the next several months would not survive. It was demonstrated that, when the inside walls of huts were treated with DDT at the rate of 2 g/m^2 of surface, effective mosquito control was obtained for periods up to 6 to 10 months. In addition the supply of DDT was unlimited and its cost minimal. The compound could be used without danger to occupants of the homes, since its toxicity by dermal exposure to human beings was quite low. Early tests showed great promise.

The island of Sardinia, with more than 78,000 cases in 1942, had only 9 cases in 1951 following several years of supervised treatments with DDT for control of *Anopheles*. Similar results were obtained elsewhere, and with great optimism the World Health Organization (WHO) in 1955 formulated plans for a worldwide eradication program for malaria. The optimism of 1955 was reinforced by early experience. In Ceylon, where 3 million cases occurred before DDT, an intensive *Anopheles* control program with DDT had reduced the number of cases to 31 in 1962 and 17 in 1963. Similar results were occurring elsewhere. World politics, jurisdictional disputes, and an anti-DDT campaign in the early 1960s intervened, however, and Ceylon discontinued its program. Resurgence in malaria was quick and relentless with 500,000 cases in 1968 and 2 million (estimated) in 1970.

Despite many problems, outstanding success was achieved. In the United States, where 6 million to 7 million cases of malaria occurred annually prior to DDT, incidence of mosquito transmission since 1953 has been only a few per year, and although almost 3000 cases were reported toward the end of the 1960s, most of these involved returning service personnel from Vietnam. Perhaps the most outstanding success story has been that of India. In 1970 only 150,000 cases of malaria occurred, in contrast to the 100 million cases prior to the use of DDT.

Many problems have developed in the antimalarial programs. It is apparent that world eradication will not be realized and that, while pesticides will continue to be essential tools in the struggle against the disease, they will not be the entire answer. In many species of malaria vectors, resistance to DDT has developed, and in many cases, this resistance has extended to related pesticides as well. In other areas of the world psychological and political resistance to the use of DDT have frustrated efficient control attempts and rendered DDT unavailable. To date no compounds have been found to replace DDT in terms of efficiency, economy, and safety. Great care must be exercised, and logic must supersede politics, or else we may lose the gains won in man's struggle against the world's most destructive disease. "In response to clear humanitarian need," the United States in December 1976 approved the sale of $450,000 worth of DDT to Cambodia for malaria control. This action was taken despite a trade embargo between the two countries (Anonymous, 1976).

3.2 YELLOW FEVER AND RELATED ARBOVIRUSES

Although yellow fever no longer poses the threat it once did, outbreaks as recently as 1959 in Ethiopia, where 15,000 died, emphasize that the disease is endemic in many parts of the world and that epidemics await only the proper conditions, conditions that are as yet incompletely known (James and Harwood, 1969). The disease arises from one of a group of viruses transmitted to man by various species of mosquitoes. These so-called "arboviruses" (short form denoting arthropod-borne viruses) have a long history, and included along with yellow fever are dengue and various encephalitides, including several endemic in North America, such as St. Louis encephalitis, eastern equine encephalitis, western equine encephalitis, and the California group of viruses.

Early accounts of yellow fever date back at least to the seventeenth century in South America (Carter, 1931). It is known to occur in many parts of the world with its greatest incidence in Africa and South and Central America. The disease is best known in the western world in relation to the building of the Panama Canal and the role it played in frustrating some of the early attempts to develop that project. The United States Yellow Fever Commission established in 1900 brought to bear on this disease the expertise of Reed, Carroll, Agramont, and Lazear and established the role of the mosquito *Aedes aegypti* Linnaeus in its transmission. This is probably the first major project where mosquito control (screening, larviciding with kerosene) was demonstrated to reduce effectively the incidence of a human disease. As a result the Panama Canal, begun in 1907, was completed 7 years later and not abandoned as was the attempt some 25 years earlier.

In the transmission of arboviruses, there is no developmental cycle in the mosquito comparable to that involved in malaria. There is, however, a high degree of specificity between the virus and its vector. To a large extent this makes it possible to concentrate on control of a limited insect population, a situation quite different from that pertaining to malaria in many world regions. Thus efforts have been made, with significant success, to eradicate *Aedes aegypti* through habitat removal and the use of pesticides. To a large extent the program has been successful.

During recent years several outbreaks of St. Louis encephalitis have occurred in urban areas in the United States with the most important (in terms of number of cases) in 1975 in the United States and part of Canada. While the specific vector or vectors in each area is not known with certainty, it is known that certain species of *Culex* are capable of transmitting the virus. In several instances large local areas in the center of the epidemic have been treated by air with insecticides to control populations of adult *Culex*. Whether or not such pesticide applications have stopped the epidemics is a matter for some speculation. It is known that in each case where application was made (Houston, Tampa Bay, Windsor) new cases ceased to occur 10 to 14 days later, a time interval consistent with the incubation period of the disease.

3.3   TRYPANOSOMIASIS—SLEEPING SICKNESS

Human sleeping sickness has been known since the fourteenth century (Willett, 1963) and is a disease endemic in much of Africa. Historically it has caused much human death, mortality approaching 500,000 in the 10 year period prior to 1906 (James and Harwood, 1969). During epidemics as many as 50% of local populations have been infected, with mortality rates as high as 60%.

The disease in man is caused by either of two species of flagellate parasites, *Trypanosoma gambiense* or *T. rhodesiense*. In either case the pathogen is ingested by a species of tsetse fly, *Glossina* sp., and the parasite undergoes a developmental cycle within the fly. Man is the reservoir host for the pathogen, which is imbibed by the tsetse as it takes a blood meal. The parasite then develops in the midgut and passes to the salivary gland, where development is completed.

There is a period of development in the tsetse that is normally about 18 days. During this time the tsetse tends to rest in rather dense vegetation on the underside of branches, and efforts toward control with pesticides have been directed, largely, toward residual sprays that act on contact with the fly in this resting location. Results have not been uniformly successful. The dense foliage preferred by the fly is not penetrated well by aerial applications of pesticide, and ground application is difficult owing to poor mobility within the vegetation. Combining pesticide application and vegetation clearing has provided a measure of control, and this plus discretionary removal of ungulates on which the tsetse can feed has reduced disease incidence in many infested areas. The battle is, however, far from won.

3.4 PLAGUE

This scourge of man has been known for centuries. In epidemics it has devastated the human population in cities and, sometimes, entire countries. During the fourteenth century, 25 million people in central Europe, one-fourth of the total population, died, and an outbreak in London in 1666 claimed 70,000 victims. In the most recent epidemic (1898–1918) 10 million people died, mostly in India and China. Whereas human epidemics have been outstanding, the disease is primarily one among rodents, associated usually with the rat, and the oriental rat flea, *Xenopsylla cheopis* (Rothschild). The disease is caused by a bacterium, *Pasteurella pestis*, and is transmitted among the rodent population through the feeding activity of fleas. Fleas, once infected, may be capable of infecting within 5 days and continue to be infective until their death (James and Harwood, 1969).

Control of plague involves control of rodents, and because many species serve as endemic foci, the potential exists for an outbreak in many parts of the world. Although the last great epidemic occurred more than 50 years ago in North China and spread throughout many of its trading nations via steamship, Southeast Asia has had increased incidence during the past 10 years. Rat control with rodenticides and the use of insecticides to control fleas in ship holds and steamship ports are important means of maintaining the disease at low levels throughout the world.

3.5 EPIDEMIC TYPHUS

This disease, often referred to as typhus fever or typhus, is associated with human populations in crowded and unsanitary conditions. Its frequent outbreaks among combat forces has earned it the name "war fever," and the name "jail fever" has a similar connotation. The disease normally has a low death rate among children but a high death rate among adults.

Typhus is known from ancient times and has been a major contributing factor to failure in a number of military campaigns in Europe. The disease was rampant during World War I, and in World War II an outbreak occurred in North Africa. In 1943 the city of Naples was under seige, bomb shelters were filthy and overcrowded, and an epidemic of typhus was imminent. The arrival of Allied forces with the "miracle DDT powder" prevented the outbreak, the first time in history than an epidemic of typhus was stopped cold.

Typhus is caused by *Rickettsia prowazeki* Da Rocha-Lima and is transmitted by the common body louse. Other species of lice (head louse, crab louse) may also be vectors, but they are not so efficient. The louse picks up the rickettsia as it takes a blood meal, and the pathogen invades and multiplies in the epithelial cells of the gut. During this development the epithelial cells are destroyed, and the rickettsia are released into the digestive tract of the louse. From here they are voided in the feces, where under humid conditions they remain infective for weeks, even months. Infected lice usually die owing to the destruction of the epithelial cells of their digestive tract.

Infection in man is the result of contamination of wounds or skin abrasions or inhalation of dust containing louse feces and the rickettsia. The itching and scratching that accompany infestations of body lice provide the necessary skin abrasions, and infrequent bathing associated with crowded conditions, especially in winter, ensures the retention of louse feces in bedding, clothing, and on the body.

Control of typhus requires a moderate degree of sanitation, good personal hygiene, and control of body lice. Modern pesticides, especially DDT, provide louse control with a minimum of effort.

Only a few of the many insect-transmitted diseases have been discussed. Those selected indicate some of the opportunities to use pesticides to break the transmission cycle but indicate also some obstacles to success. The hope of the WHO when they proposed a worldwide eradication program for malaria in 1955 was more idealistic than realistic. Certainly that goal will not be realized through pesticides alone, and though the addition of antimalarial drugs (if these can be improved and their potential realized) could improve the prospect, the likelihood of eradication is remote. Many things militate against such an achievement: a large number of species of insect vectors, each with differing habits; the ability of many insect species to develop resistance to pesticides; the political and social sensitivity with respect to the use of certain pesticides; the political instability of some of the countries where malaria is a major problem; and the shortcomings of health delivery systems in many countries, to name just a few. In addition a review of natural history suggests that eradication of any organism is perhaps an unrealistic goal. Our failure to reach an unrealistic goal should not, however, cloud the achievements won. The fight against malaria has been successful, and for each of the estimated 50 million people alive today because of its success, success has been 100%. Although we may not be satisfied, because eradication has not been accomplished, the impact of malaria control programs has been phenomenal in terms of illnesses prevented and increased labor output in malarious regions of the world. Pesticides, especially DDT, have made this advancement possible.

While the importance of the malarial control efforts has dominated the world health scene, so also has the amount of pesticide used in this effort. At the height of the campaign in 1962, malaria control programs used 130 million lb of DDT; 8 million lb of dieldrin, and 1 million lb of lindane. Most of this insecticide has been applied to the inside of huts to catch mosquitoes in their resting sites. Since 1962 there has been some reduction in the use of DDT, owing in part to development of resistance in some species of *Anopheles* and in part to public clamor for reduced usage because of environmental concerns. Since 1960 more than 1400 compounds have been evaluated for mosquito control in the malaria program. Of these only 2 approached DDT in effectiveness against the vector mosquitoes. Quite clearly DDT will have to continue as the backbone of the antimalarial program. Its effectiveness (in most areas), low cost, and safety to applicators and residents have not yet been combined in another product.

4

THE ROLE AND USE OF
PESTICIDES IN AGRICULTURE

The high quality, abundance, and variety of foods are perhaps nowhere else in the world more apparent than in the North American supermarket. This is not an accident but rather the result of effective use of all the modern technology available to the food production system. Part of that technology is pesticides, and in the United States farmers used almost 0.5 billion lb in 1971. Included in this was 170 million lb of insecticides, 42 million lb of fungicides, and 228 million lb of herbicides. Agriculture is the largest user of pesticides, especially of those products for insect, plant disease, and weed control (NAS, 1975).

4.1 INSECTS AND RELATED FORMS

More than 1 million species of insects have been described, and it is estimated that insects constitute 75% of all the species in the animal kingdom (Borror et al., 1976). Fortunately only a small percentage of these are pests of agriculture or any other of man's enterprises. Most of the insect species are either beneficial to man or are of no recognized significance. Despite this, however, the farmer is faced with pest insects on every crop he grows. Although the extent of damage varies from year to year, and from one crop to another, few crops can be produced successfully without some form of insect control. In most instances the most practical way to combat pest species is through discretionary use of insecticides.

The extent of insect damage that can be tolerated is related to the nature of the damage. Field crops grown for animal feed can be damaged by insects without rendering the produce unfit for this purpose. Harvestable yield may be reduced—in some cases to such an extent that heavy losses are sustained. For example, the northern corn rootworm and European corn borer may cause stalk breakage and lodging in field corn and affect yields through reduced plant vigor and ear loss during harvest. In field crops, however, the "loss per acre" may seem insignificant because of low per acre crop values. When, however, this loss per acre is multiplied by the vast acreage sustaining such loss, the damage and the food denied a hungry world are high indeed.

Whereas the small incremental loss is usually characteristic in field crops, this is not always the case. The locust plagues of biblical times have their counterpart in grasshopper outbreaks in western Canada and the United States that, unless controlled, destroy vast acreages of cereals. To some degree outbreaks are predictable, forecasts of grasshopper abundance being made annually. In recent years western farmers guided by these predictions have been prepared and used insecticides extensively to save their crops during outbreak years.

Similarly the armyworm has a history of great destruction. Moving as an army when food supply becomes exhausted, hordes of these larvae march across the countryside devouring practically any plant growth in their path. Cereals, including corn, are favored sites for egg laying by the moths and therefore are heavily attacked. As with the grasshoppers, heavy infestations reap a total harvest of the crop. In recent years severe loss has been avoided through the use of insecticides, occasionally applied over vast acreage by air but more commonly applied to local infestations comprising only a few thousand acres.

In the field crops in general the result of insect infestation is a reduction in yield, the level of reduction depending on the intensity of attack. Losses caused by the European corn borer during the decade preceding 1973 varied among major corn-producing states and from year to year. Estimates placed the loss at 9% with a variation from 3.4 to 16.9%. Total loss in corn production was estimated in 1968 to exceed 200 million bushels (NAS, 1975). Food value is, however, reduced only to the extent that production is reduced.

Quite a different situation pertains in fruits and vegtables. Whether the end product is destined for human or livestock food may affect the level of tolerance of insect damage, but in most fruits and vegetables postharvest spoilage associated with insect attack results in heavy storage losses.

For human consumption little insect damage is tolerated (Pimentel et al., 1977). In many countries food inspection is rigid and standards high. Food infested with insects or containing insect fragments is unacceptable. Particularly in North America, the housewife has become accustomed to high quality products free from blemishes and insects and she requires that these standards be maintained. Thus the producer must meet these requirements. To do this growers rely heavily on insecticides. For example, let us consider the production of apples in the United States or Canada and the insect pests important in New York State or Ontario. In these regions apples are attacked each year by three major insects that damage the fruit directly. These pests are the codling moth, the red-banded leafroller, and the apple maggot.

In most of New York and Ontario the codling moth has two generations each year. The insect overwinters as a full grown larva in a silken cocoon, usually in a protected place such as under bark scales on the trunk of the tree. In April the larvae change into pupae and emerge as small greysih moths in early to mid May. During warm evenings these moths become active, mate, and lay eggs on the upper surface of the newly formed apple leaves. Eggs hatch in about 2 weeks (6 to 20 days, depending on temperature), and because this is about 4 to 6 weeks after blossom, the larvae crawl to the small fruit and enter, usually by chewing a hole in the calyx end. Once entered, the larvae tunnel their way along the core of the apple, often feeding on the seeds. Most of the infested apples drop to the ground in a few weeks' time, and the codling moth larvae continue develoment in the dropped fruit. When the larvae complete development, they crawl out of the apple and move to a sheltered place, usually on the bark of the tree, spin a cocoon, pupate, and emerge as adults for a second generation. These second-generation moths are found from late July through September. By this time, however, young apple fruit are of considerable size and most of the apples entered remain on the tree until harvest. Most of the apples attacked have, at harvest, a hole at the blossom end or side, the core area tunneled, the seeds eaten, and a pinkish, brown-headed worm about $\frac{1}{2}$ to $\frac{3}{4}$ in. long happily living inside. While the entrance hole may be seen readily by one who knows what to look for, the housewife may not notice the external mark; her first indication of an infested fruit is the worm (or half worm) she encounters on the first bite.

Thus there are two phases of damage. That of the first generation results in reduced harvest whereas that of the second generation results in a wormy fruit at harvest, a fruit that must be removed before the apples can be sold. If left untreated, apple orchards would bear 20 to 90% fruit infested with the codling moth, a level of infestation that would render production impractical (Glass and Lienk, 1971).

Control of the codling moth is directed toward killing the tiny larvae as they feed (sparingly) on the apple leaves or are in the process of entering the fruit. Since egg laying and hatching extends for a considerable period, the grower seeks to maintain on the apple foliage and fruit a level of pesticide toxic to the codling moth during most of the growing season. These treatments, often referred to as "cover sprays," are now being reduced in some areas where effective monitoring programs are carried out. This monitoring determines just when the moths are laying eggs, when eggs are hatching, and when the tiny larvae are likely to be exposed.

Damage by the red-banded leafroller is seldom as severe as that of the codling moth, although in some years almost 100% of fruit in unprotected orchards may be disfigured. Like the codling moth this insect has two generations per year, these coinciding with those of the codling moth but usually about 2 weeks earlier.

Control of the red-banded leafroller goes hand in hand with that of the codling moth, although growers must begin applications a bit earlier. As with the codling moth, monitoring is being done in some areas. This insect was one of the first in which the chemical identification of the sex pheromone was determined and a system of trapping using this as a lure devised (Glass et al., 1970). As with the codling moth it appears that most of the infestation with this insect derives from an overwintering population within the orchard. Thus monitoring to detect the insect's presence and suggest the level of infestation that may result are important advances. Pesticides can then be used in a pattern consistent with the needs of the particular orchard and season and not solely on an "insurance" basis.

The apple maggot is the third and perhaps the most destructive of all apple insects in Ontario and New York. It is this insect that sets the tone for pest control on apples, for in the absence of control for only a couple of years, infestation by this insect will be present in more than 85% of the fruit (Glass and Lienk, 1971).

This insect overwinters within a puparium in the soil and emerges from late June through most of July (emergence varies somewhat from one region to another). The adult is a medium size, black and white fly. After mating, the female lays her eggs in the apple, placing them beneath the skin of the fruit with her sharp-piercing ovipositor. Larvae hatching from these eggs burrow throughout the flesh of the apple, leaving a series of winding trails that has earned for this insect the common name of "railroad worm."

In addition to the three major pests, apples may be attacked by mites (several species), aphids (several species)), various leaf-eating caterpillars, the tarnished plant bug and related species of mirids, weevils (several species), and the oriental fruit moth and other species of Lepidoptera that attack the fruit.

Thus the orchardist adopted a program of chemical sprays applied at weekly or near weekly intervals throughout the season to ensure a harvest of sound, insect-free fruit. Recent experience in pest monitoring has shown that careful field observations on insect abundance can decrease the number of applications required and reduce pesticide cost by about 30% (Tette et al., 1975). Movement in this direction is excellent, for it not only reduces the possible adverse environmental effects from insecticide use but also returns more profit to the producer.

Apples are but one of a large number of fruit and vegetable crops on which insecticides must be used if production of the crop is to continue. To a large extent the same situation exists with cole crops (cabbage, cauliflower, broccoli, brussels sprouts), onions, carrots, peaches, pears, grapes, cherries, sweet corn, and many other tender fruits and vegetables. The vulnerability of such crops lies in the fact that insects attack driectly that portion of the plant used for human food and have adapted so successfully to the crop as to destroy it completely for food pruposes.

One other crop, cotton, should be discussed, for in recent years, almost one-half of the insecticides used in agriculture in the United States are employed in its production (Pimentel, 1973). Some 13 million acres of cotton are grown throughout the south and central region of the United States, and despite the wide geographic area (east to west), wherever cotton is grown, major insect pests abound. Although the cotton boll weevil is the dominant pest over most of the production area and causes an estimated $200 million loss annually, the cotton bollworm, tobacco budworm, and plant bugs are also major pests. In some of the western areas where irrigation is practiced, insects such as the cotton leafworm and cabbage looper also cause heavy damage. Whereas it is estimated that almost half of the cotton acreage escapes heavy damage each year and therefore does not require extensive use of insecticides, other areas may require 15 or more applications. Some 70 million lb are used each year to prevent an insect loss estimated at 19% of the crop, or about $500 million (USDA, 1965).

Insect control in cotton has changed dramatically over the

past 100 years. This change has been associated with several major events. The first of these was the introduction from Mexico of the boll weevil near the end of the nineteenth century, followed by the pink bollworm. Control of these pests having become a necessity, this was achieved, rather ineffectively, in the early part of this century with sprays containing Paris Green. In the 1920s it was discovered that calcium arsenate provided effective control, and dusting by air became a standard practice that persisted until after World War II. DDT and the new crop of insecticides ushered in for the cotton grower, as for many other agricultural enterprises, a new era in insect control in terms of both effectiveness and low cost. The panacea was, however, to be short lived, for in the mid 1950s the cotton boll weevil became resistant to DDT, a resistance soon to spread from its first locus in the Mississippi Delta to practically all cotton-growing areas. Resistance did not apply only to DDT but was shown also for many of the related compounds such as dieldrin, heptachlor, endrin, and toxaphene. Growers moved to the organophosphorus insecticides (ethyl and methyl parathion, azinphosmethyl) and to some extent to the carbamate carbaryl. Whereas these materials gave good control of the weevil, control of the lepidopterous pests was less satisfactory, and mixtures containing a chlorinated hydrocarbon and an organophosphate were put into common use. The resilience of the insect population was to plague the grower again in the 1960s, when the level of pesticide resistance in some pests, and especially in the tobacco budworm in parts of Texas, made control almost impossible regardless of the pesticide used or its frequency of application. In addition, legislative decisions banning the use of DDT on cotton has further complicated the picture, and the cost of pest control on cotton has spiraled dramatically. A recent review (NAS, 1975) shows the per acre cost of insecticides in some cotton-producing areas to approach $36, a high cost indeed in a crop in which the profit margin is perilously thin.

4.2 WEEDS AND WEED CONTROL

Modern methods of weed control, primarily through the use of herbicides, have been a major contributing factor toward improving the efficiency of crop production and the value of marginal grazing land. Herbicides are, however, not the only method of weed control; cultural practices such as good seed bed preparation, tillage, crop rotation, cultivation, and hand weeding are important. Their relative importance varies with the crop situation. While the market gardener or lettuce grower may still rely heavily on cultural practices, the large producer of cereals, soybeans, or cotton depends on herbicides to ensure the production of the large weed-free acreages so common in North America. The slow, and often repeated, time- and energy-consuming cultivations have disappeared from many of these crop production systems, their place taken by selective herbicides.

Weed pests may be almost any of the 250,000 species of plants known to man. Their potential for invading our agricultural areas is great. Several million weed seeds can be found per acre in many of our agricultural soils. If, after seed bed preparation, only 10% of these weed seeds germinated, this would produce a weed population of several hundred per square foot—rough competition for any crop. These weeds compete with crop plants for nutrition and moisture. Their ability to reduce yield of cultivated crops is brought sharply into focus when we realize that one plant of wild mustard requires the nitrogen and phosphorus of two plants of oats and has the water requirement of four oat plants. One plant of ragweed, in addition to its obnoxious role toward sufferers of hay fever, has three times the water and nutrient requirements of corn.

The extent of weed damage to crop plants has been assessed in several major crops. In experiments with corn in Illinois, crop losses were estimated at 81% in unweeded plots, damage varying from 28 to 100%. Replicated studies in Minnesota over a 3 year period showed yield reductions in corn of 16 to 93% with an overall average loss of 51% (Behrens and Lee, 1966). In soybeans extensive tests in Nebraska and Minnesota demonstrated that yields in weedy plots were reduced an average of 45% over comparable plots where herbicides were applied. In these tests plots kept weed free by herbicides outyielded those where weeds were suppressed by normal cultivation practices. Weed damages in crop plants are not restricted, however, to their role as competitors. They interfere with harvesting and picking operations and harbor insect pests and plant pathogens; in seed crops, the harvest may require extensive cleaning to remove weed seeds before the crop can be sold or processed.

Weed Control in Corn

Perhaps nowhere else has the role of herbicides been more dominant than in the corn belt of the American Midwest, where about 83% of corn acreage is treated. In a recent review (NAS, 1975) 19 species of weeds are listed as "troublesome," foxtail (various species) and pigweed heading the list. The variety of weeds present in the corn production area dictates that a broad spectrum herbicide be used. As a result, five herbicides, atrazine; propachlor; alachlor; 2,4-D; and butylate (in descending order of use) account for some 95% of the 85 million pounds of herbicide applied to this crop. Approximately one-half of the acreage is treated with atrazine.

Herbicides are not the only approach to weed control in corn—tillage and cultivation are employed in most production systems, but escalations in energy costs militate against such practices. The practicality of producing corn in a no-tillage system has been demonstrated, and though yields under such a practice are usually lower than those following normal seed bed preparation, savings in labor and energy costs make the system more attractive as costs in these areas escalate.

If one were to suggest the "normal" practice for weed control in corn over the past decade, it would be to use a preemergence application of atrazine followed by one cultivation. A few growers may use no herbicide and employ three to four cultivations while an equally few may apply one or more herbicides and use no cultivations. The procedure followed is dictated by the local weed problem, the availability of labor and equipment, economics, and the particular bent of the individual producer. It is determined also by the crop rotation in which corn is produced. Since atrazine is a fairly persistent herbicide, its use, especially in a dry year, may result in a carryover in the soil the following season. Thus the grower may be faced with a situation in which last year's corn field can be planted only to crops tolerant to atrazine. Although the grower can get around this problem by using a nonresidual herbicide on corn that will be followed by a sensitive crop such as soybeans, in many instances this has contributed to a nonrotation practice in which corn is planted successively in the same field year after year. Such a practice contributes to increased problems from weed species tolerant of atrazine and a reduced yield. Hawkins and Slife (1977) found that yields following tillage practices were 22% lower than those where weed control was obtained through the use of atrazine and other herbicides. The highest corn yields were obtained where herbicides were used for weed control and a cropping system was practiced that included crops like wheat or soybeans in the rotation.

Weed Control in Small Grain Crops

Each year more than 120 million acres of land are planted to small grains in the United States and Canada. This includes 80 million acres of wheat and 40 million acres of barley, oats, and rye. Rice is grown only in California and some of the southern states with an annual seeding of 2 million to 3 million acres (USDA, 1972; *Statistics Canada*, 1970).

Compared with the corn crop, the small grain crops are all more competitive with weeds. Traditional methods of weed control such as clean seed, crop rotation, summer fallow, proper timing of seed bed preparation, seeding rate, and fertility continue to be as important today as in the "prepesticide era." In these crops the close spacing makes them more competitive with weeds but prevents cultivation for weed control after planting. Some weeds germinate with the crop, and chemical weed control is employed on more than half the acreage to ensure that the crop gets a competitive advantage over weeds early in the season. In much of the cereal-growing acreage, moreover, moisture is a limiting factor. This is particularly true in western Canada, where as much as a third of the 70 million acres devoted to cereal grains is left fallow on a rotating basis so that 3 years of moisture can be conserved for 2 years in crops. In this system weeds must be controlled also during the fallow year to conserve moisture (Friesen, 1973).

Before the development of selective herbicides, wild mustards and other broadleaf weeds were the most prevalent weed problems in small grain crops. These weeds are now so easily controlled with 2,4-D that they are no longer regarded as a major problem, the use of this herbicide in small grains production having been the most widely accepted herbicide program for more than 25 years (Klingman and Ashton, 1975). Depending on variations in climate, grain prices, and weed infestations, 50 to 80% of the small grain acreage in North America is treated with 2,4-D or MCPA, a closely related phenoxy herbicide. In 1975 more than 36 million lb of 2,4-D were used in U.S. agriculture, mostly in the cereal-producing states (Von Rumker et al., 1975). In that same year, more than 7 million lb of 2,4-D and 3 million lb of MCPA were used in the small grain-producing areas of western Canada (Daciw, 1976).

Since the introduction of 2,4-D, wheat yields have doubled in North America (Furtick, 1970). While many factors are responsible, it is generally accepted that, without chemical weed control, weeds will reduce potential small grain yields by 25 to 50% (Friesen, 1957). The application of 2,4-D at 0.25 to 1.0 lb/acre normally results in at least a 15% yield increase (Hay, 1970; Mukula, 1970).

Twenty years of repeated 2,4-D use in cereal grains has brought about a dramatic shift in problem weeds—from the sensitive wild mustards to broadleaf weeds, annual grasses like the foxtails (*Setaria* sp.), and wild oats (*Avena fatua* L.)—that are as tolerant to 2,4-D as the grain crops themselves. Wild oats, introduced from Europe less than 100 years ago, now infests more than half of the grain-growing acreage in western Canada (Alex, 1966).

These new weed problems in small grains are receiving much attention; in 1975 more than 10 million acres in western Canada were treated with new or different herbicides to control 2,4-D resistant weeds (Daciw, 1976). Unfortunately these new herbicides are more costly than 2,4-D. With the unpredictability of markets, the economics of weed control on these low "per acre return" crops is not always assured.

Weed and Brush Control on Pastures and Rangelands

Almost half of the total land area of the United States or about 1 billion acres (Klingman, 1970) plus another 40 million to 60 million acres in Canada are used as improved pasture or grazing lands. On nearly all of this land, forage production is reduced by weed or brush growth. Most of these woody species are native to North America and have spread rapidly into natural grasslands where grazing pressure from domestic animals has made the native grasses less competitive.

Herbicides such as dalapon or paraquat are used for control of existing weedy vegetation on intensive pastures, while on rangelands woody species are controlled by 2,4-D; 2,4,5-T; dicamba; or picloram, usually in combination with mechanical methods and reduced grazing. For many introduced weeds, especially those poisonous to livestock, biological control programs are being developed.

The percentage of rangeland and pasture acreage treated with herbicides is very low, mainly because of the low economic return per acre. Currently the invasion of grazing lands by woody vegetation annually exceeds the acreage on which control measures are applied (Klingman, 1970). If world demand for meat increases as expected, we shall have to reverse this trend by more intensive use of both biological and chemical methods of weed control to maintain productive rangeland.

4.3 PLANT DISEASES AND THEIR CONTROL

In North America 100,000 plant diseases are known, and while not all of these are annually important, many are; 8000 species of fungi, 500 of nematodes, 250 of viruses and 160 of bacteria cause significant diseases of our economic crops. On a crop basis 115 diseases are reported on tomato, 196 on potato, 77 on wheat, 112 on corn, and 200 on apple. In the United States the loss from plant diseases in agriculture has been estimated at $3.5 billion with an additional $100 million spent for fungicides used in control (Alder, 1970). An epiphytotic of southern corn leafblight in 1970 caused a $500 million loss in corn production.

In many cases the loss from diseases in crops is not as spectacular as that caused by insects and not as obvious as a crop overrun with weeds. More frequently the loss is insidious, manifested as a reduction in plant vigor, death of a percentage of plant seedlings, failure of blossoms to set fruit, failure of fruit to ripen evenly, and a host of other effects that reduce quantity and/or quality of the product. Diseases can, however, be dramatic in effect and rapid in onset. An attack of late blight in potatoes can sweep an area in a few days and transform lush, green vines into a mass of putrid decay.

In some ways plant diseases are similar in effect and character to the ravages of insects. As soon as the seed is placed in the soil, it is attacked by disease organisms, many of which, if successful, will prevent its germination and successful emergence as a seedling. This is especially true for large-seeded vegetables. When seeds of crops such as beans, corn, peas, lima beans, and cucurbits are placed in the soil, a high percentage will fail to produce a successful plant unless protected by a fungicide against seed-rotting organisms and seedling blights. Early planting, when the soil is cool and seed germination slow, favors the development of diseases, but most of these crops demand early seeding if the best yields are to be obtained. For many years fungicides containing mercury were used as seed protectants, and in some parts of the world, such use is being continued. Discovery of environmental and health hazards associated with mercury has resulted in the discontinuance of this practice in North America. Today most of the seed is treated with thiram, captan, or related fungicide to protect against soil-borne seed pathogens and, to a lesser extent, seed-borne pathogens. Failure to use treated seed may result in stand reductions to the point where a successful harvest will not be achieved. Seed treatments frequently include both a fungicide and insecticide, because both diseases and insects are destructive at the seedling stage. Where effective control can be had through this practice, protection is provided at minimum cost and use of pesticide. In most cases pesticides are used for seed treatments at low rates, less than 1 oz/acre and, in many cases, only a few grams. The practice of seed treatments has been so successful that most seedsmen supplying seeds for corn and other large-seeded vegetables treat the seed as a routine measure.

Ensuring a crop stand is only one of a number of steps in disease control. Leaf blights and other foliar diseases are endemic in most of our agricultural areas for many of our most important crops. In some cases, for example, rust on wheat, breeding for resistant varieties has been a successful and continuing method of control. Contrary to the history of poor success in breeding for plant resistance to insects, breeding for disease resistance has been highly rewarding. Resistance to our most important diseases is now incorporated in major cultivars of many crops, including many of the cereals, forages, and vegetables. Breeding programs for disease control are obviously the method of choice, because, once established, the annual cost is minimal. Inherent in any breeding program is, however, the requirement to monitor the pathogen constantly to detect the development of new races that may overcome the resistance incorporated. In the rust program in cereals in western Canada, this has always been a major focus of study, and the failure to predict occurrence can be dramatic. The great loss from southern corn leafblight in 1970 was due to race T, a race for which resistance in much of the corn acreage had not been incorporated (Ullstrup, 1972).

In some instances resistance to disease may be only partial. Late blight of potatoes is a good example. This disease is caused by a fungus whose spores are wind borne and germinate on the surface of susceptible foliage (many species of the Solanaceae) when temperature is favorable and moisture abundant. The ability of this disease to destroy the entire potato crop made history in the potato famine of Ireland, an event (several years) that prompted mass emigration and brought many settlers to American shores. The late blight fungus occurs throughout the world where potatoes are produced; forays to the Andean home of this vegetable have produced some genetic material with a degree of tolerance. To some extent, tolerance (or resistance, the terminology is rather arbitrary and suggests degree of susceptibility) has been incorporated in some varieties. Despite this, however, early field infection with the late blight fungus will reduce yield and, in most commercial varieties, cause tuber rot, which, in storage, is likely to spread to completely destroy the crop. The same fungus attacks tomatoes, and both foliage and fruit are destroyed.

Fortunately, fungicides are available to protect plants from invasion by this fungus. Because of the source of the infection (wind-blown spores) growers use a preventive approach with regular applications of a fungicide to keep new foliage and, in the case of tomatoes, fruit protected. No other approach has proved satisfactory.

There is one sideline to the late blight fungus that is worthy of note. It has been suggested that, in areas where late blight in potatoes is common, the incidence of spina bifida is high. This condition in the newborn in which the distal portion of the spinal column is divided is said to be correlated with consumption of late blight-infected potatoes by the mother during pregnancy (Anonymous, 1972). The evidence supporting this correlation is indeed fragmentary.

In the earlier part of this chapter we discussed the need for insecticides in the production of tender fruits and vegetables. In general these same crops need protection from diseases if they are to produce a salable harvest. As in the earlier discussion apples are again a good example. All commercial varieties of apples are susceptible to a disease called apple scab. The disease is caused by a fungus that overwinters on dead leaves in the apple orchard. In the spring, spores are discharged from these infected leaves and are wind borne to new foliage. Here, provided weather conditions are favorable, the spore germinates, penetrates the new leaf, and produces a lesion that repeats the cycle and provides spores for later infections. As the season progresses and the fruits develop, scab lesions develop on both fruit and foliage, causing vast scars on the former, distortion and abscission in the latter.

Favorable climatic factors must be present for the fungus to infect new tissue. The conditions are precisely known, tables having been developed to indicate the duration of free moisture needed at any given temperature in order for a new lesion to be formed. Favorable conditions in Ontario or New York occur several to 20 times during a gowing season.

The disease reduces tree vigor and affects the quality of the fruit. Mild infections, late in the season, produce "pinpoint" scab lesions, and fruit so affected is normally acceptable only for processing. Such fruit stores poorly, however, for the lesions enlarge in storage to cover most of the fruit surface. Apple scab infections early in the season result in badly scarred, misshapen fruit that are unsalable. Unprotected orchards, in most years, produce no marketable fruit.

Apple growers rely totally on fungicides to protect their crop. Again the fungicides must be applied largely as a protective measure to arrest spore germination and prevent penetration of the leaf or fruit surface. In Ontario and New York, growers use 6 to 15 applications of fungicide per year. The wetter the year, the more frequent the applications. As with insect control, some progress has been made in monitoring weather conditions and applying fungicides only when the moisture and temperature conditions permit spore germination. Growers' acceptance of this procedure is improving and may result in use of fewer fungicide sprays and better control of apple scab.

Thus plant diseases are important in agriculture, and though the preferred methods of control are through cultural practices, crop sanitation, and breeding, there are many situations in which these methods are not adequate. To meet these situations, some 50 million lb of fungicides are used each year in North America to control seed, foliage, and fruit pathogens.

4.4 VERTEBRATE PESTS AND THEIR CONTROL

A number of species of birds and mammals are pests of agriculture, and though pesticides are not now used in great quantities for their control, this picture may change as the demands for food become more acute and we continue to move to control anything that interferes with the quality and quantity of food we produce. Although most have recognized rodents, expecially rats and mice, as pests, few readers may realize the extent to which other mammals, and birds, are involved.

Rodents

Rodent control has been a requirement in agriculture throughout North America. Rats and mice destroy large amounts of grain in storage each year and are important pests in all areas where animals are being fed. In the orchard, both meadow mice (*Microtus* sp.) and pine mice (*Pitymys* sp.) are recognized pests, the former removing bark and girdling trees and the latter feeding underground on roots, removing small roots and the bark from larger ones. Rodent control takes various forms and must be modified to suit the particular situation. Rodenticides are used, but their use around livestock feeds and feeding stations must be controlled carefully so that feed is not contaminated. Each year some 15 million lb of rodenticides are produced in the United States, about 80% of which is warfarin (Lawless et al., 1972). Estimates on usage suggest, however, that 10% of that amount is applied by farmers, the largest use obviously being around the home (NAS, 1975). Orchardists employ a number of chemicals for control. During the 1960s, endrin was used at a high dosage rate (4 to 6 lb AI/acre) applied to the orchard floor (and especially around the base of trees) late in the season after the fruit had been harvested. Toxaphene has been used also in a similar manner. Baits (usually containing strychnine or zinc phosphide) are used widely in some areas, but results are not uniformly satisfactory.

To a large extent, rodent control both in the orchard and in buildings involves exclusion techniques, wire or plastic tree guards in the former and good construction in the latter.

Predators

Certain phases of livestock-production are plagued with predation by large animals. Whereas the sheep industry is the most vulnerable, cow-calf operations where calving occurs in the pasture also suffer loss. The predators involved vary from region to region. Coyotes and wolves are most important, but in some areas the mountain lion, bobcat, jaguar, and, occasionally, the bear are responsible. In addition, weasel, mink, fox, and skunk may ravage poultry.

For many years, and in various jurisdictions throughout North America, bounties have been paid on predators. This has been the subject of much controversy from the standpoints of both logic and effectiveness. In addition, there are many wolf and coyote admirers who claim, with some justification, that the bounty system does not discriminate between the predator that is foraging in a "natural" environment and feeding on mice and other small, wild animals (to the benefit of man) and that which is destroying sheep or other livestock.

Predator control activities against coyotes and wolves have generally been ineffective, and where they have been effective, the cost of control has greatly exceeded the loss that would have occurred in the absence of such efforts. Control activities have included shooting, trapping, use of poision baits, and lacing of fresh kills with a potent toxicant (strychnine, 1080). This last method has been the subject of much criticism, for it is unselective and poisons any animal or bird that feeds on carrion.

In addition to predation on livestock and poultry, agriculture is also plagued with skunk and bear damage in apiaries and, in some areas, damage by big game (deer, elk, antelope, moose) to cultivated cereal, vegetable, and fruit plantings. Taken on the average, the loss is minor indeed, but it may be severe to the individual producer.

Birds

Birds can be important agricultural pests, and though their depredations are usually somewhat local, this is not always the case. To a large extent, they are pests of fruit and cereal crops, but some predatory birds have been incriminated as pests of livestock and poultry. The early concept of hawks and owls as predators of the unprotected chicken flocks of our early settlers and the observed killings of lambs and kids by (most notably) the golden eagle resulted in these birds' being excluded from protection under the Migratory Bird Treaty Act of 1936. Human settlements and agricultural practices have tended to separate hawks from chickens, and the eagle predation on sheep is perhaps less than that once claimed.

A number of songbirds (including the robin), starlings, and blackbirds are damaging to tender fruits. Grapes, cherries, blueberries, peaches, nectarines, and apricots are prime targets, and damage may vary from slight to near total destruction. In many cases the taste for fruit is transient, and after a few days the birds move on to something else. In other cases, once fruit has been found, the birds return daily until the harvest is completed. Thus the extent of attack and its duration are unpredictable.

Attempts at control in fruit plantings have involved largely scare devices (scarecrows, shotguns, or timed detonations of firecrackers), but success depends on constant surveillance with alternations in scare devices as the birds become accustomed to them and are repelled no longer. In any event, success is usually local, the birds moving to adjacent plantings.

In some situations large acreages may be involved. Blueberry growers in New Brunswick have had massive invasion of plantings by robins and other songbirds. Attack has been so severe as to prompt legal action to recover losses, the aggrieved claiming that the spray programs against the spruce budworm was depriving the birds of their normal food supply (insects) and that as a result they had moved to the blueberry patch (Stevenson, 1976). Regardless of the cause the bird depredation was real. In the vast acreage involved, scare devices were of no great value (remote controlled model airplanes to simulate hawks were used among other things), but experimental results using methiocarb and other bird repellents showed some promise.

Perhaps the greatest damage in terms of economic loss is caused by blackbirds (mostly red-winged blackbirds, but starlings and cowbirds may be involved also) in corn. The population of red-winged blackbirds in parts of the corn belt (Ohio, New York, and Ontario have been hard hit) has reached enormous numbers. Whether this reflects an increased carrying capacity in the area because of improved food supply or is a transient situation is not known. Millions of these birds breed in marshy areas (especially around the Great Lakes) and, in recent years especially, the lush alfalfa plantings in the Midwest. As the juveniles are fledged during the summer, they join their parents in forays into agricultural crops and are found in massive aggregations as field corn is developing. The birds are attracted especially to corn when in the milk stage, and it is during this period that most of the damage occurs. Estimates of damage vary widely. Loss in Sandusky County, Ohio, was estimated in 1 year at $866,000 (Kottman, 1967), and a realistic appraisal perhaps approaches 1 to 2%. Damage is greatest near breeding areas but can be locally severe at locations quite removed. The birds move long distances, and once a taste for corn is established, a flock may return to the same feeding ground repeatedly. Whereas the average loss would not warrant control, local losses may be as high as 50%, and there is increasing pressure to develop a control strategy (Dyer, 1969). Scare devices have not been satisfactory and result only in moving the flock to a new location. The use of Avitrol® has received attention in recent years and is being promoted in some areas. This material is supplied as a bait in which a small percentage of the substrate particles (usually cracked corn) is treated. Birds eating the treated particle react vocally and emit distress calls that scare away the rest of the flock. There is still some question concerning the effectiveness of the method on a large-scale basis. Treatment is costly and is realistic only in areas of heavy loss. These same groups of birds that cause loss in corn in the field may also be pests at cattle feedlots and, in the case of starlings, persist throughout the year. Losses have been high. Besser et al. (1968), using cattle feed in the bird crop as an indicator, estimated the loss at $84/year/1000 starlings and $200/year/1000 red-winged blackbirds.

Losses from birds occur also in agriculture through removal of seed (especially corn and other cereals) by pheasants and crows, after planting by blackbird and starling feeding in corn cribs, and by migratory waterfowl in wheat fields in western Canada. Losses have been estimated at $50 million to $100 million in the United States, but since little research has been done in this area the accuracy of any figures must be questioned (NAS, 1970).

The whole matter of vertebrate pest control is in a state of turmoil. Some argue that agriculturalists have, through their modern technology and intensive agriculture, created their own problems and that, in the interests of conserving species that, in the main, benefit man and add aesthetically to his environment, some damage will have to be tolerated. Others maintain that food production is of such high priority that losses cannot be accepted. It is obvious that some form of middle course will prevail, but experience with insect control should not be ignored. It may be possible to develop pesticides that will provide control, but surely other approaches toward management of populations must be explored. There is now great concern for humane meth-

Table 4.1 Percentage of Losses in Major Agricultural Crops on a World Basis Due to Insects, Diseases, and Weeds[a]

Crop	Insects	Diseases	Weeds	Total
Wheat	5	9	10	24
Oats	8	9	10	27
Barley	4	8	9	23
Rye	2	3	10	15
Rice	27	9	11	46
Millets and sorghums	10	11	18	38
Maize (corn)	12	9	13	35
Potatoes	5	22	4	32
Sugar cane	20	19	16	55
Citrus fruit	8	9	4	22
Grapes	3	23	10	37
Oil crops	11	10	11	32
Vegetables	9	10	9	28

[a]After Cramer (1967). Estimates are based on various production (quantitative) tables from Cramer and are rounded to whole numbers. Thus "Total" figures may not equal the sum of the 3 preceding columns.

ods of disposing of any vertebrate pest, but we are far from being able to define what this may be. The shotgun is quick and effective when a sensitive target area is struck but may cause excessive suffering if the target is missed and a crippling wound inflicted.

Some attempts have been made to quantify pest losses in agriculture on a worldwide basis. Data compiled by Cramer (1967) represent the most nearly comprehensive treatment of the subject (Table 4.1) and indicate that world cereal losses are approximately 35% of the potential production. This represents a tremendous opportunity to improve production by applying pest control information already in hand and utilizing tools, including pesticides, now available.

5

THE ROLE AND USE OF
PESTICIDES IN THE FOREST
AND ON RIGHTS-OF-WAY

The forests of North America are vast and comprise more than one-third of the land area. In the United States (excluding Alaska) 750 million acres are classified as forest, about two-thirds of this bearing commercial timber. Canada's forests, estimated at 550 million acres, represent more than 3 acres of forest for each acre that has been put to the plow.

Forest pests are many, insects and diseases leading the list. The degree to which an organism is a pest is determined, in large part, by the use ascribed by man to a particular forest area. Most estimates of loss are presented in terms of merchantable timber. Graham (1963) summarized losses in the United States, reporting such losses as that due to tree mortality and that due to growth reduction. Losses for 1952 were 158 billion bd ft, the greatest component of which was a growth loss of 60 billion bd ft due to tree diseases. In that year diseases caused tree mortality estimated at 9 billion bd ft, while insect damage was 12 billion bd ft due to tree mortality and 9 billion bd ft of growth reduction. Thus 55% of all timber losses in that year were due to insects and diseases; 9% of the growth loss was due to domestic livestock and wildlife. To put these losses in prespective, note that 1 billion bd ft is approximately the amount of lumber required to build 100,000 American homes.

As in agriculture, losses in forests are not uniform from year to year or from one region to another. Depending on the age of tree, type of tree involved, and the mixture of timbers in a particular forest,

losses from either insects or diseases may be less or moɪ
Outbreaks of a particular insect or disease are sporadic and d
not only on the type of forest but, perhaps more importantl;
climatic factors. Outbreaks of the spruce budworm, for exaɪ
correlated with several successive seasons of dry, mild spring weather.
Outbreaks of forest pests can and have been of great economic significance. Graham (1963) lists "catastrophic" losses from outbreaks of
specific pests for the period 1900–1952 as 52 billion bd ft from six
insects' outbreaks (four insects involved, viz., spruce budworm,
mountain pine beetle, western pine beetle, Engelmann spruce beetle)
and 18 billion bd ft from two epiphytotics of chestnut blight. In New
Brunswick heavy infestations of the spruce budworm for 1 year in the
spruce-fir forests are estimated to cause a loss of 10 to 33% of merchantable standing crop (Marshall, 1975). Such estimates do not account in any way for the increased susceptibility of forests to fire after
they have been ravaged by a pest that causes high tree mortality.

Foresters have recognized for many years that certain silvicultural
practices could affect the severity of pest problems. Because many of
the tree diseases are "old age" problems, selective cutting of mature
trees reduces loss. Because the spruce budworm prefers mature fir-
spruce stands in large, contiguous blocks, harvesting practices that
minimize such susceptible stands are beneficial. The extent to which
such practices are practical may, however, be limited. Especially in
Canada, large tracts of forests exist, many rather inaccessible and, up
until now, far removed from sawmills. Under such circumstances
selective cutting is uneconomic and the only silviculture practices
available are those that nature provides. Nature's way has often been
harsh with periodic devastation by insects, ravages of fire and wind,
and regeneration to repeat the process. Through these processes stability has been maintained in the long view, but such stability has had
dramatic periods of instability and catastrophe for the life forms that
call the forest home.

The advent of the airplane permitted the forester to view the
forest from a perch better than a tree top and to survey pest damage in
a comprehensive way. It also provided a means for mass pesticide
treatments to the forest, a concept impractical with ground equipment.
Although some pesticides were applied as dusts for insect control
during the 1925–1945 period, the move to large-scale treatments
awaited the discovery of DDT. In addition to insecticides, limited use
of herbicides, fungicides, and rodenticides has been a part of modern
forest husbandry, the degree of use dependent on the role of the forest
in the (usually) local setting.

5.1   INSECT CONTROL

Since 1945 approximately 31 million acres of forest have been treated for insect control in the United States, and 80% of this acreage was treated to control two insects, the western budworm and the gypsy

Table 5.1   Aerial Application of Insecticides Against Forest Insects in the United States[a]

Year	Acreage (1000 acres) Treated for Insect Listed			
	Western budworm	Gypsy moth	Spruce budworm	Douglas fir tussock moth
1945	--	5	--	--
1946	--	61	--	--
1947	--	107	--	414
1948	4	211	--	--
1949	267	390	--	--
1950	938	583	--	--
1951	916	178	--	--
1952	669	202	--	--
1953	508	180	--	--
1954	75	1,137	20	255
1955	2,333	1,083	--	--
1956	1,367	926	--	10
1957	1,445	3,395	--	--
1958	936	516	314	--
1959	127	115	7	29
1960	119	65	242	13
1961	--	142	66	1
1962	1,027	427	57	--
1963	1,220	414	498	12
1964	684	135	58	--
1965	8	226	--	245
1966	148	275	--	--
1967	2	203	101	--
1968	6	231	10	--
1969	8	91	18	--
1970	--	223	212	--
1971	9	391	10	--
1972	--	177	500	--
1973	--	189	473	70
1974	--	171	434	466
TOTAL	12,816	12,684	3,020	1,260

[a]From Appendix D, Nat. Acad. Sci., 1975a.

moth (NAS, 1975a) (Table 5.1). At some time during this 30 year period, however, aerial applications of pesticides have been made
against 33 other insect species; 2 of the 35 insects are discussed here,
since treatments for their control aroused much public controversy.

Pesticide applications to control the gypsy moth both in New
England and neighboring states have been the subject of much debate, and their condemnation was a main thrust of Rachel Carson's
treatment of the pesticide issue (Carson, 1962). The gypsy moth is an
imported species, having been brought to America from its Eurasian
home in 1869 in an attempt to develop a silk-producing species that
would thrive under the Massachusetts climate. Thrive it did, but its
silk-producing potential was limited, the insect being interested in
supplying only enough for its own survival. Escaped moths established well, and in less than 20 years their offspring were defoliating
the beautiful deciduous trees, so prominent a feature of the New England landscape.

Vast amounts of money have been spent to control the gypsy moth
in New England and to prevent its spread to other areas. The fact that
up until recently it was confined to the Northeastern United States (a
bit in Canada) suggests that these efforts have been at least successful
in part. In recent years it has invaded new territory to the south and
west.

As with so many other insects the gypsy moth seems to survive in
low levels for many years and then for an unknown reason erupt to
epidemic numbers. Populations within this epidemic phase of the
insects' existence are not constant, and may fluctuate widely, with
the resulting numbers reflected in severity of defoliation. Studies on
defoliation in relation to tree survival have not been in agreement. Oak
mortality in Massachusetts ranged from 10 to 80% after 3 years of
"heavy" defoliation (Tierney, 1947), whereas studies in Connecticut
suggest that defoliation on this same species results in only 5% mortality and that mortality in years after a severe defoliation is less than
normal (Collins, 1961). More recent studies make it clear that tree
mortality from gypsy moth defoliation is a major loss factor and that
trees under stress from poor soil-moisture conditions sustain much
higher mortality than those growing in more favorable environments
(Anderson and Gould, 1974). Although evergreens are not preferred
hosts for the gypsy moth, attacks can result in high tree mortality,
especially in certain species of hemlock.

The extent of heavy defoliation by the gypsy moth is indicated, in
part, by the figures on acreage aerially treated for control (Table 5.1),
for treatments have been applied to areas where egg counts indicated
extensive potential defoliation. Defoliation in Connecticut, 1969–
1974, has been documented (Anderson and Gould, 1974), and figures

for the entire infested area of the United States for 1971 and 1972 are available (NAS, 1975a). The latter data indicate slightly less than 1.5 million acres infested in each of these years.

Pesticides have been used extensively for gypsy moth control. Lead arsenate was the first used on a wide scale, but DDT was more effective and replaced it after 1945. Aerial application of this compound at $\frac{1}{2}$ to 1 lb/acre provided outstanding control. More recently (since 1966) carbaryl has been used and, to a much lesser extent, insecticides derived from the bacterium *Bacillus thuringiensis* Berliner.

Whereas the gypsy moth has attracted most attention in the east, controversy in western circles has been directed in recent years to the tussock moth and its attack on the highly valued firs of the Pacific Northwest. The tussock moth is a native insect that periodically erupts to cause extensive damage. Five outbreaks have occurred during the past 40 years, and for the first four of these attacks DDT was used for control with the result that the outbreaks were arrested and extensive timber loss prevented. The insect feeds on the foliage, and this causes tree mortality. In addition it sheds hairs to which many people are allergic. This hinders logging operations and reduces recreational enjoyment in infested areas. The most recent outbreak began in 1971 and is detailed in the Report of the National Academy of Sciences (1975a). What follows is drawn heavily from that account.

Scattered infestations in the Wenatchee and Okanogan national forests in 1971 caused some tree mortality, but much of the dying timbers were harvested during the next year. A virus had struck the insect population with resulting high mortality. In 1972, however, new outbreak centers appeared and the infestation became more extensive. Insecticides other than DDT had not been evaluated against the insect, and testing of two new compounds did not demonstrate effective control. The result was that treatment for control was not employed, and 30 million bd ft of timber were lost.

In 1973 the United States Forestry Service and several other parties concerned with the timber industry sought permission to use DDT to control the tussock moth on 0.5 million acres of infested forest. The request was denied, and an estimated 600 million bd ft of timber was killed. About 100 million bd ft of this could not be salvaged.

The severe losses sustained in 1973 intensified the pressure on the Environmental Protection Agency to relax their prohibition on DDT use. This they did, and aerial application of DDT began in Washington early in June (NAS , 1975a).

The controversy over the use of DDT to control the tussock moth underscores the difficulties in assessing a benefit-risk equation (Chapter 2). The interpretation of values to assign on each side of the equation cannot be stated in economic terms only when environmental, political, and social factors defy monetary conversion. No matter what decision is taken, divergent opinions will persist among the public, scientists, and politicians. The superficial economics of DDT use for tussock moth control clearly supported its use in 1974 (and would have for 1973). Whether or not there are costs not now apparent remains to speculation.

To date the use of pesticides for insect control in United States forests has been minimal, averaging only 1 million acres/year (over the past 30 years) on an inventory of 500 million acres of commercial timber. Although minimal, such use has been critical; despite arguments to the contrary it is apparent that destructive insect outbreaks have been halted and the spread of the gypsy moth has been delayed.

Forest insect control has been a more prominent recent feature in Canada than in the United States. Outbreaks of forest insects are not new to Canada; destruction by pests such as the spruce budworm, the larch sawfly, the European spruce sawfly, and the pine sawfly are a matter of record. Such losses in the past were not of great significance, for the vast lumber supplies exceeded by far the annual harvest and much of the acreage was not readily accessible to local sawmills. But times have changed; today all provinces recognize the finiteness of their timber reserves, at least in key species essential to their markets. Perhaps nowhere is this more apparent than in New Brunswick, where forestry represents 20% of the economy and is to a large extent dependent on balsam fir and spruce for its sustenance.

Although the insect pests of Canadian forests include several species of sawfly, looper, and weevil, the nemesis in the past, present, and for the foreseeable future is the spruce budworm, and it is against this pest that massive aerial spray programs have been directed.

The spruce budworm, native to North America, exists in several races, all quite similar in general habits but relatively isolated geographically. In the vast forests of eastern Canada, the eastern spruce budworm is dominant. This insect overwinters as a small dormant larva in silken shelters embedded among the needles of its host. In early spring (late April to early May) the larvae complete their diapause and begin to feed on old and new vegetation. There are six larval instars, the last one being most destructive. Upon completion of the sixth larval stage, the insect pupates, and about 10 days later emerges as a small, brownish moth. These moths (in New Brunswick) appear in July and mate, usually on the first day of adult life. Females are unable to fly until 2 or 3 days old, and during this time they deposit a few egg masses on tree foliage near the point where they emerge. Given adequate food supply for the larvae from such egg masses, this routine ensures a surviving population at home base. After 2 days, the female takes flight, a flight that follows a consistent pattern and guarantees dispersal. The moths take to the air in late evening and ascend to altitudes of 1500 to 3000 ft, the height of ascent determined by the location of a warm air mass over the forest canopy. Having attained this goal, the moths fly for about 2 hours in the general direction of the wind currents (with the wind) but at an angle of some 30°. Such flights may carry the moths 50 or more miles in a single night, and flights may be repeated for several nights. Return to the forest canopy seems equally ordained as the ascent, a positive, directional descent being executed.

Outbreaks of the spruce budworm require the availability of a suitable food supply over a large area and a succession of years in which spring weather is warm and dry. This happy coincidence preceded the massive outbreak currently under way in eastern Canada and is believed to have occurred to initiate at least six other outbreaks since 1700 (Baskerville, 1975). Baskerville (1975) outlines the synchrony that has developed between the spruce budworm and the fir-spruce-birch forests of eastern Canada and Maine by recording the pattern of budworm infestations in the forest of Green River, New Brunswick. According to this account large areas of mature and overmature forest prevailed in this region in the mid 1870s. The forest was dominated by balsam fir and within its boundaries contained areas of dense younger trees with balsam fir again the main tree species. In 1877 an outbreak of the spruce budworm occurred that killed most of the fir in the mature and overmature forest, leaving behind the more tolerant spruce and the nonsusceptible birch. In addition the forest floor in this area contained a "thick mat of fir and spruce regeneration." In the immature stands most of the trees survived. By the early part of this century the areas that in 1877 were young, dense stands dominated by balsam fir were mature stands still dominated by balsam fir but containing a fair percentage of large spruce and birch. The understory fir and spruce surviving in the mature forests of 1877 were now a dense, young forest. Thus in 1912 the food supply was again ideal for the budworm, and several consecutive warm, dry springs resulted in an outbreak that by 1919 had "wiped out the fir component of the older stands."

As in the 1877 attack, the fir component of the younger forest was thinned during the 1913–1919 attack, and a regeneration of fir and spruce survived the attack in the mature forest. By 1949 the immature fir-spruce forest of 1913 was mature, each tree carrying a crook in its stem some 20 to 35 feet from the ground as evidence of the top killing that had occurred in the budworm outbreak 36 years earlier. The understory fir and spruce of the mature forest of 1913 was now a dense stand. Following favorable weather, an outbreak was initiated in 1949 and by 1958 most of the fir component in the mature stands that were not treated for control had been killed (Baskerville, 1975). That portion of the mature forest that was treated survived. Thus the spruce budworm, if left unchecked by man, has a cycle in New Brunswick in which it "harvests" mature fir (about 75 years old) and permits sufficient survival of younger trees (30 to 40 years) and understory (2 ft growth) to ensure its future food supply.

The spruce budworm outbreak that began in 1949 was to have a different conclusion than that of the earlier attacks. In the late 1940s, experiments in Ontario had demonstrated that DDT would provide control of this insect and prevent heavy defoliation. Thus, when New Brunswick forests were threatened with high fir mortality, a decision was made to apply DDT by air. Starting in 1952 with the treatment of 200,000 acres at approximately 1 lb of insecticide/acre, treatments continued during the next 6 years. During this period more than 7 million lb of DDT were applied to slightly more than 6 million acres (more than 14 million acres were sprayed but some acreage received several applications) (Table 5.2). Using DDT at the high rate of 1.0 lb of active ingredient/acre, the foresters hoped that budworm mortality would be so high that the outbreak would be stopped. Despite high percentage mortality of larvae (approaching 99%) this did not occur, and in subsequent years, the dosage was decreased, the concept being to prevent severe defoliation and tree mortality (Kettela, 1975).

Whereas the most prominent effect of the DDT spraying in New Brunswick was the preservation of the forest, it was learned early that aquatic life was being affected, high mortality being noted in several species of fish (see Chapter 15). To reduce this damage, the dosage of DDT was reduced to 0.25 lb/acre in the early 1960s, but at this level two applications were required for effective control of the budworm. In addition phosphamidon was substituted for DDT for treatment of forested areas within ¼ mile of major streams. This procedure reduced DDT contamination of waterways, and though phosphamidon did not cause problems in stream biology, some mortality of songbirds occurred. More recently the use of DDT has been discontinued and fenitrothion at the rate of 3 oz active ingredient/acre substituted. Between 1952 and 1973 about 50 million acres of New Brunswick forest was sprayed for spruce budworm control. This involved the use of 15 million lb of DDT and more than 3 million lb of fenitrothion (Miller and Kettela, 1975).

Table 5.2   Insecticide Treatments for Control of the Spruce Budworm in New Brunswick, 1952–1973[a]

Year	Acreage Treated[b]	Pesticide Used	Rate Ai/Acre
1952	0.2	DDT	1.0
1953	1.8	DDT	0.5
1954	1.1	DDT	0.5
1955	1.1	DDT	0.5
1956	2.0	DDT	0.5
1957	5.2	DDT	0.5
1958	2.6	DDT	0.5
1959	0.0	--	--
1960	2.6	DDT	0.5-0.75[c]
1961	2.2	DDT	0.5
1962	1.4	DDT	0.5
1963	0.7	DDT, phos.[d]	0.5
1964	2.0	DDT, phos.[d]	0.7
1965	2.1	DDT, phos.[d]	0.5
1966	2.0	DDT, phos.[d]	0.5
1967	1.0	DDT, phos.[d] fenit.[d]	various
1968	0.5	DDT, phos.[d] fenit.	various
1969	3.1	fenitrothion	0.15
1970	4.2	fenitrothion	0.15
1971	6.0	fenitrothion	0.15
1972	4.6	fenitrothion	0.15
1973	4.2	fenitrothion	0.15

[a] From Miller and Kettela, 1975.
[b] Millions.
[c] From 1960 to 1968 DDT was applied (in most cases) twice each year at a dosage rate of 0.25-0.33 lbs per acre per application.
[d] Phosphamidon used adjacent to watercourses, fenitrothion also used.

Total spray programs in eastern Canada (primarily Quebec and New Brunswick) in 1975 and 1976 have involved almost 15 million acres each year. Whereas fenitrothion is the insecticide used most widely, trichlorfon, dimethoate, and phosphamidon are also being used. The infestation of spruce budworm is massive and rated as moderate to severe in 25 million acres in Ontario, 85 million in Quebec, and 15 million in New Brunswick. Severe defoliation is also occurring in parts of Manitoba and in Nova Scotia, Prince Edward Island, and Newfoundland.

Early hopes that insecticide sprays would end an outbreak of the spruce budworm have not materialized, and forest managers have been faced with the reality that only year-to-year defoliation can be prevented. Given this fact, cost-benefit analyses with respect to New Brunswick forests indicate clearly that the spray programs make good economic sense (Marshall, 1975).

For the first time in history the fir-spruce-birch forests of eastern Canada have survived an outbreak of the spruce budworm, and it is apparent that insecticides provide the means of continuing this survival. What is not clear is whether or not forest management practices involving massive pesticide use of the past 25 years have changed a sporadic outbreak into an annual attack. Forest managers are uneasy and with good cause. The forests are green, but the price of keeping them green is escalating and the budworm shows little inclination to retire (Baskerville, 1976).

5.2 TREE DISEASES

With two exceptions, white pine blister rust and Dutch elm disease, pesticides have played a minor role in diseases of forest trees. Even with these diseases pesticide use has been limited and somewhat indirect. White pine blister rust is caused by a fungus that requires two hosts to complete its life cycle. Spores of the disease are produced on various species of *Ribes* (currents and gooseberries) and upon discharge are carried by the wind. The spores survive only a short time in moving air and thus only trees within a short distance (a few hundred ft) of the source become infected. The disease attacks five-needle pines, the white and sugar pines valued highly as a commercial construction and cabinet lumber. The disease is especially destructive to young pines, with older trees somewhat resistant. The fungus grows within the needles, and as it progresses, cankers containing fungal fruiting bodies are produced on the bark. Spores from these fruiting bodies are blown to nearby species of *Ribes* to complete the life cycle. The disease first caused extensive damage about 1918–1919, mortality of young pine approaching 100% in some areas. Foresters attacked the problem by attempting to eradicate *Ribes* mechanically; especially during the depression years of the 1930s "blister rust crews" worked white pine areas in both eastern and western United States. Whether because of these efforts or some other factors, blister rust losses were reduced and the program was believed to be successful (Martin, 1944).

As labor became more expensive so did the cost of the *Ribes* eradication program, and the hope for effective control and eradication by mechanical means became dim. Concurrent with increased labor costs, herbicides effective for *Ribes* control became available, and to some extent these were substituted for the mechanical methods used earlier. Herbicide usage for this purpose has not been great. The concept of controlling the disease by *Ribes* eradication has been questioned, and in most areas the effort has been abandoned (NAS, 1975a).

Dutch elm disease is caused by a fungus introduced to North America about 1930. It attacks practically all varieties of elm; during the past 35 years it has killed about 40% of the 25 million elms so much a part of our urban and suburban environment. Elms are being killed at the rate of 400,000/year, the cost for removal of dead elms being estimated at $100 million (NAS, 1975a).

The fungus that causes Dutch elm disease is transmitted by any of three species of bark beetles, one native and two introduced. The relationship between the fungus and the beetles is one of mutual benefit. The insect prefers dead or dying wood (with reduced sap flow) to develop its larvae beneath the bark of the elm, and the fungus requires help in getting from tree to tree.

In the early spring, beetles that have developed in infected trees fly to healthy elms and feed on twigs and tender bark in the axils of branches. Coming from infected trees, they bring with them spores of the fungus that adhere (inadvertently?) to the insect body. As the beetles feed, tissue in the healthy tree is damaged and the fungal spores germinate in the feeding wounds. As the fungus grows, mycelia ramify throughout the vascular tissue of the branches and eventually clog these tissues so that the sap flow is denied the terminal portion of the branch. This explains the characteristic, progressive death of the elm tree, starting with one or a few branches and extending to main limbs and the tree trunk as the fungal invasion continues.

Until recently fungicides effective against the fungus were not available and control attempts consisted of good cultural practices and prompt removal of dead and dying trees. Concurrent with this (and sometimes in lieu of it) insecticides were used to kill the adult beetles. DDT was used extensively during the 1948–1963 period; when applications were timed properly, applied well, and combined with prompt removal of infected trees, a measure of control was obtained. Problems arose, however, because of DDT contamination of soil beneath the elms, accumulation in earthworms, and death of robins using these as a main source of food (Chapter 18).

Urban-suburban "forests" in America are extensive and estimated at almost 35 million acres (NAS, 1975a). The fact that many communities adopted DDT-spray programs to protect their elms resulted in heavy usage of DDT for this purpose. Communities were not equally conscientious regarding tree removal, and the results of the control program were variable (Carson, 1962). In more recent years many of the spray programs have been discontinued; those now employing such techniques are using methoxychlor, lead arsenate, or carbaryl.

During the past 10 years, some progress has been made in developing fungicides effective against the Dutch elm disease fungus. The problem has been to deliver the fungicide to the site of fungal growth within the tree. Benomyl is somewhat effective when applied as a foliar treatment to young trees and will prevent the fungus from growing provided it can be carried in the sap stream. Techniques of applying this material by injection into the trunk and main branches or through the root system have been partially successful. More recently the activity of benomyl for this purpose has been improved by using certain salts of the parent compound that are more water soluble and translocated within the tree to a far greater degree.

5.3 WEED CONTROL IN THE FOREST

Although herbicide uses in the forest are minor, they are used in some management systems. In forest nurseries and Christmas tree plantations, herbicides such as simazine, amitrole, and paraquat are used to reduce weed competition and thus improve annual growth. In intensive conifer reforestation projects, nonselective, short-residual herbicides may be used in site preparation. In the first few years after planting, aerial applications of 2,4,5-T and other similarly selective herbicides can be used to kill deciduous hardwoods in the overstory. In more established forest stands mechanical devices can be used to kill trees by "herbicide injection." Cacodylic acid is used in this manner to kill undesirable trees in mixed stands or to thin out the stand of one particular species.

5.4 VEGETATION MANAGEMENT ON ROADSIDES AND UTILITY RIGHTS-OF-WAY

North American society depends on the safe, efficient, and rapid transport of people, food, energy and communications over intensively maintained strips of land collectively referred to as "rights-of-way." These include highways or roads, electrical transmission lines, telephone and telegraph lines, railroads, pipelines, airports, and even strips of land of legal significance such as the International Boundary between the United States and Canada. Maintaining these rights-of-way requires control of vegetation that continuously encroaches on these transportation links. These narrow rights-of-way (20 to 200 ft in width) extend millions of miles through urban, agricultural, and wilderness environments, and herbicide use for maintenance purposes has been the subject of public debate and controversy.

Roadsides

Reasons for herbicide use to control vegetation on roadsides are two-fold: safety and reduced maintenance costs. On these rights-of-way the objectives may be the following:

1. To extend the life of the pavement by applying a soil sterilant before the asphalt is laid.
2. To prevent breakup of the pavement edge by controlling all vegetation on the road shoulder.
3. To prevent obstruction of guide rails and traffic signs by controlling all vegetation beneath them.
4. To maintain adequate sight distances, particularly at intersections, by controlling tall growing vegetation and brush.
5. To maintain a neat, attractive primarily grass-covered roadside that is resistant to erosion, that is not an excessive fire hazard during dry periods, and that does not create snow drifting or smow removal problems in winter.

For these purposes herbicides are generally cheaper, more effective, and much more selective than strictly mechanical methods. Much roadside vegetation could be controlled by three or four mowings per season, but in most situations an annual or semiannual application of herbicides to control tall growing weeds in conjunction with one mowing has been found to be more efficient and much less expensive.

Herbicides used on roadsides for total vegetation control include TCA, MSMA, ammate, chlorate-borate mixtures, simazine, or diuron with annual use in the United States of more than 12 million lb (Von Rumker et al., 1975). For selective control of broadleaf weeds and brush, herbicide mixtures are frequently used—2,4-D; 2,4,5-T; fenoprop; and picloram are the major herbicides for this purpose, with more than 3 million lb used annually in the United States (Von Rumker et al., 1975).

Power Lines

On power line rights-of-way control of tall growing vegetation is essential to reduce power interruptions, more than 50% of which are vegetation related (Ontario Hydro, personal communication), and to facilitate access for maintenance and repair. Vegetation approaching within 10 ft of high voltage lines may cause "grounding out," and tall trees adjacent to such lines may present problems during wind storms. Herbicides are used to prevent tree growth in rights-of-way and to maintain regrowth in the shrub stage.

Normal practice is first to clear a power line mechanically by cutting. Chemicals may also be used at the clearing stage as stump treatments to prevent resprouting. Then 3 to 5 years later an aerial or ground application of selective herbicides is employed to reduce the overall population of woody species. Subsequently herbicides are applied every 5 to 7 years on a selective basis to fast growing or tall woody species (ash, maple, poplar, oak, or spruce). Since woody species are the main problem on power lines, 2,4-D; 2,4,5-T; picloram; and related chemical brush killers represent more than 90% of the more than 5 million lb of herbicide used annually (Von Rumker et al., 1975).

Railroads

Chemicals to prevent the rotting of wooden railroad ties represent the greatest use of pesticides on railroads. Creosote, the main chemical used for this purpose, is by far the most heavily used pesticide in North America with an annual use in excess of 1 billion lb (Von Rumker et al., 1975).

In addition to creosote, herbicides are used on railroads for essentially the same reasons as for roadsides. One major difference is, however, that in railroad maintenance the area of crushed rock ballast for the roadbed must be kept virtually free of all vegetation so that the fire hazard from sparks is kept to an absolute minimum.

Total acreage treated by the railroads amounts to 1.5 to 2.0 million acres annually, and 20 million lb of herbicides are used. Chlorate-borate mixtures and other nonselective herbicides account for more than 75% of this use. In addition 2 million to 3 million lb of 2,4-D; 2,4,5-T; and mixtures of related herbicides are applied for selective brush and weed control (Von Rumker et al., 1975).

6

THE ROLE OF PESTICIDES
FOR THE HOMEOWNER

$\mathbf{A}$lthough agriculture is the largest user of pesticides in North America, use in and around the home is a routine part of modern living and consumes large amounts. Accurate estimates of the amount of pesticides used are difficult to develop, since most of the pesticides produced for this market are available through a wide variety of outlets and may be applied by the homeowner himself, by a pest control operator (PCO), or, as in the case for mosquito control, control of pests of ornamental and shade trees, and for insects under quarantine, by municipal or other government agencies.

Cost figures for 1970 show that 19% of pesticides sold were directed to the homeowner. In most instances homeowner products are marketed in small packages and often in dilute concentrations. Thus the 19% figure based on value overestimates this market. Added to this, however, must be a good portion of the pesticides used by PCOs ($30 million worth of pesticides by interior PCO in 1973) both interior and exterior. These firms (11,000 interior PCOs and 15,000 exterior PCOs) did a $600 million business in 1973. Termite control alone accounted for the use of approximately 800,000 lb of insecticides, almost 600,000 lb of which was chlordane (NAS, 1975).

Interior pest control operators in the United States list the German cockroach, the house mouse, Norway rat, and subterranean termite as the most important pests, but many other pests are also involved. Other species (American, oriental, brown-banded) of roaches are troublesome, along with ants, fleas, silverfish, the brown dog tick, and housefly in the category of general household pests. Along with the suberranean termite, the carpenter ant and powder post beetles are important structural pests.

A number of insects, including the common and black carpet beetles, the webbing clothes moth, and the case making clothes moths, destroy fabrics of various kinds. In addition to the pesticides used for control of these pests in the home, many manufacturers, especially of carpets and furniture, impregnate their fabrics with an insecticide during manufacture.

Although roaches, ants, and silverfish may be the most obvious of pests in the kitchen and pantry, packaged foods may be infested with insects such as the saw-toothed grain beetle, confused flour beetle, yellow mealworm, or Indian meal moth. Infestations are more frequent than is normally believed. A survey of households in Winnipeg indicated that 18% of the homes surveyed had infestations (Loschiavo, personal communication).

Outdoor pests are also important and include wasps and hornets, clovermites, bees, crickets, ants, sowbugs, millipedes, earwigs, mosquitoes, and flies. In addition the home lawn may be damaged by white grubs, Japanese and asiatic beetles, weevils of various kinds, webworms, and chinch bugs. Moles, mice, rabbits, and pigeons and other birds may be a nuisance to the homeowner and at times destroy plants or impair the quality of the home environment.

While fungi are of only minor importance in the home (except in mildew in damp situations, diseases of house plants), they are significant outdoors, and fungicides are used for their control. This use is increased greatly in areas where home gardens (flowers, vegetable, or fruit) are abundant.

Because of the outdoor needs of the home gardener, a large volume of business has developed in multipurpose mixtures of pesticides. These contain usually one or more insecticides along with a fungicide and are designed as an all-purpose control for garden pests. For lawn maintenance a different type of product has emerged containing an insecticide (sometimes more than one) and one or more herbicides incorporated in a granular formulation with fertilizer. These developments add to the convenience of the homeowner but undoubtedly result in the use of pesticides at times when they are not needed. Overdosage by the homeowner is a common fault when it comes to pesticides, and the multipurpose mixtures aggravate the situation.

A variety of pesticides are used by the homeowner. Indoor control of many insects is based on diazinon, chlordane, or malathion, but for ant control chlordecone has been used in baits, and propoxur, for both roach and ant control. Outside the home chlordane is used for soil grubs, but large quantities of chlorpyrifos, diazinon, and carbaryl are used for control of weevils, chinch bugs, and webworms. In the garden, carbaryl, methoxychlor, and malathion are now the dominant materials replacing the DDT of two decades ago. To a significant extent rotenone and pyrethrins are used also in the home gardens.

To a large extent the homeowner has used the bis-dithiocarbamate fungicides (Chapter 8) or captan in protecting his garden plants. Recently there has been concern about ethylenethiourea, a breakdown product and a sometimes contaminant in the bis-dithiocarbamates, and at least one government (Canada) has taken steps to remove this class of fungicides from home garden mixtures. Before the advent of the bis-dithiocarbamates, copper sulfate was used by homeowners, and proposed restrictions may force users back to this general purpose, broad spectrum, persistent fungicide.

The homeowner has, in addition to insects and plant diseases, a perennial problem with weeds, and here 2,4-D is the most widely used herbicide. Some lawn weeds are not controlled with this herbicide, and manufacturers have formulated products containing several herbicides to provide more nearly complete weed control. These mixtures usually contain 2,4-D along with mecoprop, dichlorprop, or dicamba. Whereas these mixtures give excellent control of broadleaf weeds, they do not control crabgrass or other annual grasses. To control these, DCPA, trifluralin, or bensulide are used, and frequently they are incorporated in a granular fertilizer mixture for application in fall or spring.

A detailed study of homeowner use in three American cities— Lansing, Dallas, and Philadelphia—showed that 750,000 lb of pesticides were used by suburban homeowners, 122,000 lb of herbicides, 429,000 lb of insecticides, and 69,000 lb of fungicides. Most of the pesticides were used in lawn care, with insects the main problem in Dallas, weeds in Lansing and Philadelphia. In commenting on this use, the National Academy of Sciences (1975) emphasized that much of the pesticide used in lawn maintenance is applied during a 6-week period beginning the last week of April and that dosages are high, 5 to 10 lb/acre being quite common. If the survey in the three cities is representative of use by American homeowners, the fact that a population of about 3 million used 750,000 lb suggests that outdoor use of pesticides around the homes may involve some 50 million lb of pesticides, a significant amount indeed.

7

TESTING PROCEDURES
IN PESTICIDE DEVELOPMENT

All pesticides are toxicants to a greater or lesser extent. There are, however, wide variations in organisms in terms of their susceptibility to a given pesticide, and it is these differences in susceptibility that determine the role a product plays in our economy and our environment.

Thus a pesticide is said to be "selective" in toxicity if it is highly toxic to only a few related organisms or "broad spectrum" in toxicity if it has high toxicity to a wide range of organisms. In either case toxicity to different organisms is on a graded scale, the terms *selective* and *broad spectrum* being relative terms only. The biocidal nature of each pesticide must be known before its potential can be assessed, and this need has prompted the establishment of extensive testing procedures and protocols to be followed before a new compound is introduced.

There are essentially two basic conditions to be established:

1. That the product is effective in preventing damage by a pest, and
2. That use of the pesticide for the purpose intended will not cause harm to those applying the pesticide, to the product it is designed to protect, or to the environment.

A new pesticide starts out in a chemical laboratory, the product of an experimental synthesis. Sometimes these syntheses follow well-defined patterns such as the synthesis of a new chemical in a series of related compounds known to have activity as pesticides. Where this is the case, the spectrum of activity of the related known compounds gives some clue to which organisms may be susceptible and which tolerant. Such clues, while most helpful, are only guidelines, for our knowledge of biochemistry is too incomplete to predict toxicity with great accuracy. For other chemicals the product in the test tube is of completely unknown quantity, and determination of activity must await experimental results.

The pesticide company therefore maintains or has access to a testing program in which new chemicals are screened for pesticidal activity. The initial screening involves only a small number of species of pests, species that are maintained easily in the laboratory and that the toxicologist knows from past experience will provide a reliable indication of the spectrum of activity of the compound.

Included in these tests are a few of our major weeds (broad leaf and grasses), some representative plant pathogens (fungi and bacteria usually growing in culture media), and some insects, commonly the pea aphid, housefly, German cockroach, and salt marsh caterpillar. In each case the new chemical is tested in a dilution series to permit an estimation of an LD_{50} (dosage required to kill 50% of the population). This LD_{50} value is then compared to values for chemicals proved in the field to be effective and a decision made regarding the potential of the compound. Most of the new chemicals synthesized never go beyond this stage. Those chemicals that show potential in these early tests are tested on a wider variety of pests; if they appear promising, toxicological studies are begun on mammals as a first step in determining whether or not the potential pesticide can be safely used.

The toxicological testing is a sophisticated, time-consuming, and expensive procedure and, provided data on efficacy testing support the view that the new chemical will provide an economic return to the company, includes an evaluation of toxicity by ingestion, by dermal exposure, and by inhalation. The reasons for this are clear. Any pesticide that will be used widely will create a possible exposure to man by each of these routes. It is imperative that the degree of hazard be assessed in relation to the use contemplated for the product and restrictions applied that will reduce or eliminate the hazard.

The procedures for testing have been described in detail in a number of reports. Much of what follows is taken from Oser (1971) and Kenaga (1977).

7.1 TEST ANIMALS

Since the prime purpose of toxicological tests is to determine the toxicity of the compound to man, the choice of animal for test purposes should be made with this object in mind. This choice is influenced by the suspected toxicological effect of the chemical and the knowledge of what metabolic pathways may function to increase or decrease toxicity. In general the rat and mouse are normally used for oral toxicity tests, because the rat, in particular, has a digestive system geared to the same sort of diet as man and in many respects has similar metabolic responses. There are, however, certain exceptions that should be noted. The rat has the ability to synthesize its own vitamin C and lacks a gall bladder. Other animals may be used also. The guinea pig, rabbit, dog, and monkey have been employed, but for one reason or another, preference has settled on the rat and mouse. As pointed out by Oser (1971), biotransformations of pesticides may be more or less possible in different animal species, and if the transformations in man are our goals, selection of the test species should reflect this. Thus "in the case of aromatic amines, where acetylation is the usual detoxification process, emphasis should be placed on the rat which, like man, can acetylate, rather than on the dog which cannot. On the other hand, if detoxification occurs via the glucuronide pathway the response of the dog may be more significant in terms of human predictability" (Oser, 1971).

The convenience with which test animals can be maintained and the availability of healthy, uniform stock are important considerations and favor rats over dogs and most other small mammals. Rats are docile, grow rapidly (weight gain or loss easily determined), are economic in terms of cost and space, and reproduce rapidly with a gestation period of 19 to 22 days. Detracting from them as ideal test animals is that they readily acquire pulmonary infections and are likely to develop spontaneous mammary tumors.

By contrast, housing for dogs is much more difficult, and a uniform supply is not always available. Where dogs are used, beagles are the preferred breed, but even these must be immunized against hepatitis and leptospirosis and be deparasitized prior to use. A few years ago it was felt that some of the smaller monkeys would be good test animals and respond to pesticides in a manner similar to man. Experience indicated a contrary view. For example, with the rodenticide Vacor®, 1-(3-pyridinyl methyl)-3-(4-nitro phenyl) urea, LD_{50} values for acute oral ingestion for various species of rat ranged from 4.8 to 28.1 mg/kg. In the rhesus monkey the LD_{50} value is 2000 to 4000 mg/kg. Poisonings with this compound permitted the calculation of an LD_{50} to human beings, and it was found to be 5 to 10 mg/kg for females. Thus for this rodenticide the rat gave a much better indication of human toxicity than the monkey did (Laidlaw, 1976).

7.2 ORAL TOXICITY TESTS

Oral toxicity tests can be divided into three types: acute, subacute, and chronic.

Acute Oral Toxicity Tests

The purpose of this test is to determine the dosage, usually expressed as milligrams per kilogram of body weight, that is required to kill 50% of a test population. This LD_{50} figure is used universally as one of the main methods of comparison among different compounds. It should be recognized, however, as only one of many criteria that must be used if realistic comparisons are to be made. Thus the acute oral toxicity tests give a first impression of the compound. Although the main purpose of the acute oral toxicity test is to determine at what level death results in a relatively short time, observations are made also on premortal symptoms, and routine autopsies are done. The nature of premortal symptoms may indicate the nature of the toxic action. As discussed later, symptoms of poisoning with different toxicants are characteristic, to some extent, and any extra information that can be gleaned from exposed animals should not be overlooked. In the acute oral toxicity studies autopsies seldom yield much information, gross histological or pathological changes not being induced in so short an exposure.

Young rats weighing 200 to 300 g and mice weighing 15 to 20 g are used routinely in these tests. The animals are starved 16 to 18 hr prior to testing, and the toxicant is administered in food or by gavage (through a tube directly into the digestive tract). The latter method makes accurate dose determinations easier, for it is not dependent on the sometimes unpredictable voluntary food intake by the test animal.

After the test doses have been administered, food is provided, and the test animals are observed for a specific period of time, the time related in part to previous experience with the class of pesticide under test. In the initial stages of testing a few animals may be used to get a "ball park" estimate of the toxicity range of the compound. When this has been determined, statistically valid experiments are set up. These usually involve four to six replications, each containing 6 to 20 animals for each dosage tested. Dosages (often at 0.1 log intervals) are included to bracket the toxicity range of the compound. After the LD_{50} on the rat and mouse have been determined, tests are extended to a few nonrodent species (often the dog, monkey, chicken) to detect any gross differences in species. If gross differences are detected, additional testing is done to permit an interpretation of the extent to which such differences might relate to man.

Subacute Oral Toxicity Tests

These tests, often called "short term," have three main goals:

1. To determine the maximum daily dosage of a toxicant that the test animal can survive.
2. To determine the maximum daily dosage that can be consumed by the test animal without any demonstrable effect (i.e., the "no effect" level).
3. To determine the nature of the effects of the toxicant when administered above the "no effect" level.

It is obvious that these tests are extremely important in relation to the hazard associated with a pesticide. Except for the remote (usually accidental) possibility of ingestion of a large amount of poison, normal use practice will likely necessitate a low-level exposure over a protracted period of time.

In routine practice these tests involve the ingestion of a toxicant at a predetermined dosage level over a period of 3 to 6 months. The length of the feeding period varies, depending on the nature of the toxicant and the test animal being used. The concept is to expose the animal during most of its developmental period. In the rat sexual maturity is reached in about 90 days; thus the rat is an excellent choice for such studies. Feeding tests begin when the animal weighs about 50 g. Studies are also done on the dog and/or monkey to complement the rodent tests.

The dosage range included in subacute tests can be extrapolated from the LD_{50} values in the acute oral toxicity tests. Quite frequently 50% of this dosage is a starting point with about five lower dosage levels. The United States Food and Drug Administration recommends that five dosage levels be used, the steps in the dosage series being at equal logarithmic intervals. Twenty rats of about 50 g, 10 male and 10 female, are used for each dosage. In tests with dogs or monkeys the animals are used at about 6 months of age.

The importance of the subacute tests cannot be overemphasized, for it is here that the crucial questions about the effect of pesticides on mammalian systems are determined. Thus extensive (and expensive) examinations are performed. These include daily observations on behavior and general appearance of the animals and weekly (sometimes more frequent) assessments of food and water consumption and growth. Frequent analyses are made on blood and urine. Other tests may be conducted throughout the experiment as the pattern of symptoms or "educated guesses" about the mode of action of the toxicant may suggest. At the conclusion of the test (12 or 24 weeks) some or all of the animals are sacrificed and examined macroscopically and microscopically (for some organs). Weight, general appearance, and size of the treated animals are compared with those of the control group. Organs examined include adrenals, bladder, brain, gonads, heart, intestines, kidneys, liver, lungs, lymph glands, muscle, pancreas, salivary glands, spinal cord, spleen, stomach, and thyroid. Organs or tissues indicating any abnormalities are exposed to further examination microscopically.

Chronic Oral Toxicity Tests

These tests, often referred to as "long term," are designed to estimate the effect of exposure to pesticide over the normal life span of an animal. Because the rat is the animal most frequently used and its life span is about 2 years, this is the normal duration of the chronic testing, even when other animals are involved. Of particular importance in these tests are effects on reproduction and the discovery of any teratological or carcinogenic effects.

The dosage levels employed in these tests can be based on the results of subacute tests; under most circumstances only three dosage levels are included. The procedures followed are similar to those of subacute tests except that more animals are used in each group. The number of generations of offspring observed is normally two in the rat.

The whole area of carcinogenesis and teratogenesis is understood poorly and evaluation of test results a matter of considerable disagreement. Some argue that a distinction must be made between tumors that persist and those that undergo remission when the insulting toxicant is withdrawn, the former representing the effect of a carcinogen, the latter not (Weil, 1976; Deichmann, 1972). This distinction has not been made in many of the chronic feeding tests and has resulted in confusion. Current practice in interpreting the significance of tumor production in test animals tends to regard these as a carcinogenic response; thus if this is an error in interpretation, the error is on the side of safety (*Federal Register*, 1975b).

7.3   DERMAL TOXICITY TESTS

Whereas the oral toxicity tests indicate the probable level of toxicity of a pesticide by ingestion, it is equally important to learn how toxic this compound might be if it comes in contact with the skin. Normally two factors are evaluated, toxicity by the pesticide and its potential to cause irritation.

For these tests the rabbit is most frequently used, for its long torso provides surface area for exposure and the animal is economic to rear and readily available. The trunk of the animal is clipped and fitted with a rubber sleeve 7 cm in diameter and 12.5 cm long. The animal is immobilized in a restrainer, and the toxicant to be tested is injected between the rubber sleeve and the skin of the animal. After 24 hr the sleeve is removed and unabsorbed material collected and measured and the skin wiped clean. The area of contact is observed for gross reactions, and the animals are held under observation for 14 days.

In addition to the acute dermal toxicity tests, tests are also done to determine whether or not the materials may cause irritation to sensitive skin. The rabbit is again the animal of choice; in this test the skin is abraded lightly (not to the point of bleeding). Test animals are immobilized, and a gauze patch, 2.5 cm square, to which the toxicant has been applied is placed on the abraded skin and wrapped with a rubberized cloth to hold the patch in place and prevent leakage of the material. The patch is removed after 24 hr and skin reaction determined and recorded at 24, 48, and 72 hr posttreatment.

To determine further any skin sensitivity, the pesticide may be applied to sensitive portions of the body. The conjunctiva of the eye is regularly chosen with the rabbit used as the test subject. In this test a measured quantity of toxicant is placed directly under the lid of one eye of each of six albino rabbits. The lid is held closed momentarily and the rabbit confined to a cage. Eyes are examined after 24, 48, and 72 hours to note any reaction.

7.4 INHALATION TESTS

Pesticides that are used widely may present a hazard through inhalation of their vapors. The protocols for determining inhalation toxicity are not so well defined as those used for dermal or oral toxicity tests. The prime difficulty in these tests is to ensure that the inhalation route is the only one by which exposure occurs. Thus elaborate arrangements are needed to exclude possible exposure through ingestion or by the dermal route. Chambers that permit exposure to only the head of the animal have been used. Such cages prevent the animal from ingestion of toxicant through licking of fur and minimize the dermal surface exposed. The method of presentation of toxicants for testing via the inhalation route is important. In general, particles in excess of

10 microns are excluded physically by the filtering mechanisms (hairs, mucus) in the outer reaches of the inspirational network, whereas particles of less than 3 microns reach the lungs without difficulty. Thus droplet size of the pesticide atomized into test chambers must be controlled carefully.

7.5   TESTS ON OTHER SPECIES

As more information becomes available on the effect of pesticides on nontarget organisms, protocols for toxicological testing prior to release of new compounds have become more demanding in terms of evaluations on a wider range of species. The exact procedures and requirements are being reviewed continuously, and the extent to which requirements include testing on birds, fish, and molluscs reflects the intended use pattern of the pesticide (Kenaga, 1977). It must be recognized that, with many pesticides, differences in susceptibility among species are the rule, and while this is appreciated, it is impractical from a time and cost point of view to attempt evaluation against all species likely to be exposed. Thus, regardless of the amount of testing done, it is probable that use of a pesticide in the field will lead to the discovery of toxicity or other effects on some organisms, effects that laboratory experience could not predict. It is hoped that, where this occurs, the effects will be of little consequence because the effect will be minor or the organism of such nature as to recover rapidly. Where this is not the case, for example, with DDT, it is essential that the effects be noted early and restrictions placed on the use of the pesticide so that serious damage does not occur.

Throughout the preceding discussion we have used the terms *toxicity* and *hazard*. It is important to recognize a broad distinction between these terms. Toxicity refers to the inherent capacity of a substance to cause injury or death. It is determined by the nature of the compound, its ability to reach some sensitive target in the body and cause its destruction. In contrast, hazard refers to the probability that injury will result from the use of a compound under a given set of circumstances. Thus we refer to a compound as highly toxic if it has a low numerical value for an LD_{50}. This does not imply that the use of such a compound is necessarily hazardous. This latter consideration must also take into account the amount and nature of the exposure, duration of exposure, and a host of other "use related" parameters.

7.6 TESTING METABOLITES

Many pesticides undergo biotransformations that result in a variety of compounds derived from the parent material. These "metabolites" may or may not have toxicity characteristics similar to those of the product from which they were derived. For example, in nature the insecticide aldrin is broken down, and one of the main metabolites is dieldrin, a compound much more stable and with a different spectrum of activity than that of its progenitor. A profusion of metabolic products is the rule rather than the exception with most modern pesticides. It is necessary, therefore, that each metabolite be evaluated for toxicity in much the same manner as the parent compound.

7.7 EFFICACY TESTING

While the toxicological tests are being conducted to establish parameters for safe use of a proposed pesticide, extensive tests are being conducted to determine what role the product can play in pest control. There is no standard for this determination, but if a product is to get to market, data must be available to convince the company involved that the product has a significant role in pest control, that it can be used safely, that it will not result in hazardous residues in food, and that it will return a profit on investment.

A large part of the decision of the company centers on the effectiveness of the compound in field trials. These trials are conducted within the company itself, but as soon as the potential for control of an important pest is recognized, state and federal experiment stations cooperate in evaluating the pesticides under their local conditions. These tests are very much tests of mutual interests. The pesticide company wishes to learn as quickly as possible how their new product compares with products already in the field, while the experiment station personnel are faced constantly with entreaties from growers for more effective chemicals and reduced cost for pest control.

If the pesticide is to be used on food crops, a new set of problems is presented in relation to residues at harvest. For many uses a paradox must be recognized. The persistent nature of many pests requires that pesticide be on plant foliage on a continuing basis, and yet our desire not to include pesticides in our daily diet dictates that significant amounts of the pesticides not be present on the crop at harvest. A saw-off usually results in which pesticides with limited or short persistence (biodegradable) are favoured with applications made at sufficient frequency to protect the crop when protection is needed but withdrawn from use immediately prior to harvest, so that only minimum residues remain. Determining what these practices can be requires extensive field testing under a variety of growing conditions. Climatic factors, especially rainfall and temperature, may dramatically affect the residual nature of pesticides. Thus on each crop for which the pesticide shows potential, a series of treatments must be employed to determine what dosage rates and application frequencies are needed to control the pest and what residue will remain at harvest if this treatment regimen is practiced. On the basis of these tests residue tolerances are established.

Residue tolerances reflect two primary considerations. The first of these is the residue (usually expressed in parts per million [ppm]) that subacute and chronic toxicity tests indicate represents the "no effect" level. The second consideration is what residue will result if the pesticide is used in a treatment schedule needed to effect control of the pest against which it is employed. If this latter residue exceeds the "no effect" level, such treatments will not be permitted. If the level is well below the "no effect" level (by a factor of 100 or more), a tolerance will likely be established and the product will be registered for use on the basis of a specific set of directions (Chapter 22).

8

FUNGICIDES AND THEIR USES

Fungicides follow herbicides and insecticides in terms of tonnage used in crop production and are employed mainly in the protection of vegetable, fruit, and nut crops in North America. While they play an important role in the production of rice and tobacco, most forage and cereal crops, including corn, do not require fungicide treatments except as seed treatments (not forage seeds) where they are extremely important, and their use in minute quantities (e.g., 1 oz/ bushel of seed) is an outstanding example of pesticide efficiency. Apart from seed treatments fungicides are used on only 8 million acres in the United States, the largest single crop requiring regular treatment being peanuts (Blake and Andrilenas, 1975).

The relatively small acreage treated with fungicides may suggest minor usage, but this is not so. Use is intensive with multiple applications on many fruit, vegetable, and nut crops and with important and extensive use in timber, wood products, and building industries. Slime control in wood processing, mold inhibitors in building products, and wood preservatives for the construction industry, utility poles, and railway ties are among major uses of fungicides, together with use for mildew prevention in paints and varnishes. Figures for 1971 indicate 300 million lb of fungicides (exclusive of creosote) produced annually in the United States (Lawless et al., 1974). Sales figures for Canada (1974) place the value of agricultural fungicides near $10 million (*Statistics Canada*, 1975).

Fungi are a large and diverse group of plants (mostly minute),

73

some of which cause diseases in commercial crops. They reproduce vegetatively, generally, or through the production of asexual spores but may also form sexual spores in specialized fruiting bodies. They infect plants by sending haustoria or mycelia throughout the plant cells, robbing the host of needed nutrient and/or producing toxins. Their growth results in rotting of plant foliage (late blight of potatoes or tomatoes), distortion and inefficiency of roots (clubroot of crucifers), dysfunction of the vascular system (e.g., fusarium and verticillium wilts), or rotting or disfiguring of edible plant parts (e.g., cucumber anthracnose, apple scab). Sometimes only plant vigor is reduced without the dramatic crop loss associated with the more destructive pathogens. Because fungi are plants, a high degree of selectivity must exist in fungicides so that they destroy or prevent growth of the fungus without exhibiting phytotoxicity to the host plant on which the fungus is growing. In addition many fungi are beneficial and necessary components in the environment as decomposers of organic matter or as food. Thus fungicides must be narrowly selective in toxicity or used in such a way that only target organisms are affected.

In contrast to the situation with most insect pests, where the target organism moves about on the plant surface at some stage in its cycle, fungi do not move. Thus protection of plants requires that the fungicide cover the entire susceptible surface so that inoculum arriving on a plant is prevented from penetrating. To ensure this, stress has been placed on good application techniques and the development of fungicide formulations that inherently (owing to the fungicide) or by addition of spreading agents have the capacity to "redistribute" on the foliage in the presence of dew or rain. Recent progress in the design and development of systemic fungicides represents dramatic advances in plant disease control in that they provide the opportunity for distribution of fungicide within the plant tissues.

Fungicides can be classified as systemic or nonsystemic, depending on their ability to move within the plant. The use of these terms must be interpreted in a relative sense, for many fungicides not classed as systemic are in fact taken up by plants to a minor extent. Normally fungicides are referred to as "protectants," that is, those that when present on the plant surface prevent invasion by a fungus landing on the treated surface, and "eradicants" or "therapeutants," that is, those that destroy or eradicate the fungus after infection has occurred. Whereas the term *fungicide* is used broadly to include antifungal agents, many function as fungistats, that is, they prevent fungi from growing, rather than kill the organism.

8.1 SULFUR

Although its use in North America is declining, figures for 1971 indicate that sulfur accounted for 112 of the 154 million lb of fungicide used by farmers during the year (NAS, 1975).

Sulfur is obtained from natural deposits by grinding (for dust formulations), sublimation by heat and condensation (flowers of sulfur, used as a dust), or by chemical precipitation (for wettable powders). It is obtained also as a by-product in the manufacture of coal gas. Sulfur is a yellow solid, practically insoluble in water and with low solubility in most solvents. It hydrolyzes slowly in water, reacts with alkali and with some metals (e.g., tarnish of silver). Sulfur burns at temperatures above 261°C to produce sulfur dioxide, and this feature has been used in greenhouses for insect suppression. It is used as elemental sulfur or as lime sulfur against a wide variety of plant diseases but is especially effective against mildews. It is used widely on citrus, in other areas of fruit and vegetable production, and in the production of greenhouse ornamentals and vegetables.

Sulfur is phytotoxic to many species of plants, especially at high temperatures. Its exact site of action is not known. It is reduced to H_2S, and though H_2S may cause phytotoxicity to plants (McCallan et al., 1936), this does not explain sulfur's activity against fungal spores (Miller et al., 1953). Sulfur blocks some enzymes in the respiratory pathway between acetate and citrate and probably functions as a general receptor for hydrogen in the redox reactions in fungal cells (Tweedy and Turner, 1966). Sulfur acts as a protectant fungicide. It is a persistent chemical but of such low mammalian toxicity that residues in food or in the environment have not been considered important.

8.2 COPPER

Copper sulfate was first used as a fungicide as a seed treatment for control of bunt of wheat in 1761. It is moderately soluble in water (32%) and has been used as a wood preservative for almost 200 years. It is relatively toxic for a fungicide with an oral LD_{50} to the rat of 300 mg/kg.

Copper sulfate is phytotoxic and has been used as an herbicide to control annual weeds in cereals, but its main uses are to control algal growth in ponds and, in combination with lime (Bordeaux mixture), for a wide variety of fungus diseases of plants. With the advent of the dithiocarbamates the use of copper as a fungicide has declined greatly.

The first major use of Bordeaux mixture was in 1885 as a control for downy mildew on grapes. This was a fortuitous discovery, for the intent of its application to grapes was to prevent pilfering. The addition of lime reduced phytotoxicity of copper sulfate, and its use as a general plant fungicide became widespread. Some 10 million lb of copper sulfate and related copper fungicides are still used in American agriculture (NAS, 1975). The addition of lime to copper sulfate results in the production of a number of copper salts that precipitate on standing; hence the material is tank mixed and should be applied soon after mixing. Although direct phytotoxicity is seldom expressed, this fungicide, especially in repeated applications, reduces crop growth and causes yield reductions in potatoes. It is broad spectrum in activity, acting as a protectant fungistat against many species of fungi. It prevents spore germination, but if the copper is eluted from spore suspensions treated for only a brief period, the spores will germinate and grow. Copper fungicides are believed to act as the cupric ion and have been shown to inhibit a wide variety of thiol ($-SH$)-containing enzymes (Corbett, 1974).

8.3 MERCURY

The use of mercurials as fungicides predated the twentieth century. Their first uses were as seed treatments, especially of cereals to control covered smuts, bunt, and other seed-borne fungi. Mercury (especially organic mercuries) functioned as an eradicant to kill spores and prevent development of dormant mycelia. It was learned that mercurials were broadly toxic to fungi (and bacteria), and seed treatments offered an easy method of protecting seeds and seedling, not only from seed-borne pathogens but also from soil-borne pathogens.

In the early years both mercuric chloride (corrosive sublimate) and mercurous chloride were used—the latter having the distinct advantage of reduced mammalian toxicity. Both are, however, phytotoxic, and their use as fungicides was limited to seed treatments or to soil treatments for control of diseases such as clubroot of cabbage and scab on potatoes. They were used also as slimicides.

Beginning in 1915 organomercurials appeared as fungicides; these had the advantage of reduced phytotoxicity but retained the wide range of fungicidal and bactericidal action. These were used widely as seed treatments, but some were developed as eradicants for control of apple scab and other foliar diseases as well as diseases of turf. The organomercurials included a wide range of products with the general formula RHgX, where R represents an aryl-, aryloxy-, alkyl-, alkoxyethyl-grouping and X is an anionic moiety such as chloride, acetate, hydroxy, or other structures. Among the more important were the following:

Phenyl mercuric acetate (PMA, Tag®), used primarily as an eradicant for apple scab.

$$\langle\!\!\!\bigcirc\!\!\!\rangle\text{-Hg-O-}\overset{\displaystyle O}{\overset{\|}{C}}\text{-CH}_3$$

Methoxyethyl mercuric chloride and **methoxyethyl mercury silicate** (Ceresan® in Europe), developed for slurry and dry treatments of cereal seeds.

$$CH_3\text{-O-}(CH_2)_2\text{-Hg Cl}$$

Ethyl mercuric chloride (Ceresan® in the United States), developed as a seed treatment for peanuts, cotton, peas.

$$CH_3\text{-}CH_2\text{-Hg Cl}$$

Hydroxy mercury chlorophenol (Semesan®), developed for use as a seed and bulb treatment for vegetables and ornamentals.

$$\overset{\displaystyle Cl}{\underset{}{\diagdown}}\langle\!\!\!\bigcirc\!\!\!\rangle\text{-O-Hg-OH}$$

Hydroxy mercury nitrophenol (Semesan Bel®), developed as a disinfectant for potato seed pieces.

$$\overset{\displaystyle NO_2}{\underset{}{\diagdown}}\langle\!\!\!\bigcirc\!\!\!\rangle\text{-O-Hg-OH}$$

Methyl mercuric dicyandiamide (Panogen®), developed as a seed treatment for cereals.

$$CH_3\text{-Hg-}\overset{\displaystyle H}{\underset{}{\overset{|}{N}}}\text{-C}\overset{\diagup N\text{-H}}{\underset{\diagdown N\text{-CN}}{}}$$
$$\underset{H}{\overset{|}{}}$$

The mercurials are generally highly toxic, with acute oral LD_{50} values to the rat for many of the organomercurials in the 30 to 200 mg/kg range. Mercuric chloride has an acute oral LD_{50} of 1 to 5 and mercurous chloride 210 (Martin, 1972). They owe their toxicity to the Hg^{2+} or Hg^+ ion and its strong reactivity with the sulfhydryl groups of respiratory enzymes. According to Metcalf (1971), the stability of the organomercurials and the rate at which the Hg^{2+} ion is released is determined by the alkyl or aryl portion of the molecule. Perhaps more importantly, these portions give lipophilicity to the molecule and thus aid in the penetration of the cuticle.

The mercurials, whether organic or inorganic, are extremely persistent toxicants in that they undergo changes in the soil to release mercury, which may be lost as vapor or recombined as various oxides, chlorides, or nitrates. This recycling of mercury presents the continuing danger of toxicity inherent in the Hg^{2+} ion. Mercury accumulates in living tissues, and even at low dosages (0.5 ppm) phenyl mercuric acetate was shown to produce damage in the kidney (Fitzhugh et al., 1950). Many mercurials are highly volatile and irritating to the skin and respiratory tract. They are toxic by oral, dermal, or inhalation exposure, and toxicity in man is expressed as nervousness and severe impairment of coordination. The outbreak of mercury poisoning among Japanese eating contaminated fish (Minimata disease) is the most publicized incident, but other poisonings have been reported. Poisonings have been reported in human beings where mercury-treated seed has got into domestic food supply, and poisonings have been reported also in livestock and domestic and wild birds. The use of mercurials as fungicides has been largely discontinued in North America.

8.4   ORGANO-TIN FUNGICIDES

The organo-tin compounds are a relatively recent (1950) addition to the fungicidal arsenal. They are generally phytotoxic and used for mildew control in wood, textiles, and paint. They are useful also as germicides and for slime control in paper mills. Some are less phytotoxic.

Fentin acetate (triphenyl tin acetate, Brestan®) can be used on plant foliage with much the same spectrum of fungicidal activity as copper sulfate. Fentin acetate hydrolyzes in water to form fentin hydroxide. It has an acute oral LD_{50} to the rat of 125 mg/kg.

$$\underset{(C_6H_5)_3-Sn-O-\overset{\overset{\displaystyle O}{\|}}{C}-CH_3}{}$$

Fentin hydroxide (triphenyl tin hydroxide, Du-ter®) is a closely related compound used for blights on potato, sugar beets, rice, and coffee. It has an LD_{50} to the rat of 108 mg/kg.

The organo-tin compounds have been used in hospitals, especially against *Staphylococcus* sp., where the diethyl and triethyl tin iodides have been useful as skin disinfectants, the latter compound being, however, extremely toxic to man.

As fungicides the organo-tin compounds owe their activity to the tin component's acting as an inhibitor or uncoupler of oxidative phosphorylation in the mitochondria (Chapter 9) (Corbett, 1974).

8.5 PHENOLS

The phenols include a number of compounds that are effective as general disinfectants, toxic to plant life.

Pentachlorophenol is highly phytotoxic and used in weed control (Chapter 9). It is used also as a fungicide for wood preservation and for slime control in paper mills. In addition it is effective for termite control. It is used widely for nonagricultural purposes with a production of some 50 million lb annually in the United States.

A number of compounds related to pentachlorophenol are used as slimicides and as germicides in cosmetics, soaps, polishes, and the like.

A number of nitrophenols are also powerful fungicides, but again most are highly phytotoxic.

DNOC (4,6-dinitro-*o*-cresol) is a general toxicant effective against some insects and used as an herbicide. As a fungicide its use is restricted to dormant sprays and for treatment of apple scab by application to the overwintering inoculum on the orchard floor.

Dinocap (Karathane®) (4,6,dinitro-2-(1-methylheptyl)-phenyl crotonate) is a commercial acaricide but also is used extensively for mildew control on apples and a wide variety of field and greenhouse fruits, vegetables, and ornamentals.

DNOC   dinocap

The dominant fungitoxic action of phenols is probably due to their role as uncouplers of oxidative phosphorylation. Greatest effectiveness seems related to compounds with a blend of lipophilic and hydrophobic properties that orient the molecule at a lipoid-water interface. A free $-OH$ group or readily hydrolyzable ester group is essential. Whereas it is thought that activity involves lysis of the cell membrane, there is evidence for inactivation of various enzymes, including oxidases, and *in vitro* experiments indicate they may be uncouplers of oxidative phosphorylation (Metcalf, 1971; Corden, 1969).

The phenols are moderately toxic, and some have handling problems in that they are skin irritants. LD_{50} values for the rat by oral administration are pentachlorophenol, 210 mg/kg; DNOC, 25 to 40 mg/kg; and dinocap, 980 mg/kg.

8.6   DITHIOCARBAMATES

The dithiocarbamate fungicides can be regarded as two groups: (1) the dimethyldithiocarbamates represented by thiram, ferbam, and ziram and (2) the ethylene-bis-dithiocarbamates represented by nabam, maneb, mancozeb, zineb, and metiram. The distinction is important, for as is seen later (Chapter 19), a contaminant and/or breakdown product of the latter group is important in terms of human health. It is important also to distinguish between the "carbamates" that are fungicides, those that are used as insecticides, and those used as herbicides.

Group 1

Thiram (Arasan®, Tersan®) is a colorless powder slightly soluble (30 ppm) in water and most organic solvents. It is a protective fungicide, nonphytotoxic, and is used as a seed treatment to control a number of fungi (*Pythium* sp., *Fusarium* sp., *Rhizoctonia* sp.) that cause "damping off" in seedlings and seedling blights. It is useful also as a rodent repellent when painted as a slurry on the trunk of trees or sprayed heavily on the exposed bark. It is of moderate acute oral toxicity to mammals but is reported to cause reduction in egg laying when fed to hens at 35 ppm in the diet (Waibel et al., 1955). Thiram is of moderate persistence in soil and is degraded by microorganisms (Robbins and Kastelic, 1961).

Ferbam (Fermate®) is a black powder slightly soluble (130 ppm) in water and with variable solubility in organic solvents. It decomposes on prolonged storage and on exposure to moisture. It is a general purpose fungicide against foliar diseases, especially of ornamentals, and for rust on apples, but its black deposit on foliage deters such usage. It is not very persistent, is nonsystemic, and acts as a protectant. It degrades in nature to dimethylamine and CS_2 (Owens and Rubinstein, 1964).

Ziram (Milbam®, Zerlate®) is a white powder slightly soluble (65 ppm) in water and moderately soluble in most organic solvents. It is nonsystemic and used as a protectant fungicide on fruits and vegetables. It is similar in activity to ferbam.

The mode of action of the dimethyldithiocarbamates is not understood. It has been suggested that they may act by interfering with metal enzyme catalysts (e.g., forming chelates with copper) (Corbett, 1974).

Group 2

Nabam (Parzate®, Dithane D-14®) is a colorless solid soluble (20%) in water and quite unstable under moist conditions.

Zineb (Parzate®, Zineb®, Dithane Z-78®) can be prepared in the field by mixing nabam with zinc sulfate. It is a light-colored powder, almost insoluble (10 ppm) in water and somewhat unstable when exposed to light and moisture.

Maneb (Manzate®, Dithane M-22®) is a yellow crystalline solid slightly soluble in water and insoluble in most organic solvents. It is stable under most conditions but decomposes on prolonged exposure to moisture.

Mancozeb (Dithane M-45®) is a greyish powder practically insoluble in water and most organic solvents. It is a polymer of maneb with zinc and is relatively stable.

Metiram (Polyram®) is a yellowish powder practically insoluble in water and organic solvents. It is a polymer, unstable under strongly acidic or alkaline conditions and will decompose on standing in tank mixes with some insecticides.

The dithiocarbamates are the most important of the organics and accounted for half of the 26 million lb of organic fungicides used by farmers in the United States in 1971. They are equally important in Canada and are registered for use on a wide variety of fruit and vegetable crops. There are some differences in relative effectiveness and spectrum of activity between the ethylene-bis-dithiocarbamates and the dimethyldithiocarbamates, the former being active against a wider range. There is also an apparent difference in the site of action, although it is not well defined for either group. As discussed earlier the site of action of the dimethyldithiocarbamates is not known, but they probably interfere with the enzyme system. Histidine has been shown to be antagonistic to thiram but not to nabam (Metcalf, 1971). Ludwig and Thorn (1960) demonstrated the formation of isothiocyanate from nabam through the formation of ethylenethiuram disulfide, and since isothiocyanates react readily with sulhydryl groups on enzymes involved in oxidative pathways, this may be a site of action for the ethylene-bis-dithiocarbamates. The reader is referred to Owens (1969a) for a complete discussion.

As can be seen (Table 8.1) the ethylene-bis-dithiocarbamates are of low acute mammalian toxicity, but there is much concern about one of the degradation products, ethylenethiourea (ETU), which may be a contaminant in this group of fungicides and has been shown to be produced under conditions of storage (Petrosini, 1962; Engst and Schnaak, 1967; Czegledi-Janko and Hollo, 1967). It is produced also when fruit or vegetables carrying residues of the ethylene-bis-dithiocarbamate fungicides are cooked. This conversion may represent about 20% of the parent residue on precooked produce. ETU is of concern because of its goitrogenic and other properties (Chapter 19); earlier work (Smith et al., 1953) indicated there might be differences among the bis-dithiocarbamates in this respect. However, recent experiments with maneb, mancozeb, metiram, and zineb on tomatoes demonstrate that, whereas ETU residues on uncooked tomatoes were low, residues on cooked samples were significant and reflected res-

Table 8.1 Name, Structure, and Oral LD_{50} to the Rat of Some Dithiocarbamate Fungicides

Name	Structure	LD_{50} mg/kg
thiram: tetramethylthiuram disulfide	CH_3 and CH_3 groups on nitrogens: $(CH_3)_2N{-}\overset{\overset{\textstyle S}{\|}}{C}{-}S{-}S{-}\overset{\overset{\textstyle S}{\|}}{C}{-}N(CH_3)_2$	375 −865
ferbam: ferric dimethyldi-thiocarbamate	$\left[(CH_3)_2N{-}\overset{\overset{\textstyle S}{\|}}{C}{-}S \right]_3 Fe$	17000
ziram: zinc dimethyldi-thiocarbamate	$\left[(CH_3)_2N{-}\overset{\overset{\textstyle S}{\|}}{C}{-}S \right]_2 Zn$	1400
nabam: disodium ethylene-bis-dithiocarbamate	$CH_2{-}\overset{\overset{\textstyle H}{\|}}{N}{-}\overset{\overset{\textstyle S}{\|}}{C}{-}S{-}Na$ $CH_2{-}\underset{\underset{\textstyle H}{\|}}{N}{-}\underset{\underset{\textstyle S}{\|}}{C}{-}S{-}Na$	395
zineb: zinc ethylene-bis-dithiocarbamate	$CH_2{-}\overset{\overset{\textstyle H}{\|}}{N}{-}\overset{\overset{\textstyle S}{\|}}{C}{-}S{\searrow}$ $\qquad\qquad\qquad Zn$ $CH_2{-}\underset{\underset{\textstyle H}{\|}}{N}{-}\underset{\underset{\textstyle S}{\|}}{C}{-}S{\nearrow}$	5200
maneb: manganous ethylene-bis-dithiocarbamate	$CH_2{-}\overset{\overset{\textstyle H}{\|}}{N}{-}\overset{\overset{\textstyle S}{\|}}{C}{-}S{\searrow}$ $\qquad\qquad\qquad Mn$ $CH_2{-}\underset{\underset{\textstyle H}{\|}}{N}{-}\underset{\underset{\textstyle S}{\|}}{C}{-}S{\nearrow}$	6750
mancozeb: a polymer of manganous ethylene-bis-dithiocarbamate with zinc	$\left[\begin{array}{c} CH_2{-}\overset{\overset{\textstyle H}{\|}}{N}{-}\overset{\overset{\textstyle S}{\|}}{C}{-}S\searrow \\ \qquad\qquad Mn \\ CH_2{-}\underset{\underset{\textstyle H}{\|}}{N}{-}\underset{\underset{\textstyle S}{\|}}{C}{-}S\nearrow \end{array} \right]_x (Zn)_y$	>8000
metiram: zinc-activated polyethylene thiuram disulfide	$\left[\left(\begin{array}{c} CH_2{-}\overset{\overset{\textstyle H}{\|}}{N}{-}\overset{\overset{\textstyle S}{\|}}{C}{-}S \\ \| \\ CH_2{-}\underset{\underset{\textstyle H}{\|}}{N}{-}\underset{\underset{\textstyle S}{\|}}{C}{-}S \end{array} \right)_n \left(\begin{array}{c} CH_2{-}\overset{\overset{\textstyle H}{\|}}{N}{-}\overset{\overset{\textstyle S}{\|}}{C}{-}S\searrow \\ \qquad\qquad Zn \\ CH_2{-}\underset{\underset{\textstyle H}{\|}}{N}{-}\underset{\underset{\textstyle S}{\|}}{C}{-}S\nearrow \end{array} \right)_m \right]_x$	>10000

n:m − 1:3   x − unknown

idues of parent compound on the uncooked product regardless of the fungicide used (Newsome, 1976). Routes of transformations of this group of fungicides are discussed in detail by Sijpesteijn et al. (1977).

$$
\begin{array}{ccc}
\begin{array}{c}
\text{H} \\
| \\
\text{CH}_2-\text{N} \\
| \qquad\quad \text{C=S} \\
\text{CH}_2-\text{N} \\
| \\
\text{H}
\end{array}
\longleftarrow
\begin{array}{c}
\text{H} \;\; \text{S} \\
| \quad || \\
\text{CH}_2-\text{N}-\text{C}-\text{S}-\text{Na} \\
| \\
\text{CH}_2-\text{N}-\text{C}-\text{S}-\text{Na} \\
| \quad || \\
\text{H} \;\; \text{S}
\end{array}
\longrightarrow
\begin{array}{c}
\text{H} \;\; \text{S} \\
| \quad || \\
\text{CH}_2-\text{N}-\text{C} \\
| \qquad\qquad \text{S} \\
\text{CH}_2-\text{N}-\text{C} \\
| \quad || \\
\text{H} \;\; \text{S}
\end{array}
\end{array}
$$

ethylene-
thiourea   nabam

ethylenethiuram
monosulfide

8.7 PHTHALIMIDES

The phthalimides include three fungicides, captan, folpet, and captafol, which together represent the second most important group of organic fungicides used in American agriculture. They represent about half the usage of the dithiocarbamates (about 7.5 million lb in 1971) (NAS, 1975), captan accounting for most of this volume.

Captan (N-trichloromethylthio)-3a,4,7,7a-tetrahydro-phthalimide) is a white crystalline solid almost insoluble (0.5 ppm) in water and petroleum oils and with varying (usually low) solubility in organic solvents. It is stable under most conditions but hydrolyzes in the presence of alkali. On plant foliage captan is generally nonphytotoxic, but its decomposition results in the production of hydrochloric acid, and under some conditions, rapid hydrolysis may lead to some toxicity on sensitive plants. Captan is broken down rapidly in soil (half-life 3 to 4 days, Burchfield, 1959), decomposition being more rapid in moist than in dry conditions. Although effective against a wide spectrum of plant pathogenic fungi, captan is generally ineffective against powdery and downy mildews (with some exceptions) and rusts. It is of low mammalian toxicity (acute oral LD_{50} to the rat 9000 mg/kg).

In plant cells captan is degraded to thiosphosgene, a compound known to inhibit enzymes utilizing cocarboxylase as a coenzyme. The reactivity of thiophosgene in enzyme systems with free sulfhydryl, hydroxyl, and amino groups suggests that captan may have several sites of action, an interpretation that explains its wide spectrum of activity (Lukens, 1969).

captan   thiophosgene

Folpet (*N*-(trichloromethylthio) phthalimide; Phaltan®) is a white crystalline solid practically insoluble in water and organic solvents. It is stable under dry conditions but hydrolyzes readily when moist (especially at high temperatures) or in the presence of alkali. It is incompatible with alkaline pesticides. It is generally a safe fungicide on fruits and vegetables, though it may be phytotoxic to sensitive cultivars. It is of low mammalian toxicity (acute oral LD_{50} to the rat > 10,000 mg/kg); it has a broad spectrum of activity against fungi and is believed to act in a manner similar to that of captan.

folpet   captafol

Captafol (*N*-(1,1,2,2,tetrachloroethylthio)-3a,4,7,7a-tetrahydrophthalimide; Difolatan®) is a white crystalline solid practically insoluble (1.4 ppm) in water and slightly soluble in organic solvents. It is a relatively stable compound except under alkaline conditions, and this stability has prompted its use at a high dosage on fruits, especially apples, early in the season for control of apple scab. The concept is that as the season develops the fungicide is redistributed in rain and dew to cover new foliage so that repeated applications will not be needed. The phthalimides in general and captafol especially may cause skin irritation and rash on people working with these compounds or in fields treated with them. The acute oral LD_{50} of captafol for the rat is 5000 to 6200 mg/kg. Its site of action as a fungicide is probably the same as that of captan.

8.8 OTHER FUNGICIDES

Dodine (dodecylquanidine acetate; Cyprex®, Melprex®) is a white crystalline solid moderately soluble in water and ethanol but insoluble in most organic solvents. It is an important fungicide, being especially effective against apple scab. It is a relatively stable compound with local systemic activity and with eradicant protective action against the apple scab fungus (Hamilton and Szkolnik, 1958). It is a surface active compound and redistributes readily on plant foliage. This, plus its ability to arrest development of apple scab some hours after infection (Jones et al., 1963), has made it especially valuable in apple production and accounts for wide usage (more than 1 million lb in 1971). It has the disadvantage of being phytotoxic to some apple varieties and may cause fruit russeting.

Dodine is of low mammalian toxicity (acute oral LD_{50} to the rat 1000 mg/kg) and is believed to act on fungi by disruption of cell membrane integrity with resulting cell leakage, the blocking of anionic sites on cell membranes, and inactivation of enzymes (Brown and Sisler, 1960; Somers, 1963). In some areas the apple scab fungus has become resistant to dodine.

$$C_{12}H_{25}N\overset{\overset{H}{|}}{-}C\overset{\overset{H}{|}}{-}N^{+} \cdot CH_3COO^{-}$$
$$\underset{NH\ \ H}{\overset{\|\ \ \ |}{}}$$

dodine

8.9 SYSTEMIC FUNGICIDES

Whereas a number of systemic insecticides have been available and used widely for several years, it was not until the late 1960s that real progress in this area was achieved with fungicides. Prior to that most of the fungicides functioned as protectants, unable to enter plant tissue in sufficient quantity to affect fungal growth significantly. There were some exceptions (e.g., the mercurials and dodine), but the potential for plant disease therapy through eradication or stoppage of fungal growth was limited. A number of new compounds that move systemically in plants are now available and offer new opportunities in plant disease control. They represent new classes of fungicides and include benzimidazoles (and the related thiophanates), the oxathiins, pyrimidines, and morpholines. While many are still experimental, some have been marketed for several years. It is of more than passing interest that, while these represent tremendous advances in plant chemotherapy, significant numbers of fungus species have been shown to develop resistance to some of these compounds both in the field and laboratory.

Benomyl (N-1-(butylcarbamoyl)-2-benzimidazole carbamate, Tersan®, Benlate®) is a white crystalline solid practically insoluble in water but soluble in organic solvents. It is a protective and eradicant fungicide, especially useful on fruit against apple scab and generally effective against powdery mildews on a wide variety of fruits and ornamentals. It is effective also as a dip or spray against postharvest storage rots on fruit. It is sold also as a fungicide for turf diseases and is active as an ovicide on mites. Although it moves systemically upward with the transpiring stream and into new growth in plants (Peterson and Edgington, 1975), it does not move sufficiently well to control Dutch elm disease, but it is active against the pathogen. In plants benomyl is converted to methyl 2-benzimidazolecarbamate ester (MBC) (Clemons and Sisler, 1969; Sims et al., 1969), and some success has been achieved with this compound and some of its salts for Dutch elm disease therapy (Smalley et al., 1973). For this purpose novel methods of application by injection are being developed (Prasad and Travnick, 1976).

Benomyl is a relatively stable compound, persisting on and in plant foliage either as the parent compound or as MBC for several months. It is persistent also on the orchard floor, and there have been reports of thatch buildup following heavy usage. This probably reflects its toxicity to a broad spectrum of fungi, including saprophytes as well as to decomposers such as collembola and earthworms. It is especially effective against this latter group and has been used for earthworm control in grass areas around airports to reduce hazard to aircraft from birds attracted by high earthworm populations adjacent to runways.

Benomyl is of low mammalian toxicity with an acute oral LD_{50} to the rat of more than 10,000 mg/kg. It acts in a manner similar to colchicine and owes its fungicidal action to adsorption to spindle fibers involved in cell division (Dekker, 1977).

benomyl   MBC

Carbendazim (MBC) is a fungicide in its own right as well as being the active principle of benomyl. It is less active systemically than benomyl because of poorer ability to penetrate the plant cuticle.

Two additional fungicides, thiophanate and methyl thiophanate, although not benzimadizoles, owe their activity to the generation of MBC in plant tissues (Sijpesteijn et al., 1977).

Thiabendazole was developed initially as an antihelmintic. It does not produce MBC but has a similar mode of action by interfering with spindle fibers during cell division. It has much the same spectrum of activity as benomyl.

thiabendazole

thiophanate

methyl thiophanate

Carboxin   (5,6-dihydro-2-methyl-1,4-oxathiin-3-carboxanilide; Vitavax®) is a white solid slightly soluble (170 ppm) in water and with moderate but variable solubility in organic solvents. The compound is relatively stable except under highly acidic or alkaline conditions and is effective as a seed treatment for cereals and some vegetables. It is active against smuts and bunts with toxicity limited pretty much to the Basidiomycetes (Edgington and Barron, 1967). It degrades in soil in a few weeks (Edgington and Corke, 1967).

Oxycarboxin   (5,6-dihydro-2-methyl   1,4-oxathiin-3-carboxanilide-4,4-dioxide, Plantvax®), a sister compound, is more soluble (1000 ppm) in water, stable in sunlight, and resists enzymatic oxidation in plants. Thus it is more effective as a foliar spray controlling rusts of cereals (Snel et al., 1970; Edgington and Busch, 1967). As with benomyl the oxathiins move upward in plants and toward the leaf edges (Kirk and Sinclair, 1968; Snel and Edgington, 1968).

The oxathiins are believed to act as inhibitors of respiration, the site of action being succinate oxidation (Mathre, 1971). They are of low mammalian toxicity, the acute oral LD_{50} to the rat being 3820 mg/kg for carboxin and 2000 mg/kg for oxycarboxin.

carboxin   oxycarboxin

Dodemorph (4-cyclododecyl-2,6-dimethylmorpholine acetate) is a yellow liquid miscible with water and acts as a systemic, eradicant, and protective fungicide. It is a recent entry in the fungicide field and is especially effective against powdery mildews on fruit and vegetables and may play a valuable role in greenhouses for mildews on vegetables and ornamentals. It is phytotoxic to some ornamentals. It is of low mammalian toxicity, the acute oral LD_{50} to the rat being 4500 mg/kg. A related compound **tridemorph** (2,6-dimethyl-4-tridecylmorpholine) is also an eradicant fungicide and is systemic. It has been effective for mildews on cereals.

dodemorph   tridemorph

Ethirimol (5-n-butyl-2-2ethylamino-4-hydroxy-6-methylpyrimidine, Milstem®) is representative of a small group of fungicides, the pyrimidines, now under development. It is effective against mildew and in areas of high rainfall (e.g., England) acts best as a seed treatment. From seed treatments it is taken up slowly by the plant and provides long-term protection (Martin, 1972).

Under the dry soil conditions of western Canada, seed treatments with ethirimol are not effective, and a spray formulation (Milgo E®) has been tested. Although foliar treatments are not taken up effectively by the plant, such sprays have provided good control of mildew on cereals after the disease is observed (Edgington et al., 1972).

$$
\begin{array}{c}
(CH_2)_3-CH_3 \\
CH_3 \diagdown \diagup OH \\
N \diagdown \diagup N \\
N-CH_2-CH_3 \\
| \\
H
\end{array}
$$

ethirimol

From an environmental point of view most of the fungicides have not been recognized as likely to cause problems. For the most part they have low mammalian toxicity, are biodegradable, persist for only a short time, and owing to low solubility in water are unlikely to move from the site of application and contaminate water. Many fungicides are close companions in our day to day lives as ingredients in soaps and cosmetics, household disinfectants, deodorants, fabric protectants, and mold inhibitors in paints, building supplies, and paper products. Small amounts may be present also in many of our raw fruits and vegetables and as preservatives and mold inhibitors to extend the shelflife of processed foods and bakery products. The low mammalian toxicity of the products involved makes such uses possible, for most are as nontoxic as table salt.

Some fungicides have, however, been shown to cause problems, and their use has been curtailed or banned or is under close scrutiny. Included in this group are the mercurials, the ethylene-bis-dithiocarbamates, and the benzimidazoles. The environmental significance of these is discussed later.

9

HERBICIDES AND THEIR USES

C hemicals toxic to plants are called herbicides. To be phyto-toxic they must inhibit some vital process in plants to the extent that the plants die or can no longer grow. For chemicals that are to be used to control weeds within agricultural crops or plantings of other desirable plants, there must be some mechanism (or method) of achieving selective phytotoxicity so that only the weeds are con-trolled. In areas where no vegetation is wanted, chemicals with very little if any selective phytotoxicity are preferred.

9.1 HERBICIDE ENTRY INTO PLANTS

Phytotoxic chemicals used as herbicides must possess physical and chemical properties that enable them to enter plants and move within plants to the toxic sites of action.

Foliar Penetration

The primary barrier to the entry of herbicides or other substances into plant leaves is the cuticle. The cuticle is a nonliving, noncellular, lipoidal membrane that completely covers upper and lower leaf sur-faces, surrounds herbaceous stems, and even invades stomatal cavities within leaves where gases are exchanged (Bukovac, 1976). The cuticu-lar membrane consists of an outer deposition of epicuticular waxes imbedded in an underlying layer or matrix of cutin. The cutin is bound on its underside to the cellulose cell wall by a layer of pectin.

The epicuticular waxes are mainly long chain alkanes, alcohols, ketones, aldehydes, esters, or fatty acids (Bukovac, 1976), while the underlying cutin is made up of long chain fatty and hydroxy fatty acids.

Crafts (1956) was among the first to point out that lipid-soluble compounds penetrate plant cuticles more readily than water-soluble compounds. He concluded, however, that, since both types of compounds can exert biological effects in plants, there must be both aqueous and lipoidal penetration routes through the cuticle. Once an herbicide has traveled across the cuticle, it must enter and move through the cell wall, which is nonliving, composed of cellulose, and highly hydrophilic. From this point on, some chemicals may move passively by diffusion through the nonliving cell wall continuum or apoplast to other parts of the plant. Other chemicals may be actively taken up across the membrane or plasmalemma of the nearest cell and then moved to other parts of the plant through the living cell continuum or symplast.

Root Uptake

Herbicides applied to soil may enter germinating seeds (Scott and Phillips, 1973) as well as the roots or unemerged shoots (O'Brien and Prendeville, 1972) of seedlings or established plants. Chemicals dissolved in soil water can diffuse readily into the spaces between root epidermal and cortical cells. Further penetration is less well defined, and there is not complete agreement on how water or solutes cross the endodermis, the next barrier, which is a layer of lignified or suberized water-impermeable cells. Chemicals that do cross the endodermis may subsequently enter nearby living cells of the symplast, or they may enter the nonliving cell walls or cells of the apoplast for further translocation to other parts of the plant.

Uptake Through the Stem

For herbaceous plants the importance of stems as a site of herbicide uptake is quite limited. For woody plants, however, the application of herbicides to the bark-covered stems during periods of dormancy is an accepted practice. Water-soluble chemicals or aqueous spray solutions penetrate the bark poorly. They must be injected directly into the xylem or water-carrying tissues of the stem with devices that can mechanically penetrate the bark. Oil-soluble ester formulations of 2,4,5-T and related herbicides can penetrate and enter the woody stem if applied to the bark as a highly concentrated oil-based spray. The oil and dissolved chemicals may not readily move directly through the bark cells, but they may move through lenticels or pores in the bark or through natural breaks in the bark such as growth cracks (Bukovac, 1976).

9.2   TRANSLOCATION OF HERBICIDES IN PLANTS

Symplastic Movement

The plant symplast has been defined (Crafts and Crisp, 1971) as "the total mass of living cells in the plant." The protoplasm of these living cells is actually a continuum because the individual cells are interconnected by living strands of protoplasm or plasmodesmata that extend across the nonliving cell walls. Some living cells such as the sieve tubes of the phloem are arranged end to end in interconnected columns and are highly specialized for long distance movement. The main function of the phloem is the movement of sugars from sites of photosynthesis to sites of storage or active growth. The direction of movement in phloem is most often downward except during periods of very rapid shoot growth.

Many symplastic herbicides (i.e., 2,4-D) cross the cell wall and then are actively transported across the cell membrane. From this point on they may be actively held by the symplast and moved from cell to cell via the interconnecting plasmodesmata. If the plant is rapidly moving sugars to storage in the roots or to sites of active growth, these chemicals move along through the phloem with the photosynthate to these same sites. Thus, chemicals that are mainly phloem mobile are often toxic to plant roots. Furthermore, their first visible effects on plant shoots are often in the shoot tips, young leaves, or other areas of active growth.

Apoplastic Movement

The apoplast (Crafts and Crisp, 1971) "constitutes the non-living cell wall continuum that surrounds the symplast." It includes the cell walls, intercellular spaces, and xylem elements. The entire interconnected system is permeable to water and dissolved solutes (Hay, 1976). The main function of the xylem is the upward movement of water and dissolved ions from the roots to the leaves, where water is either being utilized by the leaf cells or transpired (evaporated) from the leaves into the atmosphere. The rate of water movement through stems of plants is very great and can average more than 1 gallon/day for a single sunflower plant (Fuller, 1955).

Herbicides that are primarily xylem mobile (i.e., atrazine) move only upward in plants. If they are taken up by roots, they enter the xylem and are then rapidly translocated upward through the xylem to the leaves with water. If a xylem-mobile herbicide penetrates the leaf cuticle and enters the xylem, it will move only upward in that particular xylem strand and will not move downward and out of that particular leaf (Hay, 1976). Herbicides that move with water in the transpiration stream are not accumulated in the growing points of plants. They are first widely distributed via xylem strands extending to all transpiring leaves. Here, they diffuse out of the xylem elements and finally cross the plasmalemma of living cells. The toxic symptoms of many xylem-mobile herbicides are first observed at leaf margins, where the rate of water loss is greatest. Because of the varying tendencies of different herbicides to be retained in cell walls or to enter living cells at earlier stages, different chemicals often have unique distribution patterns after root uptake and xylem transport (Hay, 1976).

9.3 TOXIC EFFECTS OF HERBICIDES IN PLANTS

The physiology and mode of action of herbicides in plants is a broad and complex field that cannot be covered here in detail. The reader is referred to several recent texts for a more nearly comprehensive treatment of the subject (Audus, 1976; Kearney and Kaufman, 1975, 1976; Ashton and Crafts, 1973).

Herbicide action in plants is complicated by the fact that a single chemical may have many different effects at different physiological sites. Furthermore, it is often difficult to determine the primary effects that begin the irreversible process toward death of the plant. Since it is impossible to examine the effects of every herbicide or related group of herbicides on every possible physiological process, our knowledge is not complete, particularly for newer compounds and for the secondary effects of older compounds.

Contact Toxicity

Several herbicides kill living plant tissue quickly and are thus referred to as "contact herbicides" (Table 9.1). Under some conditions there can be some translocation of these herbicides through nonliving por-

Table 9.1 Herbicides with Contact Toxicity in Plant Tissues[a]

phytotoxic oils	sodium arsenite
endothal	cacodylic acid
nitrofen	bromacil
pentachlorophenol	bromoxynil, ioxynil
diquat, paraquat	fluorodifen

[a]Summarized in part from Morrod, 1976.

tions of the plant, but normally they are translocated poorly. The living plant cells are killed so rapidly on contact after cuticular penetration or root uptake that further translocation is usually prevented.

Most contact herbicides are thought to act by destroying cell membrane integrity. A rapid dissociation of the bonds between the membrane proteins and phospholipids is thought to be involved (Morrod, 1976). In addition to the herbicides mentioned here (Table 9.1), other herbicides like 2,4-D that are generally not thought of as contact herbicides can have very similar rapid effects on plants if applied at very high rates. Early effects of contact herbicides usually involve a rapid darkening of green leaf tissue, a symptom of cell leakage. This is followed by a browning and necrosis of all exposed tissue within 2 to 3 days. Since there is no downward translocation, the unexposed roots of perennial plants are not affected and new shoots are readily produced from underground buds.

Mitotic Inhibitors

In plants active growth occurs in meristems (growth zones) of root tips, young leaves, apical or lateral buds, and in the vascular cambium of stems and roots. Aberrant growth, reduced growth rates, even plant death, can occur if mitotic poisons reach these meristems or growth areas in concentrations sufficient to inhibit their normal growth.

Several types of herbicides are known to be mitotic poisons (Table 9.2), and while effects of these vary, microscopic examination of affected plant tissues has revealed the following symptoms (Figure 9.1).

1. Arrested cell division at either prophase or telophase (Figure 9.1), resulting in an elevated mitotic index. This has been observed in plants treated with either pronamide (Desai, 1972) or trifluralin (Parka and Soper, 1977).

Table 9.2   Herbicides that Inhibit Mitosis in Plants[a]

carbamates	dinitroanilines
propham	trifluralin
chlorpropham	nitralin
barban	DCPA
thiocarbamates	bromacil
EPTC	
butylate	bensulide
vernolate	pronamide
maleic hydrazide	

[a]Summarized in part from Cartwright, 1976.

2. Disruption of spindle alignment so that individual chromosomes migrate to many micronuclei. This has been observed in plants treated with propham (Canvin and Friesen, 1959).
3. Many multinucleated cells as a result of nuclear division without the formation of new cell walls. This has been observed in plants treated with pronamide (Desai, 1972) and trifluralin (Parka and Soper, 1977).

Many of the mitotic inhibitors are applied to the soil, where they may affect cell division and growth in roots or shoots of germinating plant seedlings.

Photosynthetic Inhibitors

Photosynthesis is the process by which light energy (photons) is transferred, first of all to the energy of excited electrons from chlorophyll molecules and then to high energy phosphate compounds (ATP) and reduced pyridine nucleotides (NADPH), which drive reactions for the reduction of carbon dioxide to carbohydrates. The ultimate effect is that light energy is first trapped and then converted to chemical energy in the form of glucose. (See the simplified equations for the light and dark reactions, Table 9.3.) This process represents the primary means by which solar energy is converted to a usable form of energy for our biosphere.

Photosynthesis occurs in very small subcellular organelles called chloroplasts. The light reactions occur in densely packed membranous areas of chloroplasts called grana where the chlorophyll is concentrated. Surrounding the membranes and grana is a liquid matrix called the stroma, which is the site for the dark reactions of photosynthesis.

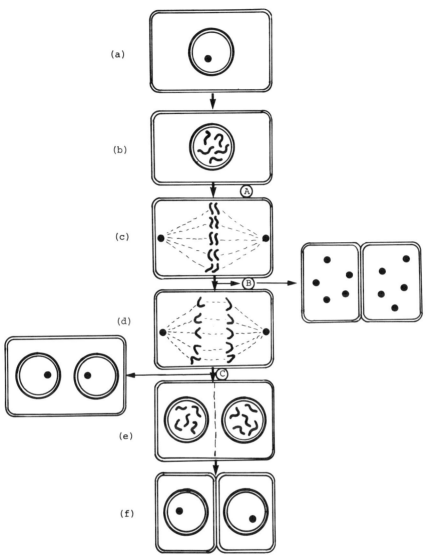

Figure 9.1 Influence of herbicides on mitosis. Stages in normal cell division include (*a*) resting cells; (*b*) prophase when the chromatin is gathered into long thread-like double chromosomes; (*c*) metaphase, when the split chromosomes are arranged in a plane across the equator of the cell and the spindle with two poles is formed; (*d*) anaphase, when the chromosome halves separate with one complete set migrating to each of the two poles; (*e*) telophase, when each new group of chromosomes is arranged in a new nucleus and a new cell wall begins to form; and (*f*) when the two new cells are complete. Points of inhibition include Ⓐ pronamide and trifluralin can arrest cell division in prophase; Ⓑ propham disrupts normal spindle development, and individual chromosomes migrate to daughter cells, forming several micronuclei; and Ⓒ trifluralin disrupts formation of the new cell wall with resultant development of multinucleated cells.

Table 9.3   Simplified Equations for the Light and Dark
Reactions of Photosynthesis

Light Reaction:

$$H_2O + ADP + Pi + NADP \xrightarrow[\text{Chlorophyll}]{\text{Light energy}} O_2 \uparrow + ATP + H^+ + NADPH$$

Dark Reaction:

$$CO_2 + ATP + NADPH + H^+ \longrightarrow CH_2O + ADP + Pi + NADP^+$$

ADP   Adenosine diphosphate
ATP   Adenosine triphosphate
NADP   Pyridine nucleotide
NADPH Reduced pyridine nucleotide
CH_2O Carbohydrates or sugar
Pi   Inorganic phosphate

There are two types of chlorophyll that participate in the light reactions; one is responsible for Photosystem I, the other, for Photosystem II (see Figure 9.2). When a photon of light is received by a chlorphyll molecule in Photosystem II, the light energy is transferred to the energy of an excited electron that is ejected from the chlorophyll molecule. Left behind is an oxidized chlorophyll molecule that receives a replacement electron from the reaction in which the water molecule (H_2O) is cleaved and oxygen (O_2) evolved. The energized, ejected electron eventually loses its energy through a series of reactions called an electron transport or cytochrome system to which is coupled a side reaction for the formation of the high energy phosphate compound ATP. After this series of phosphorylation reactions, this electron is then available to reduce the second type of chlorophyll molecule, which has previously received a photon of light and ejected an excited electron in Photosystem I. After being received by the Ferredoxin-reducing substance, the electron from Photosystem I may go on to reduce Ferredoxin, which in turn reduces $NADP^+$ to the high energy reducing substance NADPH. Alternatively, this electron may go through a second type of cytochrome system (cytochrome b) in which ATP is formed. This latter process is cyclic, and the electron can actually return to the same type of chlorophyll molecules from which it was originally ejected.

In the dark reactions of photosynthesis (Figure 9.3), a five-carbon sugar is first phosphorylated in a reaction utilizing ATP from photosynthesis. Then, in a reaction involving the energy rich reducing compound NADPH, a molecule of carbon dioxide (CO_2) is reduced and combined with the five-carbon sugar to form two molecules of a three-carbon sugar. After further phosphorylation these molecules combine to form one six-carbon sugar.

The overall process of photosynthesis can be quantified by measuring the rates of carbon dioxide fixation or oxygen evolution by plants in the light. More than half of the currently available herbicides have some inhibitory effect on photosynthesis (Moreland and Hilton, 1976), and since this process is unique to green, chlorophyll-

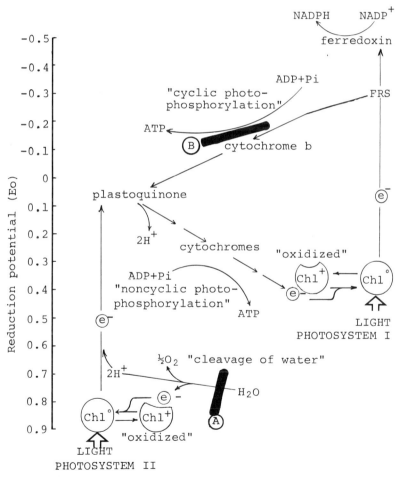

Figure 9.2 Photophosphorylation: the light reactions of photosynthesis. Many herbicides inhibit photosynthesis by blocking the cleavage of water at point Ⓐ, others "uncouple" cyclic photophosphorylation from the process by acting at point Ⓑ.

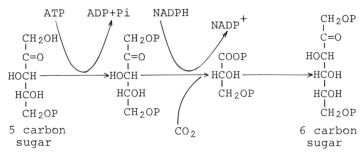

Figure 9.3 Carbon dioxide fixation: the dark reactions of photosynthesis.

containing plants, most of these herbicides are relatively nontoxic to mammals or other nontarget organisms.

Most of the herbicides that are known to be photosynthetic inhibitors block the transport of electrons from Photosystem II (Moreland and Hilton, 1976), which utilizes water as an electron donor (Table 9.4). Two other herbicides, diquat and paraquat, do not block electron flow but do compete with FRS (Ferredoxin-reducing substance) for the acceptance of electrons from Photosystem I. One herbicide, per-fluidone, is known to be an uncoupler of cyclic photophosphorylation (Moreland and Boots, 1971). To date, no commercially used herbicides are known to be inhibitors of enzymes in the dark reactions of photosynthesis.

The phenyl ureas and phenyl carbamates were the first herbicides shown to be potent inhibitors of photosynthesis (Cooke, 1956; Wessels and Vander Veen, 1956). Diuron has been so thoroughly studied that it is now routinely used by plant biochemists as a tool in studies on the basic mechanisms of photosynthesis. Diuron, atrazine, and possibly other inhibitors of Photosystem II are thought to deactivate the enzyme system responsible for cleaving the water molecule. Thus without replacement electrons from water, electron flow mediated by light can occur only until all of the chlorophyll molecules in Photosystem II are oxidized. After this, photophosphorylation (ATP production), reduction of Photosystem I chlorophyll molecules, cyclic photophosphorylation, production of NADPH, and eventually the reduction (fixation) of carbon dioxide in the dark reactions all cease.

The Hill reaction, (Hill 1937) is often used to measure Photosystem II inhibitors. In this reaction (Figure 9.4) a suspension of isolated chloroplasts is illuminated and provided with a substitute electron acceptor (usually a dye such as DPIP, dichlorophenol indophenol) instead NADP. With no inhibitor the electron accepting dye is reduced by electrons from chloroplasts exposed to light and changes from a

Table 9.4   Herbicidal Inhibitors of Electron Transport in
Photosystem II of Photosynthesis[a]

chlorinated phenyl ureas	1,2,4-triazinones
chlorinated s-triazines	azido-s-triazines
bis-carbamates	cyclopropane
uracils	carboxamides
pyridazinones	alkyl anilides
diphenyl ethers	p-alkyl thioanilides

[a]Summarized from Moreland and Hilton, 1976.

dark blue to a light blue color, a change that can easily be measured spectrophotometrically. In the presence of diuron (or another Photosystem II inhibitor) the rate of water cleavage and DPIP reduction is inhibited to varying degrees, depending on the potency and concentration of the herbicide. Inhibitors of Photosystem II can also be assessed by monitoring the rates of oxygen evolution or carbon dioxide uptake by isolated chloroplasts, leaf discs, or intact plants exposed to light.

With an uncoupler of cyclic photophosphorylation, like perfluidone, the cleavage of water and the flow of electrons through the cytochrome systems or to NADP are not inhibited, but this process is uncoupled from the process of producing ATP (Moreland and Boots, 1971). Compounds of this type might be more difficult to define except that CO_2 fixation per unit O_2 evolution should be reduced.

Bipyridylium herbicides such as diquat and paraquat exert toxic effects in plants by competing with FRS (Ferredoxin-reducing substance) for the reception of electrons from Photosystem I (see Figure 9.5). This interception of electrons by diquat or paraquat cations prevents the reduction of NADP to NADPH but does not influence noncyclic and cyclic photophosphorylation (Harris and Dodge, 1972). With a reduced supply of NADPH, plants exposed to diquat or paraquat should have a reduced ability to "fix" CO_2. Whereas this has been confirmed, these herbicides have a very rapid "contact" toxicity that cannot be explained by an inhibition of CO_2 fixation alone. Generally it is accepted that the bipyridyliums are applied to the plant in the form of cations that, when reduced by electrons from Photosystem I, become free radicals. These reduced free radicals are reoxidized by molecular oxygen, which in turn is reduced with water to form hydrogen peroxide (H_2O_2) (Calderbank, 1968; Calderbank and Slade, 1976) (see Figure 9.5). Hydrogen peroxide is a highly toxic oxidizing substance that could cause the rapid toxicity observed for paraquat or diquat by acting to increase membrane permeability with a resulting

$$2H_2O + 2\ DPIP \xrightarrow[\text{chloroplasts}]{\text{light}} O_2\uparrow + 2\ DPIP$$
$$\text{oxidized,} \qquad\qquad\qquad \text{reduced,}$$

Figure 9.4 The "Hill reaction," which is often used to measure the activity of compounds as inhibitors of photosynthesis.

complete disruption of cellular organization and function (Calderbank and Slade, 1976; Siegel and Halpern, 1965). Thus, while paraquat and diquat should be included in discussions of herbicides and photosynthesis, it is important to realize that these two chemicals are dependent on molecular oxygen and photosynthetic electron flow for their rapid contact toxicity in plants. Paraquat and diquat toxicity can also be associated with electron flow from respiration. This system does not, however, seem to be as reactive, since symptoms of phytotoxicity are reduced and plant death occurs much more slowly in the dark. The less rapid toxicity does allow for greater paraquat translocation and longer term control of perennial weeds. Putnam and Ries (1968) have shown that photosynthetic inhibitors in combination with paraquat are synergistically toxic to quackgrass.

Influence of Herbicides on Respiration

Inhibitors of photosynthesis can be correctly thought of as inhibitors of the energy trapping and storing processes in plants. In a similar sense chemicals that reduce respiration can be thought of as inhibitors of energy release and energy utilization. These compounds block por-

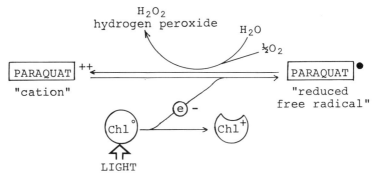

Figure 9.5 Paraquat free radical formation via electron flow from photosynthesis and its reoxidation by oxygen and water to yield the highly toxic oxidant "hydrogen peroxide."

tions of the process in which chemical energy (carbohydrates, lipids) can be converted back to high energy phosphate (ATP), pyridine nucleotides (NADPH), carbon dioxide, water, and heat. The basic process of respiration is very similar in all living organisms; thus it is not surprising that herbicides or other pesticides that inhibit respiration are quite toxic to mammals and other nontarget organisms as well.

There are three basic processes in the respiration of starch to sugars (see Figures 9.6 and 9.7): (1) glycolysis, which occurs in the cytoplasm; (2) the Krebs cycle, which occurs on the surfaces of inner mitochondrial membranes (cristae); and (3) the electron transport sys-

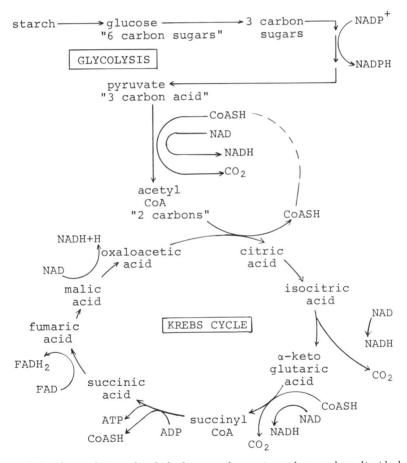

Figure 9.6 The oxidation of carbohydrates and organic acids to carbon dioxide by glycolysis and the "Krebs cycle." Benzoic acid and picolinic acid herbicides inhibit the utilization of succinic acid or α-ketoglutaric acid by this cycle (Foy and Penner, 1965).

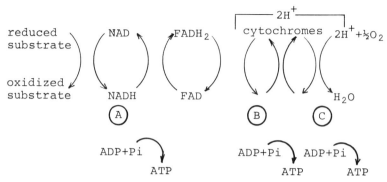

Figure 9.7   Electron flow from respiration, oxygen uptake, production of water, and the associated phosphorylation reactions. Uncouplers of "oxidative phosphorylation" are thought to "uncouple" oxidation and electron flow from phosphorylation by causing the production of unstable intermediates at points Ⓐ, Ⓑ, and Ⓒ.

tem for oxidative phosphorylation, which occurs within the membranes of the cristae.

Glycolysis

In glycolysis (Figure 9.6), starch is first hydrolyzed to form sugars. The sugars are then broken down and oxidized to the three-carbon acid pyruvic acid. The hydrogens and electrons liberated by the dehydrogenation process are utilized in the production of NADPH.

The Krebs Cycle

The "Krebs cycle" (Figure 9.6) is a cyclic series of metabolic reactions in which various organic acids are converted from one to another. One of its main functions is to regain energy from stored carbohydrates or lipids. Pyruvic acid, the end product of glycolysis, first undergoes a dehydrogenease reaction involving coenzyme ASH (CoASH) in which CO_2 is evolved and NADH plus acetyl CoA are produced. The acetyl CoA combines with oxaloacetic acid to enter the actual Krebs cycle. In the subsequent interconversion of carboxylic acids, oxaloacetic acid is regenerated to continue the cycle. In this process, however, several dehydrogenation reactions occur in which NADH, $FADH_2$, or ATP are produced, and the two carbons that entered the cycle as acetate (in acetyl CoA) are liberated as two molecules of carbon dioxide (Figure 9.6).

Electron Transport and Oxidative Phosphorylation

The reduced pyridine nucleotides, such as NADH and $FADH_2$, generated by the Krebs cycle may be utilized in other metabolic reactions in the cell to produce the high energy phosphate compound ATP. For the latter the hydrogens and electrons are transferred through a cytochrome or electron transport system (Figure 9.6) in which oxygen acts as the ultimate electron acceptor in the formation of water. At three points along the electron transport system, phosphorylation occurs as a reaction that is coupled via intermediates to the process of oxidation and electron transport. Depending on whether a pair of electrons is introduced to the cytochrome system by NADH + H or $FADH_2$, three or two molecules of ATP, respectively, are formed per atom of oxygen reduced. Thus a normal "tightly coupled" mitochondrial system would have a P/O ratio between 2 and 3.

Respiration is a physiological process that has been studied very extensively. Herbicides that are toxic to plants largely because of effects on respiratory processes in mitochondria can be classified as either inhibitors or uncouplers of oxidative phosphorylation (Table 9.5).

Respiratory Inhibitors

A compound that is regarded as a respiratory inhibitor reduces both oxygen uptake and phosphorylation (ATP formation) in mitochondria. Benzoic acid and picolinic acid herbicides do this by inhibiting one of

Table 9.5 Herbicides with Primary Toxic Effects on Respiration[a]

Inhibitors of Respiration	Uncouplers of Oxidative Phosphorylation

benzoic acids
  2,3,6-TBA, amiben,
  dinoben, dicamba,
  tricamba
	substituted phenols
	  DNP, DNOC, dinoseb
	  PCP

picolinic acids
  picloram
	hydroxy benzonitriles
	  ioxynil, bromoxynil

amides
  allidochlor, propanil
	phenoxys
	  2,4-D, 2,4,5-T

[a] Summarized from Kirkwood, 1976.

the reactions in the Krebs cycle. When succinic acid was provided as the only substrate to isolated cucumber mitochondria, Foy and Penner (1965) found that all of these herbicides inhibited both oxygen uptake and phosphorylation. When α-ketoglutarate was provided as the substrate, much higher concentrations of these herbicides were required for inhibition. Although Foy and Penner (1965) did not specify which step in the Krebs cycle is inhibited, inhibition of any step would reduce the substrate for subsequent reactions and slow down or stop the entire cycle. The net effects would be reduced CO_2 evolution, reduced NADH and ATP production, and reduced O_2 uptake, for electron transport and phosphorylation are dependent on the NADH and $FADH_2$ generated by the Krebs cycle.

Uncouplers

When a respiratory system is exposed to a chemical uncoupler, the process of oxidation continues, but it is uncoupled from the process of phosphorylation or ATP production. The Krebs cycle continues, CO_2 is evolved, NADH and $FADH_2$ are produced, electron transport through the cytochrome system occurs, oxygen is taken up, and water is produced, but the side reactions of phosphorylation occur at reduced rates (see Figure 9.7). The substituted phenols are the most notable compounds with this effect and like other uncouplers they commonly stimulate Krebs cycle activity and oxygen uptake, possibly by separating these processes from phosphorylation. DNP (dinitrophenol) is now the accepted biochemical tool for producing an uncoupled respiratory system in mitochondrial preparations. If oxygen evolution by living tissue sections or isolated mitochondria is stimulated by some experimental compound but cannot be further stimulated by additional treatment with dinitrophenol (DNP), this insensitivity to DNP may mean that the system is already uncoupled and that the experimental compound may be an uncoupler. Since phosphorylation is reduced and oxygen evolution usually stimulated, the more commonly accepted biochemical proof that a compound acts as an uncoupler is a reduced P/O ratio. In an uncoupled system fewer than two molecules of ATP may be produced for each oxygen atom reduced compared to a P/O ratio closer to 3 in a normal coupled system.

Effects on Nucleic Acid Metabolism and Protein Synthesis

In the process of protein synthesis (Figure 9.8) DNA in the cell chromatin or nucleus is thought to be shielded by nuclear proteins (nucleohistones) that may have a role in determining which sites of

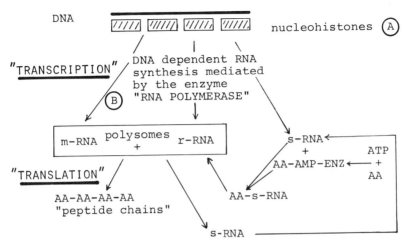

Figure 9.8 Diagrammatic scheme of protein synthesis. 2,4-D and related herbicides are thought to increase the production of RNA by influencing the action of the nucleohistone shields Ⓐ or RNA polymerase Ⓑ.

the DNA are transcribed for RNA and protein synthesis. The enzyme RNA polymerase is thought to have an important role within the nucleus in the polymerizing of nucleic acids for RNA synthesis. In the transcription process at least three types of RNA are formed: messenger RNA (mRNA), which becomes the cytoplasmic copy of the DNA code; ribosomal RNA (rRNA), which combines with mRNA to form the protein-synthesizing template; and soluble or transfer RNA (sRNA), which combines with activated amino acids (aa-AMP-enzyme complex). The translation part of the process involves the movement of amino acids in the form of the activated amino acyl-sRNA molecules to the active polysome (mRNA + ribosome complex), the assembly of these amino acids in sequence on the mRNA template, and the combination of these amino acids to form the peptide chains of protein.

Cytoplasmic factors, including metabolites or natural hormones and foreign molecules such as herbicides or other pesticides, could influence protein synthesis in various ways. Firstly, these compounds could enter the nucleus, precipitate or remove some of the histone shields, and thus uncover a portion of the DNA code for transcription. Secondly, the cytoplasmic factor could activate a specific RNA polymerase enzyme that subsequently enters the nucleus and begins the transcription process and RNA synthesis.

Other mechanisms are also possible whereby the cytoplasmic factor could alter protein synthesis by affecting specific sRNAs or ribosomes and interfering with the translation part of the process (Cherry, 1976).

Among herbicides, 2,4-D is probably the most widely studied with respect to effects on nucleic acid and protein synthesis in plants. It is generally accepted that 2,4-D acts as a synthetic version of the naturally occurring auxin indole acetic acid (IAA) and that it influences protein synthesis by similar mechanisms. The greater toxicity of 2,4-D and other synthetic auxins (2,4,5-T; dicamba; picloram; etc.) may relate to the fact that they are much less readily oxidized or deactivated by the plant than the natural auxin IAA (Table 9.6).

Chen et al. (1972) and other investigators (Cherry, 1976; Key and Hanson, 1961) have shown that 2,4-D causes a dramatic increase in the amount of RNA synthesized in susceptible plants.

In the study by Chen et al. (1972) with 2,4-D; 2,4,5-T; dicamba; and picloram, all four herbicides increased the DNA, RNA, and protein contents of tolerant wheat and susceptible cucumber roots. As the herbicide treatments were increased to toxic levels, RNA synthesis per unit DNA was further increased in the susceptible cucumber but decreased in the tolerant wheat. Furthermore, in the treated susceptible cucumber, the protein synthesized per unit RNA was sharply decreased, suggesting that the additional RNA synthesized in susceptible plants was in some way nonfunctional. In a subsequent study these same investigators (Chen et al., 1973) showed that 2,4-D or dicamba decreased the number of extractable nucleohistone proteins from susceptible cucumber but had no effect on extractable histones from wheat, suggesting that the herbicides influenced protein synthesis by acting at the histone-DNA level.

Cherry (1976) has summarized work in which most investigators agree that auxin herbicides and presumably natural hormones influence protein synthesis by acting on the DNA-RNA transcription process. He has suggested that herbicides or hormones first act on a cytoplasmic factor (hormone receptor) that in turn enters the nucleus and initiates the transcription process by affecting either the histones or the RNA ploymerase or both. The link between effects of 2,4-D at this level and death of a plant is still not clear. We do know, however, that,

Table 9.6   Herbicides Reported to Influence Nucleic Acid and Protein Synthesis in Plants

phenoxy acids (2,4-D, 2,4,5-T and others)

benzoic acids (dicamba)

picolinic acids (picloram)

after foliar penetration, 2,4-D is not evenly distributed among living cells and that, depending on concentration, 2,4-D may be stimulatory or inhibitory to plant growth. In addition, unlike natural auxins, 2,4-D concentrations in plant cells or tissues cannot be readily controlled (Van Overbeek, 1964). The overall effect in a plant treated with a toxic concentration of 2,4-D is that the growth of some cells and tissues is stimulated while the growth of others is inhibited. The result is abnormal, distorted growth characterized by swollen and twisted stems and petioles and abnormally shaped leaves. The bending and twisting of stems and "cupping" of leaves are called epinastic effects. Microscopically, phloem disruption and destruction and xylem plugging can be observed (Loos, 1975). It is most likely the disorganization and destruction of these translocation and absorption tissues (xylem, phloem, and leaves) that finally result in plant death (Hanson and Slife, 1969; Van Overbeek, 1964; Loos, 1975).

9.4   METHODS OF CLASSIFYING HERBICIDES

There are more than 150 chemically different herbicides. To simplify the study of this group of compounds, it is useful to classify them into several smaller groups of related chemicals.

A user of herbicides may find it helpful to classify compounds according to how they are applied. Preplant herbicides are applied to the soil or weeds before a crop is transplanted or seeded. Preplant soil-incorporated treatments are applied to the soil surface and then tilled into the soil, often to prevent volatility losses (see Chapter 14). Preemergence or postemergence herbicides are applied either before or after the emergence of the specified crop or weed. A user of herbicides may add mention of the crop to a classification scheme based on use and then may refer to a particular new chemical as a "postemergence" herbicide for transplanted tomatoes.

A plant biochemist may wish to classify herbicides solely according to their mechanisms of toxicity in plants and may wish to know whether a compound is selective (toxic to some plants but not others) or nonselective (toxic to most plants), contact (one that acts on contact with plant tissue) or translocated (moved within the plant via xylem or phloem). We may thus hear herbicides referred to as "inhibitors of photosynthesis, respiration, cell division or protein synthesis" or as compounds with "nonselective, contact toxicity."

For the agricultural or environmental biologist it is helpful to classify the herbicides into different groups of chemically related compounds and then to determine which generalizations are useful with respect to mechanism of action, fate in soil, mammalian toxicity, and effects on other nontarget organisms. This latter approach is employed in this text. In addition, we have distinguished between the older herbicides available before 1945 and the large number of new compounds that became available as a result of the intensive exploration phase that followed the end of World War II.

9.5 PRE-1945 HERBICIDES

History

The use of inorganic compounds as herbicides extends back hundreds of years, perhaps to the days of the Roman Empire, when ashes, common salts, and various inorganic wastes were used to control weeds on roadbeds. In 1896 Bonnet discovered that copper sulfate solutions were toxic to broadleaf weeds, particularly mustard (see Brian, 1976). This discovery led to the first chemical methods for the selective control of weeds (mustards) in crops (cereals). While most herbicides developed since 1945 are synthetic organic compounds, inorganic chemicals continue to be important, particularly for nonselective control of all vegetation on rights-of-way and in industrial areas.

Common Salts

Common inorganic salts such as NaCl can act as herbicides if applied at sufficiently high rates. Witness the damage to shrubs and trees adjacent to streets, roads, highways, or even sidewalks where common salt is used for deicing purposes. Although copper sulfate, ammonium nitrate, urea, and potassium chloride can have beneficial effects as pesticides or fertilizers, at higher rates they can be quite toxic to plants. As aqueous sprays on plant foliage, they act mainly as contact poisons that quickly desiccate and kill the plant tissue. The selective use of copper sulfate sprays in cereal crops also led to the use of dilute sulfuric acid as an herbicide. Although more effective than copper sulfate, its high corrosiveness to human skin or spray equipment hastened the search for better alternatives. The use of inorganic salts as herbicides has largely been dicontinued, though copper sulfate is still commonly used for aquatic weed control.

Chlorates and Borates

Sodium chlorate ($NaClO_3$) is a white crystalline salt that is about 30 times more toxic to plants than sodium chloride. Chlorates gained wide use as nonselective herbicides because of their effectiveness as soil applications to control deep-rooted perennial weeds (Aslander, 1926). At high rates (300 to 1000 lb/acre) sodium chlorate is used as a soil sterilant; at lower rates (1 to 4 lb/acre) it has been used as a nonselective top killer. It is also used for defoliating and desiccating crops prior to harvest. Sodium chlorate is still one of the most widely used inorganic herbicides in North America. More than 35 million lb are used annually in industrial or agricultural areas (Von Rumker et al., 1975).

Sodium chlorate is only slightly toxic to animals, with an acute oral LD_{50} of approximately 5000 mg/kg (WSSA, 1974). It can persist in soil for more than 1 year but is degraded by microorganisms and can be leached away from the soil surface, particularly in sandy soils. Persistence is much reduced in warm, moist areas.

Because sodium chlorate is a strong oxidant and desiccant, treated organic material (plant leaves, clothing, etc.) becomes highly flammable. If ignited, the fire is difficult to smother because the decomposing sodium chlorate provides the needed oxygen. For this reason sodium chlorate is always formulated with fire-retarding hydrated sodium borates.

The borates can be either beneficial as micronutrients or toxic to plants, depending on rates of application. They are not highly toxic to animals. The acute oral LD_{50}'s in mice range from 2000 to 3000 mg/kg (Weed Science Society of America [WSSSA], 1974). In soil they can persist for up to one season except when leached in high rainfall areas. They are not readily degraded by microorganisms. After uptake by plant roots they are readily moved via the xylem to leaf margins, where tissue desiccation begins. They are also highly toxic to plants in the seedling stage. They are often included in formulations with diuron, monuron, bromacil, or other soil sterilant herbicides, since their toxicity to soil microorganisms slows the degradation of these herbicides and extends their activity in the soil.

Petroleum Oils

Stove oil is widely used for weed control in carrots, because this crop has a higher tolerance to oil than most other plant species. The discovery that oils could be used to increase herbicide penetration made it possible to extend the use of atrazine in corn from strictly a soil application to use as a postemergence herbicide as well (Jones and Anderson, 1964). Fuel oil continues to have a major role in selective brush control on power lines, railroads, and roadsides, where it is used extensively as a carrier for 2,4-D; 2,4,5-T; and other brush killers. When oil is used as a carrier, ester formulations of these herbicides can be used at any time of the year, even in winter when the trees or brush are dormant.

At phytotoxic rates, oils have such a rapid contact toxicity to plants that most investigators believe physical rather than biochemical mechanisms are responsible for injury and death. While an oil film over a plant surface could certainly limit vital gas exchange, the main site of action is most likely in cell membranes. When oils actually contact and begin to dissolve in a cell membrane, the fatty acid chains of the phospholipids are forced apart and the typical symptoms of contact toxicity result (Morrod, 1976).

Effective weed control with oil requires a high dosage (60 to 80 gal/acre for weed control in carrots), and high costs of transportation have often resulted in the use of the "most available" oil. Oils vary in phytotoxicity, and there have been instances of crop damage because oil from a new source had a higher than normal aromatic compound content. For many years, used transformer oil was a convenient carrier for brush spraying on power lines. This is no longer done, because of adverse environmental effects that can occur, not from the oil, but from PCBs or other compounds added to the transformer oil to make it more resistant to heat and fire.

Phenols

Substituted phenols are possibly the oldest synthetic organic pesticides. Being respiratory uncouplers with rapid contact activity on cell membranes, they have a very broad spectrum of activity and have been used as fungicides, insecticides, ovicides, and acaricides, as well as herbicides. In 1935 DNOC (4,6-dinitro ortho-o-cresol) became the first organic chemical to be patented for selective chemical weed control (Truffaut and Pastac, 1935). DNOC is highly selective and was used for controlling broadleaf weeds in grasses. Because it is highly effec-

DNOC   dinoseb

tive at low rates and nonirritating and less corrosive, it quickly replaced the inorganic salts and sulfuric acid for weed control in cereals in both North America and Europe. PCP (pentachlorophenol), another member of this group is used mainly as a wood preservative.

In 1945 Crafts established that herbicidal activity of the phenols increased in the order DNOC (2-methyl group) < (2-ethyl) < (2-isopropyl) < (2-sec butyl = 2-sec amyl). After this, dinoseb (2-sec butyl-4,6-dinitro phenol) has become the most widely used phenolic herbicide in North America. The substituted dinitrophenols are formulated as sodium, amine, or ammonium salts for postemergence weed control or as emulsifiable concentrates for preemergence weed control. Dinoseb can be used to control seedling weeds and grasses in small grains, legumes, potatoes, corn, cucurbits, mint, small fruits, orchards, and alfalfa (WSSA, 1974). Application as a nonselective top killer to desiccate crops before harvest or to "chemically burn off" vegetation in areas that cannot be mowed are other major uses.

Dinoseb and other phenols are reported to have many toxic effects on plants (reviewed by Kaufman, 1976), but their primary mode of action is accepted to be as uncouplers of oxidative phosphorylation. While generally toxic to all green plant tissue, the water-soluble sodium salts of dinoseb can be used for selective postemergence weed control in some crops because of differences in foliar penetration. Dinoseb/oil mixtures act as contact herbicides causing rapid death of green plant tissue by destroying cell membranes.

When applied to the soil surface, dinoseb is readily leached downward to control shallow germinating weed seedlings. Deeper crop seeds are not affected unless rainfall or soil conditions result in excessive leaching. Dinoseb is not persistent in soil. Applications at herbicidal rates (6 to 9 lb/acre) provide effective preemergence weed control for only 2 to 4 weeks (WSSA, 1974). Microbial degradation has been demonstrated (Tewfik and Evans, 1966) and involves conversion of nitro groups to hydroxyls, ring cleavage, and then further degradation. Many toxic effects on microorganisms have been observed, including effects on soil respiration and nitrification. At rates of application required for weed control these effects are either very slight or very reversible (Kaufman, 1976).

DNOC and, later, dinoseb were developed for many weed control situations, but these compounds also have undesirable properties. They are nonselective and nonresidual, moderately corrosive to metal and they stain human skin and clothing. Furthermore, they are highly toxic to animals. Acute oral LD_{50}'s in the rat range from 50 to 100 mg/kg (WSSA, 1974). Symptoms of poisoning in human beings include accelerated respiration, elevated body temperature, and motor weakness.

While still registered for many situations, their use has decreased dramatically with the development of more effective, less toxic herbicides.

9.6   HERBICIDES DEVELOPED SINCE 1944

Phenoxy Alkanoic Acids

The most significant milestone in the history of chemical weed control occurred during World War II with the discovery of 2,4-D. This herbicide was found to be amazingly effective and selective at very low rates of application (0.5 to 1.0 lb/acre) for the control of broadleaf weeds in cereals. The tremendous success of 2,4-D provided much of the impetus for the search for other organic chemicals that could be used as selective herbicides. In this sense 2,4-D ushered in our modern era of selective chemical weed control. Unlike DDT, the chemical that had a similar role in the insecticide field but that is now replaced by other chemicals, 2,4-D and related herbicides continue to be among the most important chemicals for weed control.

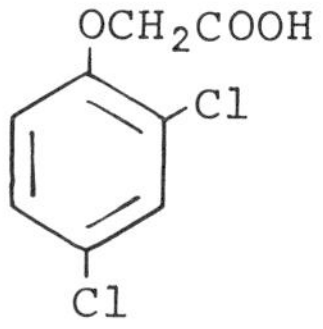

OCH_2COOH

Cl

Cl

2,4-D

Each year phenoxy herbicides account for approximately 20% of the total synthetic organic herbicide use in North America. In 1971 the total use (USA + Canada) was more than 50 million lb for 2,4-D alone (Lawless et al., 1972; *Statistics Canada*, 1972).

Chemistry

2,4-D and most other phenoxy herbicides (Table 9.7) are usually formulated as esters or amine salts to improve their solubility in oil or water, respectively. 2,4-D acid and its metal or amine salts are not very volatile, but the ester formulations vary from low to high volatility.

Table 9.7 Widely Used Phenoxy Alkanoic Herbicides

Common Name	Structural Formula
2,4-D	2,4-dichlorophenoxy acetic acid
MCPA	2-methyl-4-chloro-phenoxy acetic acid
2,4,5-T	2,4,5-trichlorophenoxy acetic acid
dichlorprop	2-[2,4-dichlorophenoxy] butyric acid
mecoprop	2-[2-methyl-4-chlorophenoxy]propionic acid
silvex	2-[2,4,5-trichlorophenoxy]propionic acid
MCPB	2-[2-methyl-4-chlorophenoxy]butyric acid
2,4-DB	4-[2,4-dichlorophenoxy]butyric acid

The amine salts are highly soluble in water, but the ester formulations are most soluble in oil.

Physiological Activity in Plants

At low application rates phenoxy herbicides can act in the same manner as IAA and can influence many natural plant processes such as stem elongation, callus formation, inhibition of lateral buds (apical dominance), root initiation, and fruit or leaf abscission. Plants do not, however, have the same ability to degrade 2,4-D or other synthetic auxins as they do for IAA. Thus at slightly higher rates these compounds become highly toxic. In general the phenoxy herbicides are much more toxic to herbaceous broadleaved plants than to grasses and much more toxic to deciduous trees than to evergreens or conifers.

Phenoxy herbicides are more active as foliar applications than as applications to the soil. Ester formulations more readily penetrate through leaf cuticles than amine or salt formulations do, but all formulations appear to be converted to the acid before entering the living cells (Wichman and Byrnes, 1970).

Phenoxy herbicides can be mobile in plants in either the xylem or the phloem (Hay, 1976). Movement out of leaves and into stems is symplastic, occurs in the phloem, and is dependent on the movement of carbohydrates. Once in the stem there is lateral movement between xylem, phloem, and cambium tissues, but the main long distance movement is upward in the xylem to other leaves in the transpiration stream. Most accumulation is in the actively growing buds and rapidly growing young leaves, since with no exit of carbohydrates from these organs, there is no mechanism for the symplastic movement of the

2,4-D back to the main stem. Under some conditions, particularly high humidity, when transpiration rates are low, there can be significant downward movement of 2,4-D to plant roots (Eliasson, 1965). Just as 2,4-D accumulates to greater concentrations in more actively growing tissues or organs, it may also accumulate to higher concentrations in the more metabolically active cells within a particular tissue. Varying concentrations of 2,4-D in adjacent cells and tissues can in turn result in stimulatory or inhibitory effects on growth via the primary mode of action, which is to disrupt normal DNA, RNA, and protein metabolism. The overall result is that leaves become abnormally shaped and cupped with twisted petioles. Developing stems are twisted, coiled, or sharply bent. Actual death of plants most likely occurs as a result of stimulated growth within stems, which results in the crushing, breaking, or plugging of the xylem and/or phloem tissues (Loos, 1975).

Degradation of Phenoxy Herbicides

Most phenoxy alkanoic acid herbicides such as 2,4-D; 2,4,5-T; or MCPA are readily metabolized, particularly in tolerant plants. Degradation occurs by alteration (decarboxylation or lengthening) of the acid chain, hydroxylation of the ring, ring cleavage, and conjugation with various plant constituents (sugars, amino acids, proteins). (See reviews by Loos, 1975; Ashton and Crafts, 1973.)

The phenoxy herbicides in general are not persistant in soil. Under warm, moist soil conditions, 2,4-D persists 2 to 3 weeks and MCPA 6 to 8 weeks. 2,4,5-T, one of the more persistent phenoxy herbicides, may vary in persistence from 2 months to a year. The phenoxy herbicides are degraded in soil by bacteria and actinomycetes that adapt to them as a carbon or food source. Microbial degradation occurs by degradation of the side chain to the phenol, ring hydroxylation and cleavage, and subsequent degradation to succinate and other naturally occurring organic acids (Loos, 1975).

In water, phenoxy alkanoic acid herbicides may be degraded by the same microorganisms and the same pathways as have been established for soil. Owing to lower numbers of organisms in water, degradation rates are slower. Crosby (1976) has reported that 2,4-D; 2,4,5-T; and MCPA undergo photolytic removal of the side chain, replacement of ring chlorines with hydroxyls, and polymerization of these polyhydroxy benzenes to form naturally occurring humic acids. As in plants, soil, and water, rates of photolysis decrease with greater chlorination of the phenoxy ring (Crosby, 1976).

Uses of Phenoxy Alkanoic Herbicides

The great physiological selectivity of phenoxy herbicides combined with their nonpersistence in the environment has resulted in extensive usage. More than 30 years after its discovery, 2,4-D is still one of the most important herbicides for the control of annual and perennial broadleaf weeds in cereals and for weed control in lawns, pastures, and aquatic situations (WSSA, 1974).

MCPA is the second most widely used phenoxy herbicide in agricultural crops. It is very similar to 2,4-D except that some cereal, legume, and flax crops are slightly more tolerant. Dichlorprop or mecoprop are often combined with 2,4-D in lawn herbicide formulations to improve the control of 2,4-D resistant weeds. Mecoprop is the only herbicide in this group that can be used to control broadleaf weeds without injury to bentgrass turf. The phenoxy butyric herbicides (2,4-DB and MCPB) are the most selective for weed control in peas, alfalfa, and other leguminous crops.

Combinations of 2,4-D and 2,4,5-T are the most important stem/foliage treatments for deciduous brush control on roadsides and power lines. Because of their inactivity on conifers and their low cost, these same herbicides have been the most important chemicals for conifer release programs in forestry. Of all the phenoxy alkanoic herbicides, 2,4,5-T is the most active on the greatest number of woody species. For this reason it is the most important chemical in brush control, particularly as a dormant/basal spray in oil during the fall, winter, and early spring. Some species of oak, ash, and hickory can be more effectively controlled with 2,4,5-TP (fenoprop) than with either 2,4-D or 2,4,5-T.

The most controversial use of any herbicide has been the use of 2,4-D and 2,4,5-T as chemical defoliants for military purposes during the war in Vietnam from 1961 to 1969. The main purpose of the defoliation program was to save lives by developing clear, vegetation-free zones around military bases and transportation routes. In total, however, one-seventh of the land area of (then) South Vietnam was sprayed (Boffey, 1971). This ignited a controversy regarding the health and environmental effects of these chemicals that has spread from the Vietnam situation to repeated criticisms of the many uses of phenoxy alkanoic herbicides in North America.

Effects on Nontarget Organisms

Without question, the most important nontarget organisms for 2,4-D; 2,4,5-T; and related herbicides are desirable plants (Chapter 14).

The toxicity of 2,4-D and related herbicides to animals is low. Acute oral toxicity for rats ranges from 300 to 1000 mg/kg for most formulations of these herbicides, and dermal and inhalation toxicities are minimal (WSSA, 1974). Whereas many different metabolites of 2,4-D have been reported in mammals, detoxification by metabolism appears to be less significant than elimination as far as explaining the low mammalian toxicity (Loos, 1975). Most investigators have reported almost total elimination of 2,4-D in the urine within 24 to 48 hours of feeding (Loos, 1975).

Picloram

Picloram (4-amino-3,5,6-trichloropicolinic acid) can be regarded as a synthetic auxin, for it has many physiological effects in plants that are similar to those of the phenoxy alkanoic herbicides and to the natural auxin IAA. Its herbicidal activity was first reported by Hamaker et al., in 1963.

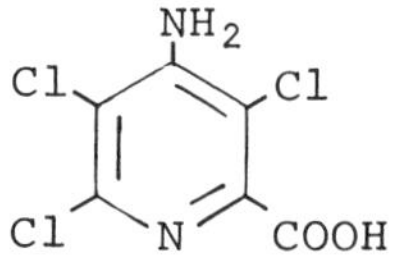

$$NH_2$$
Cl $\quad$ Cl
Cl $\quad$ N $\quad$ COOH

picloram

Chemistry and Physiology

Picloram is a white powder with a chlorine-like odor. The technical acid has a solubility in water of 430 ppm. It is often formulated as the highly soluble potassium salt for application as a spray to foliage or to the soil as a dry application of 10% active pellets. One of the most widely used picloram formulations is a mixture with 2,4-D with both as dimethyl amine salts. Ester formulations in which picloram is mixed with 2,4,5-T are also available. Picloram is a member of a family of compounds known as the pyridines. Other members of this group include N-serve (2-chloro-6-trichloromethyl pyridine), which is used commercially as a nitrification inhibitor in soil (Kapusta and Varsa, 1972), and chlorpyrifos, an important insecticide (see Chapter 10).

In comparison to 2,4-D and related phenoxy alkanoic herbicides, picloram is much more potent as either an auxin or herbicide. It is 10 to 500 times more toxic than 2,4-D or 2,4,5-T to most broadleaf plants except mustards (Foy, 1976). It is much more active and more persistent in soil, and though conifers are generally quite tolerant to 2,4-D and related herbicides, they are quite susceptible to picloram. Most grasses are as tolerant to picloram as they are to 2,4-D or 2,4,5-T.

Picloram has primarily been used for broadleaf weed and brush control on rights-of-way and roadsides. It is used particularly in areas where conifers are the dominant woody species to be controlled. Picloram is sufficiently selective for use at low rates to control broadleaf weeds in cereal crops, and in some areas of western North America use of picloram as a spot treatment in crops is permitted on a restricted basis.

The high potency of picloram seems to be related to its translocation and resistance to degradation within most plants. Picloram applied to plant foliage is readily translocated through the phloem to the stem and then via the xylem to upper portions of the shoot. Once in the stem it may also move downward in the phloem to the roots, where it may leak out into the soil or growth medium to be taken up by roots of adjacent plants (Foy, 1976). Picloram applied to the soil is leached to the root zone in most soils, where it is readily taken up by plant roots and translocated via the xylem to the leaves (Mitchell and Stephenson, 1973). Like 2,4-D, picloram is eventually accumulated in the younger, more actively growing leaves. Injury symptoms are also similar to those of 2,4-D.

Picloram reportedly exerts auxin-like effects on plants by the same mechanisms in DNA, RNA, and protein metabolism as proposed for 2,4-D and other auxin-like herbicides (Chen et al., 1972). In addition it is reported to be an inhibitor of the respiratory Krebs cycle (Foy and Penner, 1965) and it may have activity as an uncoupler of oxidative phosphorylation (Chang and Foy, 1971). Actual plant death may result from a proliferation of growth in some stem tissues with the resultant crushing of the xylem and phloem. Picloram and other pyridine derivatives often affect plants in a manner similar to the effects of the *Fusarium lycopersici* Sacc. infection of tomato plants (Gäumann, 1958). This fungus produces toxic amounts of fusaric acid, a pyridine-type compound, that causes epinastic twisting, plugging of xylem tissues, wilting, necrosis, and finally death (Owens, 1969). One example of these effects for picloram is the plugging of xylem elements, which leads to the death of red maple trees (Peterson et al., 1974).

Picloram does not appear to be readily degraded in either tolerant or susceptible plants (Meikle et al., 1966; Mitchell and Stephenson, 1973).

Fate of Picloram in Soil

In soil picloram is nonvolatile and is one of the most persistent herbicides currently in wide use. Degradation does not occur in sterile soil (Grover, 1967), and soil microorganisms do not adapt to picloram as a new carbon or nutrient source (Youngson et al., 1967). Degradation appears to be only coincidental to the microbial degradation of other types of organic matter in soil. Persistence is inversely related to soil temperature and soil moisture. Thus the half-life in soil can vary from as long as 4 years in cold, semiarid regions to as short as a few months in the warm, moist soils of the Southeastern United States (National Research Council of Canada [NRCC], 1974).

Most laboratory studies indicate that the leaching or lateral movement of picloram is greater in light textured soils than in clays or soils high in organic matter content (NRCC, 1974). Helling (1971), using comparative soil thin layer chromatography techniques, has classified picloram as a "highly mobile" herbicide. At 5 months after treatment Scifres et al., (1969) found more picloram at depths of 15 to 45 cm at the lower ends of plots with 3% slopes than could be observed in plots with 0, 1, or 2% slopes. Thus lateral movement and leaching can occur under field conditions. On a rocky power line right-of-way in Ontario, however, Suffling et al. (1974) collected and monitored runoff water from a small picloram-treated area and found that surface runoff was not as important as other factors such as vertical leaching or degradation for picloram disappearance (Suffling, 1976).

Picloram may be less persistent in water than in soil, owing to its ready photolysis in aqueous solutions exposed to either sunlight or ultraviolet light (Hall et al., 1968).

Effects on Nontarget Organisms

In animals the acute oral toxicity is low, LD_{50}'s ranging from 2000 mg/kg for rabbits to 8000 mg/kg for rats (WSSA, 1974). Picloram does not accumulate in the milk after being fed to lactating cows (Kutschinski and Riley, 1969), and most investigators (summarized by Foy, 1976) have found that it is readily excreted from animals in either urine or feces within 1 to 2 days after ingestion. Although picloram accumulates and is not readily degraded in plants, there is no evidence of further accumulation in herbivorous organisms in either aquatic or terrestrial habitats (NRCC, 1974). Owing to its low lipophilicity and relatively high water solubility, it may be ingested with plant tissues by various animals, but it is then very readily eliminated by excretory processes.

Environmental Impact of Picloram Persistence

Because of its high activity as a plant growth regulator and its persistence in soil, the greatest environmental concerns for picloram involve nontarget plants. In one study conducted in Ontario, Anderson and Smith (1969) established that picloram was still present at phytotoxic levels 5 years after an application of $2\frac{1}{2}$ lb/acre to a Guelph loam soil. If farmers were allowed to use picloram at similar rates on crop land, the problems that could result in any crop rotation scheme involving crops sensitive to picloram would be long-term problems indeed! Picloram phytotoxicity is not altered by the digestive system of cattle, and though reports of damage to sensitive plants by exposure to picloram-contaminated manure are not uncommon, few such reports have been confirmed.

Another hazard that has been reported for use of picloram in agricultural areas is the contamination of ponds used for irrigation. Kates (1965) described one such incident where picloram moved laterally into a pond and subsequently injured irrigated tobacco.

Persistence, the main property that makes picloram so hazardous to use on or near land used for agriculture, is possibly its greatest attribute for perennial weed and brush control on remote power lines and rights-of-way. In these situations its high selectivity results in effective control of tall growing brush species and in the encouragement of a grassy vegetative cover that prevents erosion.

Benzoic Acid Herbicides

The benzoic acid herbicides are yet another group of compounds that can in some respects be categorized as synthetic auxins. Plant growth-regulating properties were discovered for benzoic acids in 1942 (Zimmerman and Hitchcock, 1942) as a result of the same program that led to their discovery of 2,4-D.

TBA (2,3,6-trichlorobenzoic acid) was first evaluated as a herbicide in 1948 (see Brian, 1976) and has been developed for use at relatively high rates to control deep-rooted perennial weeds and woody species, particularly conifers. Its greatest use is on utility rights-of-way, industrial sites, and other noncrop land. By a slight alteration of TBA's structure (replacing the "2-chlorine" with a methoxy group) Richter (1961) developed a new auxin-type herbicide, **dicamba** (3,6-dichloro-*o*-anisic acid) that had greater selectivity for use in crops than TBA. Dicamba is also effective for controlling many 2,4-D resistant weeds and brush (including conifers) and is recommended for use in many crops, including small grains, corn, flax, sorghum, pastures, and sugar cane (WSSA, 1974). In combination with 2,4-D and mecoprop or other phenoxy alkanoic herbicides, dicamba is widely used at low rates for weed control in lawns. At higher rates it is also effective for controlling 2,4-D resistant perennial weeds, conifers, and other woody species.

| dicamba | chloramben | 2,3,6-TBA |

Chloramben (3-amino-2,5-dichlorobenzoic acid), another benzoic acid herbicide, differs in structure from TBA only by having an amino group at the "3" position instead of a third chlorine at the "6" position on the ring. This small difference results in a compound with dramatically different properties from those of either TBA or dicamba. All three compounds exhibit growth regulator effects in plants. In contrast, however, to TBA, which is nonselective, and to dicamba, which is selective only for broadleaf weed control in grasses, chloramben is highly selective for the control of germinating grasses and broadleaf weeds in broadleaf crops. It is widely used for preemergence weed control in soybeans, asparagus, beans, corn, peanuts, peppers, pumpkins, sunflowers, sweet potatoes, tomatoes, flowers, and ornamentals (WSSA, 1974). Because of its use in soybeans it has become one of the major herbicides in the United States. In 1971 about 20 million lb of chloramben were produced in the United States compared to only 6 and 2 million lb for dicamba and TBA, respectively (Lawless et al., 1972).

Physiological Activity and Degradation in Plants

There is not yet clear agreement with regard to the primary mechanisms of action for benzoic acid herbicides. Chen et al. (1972, 1973) established that dicamba can act as a synthetic auxin by differentially affecting DNA, RNA, and protein synthesis in tolerant and susceptible plants. Chloramben inhibits root elongation and growth, particularly in susceptible weed seedlings (Frear, 1976), but is reported not to have effects on RNA and protein synthesis (Moreland et al., 1969). TBA has been shown to affect cell elongation and apical growth, as well as root and leaf development (Frear, 1976). TBA can also disrupt the transport of IAA in plants. Another closely related growth regulator herbicide, TIBA (triodobenzoic acid), acts as a strong antiauxin, primarily by inhibiting auxin (IAA) transport in plants (Frear, 1976). Foy and Penner (1965) have reported that most of the benzoic acid herbicides are weak to moderately active inhibitors of the Krebs cycle in respiration, but these effects are not generally accepted as a primary mode of action.

In contrast to TBA, which is not degraded or detoxified in plants (Frear, 1976), dicamba is readily detoxified by ring hydroxylation and subsequent conversion to complexes with natural plant constituents (Broadhurst et al., 1966). Chloramben is also readily complexed to form an N-glucosyl chloramben metabolite (Colby, 1965; Swanson et al., 1966). TBA and dicamba can be readily absorbed by either plant leaves or roots and translocated in either xylem or phloem to accumulate in the most actively growing tissues of the plant. Chloramben is much less active as a foliar application than as an application to the soil, where it is readily taken up by roots or seeds (Knake and Wax, 1968). After uptake chloramben is very poorly translocated. Some plants may be tolerant because the chloramben is complexed with glucose and immobilized in the roots before toxic amounts are translocated to the more sensitive sites of action in the leaves (Baker and Warren, 1962).

Fate in Soil

In contrast to the phenoxy herbicides, the benzoic acids are moderately water soluble and are all highly active in soil. Dicamba and chloramben are moderately persistent and mobile, while TBA is both highly mobile and very persistent. In one extreme case Phillips (1968) reported that more than 50% of a 16 lb/acre application of TBA remained active in a silt-loam Kansas soil 11 years later. Most reported half-lives for either chloramben or dicamba are less than 1 year (Frear, 1976). Dicamba and chloramben degradation in soil appears to be microbial, and decarboxylation has been reported (MacRae and Alexander, 1965), but subsequent degradation products have not been characterized. Whereas volatility does not seem to be important, leaching appears to have a major role in the disappearance of TBA, dicamba, or chloramben from soil surfaces.

Toxicity and Environmental Effects

The benzoic acid herbicides are low in toxicity to animals. Acute oral toxicities in rats are in the range of 500 to 1000 mg/kg, 700 to 1000 mg/kg, and 3500 mg/kg for dicamba, TBA, and chloramben, respectively (WSSA, 1974). Being potent hormone herbicides with particularly high toxicity to broadleaved plants, spray drift damage to nontarget vegetation by either dicamba or TBA is an important concern. There are also some reports that dicamba can volatilize from plant leaf surfaces with resultant damage to adjacent sensitive plants by vapordrift (Behrens, 1975).

Triazines

If it is accepted that the first major milestone in the development of selective chemical weed control was the discovery of 2,4-D, the discovery of the triazine herbicides might be regarded as the second major breakthrough. This is particularly true if our terms of reference relate primarily to impacts on North American agriculture. Atrazine and to a lesser extent simazine and other triazine herbicides are probably the most important factors resulting in increases in corn yields over the last 15 years in the United States. In addition, some agriculturalists feel that, without the contributions of the triazines to weed control, corn might never have become the important crop in Canada that it is today. The herbicidal properties of **chlorazine** (Table 9.8) were first reported in 1955 (Gast et al., 1955). Since then more than a dozen symmetrical triazines and one asymmetrical triazine have become commercially important for weed control in crops or on noncrop land.

Chemistry and Uses

Chemically the triazines are heterocyclic nitrogen compounds, for they contain aromatic rings composed of both carbon and nitrogen atoms. Simazine, atrazine, and most of the commercially important triazines are symmetrical or s-triazines with the three carbons and three nitrogens evenly distributed around the ring (Table 9.8). In metribuzin, the most important asymmetrical triazine (as-triazine) de-

Table 9.8 The s-Triazine Herbicides

Common Name	Chemical Name	R$_1$	R$_2$	R$_3$	Water Solubility (ppm at 20–25 C)
chlorazine	2-chloro-4,6-bis (diethylamino)-s-triazine	Cl	-N(C$_2$H$_5$)$_2$	-N(C$_2$H$_5$)$_2$	--
simazine	2-chloro-4,6-bis (ethylamino)-s-triazine	Cl	-NH(C$_2$H$_5$)	-NHC$_2$H$_5$	5
atrazine	2-chloro-4-(ethylamino)-6-(isopropylamino)-s-triazine	Cl	-NHC$_2$H$_5$	-N.isoC$_3$H$_8$	33
propazine	2-chloro-4,6-bis(isopropylamino)-s-triazine	Cl	-N.isoC$_3$H$_8$	-N.isoC$_3$H$_8$	8.6

125

Table 9.8 *(Continued)*

Common Name	Chemical Name	Chemical Structure			Water Solubility (ppm at 20–25 C)
		R_1	R_2	R_3	
cyanazine	2-chloro-4-ethylamino-6-(1-cyano-1-methyl-ethylamino)-s-triazine	Cl	$-NHC_2H_5$	$\begin{array}{c} CH_3 \\ \| \\ -NHC-CN \\ \| \\ CH_3 \end{array}$	171
prometone	2-methoxy-4,6-bis (isopropylamino)-s-triazine	$-OCH_3$	same as for propazine		750
prometryne	2-methylthio-4,6-bis (isopropylamino)-s-triazine	$-SCH_3$	same as for propazine		48
ametryne	2-methylthio-4-ethyl-amino-6-isopropylamino-s-triazine	$-SCH_3$	same as for atrazine		185
terbutryne	2-methylthio-4-ethyl-amino-6-tert-butyl-s-triazine	$-SCH_3$	$-NHC_2H_5$	$-NHC_4H_9$	58

veloped thus far, the carbons and nitrogens are not symmetrically distributed.

Whereas chlorazine was the first triazine with documented herbicidal properties, simazine was the first to gain commercial importance in weed control because of greater selectivity in several crops. With even greater predictability for preemergence weed control, plus greater effectiveness for postemergence weed control, atrazine has now far surpassed simazine in use for weed control in corn. Because corn is such a large acreage crop in North America, atrazine has the distinction of leading all other single herbicides in volume of use during the 1970s, U.S. production and use having averaged approximately 100 million lb annually (Lawless et al., 1972; Von Rumker et al., 1975) over this period.

Atrazine. In corn, atrazine is by far the most widely used herbicide. It can be applied at three different times with little risk of crop injury: preplant, for the control of difficult weeds like quackgrass; preemergence, for the control of most germinating broadleaf weeds; and postemergence with oils or surfactants as adjuvants for improved control of some annual grasses. In addition to its use in corn, atrazine is also important for selective weed control in sorghum, sugar cane, macadamia orchards, pineapple, grass seed crops, some types of turf grasses, and in conifer reforestation. At higher rates and usually in combination with other herbicides, atrazine is widely used for nonselective weed control in noncrop land (WSSA, 1974).

Simazine is the least water soluble of all the triazine herbicides (Table 9.8). It is very immobile in soil and very ineffective as a foliar application to weeds. For preemergence weed control in annual row crops, adequate rainfall soon after application is more important for simazine than for atrazine. Because of its immobility, simazine is less effective than atrazine for preemergence weed control in corn. But this same property of immobility makes simazine more selective for use in a greater number of different crops, primarily deep-rooted perennial crops, than any other triazine herbicide. Simazine is currently registered for use in alfalfa, artichokes, asparagus, cane fruits (i.e., raspberries), corn, cranberries, currants, pineapples, most orchard crops, sugar cane, forages, bermuda grass, many species of woody ornamental nursery plants, shelterbelts, and Christmas tree plantings (WSSA, 1974). Simazine has gained wide use at higher rates for nonselective weed control in noncrop land. It also has some use in the control of aquatic weeds. In 1971 the estimated United States production of simazine was 5 million lb (Lawless et al., 1972).

Propazine. The importance of propazine (1971 production of 4 million lb, Lawless et al., 1972) is solely due to its greater selectivity for weed control in sorghum, an important crop in the United States.

Cyanazine is a symmetrical triazine herbicide that has recently been developed. As a preemergence application cyanazine is nearly as effective as atrazine for weed control in corn. It is much less persistent than atrazine, and for this reason some weeds escape control later in the season. But in situations where a triazine-sensitive crop is being planned for the subsequent year, cyanazine is gaining wide acceptance as a triazine herbicide that will give a similar spectrum of weed control with a much reduced risk of residue carryover (Anderson, 1971).

Methoxy and **methylthio triazines.** Many derivatives other than the diamino substituted chloro-*s*-triazines have been synthesized and found to have different but very useful properties as herbicides. Important among these are the methoxy triazines with a "2" methoxy ($—OCH_3$) instead of the "2" chloro substituent and the methylthio triazines with a "methylthio" ($—SCH_3$) substituent at the "2" position on the ring (Table 9.8). Nomenclature for these other triazines is conveniently systematic in that a methoxy triazine or a methylthio triazine with *N*-substituents identical to atrazine would be named atratone and ametryne, respectively. Likewise, methoxy and methylthio derivatives of simazine would be known as simetone and simetryne, respectively (Table 9.8).

The methoxy triazines are characterized by greater water solubility and greater activity as foliar or soil applications, but in comparison to the chloro-triazines, they are less selective (Knüsli, 1970).

Prometone (the methoxy analogue of propazine) has been developed for nonselective preemergence and postemergence weed control in noncrop land.

The methylthio triazines are quite varied in herbicidal activity (Knüsli, 1970). Prometryne selectively controls annual grasses and broadleaf weeds in celery and cotton and has activity as either a preemergence or postemergence application. Ametryne is a selective herbicide for pineapple, sugar cane, and bananas, whereas terbutryn can provide selective weed control for fall seeded cereal crops in some of the Northwestern states (WSSA, 1974).

Metribuzin [4-amino-6-tert-butyl-3-(methylthio)-*as*-triazine-5(4H)one], an asymmetrical triazine, was developed during the early 1970s. It has physiological similarities to the symmetrical triazines with respect to both translocation and mode of action in plants. It is, however, remarkably different in selectivity and can be used selectively in soybeans, tomatoes, and potatoes, crops that are highly sensitive to most s-triazine herbicides (WSSA, 1974).

$$(CH_3)_3C \qquad \overset{O}{\underset{N}{\parallel}} \quad N-NH_2$$

metribuzin

As a group the triazine herbicides are quite insoluble in water, solubilities ranging from 5 ppm for simazine to 750 ppm for prometone (WSSA, 1974). They are most commonly formulated on inert clay particles and sold as "wettable powders" that are suspended in water and applied as a spray. Some liquid, suspension, or granular formulations are available for some of the triazines.

Soil Interactions

The preemergence weed control activity, persistence, and possible movement of the triazine herbicides are all very much influenced by variations in the soil environment. Triazine/soil interactions have been extensively researched. Volume 32 of *Residue Reviews* is one of the most extensive reviews, with all 15 chapters devoted primarily to triazine/soil interactions (Gunther, 1970).

With the exception of cyanazine, which has a half-life of only a few weeks, most of the symmetrical triazine herbicides are moderately persistent in soil. This persistence is very much dependent on (1) rates of application, (2) soil organic matter or clay content, (3) soil moisture, (4) soil temperature, and (5) soil pH. Simazine or atrazine applied at rates less than 2 lb/acre are normally degraded to nonphytotoxic levels by the end of the growing season. With rates of application greater than 2 lb/acre, residues often carry over into the next year. The very high rates of 10 to 20 lb/acre required for total vegetation control can keep treated areas free of most vegetation for up to 3 years.

The adsorption of the triazines to soil clay or organic matter results in a temporary inactivation of a portion of the herbicide applied. As a result higher rates of application are required for effective preemergence weed control on heavy clay or organic soils (Anderson, 1971). Low soil moisture, low temperatures, and acid soil conditions (low pH) all increase persistence by increasing the tendency for triazine herbicides to be adsorbed and also reduce their effectiveness for weed control. Triazine herbicides are also less likely to leach in heavy clay or organic soils, but they are likely to be more persistent, and in the moist soil conditions of the following spring, residue carryover problems can be greater than in light sandy soils (Anderson, 1971). On soil surfaces photodecomposition may be an important factor for triazine degradation, and weed control is often poor if there is no rainfall within the first several days to leach the chemical down in the soil. It is most difficult to study photolysis of triazine herbicides while adsorbed to soil, but Jordan et al. (1964) and Pape and Zabik (1970) have documented significant photolysis of atrazine on soil under some conditions.

Chemical or microbial degradation of most triazine herbicides occurs quite slowly in soil. According to Armstrong and Chesters (1968) adsorption to soil colloids may slowly catalyze the chemical degradation of atrazine to its 2-hydroxy atrazine analogue. The microbial degradation of triazines has been extensively documented (Kaufman and Kearney, 1970), but it occurs slowly with no lag phase. It has been suggested that microorganisms may not have to become "adapted" to degrade triazine herbicides (Kaufman and Kearney, 1970). The slow rate of breakdown for most triazines may be due to adsorption, desorption, and solubility factors that make all but a small percentage of the s-triazines unavailable for microbial uptake and degradation at any particular time (Harris et al., 1968). Microbial degradation, while slow, involves conversion to 2-hydroxy and N-dealkylated derivatives, eventual ring cleavage, and liberation of carbon and nitrogen for microbial or plant growth (Esser et al., 1975). Microbial degradation of the s-triazine herbicides occurs much faster in warm, moist, low adsorptive soils (McCormick and Hiltbold, 1966). It is very interesting how a small change in the sturcture of atrazine (the addition of a —C≡N group) to form cyanazine can so dramatically reduce soil persistence from months to weeks. It is possible that both greater availability because of greater solubility and greater ease of microbial attack are involved in the shorter persistence of cyanazine (Anderson, 1971).

Metribuzin, the asymmetrical triazine, has not been studied as extensively as the s-triazines, but its activity, mobility, and persistence are also very dependent on soil organic matter. It has a half-life in soil of a few to several months like atrazine, and initial deaminated microbial metabolites have been documented (Sharom, 1974). Residue carryover into the next season has been observed for metribuzin, particularly on muck soils.

Mode of Action and Metabolism in Plants

Most of the triazine herbicides are more readily absorbed by plant roots than by leaves. After root uptake or foliar penetration, however, translocation is almost exclusively upward in the xylem. Unlike the growth regulator herbicides (phenoxys, benzoics, picloram) there is no accumulation of the triazines in the most rapidly growing portions of the plant. Once atrazine arrives in a plant leaf, it does not exit from that leaf. Thus triazines finally accumulate in leaf cells, particularly at leaf tips and margins, where water loss by transpiration is greatest. Subsequently they arrive at their ultimate site of action—the chloroplasts.

It is well documented that the symmetrical triazines (Esser et al., 1975) and metribuzin (Sharom, 1974) are all potent inhibitors of Photosystem II (cleavage of water) in photosynthesis. But because atrazine-treated plants develop injury symptoms faster and differ in visible effects from "dark starved" plants, some investigators have postulated that a directly phytotoxic substance is formed when the chlorophyll is illuminated in the presence of atrazine (Ashton and Crafts, 1973). In addition to effects on photosynthesis, atrazine has been shown to reduce transpiration by influencing stomata (Jensen et al., 1977; Wills et al., 1963). Moreover, at low rates of application, triazines are known to stimulate plant growth, possibly by enhancing nitrate uptake, nitrate reductase, and protein synthesis (Ries et al., 1967).

Largely owing to the investigations of Shimabukuro and associates (reviewed in Esser et al., 1975), more is known about the degradation of atrazine in plants than is known for most other herbicides. The very high tolerance of corn (Shimabukuro and Swanson, 1969) and the moderate tolerance of many annual grasses (Jensen et al., 1977) are due to the ability of these plants to metabolize and detoxify the atrazine molecule. There are at least three pathways that can be important (Figure 9.9).

1. The nonenzymatic replacement of the 2 chloro group with a hydroxyl to form hydroxy atrazine.
2. Dealkylation of either N-alkyl group.
3. Conjugation with amino acids at the "2" position to form the "glutathione complex."

These pathways can have varying degrees of importance in different plants, but they all seem to be fully operative in corn. Not all of the

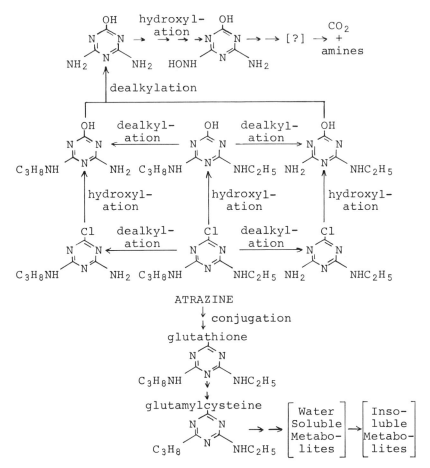

Figure 9.9 Metabolism and detoxification of atrazine in higher plants. The three main pathways—hydroxylation, dealkylation, and conjugation—are all operational in corn with eventual complete degradation to CO_2 and other naturally occurring compounds.

degradative end products are known. There are, however, several reports of radiolabeled $^{14}CO_2$ being slowly evolved after plants have been fed ^{14}C-ring labeled atrazine, simazine, propazine, prometryne, or ametryne (summarized by Esser et al., 1975), which indicates that complete degradation of the triazines in plants is possible.

Nearly all plants can metabolize triazine herbicides, but the differing rates at which this can occur in different species is generally regarded as the major explanation for atrazine selectivity. But in some areas where atrazine has been used repeatedly for weed control in corn grown for several years without rotation, some biotypes of weeds are appearing that have a new mechanism for atrazine resistance (Radosevich, 1973). One such example is lamb's-quarters (*Chenopodium album* L.). Normally a 1 to 2 lb/acre application of atrazine gives very good control of this weed, but the resistant biotypes can tolerate this chemical at rates in excess of 20 lb/acre (Bandeen and McLaren, 1976). Jensen et al. (1977) established that differences in uptake or translocation of atrazine were not involved. In subsequent studies Souza-Machado et al. (1977) showed that, contrary to the normal, Photosystem II was remarkably resistant to atrazine inhibition in chloroplasts isolated from the tolerant biotype. Radosevich and Conrad (1977) have suggested that the tolerant biotypes of lamb's-quarters and other weeds (groundsel, redroot pigweed) have always been present in low populations but that it is atrazine as a new selection factor that has made them a dominant weed over susceptible biotypes in some corn fields.

Environmental

The triazines are low in acute toxicity to animals, oral LD_{50}s ranging from 300 mg/kg for cyanazine to as high as 5000 mg/kg for simazine (WSSA, 1974). The main environmental concerns for the triazine herbicides relate to their moderate persistence in soil and their possible damage to nontarget plants. Along with persistence, movement of atrazine in the environment has to be a major concern. Certainly the movement of atrazine-contaminated soil to the greenhouse results in problems, as the junior author of this text can personally attest. But the appearance of atrazine as the most common herbicide residue in water in some corn-growing areas (Frank, personal communication) poses some important questions. Has atrazine moved laterally from treated corn fields into adjacent streams and ponds? Is this a result of sloppy sprayer loading practices at water sources with contamination resulting from back flow? At present the levels are too low for immediate concern with respect to effects on algae, other aquatic plants, or fish.

Ureas

Highly potent but selective herbicidal properties were discovered for monuron, the first of the urea herbicides, in 1951 (Bucha and Todd, 1951). In quick succession thereafter a series of urea compounds, including fenuron, diuron, and neburon, were developed for both selective and nonselective chemical weed control (Table 9.9). In the years since, at least 10 commercially important urea herbicides have been developed.

Table 9.9 Examples of the Substituted Urea Herbicides

$$\begin{array}{c} \text{H} \quad \overset{\displaystyle O}{\overset{\displaystyle \|}{\text{N—C—N}}} \diagup ^{R_2} \\ R_1 \diagup \qquad\qquad \diagdown R_3 \end{array}$$

Common Name	Chemical Name	Chemical Structure		
		R_1	R_2	R_3
chloroxuron	3-[p-(p-chlorophenoxy)-phenyl]-1,1-dimethyl urea	Cl—⟨⟩—O—⟨⟩	$-CH_3$	$-CH_3$
diuron	3-(3,4-dichlorophenyl)-1,1-dimethyl urea	Cl—⟨⟩ (Cl)	$-CH_3$	$-CH_3$
fenuron	1,1-dimethyl-3-phenyl urea	⟨⟩	$-CH_3$	$-CH_3$
linuron	3-(3,4-dichlorophenyl)-1-methoxy 1 methyl urea	Cl—⟨⟩	$-CH_3$	$-OCH_3$
metobromuron	3-(p-bromophenyl)-1-methoxy 1-methyl urea	Br—⟨⟩	$-CH_3$	$-OCH_3$
monolinuron	3-(p-chlorophenyl)-1-methoxy 1-methyl urea	Cl—⟨⟩	$-CH_3$	$-OCH_3$
monuron	3-(p-chlorophenyl)-1,1-dimethyl urea	Cl—⟨⟩	$-CH_3$	$-CH_3$
norea	3-(hexahydro-4,7-s-yl)-1,1-dimethyl urea	(bicyclic CH_2 structure)	$-CH_3$	$-CH_3$

In some respects some of the urea herbicides have use patterns quite similar to some of the s-triazines. **Diuron,** an insoluble immobile compound in the soil, is recommended for use in many of the same deep-rooted crops (sugar cane, pineapple, grapes, apples, pears, citrus) as simazine is, and it too is widely used for nonselective weed control on noncrop land (WSSA, 1974). **Linuron,** like atrazine, controls germinating weeds in corn and sorghum. It does, however, differ in selectivity, because corn cannot tolerate high rates of linuron, and yet crops such as soybeans, cotton, potatoes, carrots, and parsnips are much more tolerant to linuron than to atrazine (WSSA, 1974). More than half of the 6 million lb (Lawless et al., 1972) of diuron produced in the United States is used for weed control in industrial areas, the remainder being used in agriculture. In contrast, most of the linuron and other urea herbicides produced are used for weed control in soybeans, cotton, or other broadleaf crops. **Chloroxuron** has some postemergence activity and is used in carrots, celery, soybeans, onions, and strawberries. **Flometuron** is used in cotton and sugar cane. **Metobromuron, monolinuron,** and **norea** have some of the same uses as linuron. **Neburon** and **siduron** are used in ornamentals and in some types of turf, respectively (WSSA, 1974). **Tebuthiuron,** a new urea herbicide, is being widely developed for total nonselective vegetation control in noncrop land and at lower rates for selective weed control in sugar cane.

Fate in Soil

The adsorption, leaching, and persistence of the urea herbicides in soil are all to a large extent dependent on differences in chemical structure. Thus, among the N, N-dimethyl ureas, adsorption increases according to the following order: fenuron < monuron < diuron (Bailey and White, 1964). Leaching and persistence are often dependent on availability in the soil solution (Chapter 14); thus fenuron, which is the most water soluble, is the least persistent but the most mobile of these three urea herbicides (Geissbühler et al., 1975). N-methoxy groups (linuron) instead of N-methyl groups (monuron) and ring bromines (chlorbromuron) instead of ring chlorines (linuron) both tend to increase adsorption. Soil type is also very important; thus many investigators (reviewed by Geissbühler et al., 1975) have found adsorption of urea herbicides to be lowest in sandy soils, intermediate in clay loam soils, and highest in organic soils.

Many of the urea herbicides are moderately persistent in soil. Persistence at rates selective for crops can vary from a few months to more than a year. Damage to sensitive crops from soil residues persisting into the next season can sometimes be a problem.

Microorganisms have been isolated that are capable of using urea herbicides as a source of carbon (Hill et al., 1955); the kinetics for urea disappearance do not, however, resemble those for 2,4-D, possibly because low solubilities and adsorption phenomena reduce the availability in the soil solution and discourage rapid proliferation of "adapted" organisms. Under conditions of too little rainfall to leach the chemicals into the soil, photodecomposition of the urea herbicides on the soil surface (Ashton, 1965; Comes and Timmons, 1965) can be an important factor in the reduction of urea persistence. Photolysis of urea herbicides in aqueous solutions by dealkylation, hydrolysis of the urea linkage, and liberation of the corresponding anilines has been documented by many investigators (reviewed by Crosby, 1976).

Mode of Action and Degradation in Plants

The urea herbicides are readily taken up by plant roots but, except for chloroxuron, have only limited activity when applied to plant foliage. They are translocated primarily upward in plants through the xylem via the transpiration stream to cells in plant leaves. Here they act as inhibitors of photosynthesis (Cooke, 1956). The urea herbicides are considered to be classical inhibitors of Photosystem II in photosynthesis, but as with the triazines, some investigators postulate that toxicity to plants also involves the production of an as yet unidentified toxic substance in the light (Sweetser and Todd, 1961).

Urea herbicides are degraded in plants but at widely varying rates that in some cases can result in selective phytotoxicity. Degradation pathways usually involve successive N-dealkylation, deamination, decarboxylation, and liberation of the corresponding aniline, which is in some instances conjugated with carbohydrates. The importance of these steps in the pathway varies for different plants and for different urea herbicides (Geissbühler et al., 1975).

Toxicity to Animals

The urea-type herbicides have a low order of toxicity to animals, oral LD_{50}s ranging from 1500 to 11,000 mg/kg (WSSA, 1974). When fed orally to rats, urea herbicides are rapidly eliminated via either the urine or feces. Continuous feeding of linuron or diuron at levels of 25 to 2500 ppm to dogs or rats resulted in no significant accumulations within various organs or tissues of these animals (Geissbühler et al., 1975).

Uracils

The substituted uracil herbicides are highly active soil-applied herbicides that are potent inhibitors of photosynthesis in plants. Their chemistry and potential as herbicides were announced in 1962 (Bucha et al., 1962). **Bromacil** is used for selective weed control in citrus and pineapple but is more important for total vegetation control on railroads and in other noncrop areas. **Terbacil** is used for selective weed control in orchards, sugar cane, mint, and strawberries (WSSA, 1974).

bromacil   terbacil

The uracil herbicides are moderately persistent in soil, with an average half-life of several months. Mobility in soil is also moderate (Rhodes et al., 1970). In comparison to the urea herbicides, bromacil and terbacil are more mobile than diuron but less mobile than fenuron.

Uracil herbicides are not highly selective in toxicity to plants. They are quite readily translocated from roots to shoots by the apoplastic system (Barrentine and Warren, 1970) except in citrus, where they can be immobilized in the roots (Gardiner et al., 1969). Their phytotoxicity seems to be by a mechanism very similar to that of the triazines and ureas, namely, an inhibition of Photosystem II in photosynthesis (Hilton et al., 1964).

Owing to very rapid excretion they are low in toxicity to mammals. Oral LD_{50}s for both bromacil and terbacil exceed 5000 mg/kg (WSSA, 1974).

Dinitroanilines

The activity and selectivity of substituted 2,6-dinitroaniline compounds as herbicides were first reported in 1960 (Alder et al., 1960). **Trifluralin** is by far the most widely used chemical in the group. It was first registered for use in food crops in 1964, and because of its selectivity for use in two very important crops, cotton and soybeans, it has rapidly become one of the most widely used herbicides in North

Table 9.10 The Dinitroaniline Herbicides

R$_1$—N—R$_2$

O$_2$N, NO$_2$, R$_3$ (benzene ring structure)

Common Name	Chemical Name	Chemical Structure		
		R$_1$	R$_2$	R$_3$
benefin	N-butyl-N-ethyl-α,α,α-trifluoro-2,6-dinitro-p-toluidine	$-C_2H_5$	$-C_4H_9$	$-CF_3$
butralin	N-(sec-(butyl)-4-(t-butyl)-2,6-dinitroaniline	$-H$	$-CH(CH_3)(C_2H_5)$	$-C(CH_3)_3$
dinitramine	N^4,N^4-diethyl-α,α,α-trifluoro-3,5-dinitro toluene-2,4-diamine	$-C_2H_5$ (plus a second amino group on the ring)	$-C_2H_5$	$-CF_3$
fluchloralin	N-(2-chloroethyl)-2,6-dinitro-N-propyl-4-tri-fluoromethyl aniline	$-C_3H_7$	$-C_2H_4Cl$	$-CF_3$
isopropalin	2,6-dinitro-N,N-dipropyl cumidine	$-C_3H_7$	$-C_3H_7$	$-CH(CH_3)_2$

nitralin	4-methylsulfonyl-2,6-dinitro-N,N-dipropylaniline		$-C_3H_7$	$-SO_2CH_3$
oryzalin	3,5-dinitro-N^4,N^4-dipropyl sulfanilamide		$-C_3H_7$	$-SO_2NH_2$
penoxylin	N-(1-ethylpropyl)-2,6-dinitro 3,4-xylidine	$-H$ [plus a second methyl group ($-CH_3$) on the ring]	$-CH(C_2H_5)_2$	$-CH_3$
profluralin	N-cyclopropylmethyl-α,α-trifluoro-2,6-dinitro-N-propyl-p-toluidine	$-CH(CH_2)_3$	$-C_3H_7$	$-CF_3$
trifluralin	α,α,α-trifluoro-2,6-dinitro-N,N-dipropyl-p-toluidine		$-C_3H_7$	$-CF_3$

America. In 1971 more than 20 million lb were produced (Lawless, et al., 1972). In addition to cotton and soybeans, trifluralin is registered for use on more than 40 other broadleaf crops, particularly vegetables (WSSA, 1974). Because of its growing importance for weed control in summer fallow and for controlling annual grasses in wheat, barley, and rapeseed, trifluralin is becoming one of the most important herbicides in western Canada. More than 10 different analogues of trifluralin have now been developed as herbicides (Table 9.10). While they may differ slightly in selectivity, their properties, methods of use, and mode of action are similar.

The dinitroanilines are yellow-orange crystalline solids with solubilities in water of 1 ppm or less. They are most commonly formulated as emulsifiable concentrates, but granular formulations of trifluralin are also available.

Trifluralin and the related dinitroanilines have relatively high vapor pressures and readily volatilize from soil surfaces, particularly if the soil is warm and moist. They are also one of the most susceptible groups of chemicals to photolysis. Leitis and Crosby (1974) have documented a multitude of photolysis products after exposure of trifluralin to sunlight. To prevent reductions in effectiveness by volatility or photolysis, most of these herbicides are applied as preplant soil-incorporated treatments. Once incorporated in the top 1 to 2 in. of soil, the dinitroaniline herbicides are extremely immobile, owing to their very low water solubility and their tendency to be adsorbed. They are not usually recommended for use on muck soils. Dinitroanilines are degraded by soil microorganisms. While the pathways may differ, depending on whether the conditions are anaerobic or aerobic, reduction of the amino groups and N-dealkylation reactions are normally important early in the degradation process (Probst et al., 1975).

The dinitroaniline herbicides are most toxic to germinating plant seedlings, particularly grasses. Perennial crops or weeds, deep-seeded crops, or transplanted crop plants with established root systems are the most tolerant. They are readily absorbed from soil by the penetrating shoots and to some extent by the roots of young seedlings. They inhibit growth of the entire seedling, but their most obvious effect is the inhibition of lateral root formation. Affected roots usually swell in the meristematic region near the root tip. Microscopic investigations have revealed that trifluralin blocks cell division at metaphase and produces multinucleated cells or cells with many micronuclei (Hacskaylo and Amato, 1968) (see Figure 9.1). An interference with the production of microtubules and the formation of the mitotic spindle are also indicated (Parka and Soper, 1977).

Dinitroaniline herbicides may be taken up by roots of established plants but are poorly translocated from these sites of root uptake to upper portions of the plant (Strang and Rogers, 1971). Metabolism in plants is also very slight. Trace amounts of some metabolites have been isolated from carrot roots growing in treated soil, but whether these degradation products were taken up from the soil or produced in the roots could not be determined (Golab et al., 1967).

The dinitroanilines are toxic to fish (Parka and Worth, 1965) if placed directly in water. LC_{50}'s for trifluralin in static fish ponds range from 0.058 ppm for bluegills to 0.56 for goldfish (Probst et al., 1975). Trifluralin is of low toxicity to birds and mammals, acute oral LD_{50}'s ranging from 800 mg/kg for newborn rats; to 2 g/kg for dogs, chickens, and rabbits; to greater than 10 g/kg for adult rats (reviewed by Probst et al., 1975). Although trifluralin is toxic to fish, it is very immobile and only moderately persistent; thus its use on agricultural land is not considered to be a serious hazard to aquatic environments.

Chloroacetamides

CDAA (N,N-diallyl-2-chloroacetamide), the first chloroacetamide herbicide, was introduced in 1956 (Hamm and Speziale, 1956; Jaworski, 1956) (Table 9.11). It was first developed for preemergence control of grasses in corn, sorghum, and soybeans, plus other legume crops, vegetables, fruits, and ornamentals. CDAA is quite volatile and must be incorporated to prevent losses in most soils (WSSA, 1974). It is highly water soluble, and because of excessive leaching, effectiveness for weed control is often lower on mineral soils than on muck soils. **Propachlor** was introduced in 1965 (Baird et al., 1965), and because of lower volatility, lower water solubility, and greater activity on some broadleaf weeds in addition to grasses, it largely replaced CDAA in corn, sorghum, and soybeans on mineral soils. **Alachlor** was introduced in 1969 (Jaworski, 1975) and has rapidly surpassed CDAA and propachlor to become the most widely used chloroacetamide herbicide. Its remarkable selectivity in four of the most important field crops in the United States, namely, corn, soybeans, cotton, and peanuts, resulted in a total use of more than 20 million lb in 1971 (Lawless, et al., 1972), only 3 years after its introduction. In addition to its use on these major field crops it has also been developed for weed control in fruits and ornamentals.

The chloroacetamides are nonpersistent herbicides in soil. The average persistence of weed control activity ranges from 3 to 6 weeks for CDAA and propachlor to 6 to 10 weeks for alachlor (WSSA, 1974).

Table 9.11 The Chloroacetamide Herbicides

$$R_1-\overset{\overset{O}{\|}}{C}-N\overset{R_2}{\underset{R_3}{}}$$

Common Name	Chemical Name	Chemical Structure		
		R_1	R_2	R_3
alachlor	2-chloro-2'-6'-diethyl-N-(methoxymethyl)acetanilide	$ClCH_2-$	$-CH_2OCH_3$	CH_3CH_2 / CH_3CH_2 diethylphenyl
CDAA	N,N-diallyl-2-chloro acetamide	$ClCH_2-$	$-CH_2CH=CH_2$	$-CH_2CH=CH_2$
propachlor	2-chloro-N-isopropyl acetanilide	$ClCH_2-$	$-CH(CH_3)_2$	phenyl

Some soil fungi have been shown to degrade alachlor, but the degradation pathways are not fully known (Jaworski, 1975).

In plants CDAA has been shown to inhibit cell division in root tips of barley (Canvin and Friesen, 1959) and respiration in germinating ryegrass seedlings (Jaworski, 1956). Duke has proposed that propachlor inhibits cucumber root growth by inhibiting protein synthesis (Duke et al., 1967). Alachlor is known to inhibit growth of shoots and roots in germinating seedlings (Keeley et al., 1972).

Chloroacetamide degradation occurs at a fast rate in tolerant plants (Smith et al., 1966). With CDAA both the chloroacetyl and allyl moieties are metabolized. Propachlor degradation is very rapid in corn and soybeans with no undegraded herbicide detectable 5 days after treatment (Jaworski, 1969).

One major undesirable property associated with CDAA is that it is very irritating when contacted by the skin or eyes. This has most likely been a major factor in the shift from the use of CDAA to propachlor, alachlor, or other equally effective but less irritating compounds. As a group the chloroacetamide compounds are low in toxicity to mammals, acute oral LD_{50}'s for the rat ranging from 700 to 2000 mg/kg (WSSA, 1974).

Thiocarbamates

The thiocarbamates are a relatively large group of closely related herbicides that are sulfur derivatives of carbamic acid. The most widely used chemicals in this group are **butylate, cycloate, diallate, EPTC,**

$$
\begin{array}{ccc}
\overset{\text{O}}{\underset{\parallel}{\text{NH}_2\text{-C-OH}}} & \overset{\text{O}}{\underset{\parallel}{\text{NH}_2\text{-C-SH}}} & \overset{\text{S}}{\underset{\parallel}{\text{NH}_2\text{-C-SH}}} \\[1em]
\text{carbamic} & \text{thiocarbamic} & \text{dithiocarbamic} \\
\text{acid} & \text{acid} & \text{acid}
\end{array}
$$

molinate, pebulate, triallate, and **vernolate** (Table 9.12). EPTC was the first of the group to be developed commercially and is particularly effective for controlling nutsedge (*Cyperus* spp.), quackgrass, and other difficult grasses. It is registered for use in alfalfa, a number of bean crops, corn, cotton, potatoes, beets, peas, fruits, and ornamentals (WSSA, 1974). Diallate and triallate are used widely for the control of wild oats in a number of crops (beets, barley, corn, legumes, peas, potatoes, root crops, fruits, and ornamentals), but triallate is more selective for the control of wild oats in wheat. Because of its greater

Table 9.12 The Thiocarbamate Herbicides

$$R_1 \diagdown \underset{R_2}{\diagup} N-\overset{\overset{\textstyle O}{\|}}{C}-S-R_3$$

Common Name	Chemical Name	Chemical Structure		
		R_1	R_2	R_3
butylate	S-ethyl diisobutylthio-carbamate	$CH_3-\overset{\overset{\textstyle CH_3}{\|}}{C}H-CH_2-$	$CH_3-\overset{\overset{\textstyle CH_3}{\|}}{C}H-CH_2-$	C_2H_5-
cycloate	S-ethyl N-ethylthio-cyclohexane-carbamate	C_2H_5-	(cyclohexyl ring)	C_2H_5-
diallate	S-(2,3-dichloroallyl) diisopropylthiocarbamate	$CH_3-\overset{\overset{\textstyle CH_3}{\|}}{C}H-$	$CH_3-\overset{\overset{\textstyle CH_3}{\|}}{C}H-$	$Cl-\overset{\overset{\textstyle H}{\|}}{C}=\overset{\overset{\textstyle Cl}{\|}}{C}-CH_2-$
EPTC	S-ethyl dipropylthio-carbamate	C_3H_7-	C_3H_7-	C_2H_5-
molinate	S-ethyl hexahydro-1 H-azepine-1-carbothioate	(azepine ring) $N-a$		C_2H_5

pebulate	S-propyl butylethylthio-carbamate	C_2H_5-	C_4H_9-	C_3H_7-
triallate	S-(2,3,3-trichloroallyl) diisopropylthiocarbamate	$CH_3-\overset{\displaystyle CH_3}{\underset{\textstyle \|}{CH}}-$	$CH_3-\overset{\displaystyle CH_3}{\underset{\textstyle \|}{CH}}-$	$Cl-\overset{\displaystyle Cl}{\underset{\textstyle \|}{C}}=\overset{\displaystyle Cl}{\underset{\textstyle \|}{C}}-CH_2-$
vernolate	S-propyl dipropylthio-carbamate	C_3H_7-	C_3H_7-	C_3H_7-

[a]The nitrogen atom in the molinate ring structure is the nitrogen atom of the parent thiocarbamic acid molecule; there is only one nitrogen atom in molinate.

145

selectivity in corn, butylate is used in the greater volume in North America (more than 6 million lb in 1971, Lawless et al., 1972) than any other thiocarbamate herbicide. Molinate is used primarily for the control of grassy weeds in rice. Pebulate is registered for use in sugar beets, tomatoes, and tobacco. Vernam is used in soybeans, peanuts, tobacco, and sweet potatoes, and cycloate is used in spinach and beets (WSSA, 1974). Because of their effectiveness for controlling annual grasses and nutsedge and their selectivity in a wide range of crops, the thiocarbamates have become one of the major groups of herbicides in North American agriculture.

Most of the thiocarbamates are either colorless or yellow, oily liquids with moderate to low solubilities in water. They are generally quite volatile, and in most situations they must be applied as preplant soil-incorporated treatments. Because of their volatility they are least effective when applied to warm, very moist soils. The thiocarbamate herbicides are readily degraded in soil, and although the various degradation products have not been defined, radiolabeled CO_2 is evolved after soil treatment with ^{14}C-EPTC (MacRae and Alexander, 1965). Molinate usually persists in soil for a few weeks, but most other thiocarbamate herbicides persist for several weeks. Effectiveness of thiocarbamates in soils with moderate, compared to very low, amounts of organic matter is often greater, since adsorption prevents losses by volatility and leaching and keeps the herbicide in the zone of weed seed germination (Ashton and Sheets, 1959).

While the thiocarbamates are known to have a multitude of different effects on plants, their actual toxic mechanism of action has not been established. Thiocarbamates promote the germination of some weed seeds in soil, but susceptible weed seedlings are killed (Fawcett and Slife, 1975). This might explain the common occurrence of season long weed control with such short residual herbicides. The most important site of plant uptake for these herbicides seems to be the coleoptile (shoot) of susceptible grass seedlings as it pushes through the treated soil prior to emergence (Parker, 1963). Translocation occurs readily in the xylem, and the emanation of EPTC vapors from plant leaves has been reported (Chang et al., 1974). The growth of treated susceptible grass shoots is severely retarded with bent and twisted leaves. The leaves are often so tightly rolled that younger leaves cannot emerge.

These effects could involve, as suggested by Chen et al., (1968), antagonistic interactions with hormones, particularly gibberellic acid, in the intercalary meristem at the base of grass shoots. EPTC is readily metabolized in corn plants, first to EPTC-sulfoxide (Lay et al., 1975) and subsequently to CO_2 (Chang et al., 1974; Lay and Casida, 1976). Lay and Casida (1976) have proposed that the sulfoxide metabolites are the phytotoxic moieties and that they may act by complexing with important thio (—SH)-containing compounds in the plant, such as glutathione-SH and coenzyme A-SH. Some formulations of butylate and EPTC now contain small amounts of diallyl dichloro acetamide, which acts as a physiologically selective antidote to reduce thiocarbamate injury to corn but not to other germinating grass species (Stephenson and Chang, 1978).

Fang et al. (1964) have established that thiocarbamates are readily metabolized in animals to natural compounds that enter the metabolic pool. After feeding rats ^{14}C-pebulate, radioactivity was eliminated as $^{14}CO_2$ in expired air. EPTC and butylate are of low toxicity to animals with oral LD_{50}'s of 1600 mg/kg and 4000 to 5000 mg/kg respectively. The LD_{50}'s for all other thiocarbamates are in the range from 500 to 4000 mg/kg (WSSA, 1974).

Dithiocarbamates. CDEC and **metham** are two compounds with herbicidal properties that are derivatives of dithiocarbamic acid. They are both volatile, nonpersistent in soil, and of low toxicity to animals. Acute oral LD_{50}'s are more than 800 mg/kg. CDEC can be used for weed control on many different vegetable crops (WSSA, 1974). It is used as either a "watered in" or soil-incorporated treatment to control germinating weeds. Metham is used as a temporary soil fumigant to control germinating weed seeds, nematodes, soil insects, and disease organisms (WSSA, 1974). It has also been used to kill tree roots in sewers (Ahrens et al., 1970).

$$CH_3CH_2 \diagdown \underset{\overset{\|}{N}-\overset{\overset{S}{\|}}{C}-S-CH_2\overset{\overset{Cl}{|}}{C}=CH_2}{}$$

$$CH_3CH_2 \diagup$$

$$CH_3NH-\overset{\overset{S}{\|}}{C}-S-Na$$

CDEC   metham

Bipyridyliums

Diquat and **paraquat** are two heterocyclic cationic compounds that are referred to as either bipyridyliums or quarternary ammonium compounds because of the positively charged "four bonded" nitrogens in their structure. Brian et al. First disclosed the herbicidal properties of this family of compounds in 1958 (Brian et al., 1958).

Both diquat and paraquat act as nonselective contact herbicides with essentially no activity in' soil. Diquat is used primarily as a crop desiccant in seed crops and as an aquatic herbicide (WSSA, 1974). Paraquat is used extensively as a nonselective herbicide for noncrop land and as a directed spray for weed control in orchards, nurseries, and landscape situations. The use of paraquat to "burn off" existing vegetation nonselectively in combination with residual selective herbicides has led to the development of "zero tillage" methods for the production of corn and other crops. Paraquat is also used as a crop desiccant and as a defoliant in cotton (WSSA, 1974).

diquat   paraquat

Both diquat and paraquat are inactivated immediately on contact with soil by the nearly irreversible binding at cation exchange sites on clay or organic matter. Microbial degradation and photolysis (Bozarth et al., 1965; Funderburk et al., 1966) are thought to be important means for diquat or paraquat degradation in solutions or on plant surfaces, respectively, but not in soil. Chemical, microbial, and photodegradation, as well as biological activity for microbes or plants, are almost totally prevented by adsorption of diquat or paraquat to soil colloids. Baldwin and other investigators (reviewed by Calderbank and Slade, 1976) have found residues of paraquat or diquat to persist in soil for many years with evidence of only slight degradation. In water, diquat is readily taken up by or adsorbed to aquatic plants. Some microbial or photodegradation while in solution is possible, but inactivation occurs primarily by adsorption to suspended soil or organic particles and eventual long-term persistence in bottom sediments (Calderbank and Slade, 1976). Diquat and paraquat have been partially desorbed or displaced from adsorption sites by treatment of soils with ammonium chloride, but complete release requires boiling of the soil with concentrated sulfuric acid (Tucker et al., 1967). Whether or not certain soils can be saturated with diquat or paraquat ions and what the possible biological significance is of a build-up of these undegraded but inactivated (?) herbicidal molecules are environmental questions that deserve continued research.

In plants diquat and paraquat are so rapidly toxic that translocation is limited. Light, oxygen, and the flow of electrons from photosynthesis are required for rapid toxicity. Cations of paraquat or diquat are reduced to free radicals by electrons from photosynthesis and in turn react with water and oxygen to produce hydrogen peroxide, which is thought to be the main toxicant. Treatment of plants with paraquat during periods of low light or in the presence of photosynthetic inhibitors (Putnam and Ries, 1968) has resulted in reduced contact activity, greater translocation, and longer lasting activity on perennial weeds. Toxic effects in plants in the dark or in nonphotosynthetic organisms are thought to be due to free radical production by electrons from respiration. Toxic effects associated with respiration are not, however, thought to be as rapid as those associated with photosynthesis (Calderbank and Slade, 1976). Except for the reversible conversion to reduced free radicals, there is no evidence that diquat and paraquat are degraded in plants (Funderburk and Lawrence, 1964).

Paraquat is one of the more toxic organic herbicides in animal systems. Oral LD_{50}'s for rats are between 100 and 150 mg/kg (WSSA, 1974). Diquat has an oral LD_{50} in excess of 200 mg/kg. These differences in mammalian toxicity may be due to the fact that diquat is readily metabolized in mammalian systems whereas paraquat is not (Calderbank and Slade, 1976).

Many human fatalities have resulted from accidental and intentional ingestion of concentrated paraquat liquid (Staiff et al., 1973). Progressive fibrosis (hardening of alveolar membranes) of the lungs has been associated with most of the fatalities (Sinow and Wei, 1973). With normal spraying practices adverse effects of paraquat on human beings have not been observed. Even continual daily contact with sprays of paraquat for long periods of time in tropical areas has resulted in only minor skin irritations, particularly near other wounds, and in the increased occurrence of nose bleeds (Swan, 1969).

Organic Arsenicals

The monosodium (**MSMA**) or disodium (**DSMA**) salts of methane arsenic acid ($CH_5A_5O_3$) were first patented as herbicides for controlling annual grasses in turf in 1951 (Schwerdle, see review by Woolson, 1976). Currently, their most extensive use is as selective postemergence herbicides in cotton. They are also used as nonselective herbicides in noncrop areas and are combined with 2,4-D/2,4,5-T mixtures to improve the control of conifers in brush control operations (WSSA, 1974). **Cacodylic acid** ($(CH_3)_2A_5O_2H$), a closely related organic arsenical, is used as a nonselective contact herbicide or as a tree injection to thin out stands or to kill undesirable trees in forestry situations.

Largely because of the use in cotton, more than 35 million lb of DSMA and MSMA were produced and used in the United States in 1971 (Lawless et al., 1972).

$$CH_3As\underset{OH}{\overset{\overset{O}{\parallel}}{<}}O\text{-}Na \qquad\qquad CH_3\text{-}\underset{\overset{\parallel}{O}}{\overset{\overset{CH_3}{|}}{As}}\text{-}OH$$

MSMA   cacodylic acid

Methane arsenic acid is readily adsorbed and loses its herbicidal activity once it reaches the soil. The sodium salts lose their activity within several weeks (Schweizer, 1967). Microorganisms can apparently adapt to the organic portion of these compounds as a partial source of carbon (Dickens and Hiltbold, 1967), degradation products being CO_2 and arsenate. The arsenate ion then becomes a rather permanent inorganic component of the soil.

The organic arsenicals can be translocated in plants but only in the xylem. Their mode of action is poorly understood, but they may act either as mitotic inhibitors (Woolson, 1976) or as uncouplers of oxidative phosphorylation (Ashton and Crafts, 1973). MSMA and DSMA may be converted to amino acid complexes, but there is no evidence of cacodylic acid degradation in plants (Woolson, 1976).

Although these organic herbicides contain arsenic, they are surprisingly nontoxic to animals. The oral LD_{50}'s in rats are more than 800 mg/kg and 1800 mg/kg for cacodylic acid and methane arsenic acid, respectively (WSSA, 1974). Environmentally, there is concern that repeated applications to soil could lead to accumulation of inorganic arsenic residues sufficient for toxic effects on soil organisms or plants.

Dicarboxylic Acids

Endothal is the common name for 7-oxabicyclo [2,2,1] heptane 2,3-dicarboxylic acid. Its various uses include preemergence and postemergence weed control in crops such as beets, control of annual bluegrass in turf, and preharvest desiccation of seed crops (i.e., alfalfa, soybeans). It is also an important aquatic herbicide (WSSA, 1974). More than 2 million lb were produced in the United States in 1971 (Lawless et al., 1972). In soil or water, endothal persists for only a few weeks (Audus, 1964; Yeo, 1970). Endothal is a fast acting contact herbicide in plants (Ashton and Crafts, 1973), and very little is known about its metabolism in plant systems. Hiltibran (1963) found that endothal toxicities to several aquatic plants could vary dramatically, depending on whether liquid or granular formulations were applied.

endothal

The pure acid of endothal is highly toxic to mammals (oral LD_{50} less than 100 mg/kg), but when it is formulated as the amine salt, toxicity is reduced (>200 mg/kg for the 66.7% formulation) (WSSA, 1974). Walker (1963) found that one major advantage of its use as an aquatic herbicide was its low toxicity to fish.

Amitrole

Amitrole (3-amino-s-triazole), a nonselective herbicide, was first introduced by Hall et al. in 1954. It is taken up by plant foliage and readily translocated throughout plants in xylem or phloem. It is one of the most effective herbicides for the control of many perennial weeds such as quackgrass, milkweed, Canada thistle, poison ivy, cattails, and many aquatic weeds. Originally it was developed and registered for many uses on crop land such as for small fruits and orchards and for the control of quackgrass in fields prior to planting of crops. In 1959 it was proposed that it could induce thyroid tumors in rats (see Carter, 1975), and this led to the cancellation of all registered uses on food crops in the United States. In Canada its use on food crops has not been similarly restricted. Amitrole uses on noncrop land and as an aquatic herbicide have been continued in both countries.

amitrole

Amitrole, is rapidly degraded in soils with a half-life of only 2 to 3 weeks (Kaufman et al., 1968). Degradation appears to be chemical but occurs more readily in soils with normal levels of microbial activity.

Short persistence in soil and good nonselective activity on perennial weeds have made amitrole important as a preplant treatment to control difficult weeds, particularly in Canada, where use on crop land has been allowed. In this use amitrole is applied to emerged quackgrass or other perennials early in the spring. The field can be plowed or tilled after 1 to 2 weeks, and after 4 to 5 weeks amitrole is degraded in the soil and it is safe to plant the crop.

In plants, amitrole is metabolized primarily to complexes involving amino acids or sugars, but the triazolyl ring is quite stable (Carter, 1975). The most visible effect of amitrole treatment on plants is the bleached, white, albino appearance of foliage that is produced after treatment. Some investigators have suggested that amitrole first blocks carotenoid synthesis in newly developing leaves and that, in the absence of carotenoids, the unshielded chlorophyll is photooxidized (reviewed by Carter, 1975). This leaves a white leaf void of the yellow-orange carotenoids and the green chlorophyll. Growth soon becomes limited owing to a lack of photosynthetic activity.

Amitrole is not acutely toxic to animals. Its oral LD_{50} of more than 24,600 mg/kg is among the highest of all pesticides (WSSA, 1974). However, long-term feeding studies with rats commonly result in enlarged thyroid glands (WSSA, 1974; see Carter, 1975), and as a result its use has been sharply limited.

Glyphosate

Glyphosate (N-phosphonomethyl glycine) is a new herbicide introduced in 1971 (Baird et al., 1971). In many respects the activity of glyphosate is similar to that of amitrole. It is absorbed by plant foliage and readily translocated in plants. It is very nonselective and has high activity for the control of many perennial weeds, including quackgrass and field bindweed. Persistence and activity in soil are less than for amitrole; it is deactivated on contact by adsorption and then degraded within weeks by microorganisms (Baird et al., 1971; Sprankle et al., 1975). Some of the registered or proposed uses for glyphosate include preplant or postharvest control of perennial weeds for many crops, chemical mowing as a directed spray in landscape situations or in fruit or plantation crops, brush control, conifer release in forestry programs, and aquatic weed control.

Its mode of action in plants is poorly understood. Jaworski (1972) has shown that it disrupts normal amino acid synthesis in duckweed (*Lemna minor* L.), but the significance of these effects in higher plants is not yet known. Although it is readily translocated in phloem, there is little evidence that it is metabolized in plants.

$$CH_2COOH$$
$$|$$
$$NH \quad O$$
$$| \qquad ||$$
$$CH_2-P-OH$$
$$|$$
$$OH$$

glyphosate

Glyphosate has low acute and chronic toxicity to animals. Oral LD_{50}'s for rats are in excess of 4000 mg/kg, and 2 year feeding studies with rats or dogs with dietary levels as high as 300 ppm have not resulted in adverse effects (WSSA, 1974).

Because of its nonselective toxicity to many herbaceous plants and brush, short persistence in soil, and low toxicity to animals, a very broad use pattern could be developed for glyphosate. Its use as an herbicide could become important in agriculture, in landscape situations, around the home, on industrial sites or utility rights-of-way, and in forestry or aquatic areas.

9.7   SELECTIVE TOXICITY OF HERBICIDES IN PLANTS

Herbicides are, by definition, toxic to vital processes in plant tissues. There are, however, many mechanisms by which herbicides can be used to control certain plant species selectively without injury to others in a mixed population. It is important to realize that, at very low rates of application, all plants may be completely tolerant to most herbicides and conversely that no plant is completely tolerant to very high rates of any herbicide. Selective herbicidal activity is very much dependent on the amount of the herbicide applied. For most herbicides selectivity is dependent on how much of a toxic chemical actually accumulates at site(s) of action within plant cells.

At normal herbicidal rates of application a multitude of factors determine how much of a chemical eventually reaches potential site(s) of action in plants (Figure 9.10). For a chemical that is applied to plant foliage, not all of the spray is intercepted by the leaves—some may run off, some may evaporate or volatilize, or some may be adsorbed to the cuticle or cell walls. Foliar penetration studies have revealed that when most herbicides are applied to leaves less than 50% of that applied actually penetrates the cuticle and enters the plant leaf cells. Even after the chemical enters plant leaf cells, other processes prevent toxic concentrations from reaching the sites of action. The chemical may not be very mobile within plant cells or it may be adsorbed at inactive sites. After entering cells it may be metabolized by plant

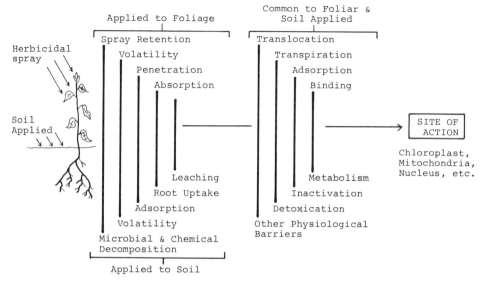

Figure 9.10 Obstacles that determine the concentration of phytotoxic material at the site of action in plants.

enzyme systems to nontoxic metabolites. When herbicides are applied to the soil, losses due to volatility, adsorption to clay or organic matter, leaching, or degradation processes result in only a fraction of the applied amounts being available for root or shoot uptake. When we consider all of the possible losses, it is probable that far less than 1% of an herbicide application ever reaches sites of action in plants growing in the treated area.

In most instances where an herbicide is differentially toxic to two different plants, it is because chemical, physical, morphological, or physiological factors interact to result in different concentrations of that herbicide's reaching the sites of action (chloroplast, nucleus, etc.) in those two plants.

10

INSECTICIDES AND THEIR USES

Pesticides used to control insects are called insecticides, and within this broad category one may find reference to materials that are especially effective against a particular stage in the insect's development. Thus, ovicides, larvicides, and adulticides are used frequently to indicate insecticides directed against the egg, larval, and adult stages, respectively. In addition, acaricides, toxicants effective against mites (acarines), are usually included along with insecticides. Insecticides are categorized also in relation to the manner of entry into the insect. Thus we refer to "stomach poisons," those that are ingested by the insect; "contact poisons," those that act upon contact with the insect; and "fumigants," those that enter the insect in a gaseous state via the insect's respiratory system.

Insecticides kill by disrupting some vital system or action within the insect. Two things are necessary. Firstly, there must be a sensitive site in the insect; for many of the present insecticides this is the cholinesterase enzyme that breaks down the transmitter substance in the nerve synapse. Secondly, there must be a method by which the insecticide can contact this sensitive site. It is important to understand that both of these features must be present for toxic action to occur. Frequently the relative toxicity of an insecticide to two different species is determined by barriers to penetration, barriers to entry in the first place or barriers through detoxification at a rate so rapid that the material does not reach the site of action in its toxic form.

10.1   THE NATURE OF INSECTS

In a simplistic view the insect can be thought of as a tube within a tube. On the outside is the exoskeleton or cuticle. The nature of the cuticle is important in relation to insecticides, for it is a highly complex structure with differential permeability. The cuticle is composed of several layers. The outermost, the epicuticle, is 1 to 2 microns thick and composed of four layers, more or less distinguishable. The outer layer is a thin, protective coating of a cement-like substance highly resistant to chemical attack. Underlying this is a wax layer impermeable to water and specific to each insect. Beneath this is a thin layer of polyphenols and protein, and under this the cuticulin, a layer of polymerized lipoprotein. Beneath the epicuticle is the exocuticle, a much thicker structure (10μ) than the epicuticle and made up of alternating layers of chitin and arthropodin—the former a polymerized glucosamine, the latter a protein. Underlying the exocuticle is a layer of similar nature and thickness but less densely tanned called the endocuticle and beneath this the hypodermis (epidermis). The hypodermis is a single layer of cells supported on a basement membrane. This multilayered nature of the cuticle presents a barrier to water and provides protection against many external conditions. Its waxy nature provides, however, rapid entrance for lipophilic chemicals, a description that fits many insecticides.

The inner tube in the insect body is the alimentary tract extending from the mouth to the anus. The anterior and posterior portions of this system are derived from an infolding of the outer wall and thus resemble the multilayered cuticle just described, except that tanning is generally absent and the layers are reversed. Joining the fore and hind parts of the system is the midgut, derived from endodermal tissue. Thus it lacks the cuticular lining and consists of a thin-walled cellular structure with an apparent high degree of permeability to many materials, including pesticides.

Between the exoskeleton and the alimentary tract lies the hemocoel, filled with blood to bathe the various organs and tissues lying within the hemocoel.

Thus it should be apparent that, if an insecticide penetrates the insect cuticle, it can be carried by the blood to vital organs. During this transmission, however, and during the passage of an insecticide through the cuticle, the chemical nature of the compound may be changed by the action of enzymes, and these changes may either decrease or enhance its toxicity.

The Nervous System

The postwar insecticides, with some significant exceptions to be discussed later, act on the nervous system. Thus an understanding of that
system is important to the discussions that follow. The nature of the
nervous system and how it functions is discussed with great clarity by
O'Brien (1967), and much of what follows is taken from his
description.

The nervous system has as its basic unit the "neuron." This is a
specialized cell containing a nucleus, cytoplasm, and a host of other
defined components and extending into processes, one or more (usually short) "dendrites" and a much longer "axon." The dendrites function to receive a message or impulse from an adjacent tissue and relay
it to the neuron body, and the axon serves to carry a message out to
another neuron or to an "effector," usually a muscle or a gland. The
ends of these processes are adapted with specialized surfaces to serve
for reception (dendrites) and transmission (axons). Thus, with respect
to the neuron body, the dendrites are efferent, the axon afferent.

Within the animal the neurons occur peripherally and internally
as single cells or small groups and also in larger, internal masses to
form "ganglia" and a larger mass, the brain. Interconnecting the peripheral nerves, the ganglia and the brain are nerve cords made up of
large numbers of intertwined axons. Thus impulses registered by sensory neurons are transmitted to the brain for interpretation, and a
response is transmitted back to the point in the body where an appropriate action is carried out. Within the animal other neurons are monitoring internal functions and reporting these to the brain. Thus we have a
"sensory nervous system" that functions as a body sentinel and a
"central nervous system," the brain, to direct the actions and responses of the body. In a simplistic view the ganglia can be interpreted as "junction boxes" to relay messages from the brain to the
location (smooth muscle, gland) where action is required. Ganglia
function also as a coordinator system independent of the brain.

The transmission of an impulse involves two steps: (a) transmission within the neuron; (b) transmission across the junction
("synapse") between neurons or between a neuron and an effector.
Transmission within the neuron is believed to be electrical and is
effected through a wave of changing polarity as an impulse passes
down the axon. When at rest the axon carries a negative electrical
potential inside with respect to the outside of the nerve membrane.
This is due to the relative ratio of Na^+ to K^+ ions with the concentration

of Na^+ ions low inside, high outside, the reverse situation applying to K^+. This balance is maintained by the sodium-potassium pump, whose anatomy is not known, but it somehow functions in concert with a nerve membrane differentially permeable with respect to the two ions. To carry an impulse down the axon, sodium rushes in through what is termed, picturesquely, the "sodium gate" to cause a rapid wave of changed polarity down the nerve axon. At the same time potassium ions burst from within the axon through the "potassium gate" to restore the electrical potential to its former state. The sodium-potassium pump then operates to restore the resting position. Thus the transmission of the impulse is a rapid reversal of Na^+ to K^+ ions within the nerve with a wave of changing polarity. The restoration of the resting state involves a less rapid "pumping" action.

Transmission from the axon of one neuron to the dendrite of another or from the axon to the organ to be activated is believed to be chemical, rather than electrical. The junction across which the impulse must pass is called a synapse. This can be thought of in terms of five components.

1. The bulbous end of the axon from which the impulse is emitted, the presynaptic membrane.
2. The opening between the presynaptic membrane and the receptor, the synaptic cleft, a width of about 500 Å.
3. The receptor, a dendrite or a tissue such as smooth muscle, gland, or the like, the postsynaptic membrane.
4. The transmitter substance, a chemical such as acetylcholine.
5. A second chemical capable of acting on the transmitter substance to return it to an ionic form after the impulse has been transmitted.

As an impulse is released at the end of an axon, the transmitter substance on the surface of the presynaptic membrane is diffused across the synaptic cleft to contact the postsynaptic membrane and effect a response. How this is done remains obscure, but the result is a passage of the impulse to another neuron (if the receptor is a neuron) or an action by a muscle or gland (if the receptor is one of these). Two transmitter substances are known, acetylcholine and norepinephrine (synonymous with noradrenaline). Synapses in which acetylcholine acts as the transmitter substance are said to be cholinergic, and those utilizing norepinephrine are referred to as adrenergic. When the message is passed across the synaptic cleft, an enzyme (acetycholine-sterase in cholinergic junctions; monoamine oxidase in adrenergic junctions) acts upon the transmitter substance to restore it to the "ready" position for the next transmittal.

For the most part, the vital activities of insects, namely, respiration, nutrition, sensation, reproduction, and protection, are similar to those of other members of the animal kingdom. Thus materials that are toxic to insects are likely to be toxic to other animals, differences in degree of toxicity being related to the extent to which these various animals are able to prevent the chemical from entering their system or are able to metabolize it on its way to the site of action. In recent years attempts have been made to utilize a process almost unique to insects—the process of molting. This is discussed at more length in the section dealing with insect growth regulators and their mimics.

In the sections that follow, we discuss some of the insecticides in major use in North America. They are discussed in terms of their use pattern, persistence, solubility, and toxicity. The mode of action and route of metabolism are given for some representative examples, in an effort to permit the reader to get some indication of what environmental effects might be anticipated. No attempt is made to include all insecticides. Instead we describe high use compounds and those that are of greatest concern environmentally. In the discussion that follows we have not cited individual references for general statements about the characteristics of insecticides. Much of our information on general properties is taken from Martin (1972) and Spencer (1973) and on uses and use estimates for older compounds from Metcalf et al. (1951).

10.2 PRE-1945 INSECTICIDES

For convenience of presentation we distinguish between the relatively few older insecticides, many of them inorganic, and the profusion of new compounds developed after World War II.

Arsenical Insecticides

A number of compounds containing the element arsenic have insecticidal properties; as much as 100 million lb of these products were used annually in North America during the 1930s and 1940s. The earliest important arsenical insecticide was **Paris Green**, a vibrant pigment discovered in 1865 to be highly toxic to the Colorado potato beetle. This compound contains about 40% arsenic and is water soluble to the extent of 3%. It is highly toxic to mammals (acute oral LD_{50} to the rat

22 mg/kg) and, because of its water solubility, to plants. The discovery of other, less phytotoxic arsenicals has reduced the uses of Paris Green to two. A small amount is used for control of mosquito larvae (note high water solubility), but it finds its greatest use as a bait mixed with bran and molasses for control of cutworms, crickets, grasshoppers, and armyworms. Certain other high arsenic-containing compounds (e.g., white arsenic, As_2O_3) may also be used in baits. By far the largest use of arsenicals has involved the two compounds lead arsenate and calcium arsenate.

Lead arsenate, $PbHAsO_4$, was developed about 1892 for control of the gypsy moth. It contains about 22% arsenic and has a relatively low (0.25%) solubility in water. It is relatively safe for use on plants but may cause some foliar burn under humid conditions. Lime or zinc sulfate can be added to reduce foliar burn, or basic lead arsenate, $Pb_4(PbOH)(AsO_4)_3$, with 14% arsenic can be used.

Lead arsenate has been used widely on plant foliage for many insect pests, particularly the codling moth and apple maggot on apples. It is toxic to mammals (acute oral LD_{50} to the rat about 50 mg/kg).

Calcium arsenate was developed shortly after 1900 as a less expensive substitute for lead arsenate. It consists of a mixture of $Ca_3(AsO_4)_2$ and $CaHAsO_4$ and contains 30 to 40% arsenic. The early preparations were quite phytotoxic, and although calcium arsenate was generally more effective than lead arsenate as an insecticide, its phytotoxic properties restricted its use. A less phytotoxic preparation was developed about 1915. Calcium arsenate was applied aerially for cotton boll weevil control prior to 1945, and although it was replaced largely by more recent pesticides, it is still used to some extent in areas where the boll weevil has become resistant to the post-1945 insecticides. Calcium arsenate has an acute oral LD_{50} to the rat of 40 to 100 mg/kg.

The arsenicals continue to be important insecticides. Their effectiveness as insecticides is related to their percentage content of arsenic, and their phytotoxicity is correlated with their solubility in water. The arsenates are less water soluble than the arsenites and are more useful as insecticides. The arsenites are used in baits and as herbicides (7 million lb in 1972) (NAS, 1975).

Arsenicals are stomach poisons and as such are somewhat selective in insects. Their solubility in the digestive tract of the insect aids their penetration of the insect system, and insects poisoned show pathological symptoms in the cytoplasm and nuclei of the midgut epithelium. It is believed that the arsenic ion reacts with the thiol (−SH) groups of various enzymes and thus inhibits respiration. Because a number of enzymes are inhibited (Webb, 1966), the general biocidal nature of the arsenicals can be appreciated, and because respiration is involved, their toxicity to plants as well as animals is apparent.

An interesting aspect of the arsenicals is the finding (Pickett and Patterson, 1963) that, where lead arsenate is used in orchards for insect control, reproductive maturity in the female apple maggot is delayed.

The arsenicals are persistent chemicals that build up on plant or fruit surfaces exposed to repeated applications and accumulate in the soil. High residues are found in orchard soils (Chapter 14)—small wonder when it is noted that in the early 1940s spray calendar recommendations in eastern apple orchards called for up to 100 lb of lead arsenate/acre/season.

Fluoride Insecticides

The fluorides came into use in the early part of this century with the discovery that **sodium fluoride** was an effective stomach poison for cockroaches. About 1915 this compound was used in the United States as a dry powder for cockroaches and ants and for chewing lice on poultry. Sodium fluoride has also been used in baits, but the big use of fluorides came with the search for alternative insecticides for arsenicals (because of residues on fruit and vegetables) and the discovery of cryolite in 1928.

The fluorides are found in nature in fluorspar, CaF_2, and **cryolite,** Na_3AlF_6, the latter present in large deposits in Greenland. They can also be synthesized, and these usually have better physical properties as insecticides. Whereas NaF is quite soluble in water (4.3%) and is highly phytotoxic, cryolite is not and can be used safely on plants. Thus in the 1930s cryolite, with a range of activity against insects approaching that of lead arsenate, was used to replace some of the arsenicals in insect control programs on fruit and vegetables. In 1944 almost 20 million lb of fluoride insecticides were used, 75% of this being cryolite. The fluorides are not especially good insecticides, and their use has declined greatly. Whereas NaF is moderately toxic (acute oral LD_{50} to the rat 200 mg/kg), cryolite is of low toxicity (acute oral LD_{50} to the rat 13,500 mg/kg).

The fluorides are believed to be enzyme inhibitors acting on those that use iron, calcium, or magnesium as a prosthetic group. Relatively high concentrations must be present in the insect to cause death (O'Brien, 1967), leading Corbett (1974) to suggest that toxicity is due to "simultaneous inhibition of several enzymatic processes."

Insecticides of Botanical Origin

Insecticides derived from certain plant species were much more important during the pre-1945 era than they are at present, their use dropping from well over 20 million lb to less than 5 million. Although difficulties in obtaining adequate supplies contributed to this decline, the fact that they are, in general, less effective than many of the newer pesticides was a dominant factor.

Nicotine and the related alkaloids nornicotine and anabasine are derived from a number of species of *Nicotinia* (*tobaccum, rustica, glauca*) and certain other plants. The activity of extracts from tobacco against aphids was recognized early, and recommendations for tobacco "teas" can be found as early as 1763. In the early 1800s the nature of the toxicant was discovered and commercial products derived from tobacco waste were placed on the market.

Nicotine is readily soluble in water, and although it acts as a contact insecticide, its activity is restricted to aphids and some species of mites and ticks. It is highly volatile and as such is a useful fumigant, finding application in greenhouses and as a roost paint applied to roosting areas in poultry houses, the vapor serving to kill ectoparasites on poultry.

The alkaloids are broken down rapidly on exposure to light. This has been a deterrent to their use, and attempts have been made to develop "fixed nicotines" that would persist on plant foliage and function as a stomach poison. Some limited success was attained, nicotine bentonite being the most satisfactory and used to some extent for control of the codling moth.

Nicotine is highly toxic to mammals (acute oral LD_{50} to the rat 30 mg/kg) by the oral route and almost equally toxic dermally. Nicotine affects the nervous system, acting on the ganglia of the central nervous system in insects. Low dosages of nicotine stimulate synaptic transmission, and high concentrations induce blockage. The site of action is believed to be the acetylcholine receptor (see discussion under organophosphorous insecticides), and the ability of the nicotine to reach that site is believed to be the basis for selectivity in insects.

Usage of nicotine exceeded 1 million lb during the early 1940s but is probably less than 10% of that amount now.

Pyrethrum. For many years it has been known that extracts from the flowers of *Chrysanthemum cinerariaefolium* and *C. coccineum* had insecticidal properties; in parts of Asia powders of these flowers have been used for louse and flea control for almost two centuries. Clinical investigations showed that the commercial product, referred to for many years as pyrethrum, was a mixture of five esters, the pyrethrins I and II, cinerins I and II, and jasmoline II.

These pyrethrins have been used extensively as dusts for plant pests, as oil solutions for control of flies on cattle, and as aerosols for control of household pests.

The pyrethrins decompose rapidly when exposed to light, air, and moisture and are thus suitable for use on crops close to harvest. They are contact insecticides characterized by a rapid "knock down." They are detoxified readily in the mammalian system and also in insects. As a result insects, although momentarily paralyzed, may recover quite quickly. There is normally about a threefold difference in dosage required to cause death over that required to cause knock down. The ability of insects to detoxify pyrethrins is believed due to "mixed function" oxidases, and this is inhibited by certain synergists. The effect is so striking that most pyrethrin aerosols also contain piperonyl butoxide, propyl isome, MGK264, or some other synergist. In practice the synergist is added at about 10 times the concentration of the pyrethrins.

The site of action is the nervous system. Both the peripheral and central nervous systems are affected, and the pyrethrins have a negative temperature coefficient, that is, they are more effective at low than at higher temperatures (Corbett, 1974).

The discovery of the chemical nature of the pyrethrins and their high cost led to attempts to synthesize these "natural" insecticides. Allethrin was the first of these synthetic pyrethroids, but others soon followed. In recent years English and Japanese workers have succeeded in synthesizing a large number with outstanding insecticide activity. These are still in the experimental stage, but their high toxicity to insects, low mammalian toxicity, and wide range (among different synthetic products) of residual properties have generated a great deal of interest in this class of compounds (Bader, 1976).

The pyrethrins have low mammalian toxicity (acute oral LD_{50} to the rat 820 mg/kg), and a wide range of toxicity is associated with some of the synthetic pyrethroids. The LD_{50} value (rat) for allethrin is 920 mg/kg; for dimethrin (a compound sometimes used for insect control in potable water) it is 40,000 mg/kg. Some of the new experimental compounds have LD_{50} values as low as 25 mg/kg.

Rotenone. The ground-up roots of *Derris* sp. have been used as fish poisons in South America for centuries, but their first use for insect control was in 1848. Rotenone (also called cubé or derris) is prepared either as the powdered roots of *Derris elliptica* and *D. malaccensis* from Malaysia or *D. urucu* or *Lonchocarpus utilis* from South America or as extracts from these roots. Rotenone is the most active insecticide in these preparations, but at least five related compounds (deguelin, elliptone, malaccol, sumatrol, toxicarol) also occur, each having some insecticidal activity.

Rotenone and the related extractives act as contact and stomach poisons. They have been used as sprays or dust for plant pests and are moderately effective against some species of aphids and most species of lepidopterous larvae. They are nonphytotoxic and have moderate toxicity to mammals (acute oral LD_{50} to the rat 132 mg/kg). Since the commercial preparations are formulated with a low concentration of the active ingredients, these products are recommended widely for home garden use. Rotenone has been used extensively for control of cattle grubs. The rotenoids are broken down to nontoxic products quite rapidly when exposed to light and air and are thus good insecticides for use on crops close to harvest.

Rotenone acts as both a stomach and contact poison in insects, but its effect is not on the nervous system. Poisoned insects have reduced oxygen consumption and die quite slowly, eventually becoming paralyzed. The site of action is believed to be the electron transport mechanism in the tricarboxylic acid cycle, but the biochemical lesion that effects this interference has not been demonstrated.

Rotenone is still in demand for home garden products and cattle grub control, but its use is declining from the 12 to 15 million lb level of the early 1940s. It is used also as a fish toxicant for pond reclamation (Chapter 12).

Other botanical insecticides include **Ryania** and **Sabadilla,** the former effective against lepidopterous larvae and the latter against some hemipterans and thrips. Neither is used extensively.

Dinitrophenols

The dinitrophenols were the first of the synthetic pesticides. First produced in Europe in 1892, these products are more important as herbicides (Chapter 9) than as insecticides. Some of the dinitrophenols are also effective fungicides. As a group the dinitrophenols are phytotoxic (hence their use as herbicides), and their use on plants derives from the fact that some of the esters are not water soluble.

From an insect control point of view the important products are **DNOC,** used as an ovicide for control of mites, aphids, and scale insects; **dinocap;** and **binapacryl.** Because of phytotoxicity DNOC can be applied only when plants are dormant. Dinocap is effective against mites and also for control of powdery mildew. It can be applied to plant foliage as can also the acaricide binapacryl.

The dinitrophenols act to uncouple oxidative phosphorylation (see section on herbicides). This explains their biocidal nature, for the respiratory process is universal in living organisms.

Mammalian toxicity is variable, acute oral LD_{50} values to the rat for DNOC, dinocap, and binapacryl being 26 to 65, 980 to 1190, and 136 to 235 mg/kg, respectively. Symptoms of poisoning include fever, increased breathing rate, gastric distress, nausea, perspiration, and cyanosis followed by coma.

Petroleum Oils

Many kinds of petroleum oils have been used in insect control both for control by themselves and as carriers for other pesticides. The history of research on petroleum oils as insecticides is extensive. Kerosene was employed for control of mosquito larvae almost a century ago, and the use of oils for control of plant pests began as early as 1763. Petroleum oils are highly effective ovicides but, in general, have rather low toxicity to most other insect stages. Their phytotoxicity has been a problem, and extensive research in New York and California has resulted in rigid specifications for spray oils that can be used on plant foliage. These "superior" oils have been used in orchards for control of mites, aphids, scales, and other insects whose eggs are placed on exposed plant surfaces.

The mode of action of oils remains a mystery. Those that are phytotoxic to plants seem to owe this herbicidal effect to a disruption of cellular membranes and subsequent leakage of vital water from the cell. The damaged tissue becomes flaccid and dies. Whether or not an effect on cellular membranes in insects occurs has not been established, but it is generally accepted that at least one major effect of petroleum oils is a physical blockage of oxygen. One of the characteristics of oils is their ability to "creep" over a surface. This creeping action may provide such a complete oil film on insect eggs that the diffusion of oxygen through the egg chorion is prevented, resulting in the asphyxiation of the embryo.

Sulfur

Sulfur, in a wide variety of forms, has been used as an insecticide for almost 100 years. When used alone as a dust, sulfur is effective in controlling mites, chiggers, ticks, and thrips and has found wide usage in greenhouses. It has the added advantage of being a good fungicide against a wide variety of diseases (Chapter 8).

10.3   POST-1945 INSECTICIDES

The large number and diversity of insecticides developed during and after World War II provided insect control at a level much greater than was possible with the older insecticides. Many of these new materials act on the nervous system as contact insecticides and are effective at low dosages. Their effectiveness and, in the early days, low cost brought a much expanded use of these toxicants and with wide usage some became general contaminants of the environment.

DDT and Related Compounds

Any discussion of modern insecticides must begin with DDT. The effectiveness of this chemical in controlling a wide variety of insect pests ushered in a new way of thinking about insect control and evoked in the minds of some entomologists the idea of eradication of major pests. No other insecticide has received so much praise or so much condemnation. Hailed by Sir Winston Churchill as that "miraculous DDT powder" because of its unprecedented effectiveness in halting an outbreak of louse-borne typhus among allied troops in Italy in 1943, less than 20 years later it would be reviled as the "elixir of death" by Rachel Carson (Carson, 1962). Because of environmental concerns its use in North America has been virtually eliminated, and this has also been the case in many of the other "developed" countries of the world. It still plays a key role in WHO campaigns against insect-borne diseases and in agricultural production in the developing nations, although this latter use is decreasing owing to insect resistance.

DDT is the name given to the chemical 1,1,1-trichloro-2-2-bis (parachlorophenyl) ethane. It was first synthesized by Zeidler in 1874, but its insecticide properties were not suspected until 1939, when it was tested by Müeller against clothes moths (Metcalf, 1973). As mentioned previously, it was used for louse control in 1943–1944 and after the war became established in agriculture and forestry. In 1948 Müeller was awarded the Nobel Prize in medicine and physiology for his developmental work with DDT. As early as 1950 its persistence in soil was recognized and there were hints that it might be harmful to certain nontarget species (beneficial insects and fish). By 1952 certain strains of the housefly in Italy had become resistant to this insecticide, a development that was to be repeated in more than 150 other insect species (Brown, 1971).

The great effectiveness of DDT is apparent in the fact that by 1961 more than 1200 formulations were available and in the United States it was registered for use on 334 crops against 240 species of agricultural pests. In addition it was used widely in forestry, against household and fabric pests, against pests of lawns and ornamentals, against pests of livestock, for control of black flies and mosquitoes, against the tsetse fly in Africa, and for control of mice (as a tracking powder) and bats. In the peak of its production (1963) 180 million lb were produced in the United States annually.

The development of insect resistance to DDT, coupled with the concern that it might cause widespread and irreparable damage to the environment, resulted in a drastic decline in use of this insecticide during the 1960s, and by the end of that decade the chemical had been severely restricted or banned in much of North America. As already discussed, the effectiveness of DDT against many insect pests is as legendary as its environmental impact, and when the tussock moth threatened western forests, petitions were raised in 1973 for the reinstatement of this insecticide, a request granted the following year (Graham, 1974).

DDT is almost insoluble in water (< 1 ppm), has low vapor pressure, and is relatively unaffected by light and air. Thus, it is one of the most persistent insecticides known. It is soluble in most organic solvents and petroleum oils. It dissolves readily in animal fats, a fact of great significance, since it accounts for its accumulation in animal tissues.

Under alkaline conditions it is dehydrochlorinated to DDE, a product that is not insecticidal but that is important as an environmental contaminant. A similar dehydrochlorination occurs in insects by enzymatic action. In addition, in insects hydroxylation occurs, converting some DDT to dicofol, a commercial acaricide (Menzie, 1969).

Many studies have been conducted on the metabolism of DDT in living systems. It is clear that mammals, birds, fish, and a host of microorganisms are capable of breaking down DDT to DDE and several other metabolites. What is not clear is which of these metabolic products are innocuous and which might in some way impair biological processes. Some of the metabolic products are indicated in Figure 10.1.

The persistence of DDT has been a strong factor contributing to its effectiveness as an insecticide. In antimalarial compaigns an application of 1 to 2 g/m² on the inside of huts remained effective against mosquitoes for 6 months to 1 year; soil applications (10 to 25 lb/acre) to permanent sod (lawns, etc.) gave protection against soil-inhibiting grubs (Japanese beetle, European chafer) for 2 to 6 years; and field

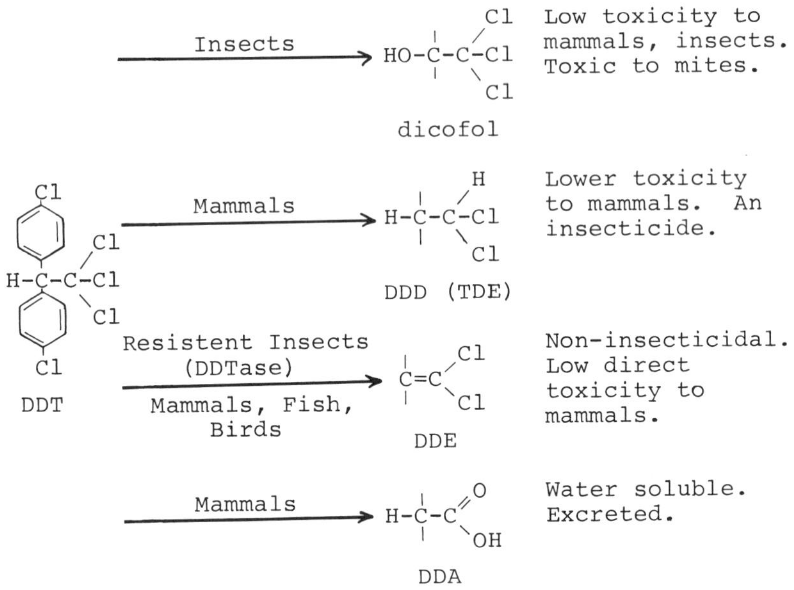

Figure 10.1 Some of the metabolites of DDT. For more nearly complete information, see Menzie, 1969.

applications (0.25 to 1.0 lb/acre) to plant foliage provided control of leaf-eating insects for several weeks. The result was that, with DDT, fewer applications were required than with most other insecticides, and thus savings were effected in the labor involved in insect control.

Although researched extensively, the exact biochemical lesion that causes toxicity of DDT is not known. Toxic symptoms in insects are expressed as hyperactivity; tremors, evident in the appendages; and finally paralysis. The lipophilicity of DDT permits it to penetrate the cuticle rapidly, and it acts as both a contact and stomach poison. The initial symptoms of poisoning suggest an effect on the peripheral nerves to trigger involuntary muscle action. Experiments with insect nerves show that DDT treatment results in repetitive firing of impulses, and this is believed due to an interference with the ionic balance within the nerve. DDT has a general affinity for membranes, and it would appear that it may bind with lipoprotein in the nerve membrane and alter the permeability of that membrane to Na^+ and K^+ ions. Holan (1969, 1971) experimented with DDT and a group of its analogues, some insecticidally active, others less so. He found a relationship between the shape of molecules and their activity. If one can visualize the sodium gate in a literal sense, then Holan postulates that the apex (the—C Cl_3 moiety) of a wedge-like DDT molecule may hold the gate ajar with the base of the wedge bound to the membrane. In any event the initial effect of DDT is to upset axonic transmission, and the resulting hyperactivity brings on exhaustion and paralysis. In the process toxins of unknown chemistry are found in the insect blood, toxins that, when transferred to healthy insects, cause death. The production of this toxin(s) is probably secondary, for it has been shown that cockroaches under other types of stress conditions (i.e., tied) also develop toxins so that the blood of these is toxic when injected into other cockroaches (O'Brien, 1967).

In addition to its effect on axonic transmission via membrane permeability, DDT is believed to inhibit a specific ATPase important in ion transport in nerves (Matsumura and Patil, 1969) and to increase activity of microsomal enzymes. It has also been reported to inhibit carbonic anhydrase, an enzyme believed important in eggshell production in birds (Peakall, 1970). The importance of this is discussed in more detail in a later section (Chapter 18).

Although DDT is generally not phytotoxic, it has been shown to inhibit oxidative phosphorylation in mitochondria and the Hill reaction in chloroplasts. Failure to cause damage has been interpreted as an inability of the compound to reach the site of action.

DDT is of moderate toxicity to mammals by the oral route (acute oral LD_{50} to the rat 200 mg/kg) and of low dermal toxicity (LD_{50} to the rat about 3000 mg/kg). By contrast its LD_{50} to the American cockroach is near 8 mg/kg and to the housefly 2 mg/kg. In the case of insects the LD_{50} values are about the same regardless of whether the exposure is dermal or by ingestion.

Its acute toxicity to birds is low (LD_{50}'s higher than for mammals), but fish are extremely sensitive, some species unable to survive in environments containing 0.01 ppm or more. DDT shares with the pyrethroids the characteristic of having a negative temperature coefficient; that is, it is more toxic at low than at high temperatures.

The most extensive present-day use of DDT is for vector control in developing countries, principally in Asia and Africa. Resistance of some vector species to DDT and more selective application methods have reduced the usage in these programs from a peak of almost 100 million lb to slightly less than half that amount. In another area of human health DDT is still used by the medical profession for control of human lice.

Related Insecticides

A number of insecticides closely related to DDT have been important products, and though the chemical differences appear minor (Figure 10.2), the biological properties are quite distinct.

TDE (DDD). 2,2 bis(p-chlorophenyl)-1,1-dichloroethane was marketed for a number of years under the trade name Rhothane®. It is much less toxic than DDT to mammals (acute oral LD_{50} to the rat 3400 mg/kg, dermal > 10,000 mg/kg) but is more toxic to hornworms on tomato and tobacco and to some species of leafrollers. It is believed to present similar environmental hazards as DDT, and its use has been discontinued.

Perthane®. 1,1-dichloro-2,2-bis(4-ethylphenyl)ethane has had minor use. Insecticidally it is less active than DDT and has low toxicity (acute oral LD_{50} to the rat 8000 mg/kg, dermal near 10,000 mg/kg). Perthane® has never been used widely, but it is especially effective against the pear psylla and on leaf hoppers. It is not as persistent as DDT.

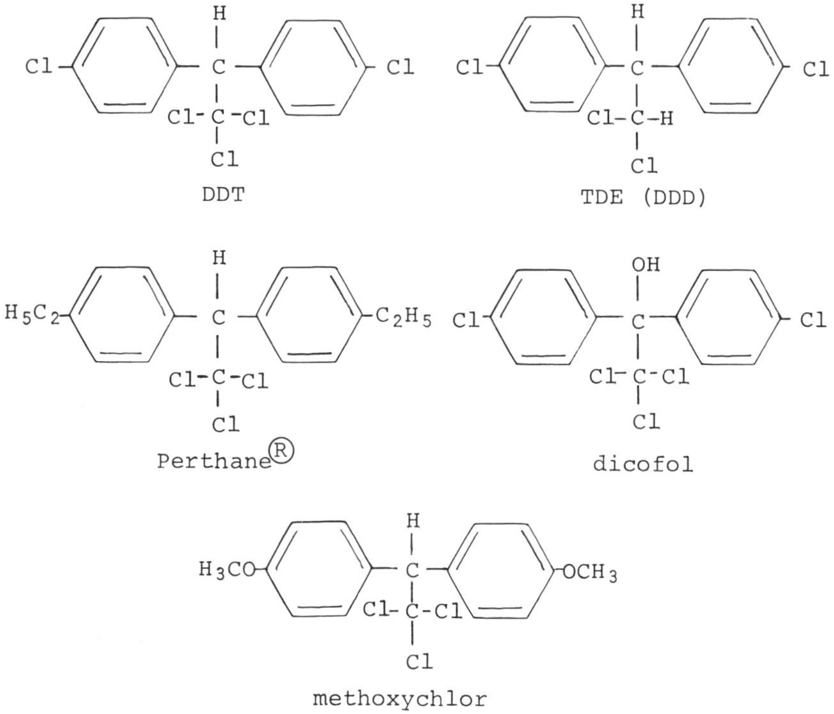

Figure 10.2 DDT and related insecticides.

Methoxychlor. 1,1,1-trichloro-2,2-bis(p-methoxyphenyl) ethane is a widely used and important insecticide. It is relatively nonpersistent, though it is not decomposed readily by heat or oxidation. It does not accumulate to a significant extent in tissues of animals. When fed to cattle, it is not secreted in milk, apparently forming a number of water-soluble metabolites that are excreted in the urine. Although most species of animals metabolize methoxychlor quite rapidly, this is not a universal phenomenon. Higher animals and birds detoxify methoxychlor apparently by the action of mixed function oxidases, and because the activity of these is low in insects, it has been suggested that this explains the selective toxicity toward the latter group (Metcalf, 1971). Methoxychlor does accumulate in some animals, the snail, *Physa*, being an example where this occurs (Kapoor et al., 1970). Molluscs and crustaceans, in general, may not metabolize methoxychlor as rapidly as higher animals, with the result that accumulations may occur in their tissues (National Research Council [NRC], 1975a). Although fish may concentrate methoxychlor above the level found in their food and environment, accumulation is not nearly so pronounced as for DDT, and the residue is lost soon after exposure is discontinued. Methoxychlor has low toxicity to mammals (acute oral LD_{50} to the rat 6000 mg/kg, dermal toxicity in the same range). It has similarly low toxicity to birds but, like DDT, is highly toxic to fish. In general it has the same spectrum of activity against insects as DDT does but is less effective, requiring in general two to four times the dosage to achieve comparable immediate control. Its lack of persistence means that repeated applications at short intervals are often needed.

The lack of prolonged persistence of methoxychlor, together with its failure to accumulate in most biological tissues, has made it an attractive insecticide for use on products near harvest, in the dairy barn for housefly control, and as either a larvicide or adulticide against black flies and mosquitoes. About 10 million lb are produced annually in the United States. Usage figures suggest less than 30% of this is used by farmers and much of their use is on livestock. It would appear that most of the methoxychlor produced is used in home products and for biting fly control (NAS, 1975).

Dicofol. 1,1-bis(p-chlorophenyl)-2,2,2-trichloroethanol is an acaricide that from an environmental viewpoint shares many properties with the closely related DDT. It is insoluble in water and persistent, being relatively unaffected by light or moisture. It persists in soil for more than 1 year. Its breakdown products have not been investigated extensively; however, in the presence of alkali it forms the biologically inactive dichlorobenzophenone and chloroform.

It is used for mite control on a wide variety of agricultural and ornamental crops. Annual production in the United States is about 4 million lb. It has low mammalian toxicity (acute oral LD_{50} to the rat 1000 mg/kg, dermal toxicity > 4000 mg/kg).

As can be seen from the foregoing discussion, one cannot generalize on the biological characteristics of insecticides, even within such a narrow group as DDT and closely related compounds. While all share the common features of being chlorinated hydrocarbons, almost insoluble in water, they differ markedly in their toxicity to vertebrates and invertebrates, in their persistence, and most importantly, in their tendency to concentrate in lipid animal tissues.

The Cyclodienes

The cyclodienes are an important group of chlorinated cyclic hydrocarbons containing an endomethylene bridge. Discovered in 1945, the first commercial product was chlordane. The commercial preparation of this insecticide contained a mixture of isomers based on the chlordane molecule. A number of related and "purer" (from a chemical point of view) insecticides in this class were developed during the next few years.

Chlordane. In its early stages of production technical chlordane was reported to be a mixture of the cis (α) and trans (γ) isomers of chlordane (60 to 75%) containing in addition some 25 to 40% related isomers. This characterization was based on the analytical competence of the times, but it has since been shown that at least seven major components are involved, including isomers of chlordane (20%), α and γ chlordane (40%), heptachlor (10%), and transnonachlor (7%). The exact percentages are still somewhat debatable. Suffice it to say that the compound contains many components, one study distinguishing 26 peaks on an analysis by gas chromatography (NRC, 1974a). In this discussion we refer to the product as technical chlordane, recognizing that generalizations may not apply equally to each of the components.

Technical chlordane is a brownish, viscous liquid nearly insoluble in water (5 ppm) but soluble in most organic solvents and petroleum oils. It is stable at normal temperatures and only slowly degraded when exposed to strong light. It is dehydrohalogenated under alkaline conditions.

Perhaps because of the large number of products involved, little has been reported on the metabolites of technical chlordane. In rat studies most of the ingested dosage is excreted in the feces and urine as hydrophilic compounds (Poonawalla and Korte, 1964) and there is no evidence for biomagnification. When fed to cattle, dogs, goats, and pigs, only low levels of oxychlordane can be detected in animal tissue and milk. This is surprising, since heptachlor is a significant component of technical chlordane, and heptachlor epoxide has been shown to deposit in animal tissues (Menzie, 1969).

In agriculture chlordane is used mainly as a soil insecticide, and there is little evidence that it is translocated into aerial parts of plants, though such has been reported in alfalfa (Boyd, 1971; Wilson and Oloffs, 1974) and low levels have been found (infrequently) in milk. Root crops growing in chlordane-treated soils do contain residues, residues quantitatively related to the chlordane content of soils in which they are grown. The low solubility of chlordane and its tendency to adhere to soil particles render it immobile in soils, and its transfer to rivers and lakes has not been significant.

chlordane

Chlordane is persistent in soils, one study on sandy loam soil reporting 16% of a 14 kg/ha application remaining after 15 years. There is an initial sharp decline in residue during the first month after application followed by a very slow decline. This initial decline may reflect, in part, the tendency for some components of the technical chlordane to volatilize. Biological activity suggests that chlordane provides control of susceptible soil invertebrates for several years, and studies on root crops often detect residues when these are grown in treated soil up to 4 years after treatment. Chlordane is effective in controlling wood-infesting borers, carpenter ants, and termites. Treatments of timbers are effective for many years.

Chlordane is a contact and stomach poison (acute oral LD_{50} to the rat of about 400 mg/kg and a dermal LD_{50} value near 800 mg/kg). Several of the components of technical chlordane are much more toxic. It is toxic to a wide range of insects. Symptoms of poisoning include hyperactivity and violent convulsions; in larvae a portion of the hindgut is often extruded through the anus. Toxicity is believed to occur in the ganglia rather than in the peripheral nerves, but the exact site of action is not known.

Chlordane has been used extensively as an agricultural soil insecticide, for cockroach and ant control around the home, and for control of wood-infesting pests. Usage figures indicate 1.9 million lb used by agriculture in the United States in 1971, with a production of 25 million lb (NAS, 1975). Nonagricultural uses obviously accounted for the major part of the production.

Heptachlor. Much of what has been said about chlordane applies also to heptachlor, but there are some outstanding differences. Heptachlor was isolated from technical chlordane and introduced in 1948. It is a white solid insoluble in water but sharing with chlordane ready solubility in organic solvents. It is more active insecticidally than chlordane and in practice is applied at from one-half to one-fourth the dosage. Its also has a higher toxicity to mammals, with an acute oral LD_{50} to the rat of about 100 mg/kg and a dermal toxicity near 200 mg/kg. When applied to soil, a large proportion volatilizes, but over time some degrades to heptachlor epoxide. Heptachlor is metabolized also by birds, insects, microorganisms, mammals, and plants to form the epoxide (Brooks, 1969), a product as toxic as the parent compound and with the undesirable characteristic of bioaccumulation in animal tissues. The epoxide may be formed also through photodecomposition. Conversion to heptachlor epoxide is not the only route by which heptachlor is altered. In fact, soil treatments with heptachlor result in only about 1% conversion to the epoxide (NRC, 1974a). Regardless of this observation, the epoxide has repeatedly been found in both mammals and birds in areas where heptachlor has had extensive usage.

Heptachlor has had an important use in agriculture against soil insects, but although about 6 million lb were produced in 1971, only 15% of this was used by farmers in the United States that year (NAS, 1975). It has been used extensively for fire ant control. In Canada concern for heptachlor epoxide residues in milk and anxiety for possible effects on birds resulted in a ban on the use of heptachlor in 1969.

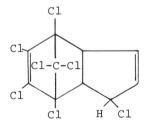

heptachlor   heptachlor epoxide

Aldrin and **Dieldrin.** Aldrin and its epoxide dieldrin were developed in the late 1940s as insecticides with much of the same spectrum of activity as heptachlor and chlordane but effective at somewhat lower dosage rates than chlordane. As well as being more toxic to insects, they have greater mammalian toxicity (acute oral LD_{50} values to the rat being about 50 mg/kg). Corbett (1974) refers to work by Wang et al. (1971) in which it was shown that, when aldrin trans-diol was applied to the cockroach ganglion, an immediate response was elicited. This was in contrast to the delayed response following dieldrin application. It is suggested that, since aldrin is epoxidized to dieldrin and dieldrin to aldrin trans-diol in some insects and mammals, aldrin trans-diol may be the active toxicant (Corbett, 1974).

Both aldrin and dieldrin have had extensive use as soil insecticides for control of grubs, wireworms, root maggots, and corn rootworms, among others. They have been used also as foliar insecticides, especially against Coleoptera. Dieldrin and, to a lesser extent, heptachlor have been useful as seed treatments to control root maggots and wireworms. For example, bean seed treated at $\frac{1}{4}$ oz dieldrin/bushel of seed is protected effectively. Many other large seeded vegetables and cereals have been treated in this manner, but the development of resistance, especially in the seed-corn maggot, has reduced this practice. In addition treated seed proved toxic to a number of seed-eating birds that follow agricultural operations and pick up any seed left uncovered at planting.

Dieldrin is highly effective against ants and termites. It was used extensively in an attempt to eradicate the fire ant in southeastern United States, a use that was criticized severely by "environmentalists." It has been used also in eradication programs against the Japanese beetle. On foliage aldrin residues are short lived, but dieldrin persists for several weeks. When aldrin is applied to the soil, it disappears quite rapidly, per se, through volatilization and epoxidation to dieldrin, the latter an extremely persistent metabolite. Application of dieldrin to soil results in persistent residues with insecticidal activity for several years.

Dieldrin accumulates in animal tissue and is eliminated slowly and in dairy animals may be secreted in milk for several months following exposure. Dieldrin has been found in tissues of many birds, in some cases in toxic quantities (Chapter 18). Dieldrin has been found in human tissue and in mothers' milk (Chapter 19).

Aldrin and dieldrin are metabolized by plants, animals, and microorganisms to a number of hydrophilic compounds. In animals the rate of excretion is related to the rate of intake. Some of the ingested compound is passed unchanged, this percentage increasing at higher dosage. Apparently an equilibrium is established in the animal system and, within limits, can be maintained despite increasing dosages (Robinson, 1970). Metabolism is accomplished by enzymes in the liver microsomes, but despite such metabolism, intake of these insecticides has been associated with permanent liver damage.

Production in the United States (1971) was estimated at 10 million lb for aldrin and about 1 million lb for dieldrin. Much of the aldrin production was used for control of corn rootworms in the Midwest (NAS, 1975).

aldrin

dieldrin

Endrin is the endo-endo isomer of dieldrin, differing only in the spatial arrangement of the two rings (Metcalf, 1971). It is the most toxic of the commercial cyclodienes with an acute oral LD_{50} to the rat of about 7 mg/kg and a dermal toxicity near 10 mg/kg. Endrin has an even broader spectrum of activity than most other cyclodienes. Its main use was as a foliar insecticide, especially effective against lepidopterous larvae, Coleoptera, Hemiptera, and Homoptera. Its high mammalian toxicity made it useful as a rodenticide for mouse control in orchards—as a spray applied under the tree canopy and around the trunk. It is persistent and highly toxic to fish.

At one time endrin was used extensively in cotton, tobacco, and potatoes. Cyclodiene resistance in insects and the persistence and high mammalian toxicity of the compound have drastically reduced its use. It is no longer recommended in Canada. Slightly more than 1 million lb were used in agriculture in the United States during 1971.

dieldrin

endrin

Mirex and chlordecone. Mirex and the closely related metabolite chlordecone have found a small but significant role as stomach poisons in ant and cockroach baits. At one time these chemicals were viewed as ideal for this use, for they were rather selective in action and effective at such low concentrations (0.15 to 0.075% in baits) that their use seemed to present no hazard. They are of moderate toxicity (acute oral LD_{50} to rat of mirex about 400 mg/kg, of chlordecone 100 mg/kg). Both chemicals are extremely persistent and lipophilic. They are practically insoluble in water.

The concept of mirex and chlordecone as "good" insecticides was jolted with the revelation of severe, debilitating illness in a large number of workers in the factory where chlordecone was produced. Recent discovery of high pollution from this same chemical in the James River because of careless dumping of waste product into one of its tributaries has resulted in a ban on fishing in that river, and the possibility exists that contamination of fish may be critical also in Chesapeake Bay (Wilbur, 1976).

Ant baits containing mirex have been effective in the fire ant program and for ant control in pineapples and sugar cane. Chlordecone wafers were finding a role in termite control, but with the revelation of human illnesses and environmental contamination from these chemicals, it is unlikely that their use will be continued. Under field conditions mirex bait provided ant control at a dosage of only 3 g toxicant/acre.

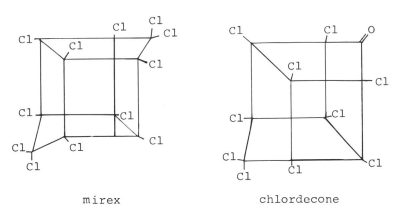

mirex   chlordecone

Endosulfan (Thiodan®) is a brown crystalline solid soluble in organic solvents and only slightly (20 ppm) soluble in water. It is unaffected by light but decomposes when exposed to water, forming SO_2 and endosulfan alcohol. It is moderately persistent and used mainly as a foliar insecticide. On plant foliage residues are reduced to low levels within a few weeks, disappearance believed due primarily to volatilization. The technical product contains α and β isomers. These differ markedly in terms of their persistence. Although α endosulfan disappears in less than 100 days, residues of the β isomer have been detected for more than 2 years. In soil both isomers are metabolized to endosulfan sulfate, and this metabolite is highly persistent (Stewart and Cairns, 1974).

Endosulfan is of moderate mammalian oral toxicity (acute oral LD_{50} to rat about 75 mg/kg) and only slightly less toxic by the dermal route. Toxicity to birds is of the same order, but fish are highly susceptible, and contamination of water has resulted in striking fish kills. The insecticide is metabolized readily by mammals to endosulfan diol and other hydrophilic metabolites that are excreted. Endosulfan sulfate has been detected in animal tissues but only for brief periods after exposure (NRC, 1975).

Endosulfan is used in agriculture against a wide variety of insect species, including lepidopterous larvae, beetles, and aphids. It has been especially effective against certain wood and stem borers such as the peach tree borer, bronze birch borer, and locust borer. Its high toxicity to fish makes it an effective chemical for removal of trash fish in small lakes, and because it degrades quite rapidly in water, restocking can be done after a reasonable waiting period. Production in the United States was reported as 2 million lb in 1971 with about one-half this amount used by farmers. Canada used about 200,000 lb in 1974.

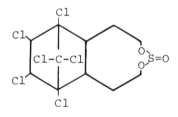

endosulfan

As a group the cyclodienes share the common characteristics of low water solubility and persistence. They act on the ganglia of the central nervous system. The relative rates by which they are metabolized in vertebrate systems and the toxicity of the metabolic products differ drastically. Thus endosulfan does not bioaccumulate, in contrast to dieldrin and heptachlor (as its epoxide), which do.

In recent years severe restrictions have been placed on the use of aldrin, dieldrin, endrin, and heptachlor in Canada, and the United

States has restricted and is continuing to restrict these insecticides. Chlordane is also under review, and production of mirex and chlordecone has been discontinued.

Chlorinated Terpenes

The chlorination of bicyclic terpenes produces insecticidal compounds, the most important of which is **toxaphene,** a mixture of camphenes chlorinated to 67 to 69%. Toxaphene is a yellowish wax readily soluble in organic solvents and relatively insoluble (3 ppm) in water. The technical product contains a mixture of isomers whose chemistry is not well understood. It is unstable in strong sunlight and in alkali and under high temperatures releases HCl and becomes less insecticidal (Metcalf, 1971).

The literature on toxaphene is not informative regarding its mode of action, but it is believed to be similar to that of the cyclodienes. It is moderately toxic (acute oral LD_{50} to rat about 100 mg/kg with a dermal toxicity in the 600 to 1000 mg/kg range). Residues have been detected in animal tissues, and it is considered to be a persistent insecticide. Its main use has been as a foliar insecticide on cotton, but it has been used also for control of grasshoppers and crickets in field crops. It is extremely toxic to fish and has been used for control of trash fish prior to restocking of small lakes. Porduction figures (1971) indicate 50 million lb in the United States, and usage figures for the same year show that farmers used 37.5 million lb, 4.6 million of this on livestock (NAS, 1975).

toxaphene

Organophosphorous Insecticides

The group of insecticides having phosphorus as the active nucleus and referred to commonly as the "organophosphates" is the largest and most diverse group of insecticides known. To refer to them as "organophosphates" is not correct, since one or more of the oxygens in the phosphoric acid moiety may be replaced, and proper nomenclature reflects these substitutions. The reader is referred to O'Brien (1967) or Corbett (1974) for a discussion of this aspect.

As pointed out by O'Brien (1967), these insecticides are esters of alcohols with phosphoric acid or anhidrides of phosphoric acid with another acid. Thus they can be expected to be subject to hydrolysis and be chemically reactive.

Writers who are antipesticide and wish to appeal to emotion in their readers to press home the dangers of insecticides often report that the organophosphorus insecticides are a product of the research on "war gas" during World War II. To an extent this is true, for nerve gases such as toban, sarin, and soman developed during that period act in a manner similar to that of this class of insecticides. The insecticide research was, however, a distinct entity necessitated in Germany by the difficulty during wartime of obtaining nicotine, an aphicide essential to the sugar beet industry. This research led to the discovery by Schrader of Bladan in the late 1930s, a compound that owed its insecticidal activity to TEPP, one of the most potent aphicides ever developed. TEPP was narrow in spectrum and unstable, especially under alkaline conditions, and because of this was not suitable for inclusion in sprays of Bordeaux mixture (copper sulfate and lime) used for control of fungus diseases on many crops. Further research by Schrader resulted in the discovery of parathion in 1944 and the systemic aphicide schradan. After the Second World War, the chemistry behind these insecticides became known, and since then an estimated 100,000 organophosphorous compounds have been synthesized and tested against insects. Kenaga and End (1974) include 94 of these in their listing of commercial and experimental organic insecticides. Perhaps in no other group of insecticides is such diversity found in toxicity to animal life, degree of persistence, selectivity, or use patterns. Only a few will be discussed. Included are those used in multimillion lb volume and some others that, though not of such extensive use, illustrate important features. For more nearly complete listings the reader is referred to Martin (1972), Corbett (1974), or Spencer (1973).

Mode of Action

Organophosphorous insecticides act on the nervous system by inhibition of acetylcholinesterase at the synapse. In this they share a common site with the carbamates, discussed later. The exact mode of action, kinetics of the reactions involved, and the experimental evidence supporting the conclusions concerning this mode of action are detailed by others (O'Brien, 1967; Corbett, 1974) and are reviewed here only in simple form.

As pointed out earlier, messages are transmitted across the nerve synapse by a transmitter substance, and acetylcholine serves that role in cholinergic junctions. When the impulse is passed, acetylcholine-sterase acts on the acetylcholine to return it to an active ionic state. The action can be visualized as follows (Hellenbrand and Krupta, 1970):

$$EH\ +\ AcCh \rightleftharpoons EH \cdot AcCh \xrightarrow[ChH]{} EAc \xrightarrow[HOH]{} EH\ +\ AcOH$$

Thus in the first step the enzyme (EH) combines with the acetyl-choline (AcCh) to form a reversible and transitory complex. This complex acetylates the enzyme (EAc) with the release of choline (ChH). The addition of water serves to deacetylate the enzyme to give free enzyme and acetic acid (AcOH). In relation to the toxicity of organophosphorous insecticides the important impetus for the initial steps in this reaction is the nucleophilic (attraction to a positive center) hydroxyl group of a serine component in acetylcholinesterase attacking the positively charged nitrogen atom of acetylcholine.

$$CH_3-\overset{\overset{O}{\|}}{C}-O-\overset{\overset{H}{|}}{\underset{\underset{H}{|}}{C}}-\overset{\overset{H}{|}}{\underset{\underset{H}{|}}{C}}-\overset{+}{N}\overset{\diagup CH_3}{\underset{\diagdown CH_3}{-CH_3}} \qquad HO-\overset{\overset{H}{|}}{\underset{\underset{H}{|}}{C}}-\overset{\overset{H}{|}}{\underset{\underset{NH_3}{|}}{C}}-COO-$$

acetylcholine   serine

Within the organophosphorous insecticides the P=O nucleus has a strong electronic pull on the hydroxyl group of acetylcholinesterase with the result that, when one of these insecticides penetrates the synapse, it is attacked by the acetylcholinesterase, bonds with it, and thus ties up the enzyme so that it is not free to act on the transmitter substance. In the case of the organophosphorous insecticides the bonding is strong and the addition of water releases the enzyme very slowly. Thus some literature on the subject refers to this as an irreversible step, in contrast to the situation with carbamate insecticides, in which the bond is much less "tight." Continuous exposure being barred, the concentration of enzyme returns rapidly to preexposure levels.

The strength of the positive pull on the P=O group on the enzyme is altered when oxygen is replaced in this nucleus by sulfur (P=S), and the pull is also affected by other components of the insecticide molecule adjacent to the P=O center. In fact P=S has little drawing power and organophosphorous insecticides of the–thionate configuration must be oxidized to the corresponding P=O compound before they become active. This oxidation occurs readily in both animals and plants, a step referred to frequently as "lethal synthesis" (O'Brien, 1967).

Organophosphorous poisoning is thus brought about by a phosphorylation of acetylcholinesterase and a resultant accumulation of acetylcholine in the nerve synapse. When this occurs, normal transmission of nerve impulses is prevented; in insects this is expressed in hyperactivity, tremors, convulsions, paralysis, and death. These symptoms are consistent with effects in the central nervous system, where cholinergic junctions undoubtedly occur.

In higher animals susceptible cholinergic synapses occur throughout much of the nervous system, symptoms of organophosphorous poisoning including nausea, salivation, giddiness, tremors, tearing, myosis, coma, convulsion, prostration, and death. Contrary to the situation in insects, where respiration through the spiracles is largely passive, respiration in higher animals requires controlled muscular activity. Disruption of this activity by poisoning the system by which it is controlled results in asphyxiation.

Widely Used Organophosphorous Insecticides

Although many organophosphorous insecticides are being marketed, the major use has involved a relatively small number of compounds. These will be discussed, for their volume of use may place them in the environment in a significant way.

Parathion (*0,0*-diethyl *0*-nitrophenyl phosphorothionate) is one of the most toxic and widely used insecticides in North America. It is a dark brown liquid miscible in benzene and many organic solvents but insoluble in petroleum and mineral oils. It is practically insoluble in water (less than 1 ppm). It is resistant to acid hydrolysis but is hydrolyzed by alkali to form diethyl phosphorothioic acid and *p*-nitrophenol, both of which are of no significance as toxicants.

Parathion is converted in animal and plant tissues to the powerful anticholinesterase paraoxon. Conversion was shown to occur in the liver of higher animals, brook and brown trout, the frog, and pigeon (Potter and O'Brien, 1964). This occurs also in other animal tissues. Activation to paraoxon and degradation to nontoxic metabolites occur simultaneously and at about equal rates in liver microsomes (Johnson and Dahm, 1966). A large number of metabolites have been identified, some the result of dealkylation. Conversion to paraoxon involves both oxidases and $NADPH_2$. Metabolic degradation also involves reduction of the nitrophenolic moiety to the corresponding aminophenol. The metabolic pathways and products are depicted by Menzie (1969). In man parathion is excreted in urine as p-nitrophenol (Funckes et al., 1963), and this is a useful method for determining human exposure. In ruminants a conjugated p-aminophenol is excreted, probably owing to organisms in the rumen that bring about reduction of the parathion to amino parathion.

The conversion of parathion to paraoxon activates the compound as an anticholinesterase, and this occurs in plants and on plant surfaces, as well as in animals. This conversion on plant surfaces is of special importance in hot, dry agricultural areas where the conversion to paraxon is not accompanied by an equally rapid breakdown of this product. As a result paraoxon accumulates on the dry, dusty foliage and presents a hazard to field workers, a hazard that may persist for many weeks after application and result in illness (McEwen, 1977).

Although parathion does not normally persist for long periods when applied to plant foliage, it can be quite persistent in soil, with measurable residues up to 1 year after application. Its low solubility in water militates against its being a major contaminant in that environment, and it is rarely detected in streams or lakes except immediately after applications to adjacent crops.

Parathion is toxic to all animal life, with an acute oral LD_{50} to the rat near 10 mg/kg and with equal toxicity by dermal exposure. As a result of its high mammalian toxicity by the dermal route and its use in high volume, it and the closely related methyl parathion have been responsible for many of the illnesses and deaths reported from pesticide poisonings.

Parathion is broad spectrum as an insecticide and used widely in agriculture against foliar pests and the adult stage of root maggots. It has also been used in some countries for bird control (Quelea birds in Africa) by spraying of sites where the birds congregate. In some instances (e.g., cotton pests), resistance has developed to parathion, and the insecticide is no longer effective. Figures for the United States indicate 15 million lb produced in 1971 with 9.5 million lb used in agriculture (NAS, 1975).

parathion   paraoxon

Methyl Parathion. The closely related methyl parathion (*0,0*-dimethyl *0-p*-nitrophenyl phosphorothionate) shares with parathion many chemical properties, but it isomerizes and hydrolyzes more readily and is thus less persistent as a foliar residue and less stable in storage. Similarly, it is much less persistent in soil (Lichtenstein and Schulz, 1964). Methyl parathion is a white, crystalline powder soluble in most organic solvents, slightly soluble in petroleum distillates, and more soluble in water (60 ppm) than parathion. It is highly toxic (acute oral LD_{50} to rat near 20 mg/kg with dermal toxicity near 60 mg/kg). Evidence suggests that it is metabolized in a manner similar to that of parathion, being first activated to methyl paraoxon and then degraded to nontoxic products in life systems. Although it has the same general spectrum of activity against insects as parathion has, it is more effective against aphids and beetles. This, plus its slightly lower mammalian toxicity, has resulted in its being used instead of parathion in many agricultural programs, especially in cotton and citrus, 45 million lb being produced in the United States in 1971, with more than 27 million lb used by farmers (NAS, 1975).

$$CH_3O \diagdown \underset{CH_3O \diagup}{\overset{\overset{\textstyle S}{\|}}{P}} {-}O {-} \langle \text{benzene ring} \rangle {-} NO_2$$

methyl parathion

Diazinon (*0,0*-diethyl *0*-2-isopropyl-4-methyl-6-pyrimidyl phosphorothionate) is a reddish-brown liquid miscible in most organic solvents, soluble in petroleum oils, and slightly soluble (40 ppm) in water. It is subject to oxidation and decomposes at high temperatures. Diazinon hydrolyzes rapidly in strong acid or alkaline conditions but only slowly where conditions are near neutral. In both animal and plant tissues diazinon is converted to the active diazoxon, a product much more susceptible to hydrolysis than the parent compound. Diazinon is about as persistent in soil as parathion, being degraded by microorganisms and by hydrolysis.

On plant foliage diazinon is moderately persistent, providing protection from insects for 7 to 10 days.

Diazinon has been detected at low levels in water and, even at low levels, has been shown to cause effects, lethal and sublethal, on fish and aquatic invertebrates (Chapter 15). In higher animals diazinon is degraded by microsomal enzymes in the presence of $NADPH_2$ and excreted (Menzie, 1969).

Diazinon is much less toxic than parathion with acute oral LD_{50} values reported for rats as 66 to 600 mg/kg. Dermal toxicity is 379 to 1200 mg/kg (Kenaga and End, 1974).

Diazinon is used as a soil and foliar insecticide and is effective for a broad range of insect pests. It is a good residual treatment for control of flies in barns (where resistance has not developed) and is used also in household sprays and dusts for ant and cockroach control. Production in the United States (1971) was reported at 10 million lb, with one-third of this used in agriculture (NAS, 1975).

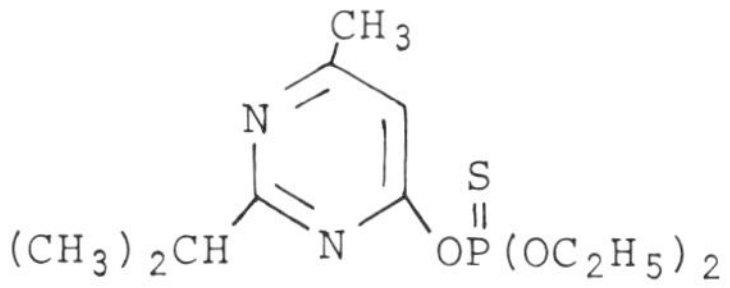

diazinon

Azinphosmethyl (*0,0*-dimethyl *S*-4-oxo-1,2,3-benzotriazin-3-(4-H)-yl-methyl phosphorodithioate), better known as Guthion®, is a white solid soluble in organic solvents and slightly (30 ppm) soluble in water. It is a broad spectrum insecticide used against foliage-feeding insects and is relatively persistent, giving effective protection for 2 or more weeks. It is readily hydrolyzed by acid or alkali.

Azinphosmethyl has high mammalian toxicity, the acute oral LD_{50} to the rat being near 15 mg/kg, with dermal toxicity near 225 mg/kg. Because of its relatively lower dermal toxicity to mammals, it is preferred over parathion and finds extensive use in orchard and vegetable production. It has been used also quite widely in cotton but has not been as effective there as methyl parathion has. Its use in vegetable and fruit production has resulted in occasional contamination of adjacent streams with resulting local fish kills. Production figures for the

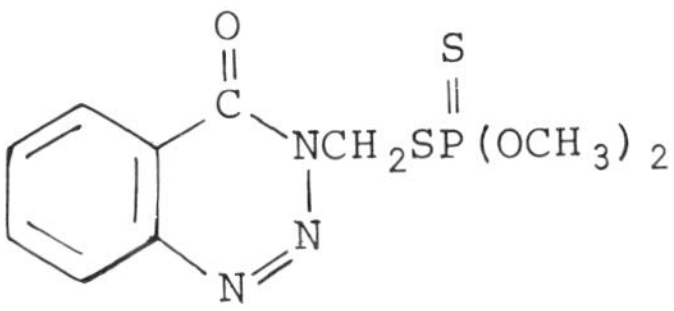

azinphosmethyl

United States (1971) indicate 4 million lb with estimated use in agriculture for the same year at 2.7 million lb (NAS, 1975).

Malathion. The discovery of malathion (*0,0*-dimethyl S-(1,2-dicarboxyethyl) phosphorodithioate) ushered in a new era in organophosphorous insecticides, for this product is toxic to insects but has low toxicity to mammals. Malathion is a clear, amber liquid miscible in organic solvents, moderately soluble in light petroleum distillates, and slightly soluble in water (145 ppm). Malathion is fairly stable in aqueous solutions in the pH range 5.0 to 7.0 but hydrolyzes rapidly in more acidic or alkaline solutions. While its toxicity to insects approaches that of parathion, it has low mammalian toxicity, acute oral LD_{50} to the rat being about 2000 mg/kg with dermal toxicity in the 4000 mg/kg range.

The explanation for the wide selectivity between insect and mammalian toxicity lies in the ease with which malathion is degraded to nontoxic metabolites by carboxylesterases. These enzymes are not nearly so abundant in insects as in vertebrates, with the result that in insects the oxidases convert malathion to the active malaoxon much more rapidly than the latter compound is degraded. In mammals, by contrast, while the conversion to malaoxon does occur, it is degraded before it can reach the site of action. In addition to degradation by carboxylesterases, malathion is inactivated also by phosphatases in both insects and mammals (Figure 10.3). Again there is a rate difference favoring less toxicity to mammals.

Thus malathion presents several foci for enzymatic hydrolysis (Figure 10.3), with a wide variety of metabolites formed and final degradation to inorganic phosphate (Menzie, 1969).

Malathion is effective against many species of insects and mites and because of its relative safety to mammals is used widely in world health, household, garden, and nuisance pest control. It is used also around livestock for fly control and has been one of the preferred

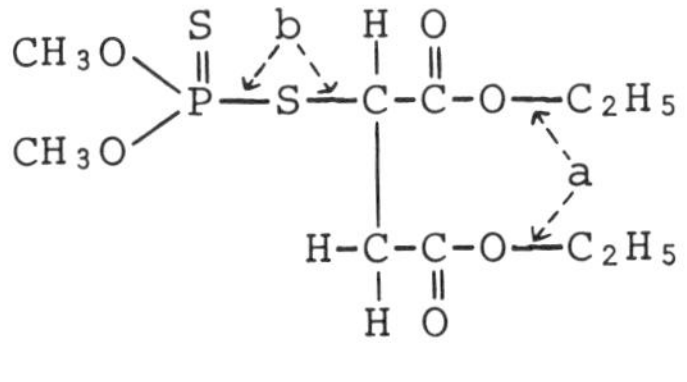

malathion

Figure 10.3 Diagrammatic representation of malathion indicating points at which hydrolysis occurs by carboxylesterases (*a*) and phosphatases (*b*).

insecticides for control of adult mosquitoes. Some insects, notably houseflies, have become resistant to malathion in some areas.

Although malathion is effective against many pests in agriculture, it must be used at higher dosages than materials like parathion and azinphosmethyl, with a resulting higher cost. Thus the use of this compound in agriculture has not been as extensive as some other organophosphorous insecticides. Figures for the United States (1971) indicate a production of 3.5 million lb, only about 10% of which was used in commercial agriculture (NAS, 1975).

Fenitrothion (0,0-dimethyl 0-3-methyl-4-nitrophenol phosphorothionate) is a brownish liquid soluble in most organic solvents but with low solubility in water (20 ppm). In common with the organophosphorous insecticides discussed previously, it is converted by oxidases to the active fenitrooxon in mammals, plants, and insects. Feeding studies show that it is degraded rapidly in mammals to a number of metabolites, including p-nitrocresol, demethyl fenitrothion, aminofenitrothion, and dimethyl phosphorothioic acid (NRC, 1975b). Metabolites are readily excreted and residues are rarely detected in milk.

Although fenitrothion is used widely in Asia and Europe (called sumithion there) as a general purpose agricultural insecticide, its use in North America has been limited almost entirely to forest spraying against the spruce budworm. In this program it has been used at 2 to 4 oz/acre and applied by air to millions of acres of forests in New Brunswick and Quebec. Contamination of water occurred during spray operations, but this contamination was short lived. In natural waters, fenitrothion degrades rapidly through hydrolysis to form p-nitrocresol and other polar metabolites (Kovaciova et al., 1973).

Fenitrothion is moderately toxic, with an acute oral LD_{50} to the rat of 250 to 670 mg/kg and with dermal toxicity reported as 200 to 730 mg/kg (Kenaga and End, 1974).

Fenitrothion is not produced in North America, and use figures are not available. In Canada several million pounds were imported in 1975, most of which was used in spruce budworm control.

fenitrothion

Other Organophosphorus Insecticides

Whereas the organophosphorous insecticides already discussed are those in major use in North America, a large number of related compounds have limited use in specific circumstances. The names, structural formulae, and toxicity of some of these are included in Table 10.1.

TEPP was the first of the commercial organophosphate insecticides, discovered about 1938 by Schrader. It is miscible in water and hydrolyzes readily to nontoxic compounds. On plant foliage it dissipates rapidly through volatilization and hydrolysis. It is selectively toxic to aphids and mites and is used in aerosols in greenhouses. Because of its short persistence (less than 1 day) on plant foliage it is useful for aphid control on food crops immediately prior to harvest.

Dicapthon, although closely related to methyl parathion is much less toxic, especially by dermal exposure. It is useful for fly control and as a household insecticide.

Trichlorfon (Dipterex®, Dylox®) is a broad spectrum insecticide of low mammalian toxicity effective as a stomach and contact insecticide. It is used as a foliar spray against lepidopterous larvae and finds use in the dairy barn as a bait for fly control. Dichlorfon is converted readily to dichlorvos, a conversion believed responsible for much of its activity.

Dichlorvos (Vapona®) is an interesting insecticide with a characteristic rapid knockdown of insects when used as a spray or bait. This compound is in the "active" form and thus does not require "lethal synthesis" within the insect. Dichlorvos is used extensively in Vapona® strips, a preparation in which the insecticide is impregnated in a resin and volatilizes at a fairly uniform rate over an extended period of time to give control of household pests, especially those that fly. Although the LD_{50} values indicate that the insecticide is highly toxic, the rapid rate at which it is metabolized in the mammalian system permits its use in areas where human exposure occurs. Dichlorvos is effective against ectoparasites and is used in flea collars for dogs and cats and a number of other veterinary applications. There is some question concerning the effect of dichlorvos exposure on the ability of companion animals to withstand stress.

Naled (Dibrom®) is closely related chemically to trichlorfon and has a similarly low mammalian toxicity. It is quite volatile and because of this has been used as a "pour on" on heating pipes in greenhouses. This provides an effective fumigation for control of aphids and mites. In addition to its use in greenhouses, naled is effective against adult Diptera and has been used to control these in agriculture and in mushroom houses. About 2 million lb were produced in the United States in 1971 (NAS, 1975).

Ethion is a persistent (several months in soil) insecticide and acaricide. It is practically insoluble in water and is used frequently in combination with oil as a dormant spray for control of aphids, mites, and scale insects in orchards. It is effective also when used without oil

Table 10.1 Name, Structure, Acute Oral Toxicity to the Rat, and Dermal Toxicity of Some Organophosphorous Insecticides

Name	Structure	Acute Toxicity* (LD_{50} to Mammal)	
		Oral	Dermal
TEPP: bis-0,0-di-ethylphosphoric anhydride	C_2H_5O \\ P-O-P / C_2H_5O, O, O, OC_2H_5, OC_2H_5	0.2 -2	2 -20
dicapthon: 0,0-dimethyl 0-2-chloro-4-nitrophenyl phosphorothionate	CH_3-O \\ P-O-⟨ring⟩-NO_2 / CH_3-O, S, Cl	220 -400	>2000
trichlorfon: 0,0-dimethyl (1-hydroxy-2,2,2-trichloroethyl phosphonate	CH_3-O \\ P-C-C-Cl / CH_3-O, O, OH, Cl, H, Cl, Cl	450 -469	>2000
dichlorvos: 2,2-dichloro-vinyl dimethyl phosphate	CH_3-O \\ P-O-C=C / CH_3-O, O, H, Cl, Cl	25 -170	59 -900
naled: 1,2-dibromo-2,2-dichloroethyl dimethyl phosphate	CH_3-O \\ P-O-C-C-Cl / CH_3-O, O, Br, Br, H, Cl	430	1100
ethion: 0,0,0,0-tetra-ethyl S,S-methy-lene bis phos-phorodithioate	C_2H_5O \\ P-S-C-S-P / C_2H_5O, S, H, S, H, OC_2H_5, OC_2H_5	27 -119	62 -245

Table 10.1 *(Continued)*

Name	Structure	Acute Toxicity* (LD$_{50}$ to Mammal)	
		Oral	Dermal
fenthion: 0,0-demethyl 0-[-4-(methylthio) m tolyl] phosphorothioate		225 -298	330
fensulfothion: 0,0-diethyl 0-[p-(methylsulfinyl) phenyl] phosphorothioate		2 -10	3 -30
chlorpyrifos: 0,0-diethyl 0-(3,5,6-trichloro--2-pyridyl) phosphorothioate		97 -276	2000
phosmet: 0,0-dimethyl S-phthalimido-methyl phosphorodithioate		147 -299	3160
phosalone: S-[(6-chloro-2-oxo-3 benzoxazolinyl methyl] 0,0-diethyl phosphorodithioate		125 -180	1500
chlorfenvinphos: 2-chloro-1-(2,4-dichlorophenyl) vinyl diethyl phosphate		12 -56	31 -108

during the growing season. Ethion has been used for many years as a soil application in the seed furrow for maggot control on onions. In the soil it is longlasting and has been shown to accumulate to significant levels (>5 ppm) in organic soils (Miles et al., 1978). About 3 million lb were produced in the United States in 1971 (NAS, 1975).

Table 10.1   (*Continued*)

| Name | Structure | Acute Toxicity* (LD$_{50}$ to Mammal) | |
		Oral	Dermal
stirofos: 2-chloro-1-(2,4, 5-trichloro- phenyl) vinyl dimethyl phosphate		4000 -5000	>5000
fonofos: 0-ethyl S- phenyl ethyl- phosphonodi- thioate		8 -16	147
methamidophos: 0,S-dimethyl phosphoramido- thioate		13 -30	110
monocrotophos: dimethyl cis-1- methyl-2-methyl- carbamoylvinyl phosphate		21	354
phosphamidon: 2-chloro-N,N- diethyl-3- hydroxycrotona- mide dimethyl phosphate		15 -33	125 -150
temephos: 0,0-dimethyl phosphorothioate 0,0-diester with 4-4'-thio diphenol		1000 -3000	4000

Fenthion (Baytex®) is a persistent (several months) insecticide of moderate toxicity with a fairly broad spectrum of activity but is especially effective against Diptera. It finds use in controlling fruit flies, house flies, and mosquitoes, the last both as an adulticide and a larvicide. It is used against ectoparasites on livestock (Tiguvon®). It is

Table 10.1   (*Continued*)

Name	Structure	Acute Toxicity* $(LD_{50}$ to Mammal)	
		Oral	Dermal
leptophos: 0-(4-bromo-2,5-dichloro phenyl) 0-methyl phenyl phosphonothioate		43 -53	10000

*Toxicity figures are from Kenaga and End, 1974. In most cases acute oral LD_{50} figures are based on the rat, dermal LD_{50} figures on the rabbit.

highly toxic to birds and has been used in Africa against weaver birds.

Fensulfothion (Dasanit®) is very similar in structure to fenthion but highly toxic. It is persistent (few months) and an effective soil insecticide against root maggots in vegetable crops and rootworms in corn. It is effective also against free-living cyst and rootknot nematodes. About 4 million lb were produced in the United States in 1971 (NAS, 1975).

Chlorpyrifos (Dursban®, Lorsban®) is a moderately toxic broad spectrum insecticide, of greatest interest from an environmental standpoint because of its use as a mosquito larvicide. Chlorpyrifos is relatively stable under acidic conditions but hydrolyzes when exposed to alkali. Many metabolites are formed in both animal and plant tissues, and when it is applied in an aquatic environment (e.g., as a mosquito larvicide), the residue is short lived and found mostly in the phytoplankton (Frank, personal communication). It has low solubility (2 ppm) in water.

Chlorpyrifos is somewhat persistent in soil (weeks) and is effective against root maggots in vegetable crops, chinch bugs, and white grubs in lawns. It is effective against cutworms but may be phytotoxic to some plants (e.g., head lettuce).

Phosmet (Imidan®) and **phosalone** (Zolone®) are general purpose insecticides used to a significant extent on deciduous orchards. They are of moderate toxicity orally but of low toxicity to mammals by dermal exposure.

Chlorfenvinphos (Birlane®) and **stirofos** (Gardona®) are two closely related insecticides but with vastly different biological activity. Chlorfenvinphos is highly toxic to most mammals (the dog being an exception) and has a narrow spectrum of activity as a soil insecticide against root maggots and corn rootworms. By contrast, stirofos is of low toxicity to mammals and is selectively toxic to Diptera, Coleoptera, and Lepidoptera. Whereas chlorfenvinphos is a persistent (several months) soil insecticide, stirofos is not.

Fonofos (Dyfonate®) is a highly toxic, persistent (months) general soil insecticide with a broad spectrum of activity.

Methamidophos (Monitor®) is a broad spectrum, highly toxic, persistent (weeks on foliage) insecticide and acaricide. It is readily soluble in water and has some systemic activity. It is an outstanding aphicide and has provided excellent control of lepidopterous larvae that have developed resistance to many insecticides (e.g., cabbage looper, cotton bollworm).

Monocrotophos (Azodrin®) is also a broad spectrum, highly toxic, persistent (months) insecticide. Although it provides good control of many vegetable and cotton pests, its high toxicity and persistence on foliage have been associated in a few instances with illness among field workers.

Phosphamidon is a highly toxic, general purpose insecticide with systemic activity. It has high toxicity to birds and when used in the spruce budworm program caused some mortality in songbirds (Chapter 18).

Temephos (Abate®) is an important insecticide for control of larvae of mosquitoes, black flies, biting midges, and sand flies of public health or nuisance importance. Because of its widespread use in the aquatic environment, temephos has been studied extensively with regard to its effects on nontarget organisms.

Leptophos (Phosvel®) is a broad spectrum insecticide, especially effective against lepidopterous larvae. It is of environmental interest because of its reported "demyelinating" effect, this being observed first in water buffalo that drank contaminated water (Shea, 1974). More recently chronic poisoning from leptophos among workers exposed during its manufacture has been experienced. Leptophos was registered for cutworm control in Canada, but this registration has been withdrawn (Chapter 19).

Systemic Organophosphorous Insecticides

Included among the organophosphorous insecticides are a number that are taken up systemically by plants, are converted into active forms, and kill insects. Most of these are primarily effective against sucking insects, but some by virtue of high inherent toxicity and/or high concentration in plant tissues are effective also against lepidopterous larvae (especially stalk borers), leaf eating beetles, and leaf-mining larvae.

The value of these insecticides is indicated by the fact that more than 20 million lb were produced in the United States in 1971 (NAS, 1975).

Schradan, developed by Schrader in the late 1940s, was the first of this group of plant systemics. It is a mixture of octamethyl pyrophosphoramide and decamethyl triphosphoramide and is miscible in water. It is highly toxic to mammals (acute oral LD_{50} to the rat of about 10 mg/kg and a dermal toxicity of about 15 mg/kg). As a technical product Schradan has little toxicity to insects. Toxicity appears when the compound is converted in plants to the monophosphoramide oxide (Casida et al., 1954).

schradan

Schradan is used primarily as a foliar spray against sucking insects and is effective for several weeks after application. It is not used in North America.

Demeton (Systox®) (mixture of $0,0$-diethyl S-(and 0)-2-[(ethylthio) ethyl] phosphorothioates) and oxydemeton methyl (S-[2-(ethyl-sulfinyl) ethyl] $0,0$-dimethyl phosphorothioate) (Meta-Systox-R®) are closely related systemic insecticides effective against sucking insects when applied as foliar sprays or in the soil. Demeton is a mixture of demeton-0 and demeton-S. They are slightly soluble in water (demeton-0, 60 ppm; demeton-S, 2000 ppm) and are taken up rapidly by plants, the demeton-S compound being translocated more rapidly. Both compounds are highly toxic (acute oral LD_{50} to the rat for demeton-S, 30 mg/kg and, for demeton-0, 6 to 12 mg/kg) and are converted in plants to the sulfoxide and sulfone, metabolites much more toxic than the parent compound.

Oxydemeton methyl is less toxic to mammals (acute oral LD_{50} to rat about 75 mg/kg and dermal toxicity near 200 mg/kg). This compound has some contact activity.

Both demeton and oxydemeton methyl are used in fruit and vegetable production.

demeton (thiol isomer)   oxydemeton methyl

Disulfoton (Di-syston®) (0,0-diethyl S-[2-(ethylthio) ethyl] phosphorodithioate) is a more persistent systemic than demeton and is used primarily as a granular application in the soil. It is effective against sucking insects and also against corn rootworms. It is highly toxic with an acute oral LD_{50} to the rat of 2 to 12 mg/kg and a dermal toxicity in the 20 to 50 mg/kg range. It is slightly soluble in water (66 ppm) and when applied in the granular form is taken up by plants over an extended period. In plants it is converted to the sulfoxide and sulfone, metabolites identical with those produced from the thiol isomer of demeton. About 8 million lb were produced in the United States with half that amount reported used by farmers in 1971 (NAS, 1975).

disulfoton

Phorate (Thimet®) (0,0-diethyl-S-[(ethylthio)methyl] phosphorothioate) is a widely used systemic effective against sucking insects, larvae of the corn rootworms, and leaf-eating beetles. It is highly toxic, with an acute oral LD_{50} to the rat of 1 to 5 mg/kg and a dermal toxicity reported as 2 to 300 mg/kg (Kenaga and End, 1974). Because of its high mammalian toxicity it is marketed only as a granular product. In common with other organophosphorous insecticides, phorate does not accumulate in animal tissues, it being hydrolyzed and excreted in urine and feces (Menzie, 1969). Phorate had about the same volume of use as disulfoton in 1971.

phorate   Counter®

Counter® (S-[(tert-butylthio)methyl]0,0-diethyl phosphorodithioate) is closely related to phorate, a newer systemic with similar
characteristics.

Dimethoate (Roger®, Cygon®) (0,0-dimethyl S-(N-methyl-
carbamoylmethyl) phosphorodithioate) is effective as a systemic and/or
contact insecticide against a wide range of insects and mites. In contrast
to the high toxicity of most organophosphorous systemics, dimethoate is
of moderate toxicity with an acute oral LD_{50} to the rat near 500 mg/kg by
dermal or oral exposure.

Dimethoate is used on a wide range of plants for control of mites
and both sucking and leaf-feeding insects. It is effective also as a
residual treatment for fly control in livestock pens. It is highly effective against dipterous larvae and is used for fruit fly control and for
control of leaf miners on fruit, vegetables, and ornamental trees. Concentrated preparations can be "painted" on the trunk and main limbs
of large trees, for example, birch, to provide control of the birch leaf
miner. It has been used also for spruce budworm control.

$$CH_3O \underset{CH_3O}{\overset{S}{\underset{\diagup}{\overset{\diagdown}{P}}}} - S - \overset{H}{\underset{H}{C}} - \overset{O}{\overset{\|}{C}} - \overset{H}{\underset{}{N}} - CH_3$$

dimethoate

Mevinphos (Phosdrin®) (methyl 3-hydroxy-alpha-crotonate, dimethyl phosphate) is a short-lived systemic and contact poison useful
on fruits and vegetables where insect control is required near harvest.
Mevinphos is highly toxic by oral or dermal exposure (acute oral LD_{50}
to the rat 3 to 7 mg/kg, dermal, 3 to 90 mg/kg) (Kenaga and End, 1974).

$$CH_3O \underset{CH_3O}{\overset{O}{\underset{\diagup}{\overset{\diagdown}{P}}}} - O - \overset{CH_3}{\underset{}{C}} = \overset{H}{\underset{}{C}} - \overset{O}{\overset{\|}{C}} - O - CH_3$$

mevinphos

Dicrotophos (Bidrin®) (3-hydroxy-N,N-dimethyl-cis-crotonamide
dimethyl phosphate) is a water-miscible systemic of interest owing to
its unique method of application. When injected into tree trunks, it
provides control of a wide range of sap borers and leaf miners. It is
highly toxic, with an acute oral LD_{50} to the rat of 22 to 75 mg/kg and a
dermal toxicity near 200 mg/kg.

$$CH_3-O \quad O \quad CH_3 \; H \; O \quad CH_3$$
$$P-O-C=C-C-N$$
$$CH_3-O \quad \quad \quad \quad \quad \quad \quad CH_3$$

dicrotophos

Organophosphorous Systemics for Animals

Livestock are infested with various ticks, lice, fleas, mites, and flies as ectoparasites and with a number of helminths and other internal parasites. Several insecticides have been discussed in relation to their role in ectoparasite control, and the role of rotenone for control of cattle grubs has been cited. Among the organophosphorous insecticides are several that can be applied externally to cattle or taken internally with food or as a bolus for control of parasites. These "animal systemics" are important in livestock production, especially for control of cattle grubs.

Crufomate (Ruelene®) (4-tert-butyl-2-chlorophenyl methyl methyl phosphoramidate) is effective against cattle grubs and some helminths. It is a white crystalline solid practically insoluble in water, with a high toxicity to cattle grubs but a low toxicity to the host. The acute oral LD_{50} of crufomate to the rat is 900 to 3000 mg/kg, with dermal toxicity in the same range. The insecticide is effective when applied as a spray to the back of the animal or when included in the food at 20 to 25 mg/kg of body weight. This is approximately 20% of the dosage where slight depression in cholinesterase activity of the host animal is detectable. Within the animal tissue dosages toxic to the grubs occur, but these are degraded rapidly to nontoxic components and excreted or used in the animal's metabolism (e.g., inorganic phosphate) (Menzie, 1969). About 2 million lb were produced in the United States in 1971 (NAS, 1975).

crufomate

Ronnel (Korlan®) (0,0-dimethyl-0-2,4,5-trichlorophenyl phosphorothioate) is about equal in toxicity to crufomate and provides control of cattle grubs when administered in the food at 100 mg/kg body weight or when given as a single 15 g bolus. In the animal it is more persistent than crufomate but is metabolized and excreted, peak levels occurring in the urine 18 to 32 hours after administration (Plapp and Casida, 1958). About 3 million lb were produced in the United States in 1971 (NAS, 1975).

ronnel

Although crufomate and ronnel are the major animal systemics used in North America, others include **crotoxyphos** (Ciodrin®), used primarily as a spray for control of ectoprarsites; **coumaphos** (Co-ral®), applied as a spray for cattle gurb control; **cythioate** (Proban®); and **famphur** (Warbex®).

crotoxyphos   coumaphos

cythioate   famphur

Although the discussion of organophosphorous insecticides given here is not exhaustive and many important compounds are omitted, it is clear that this group of compounds is widely diverse in biological activity. It includes some of our most toxic insecticides such as parathion, phorate, and mevinphos and also some such as malathion, stirofox, and temephos whose mammalian toxicity is in the same order of magnitude as that of table salt. Among the organophosphorous insecticides are some like fensulfothion and ethion that persist in soil for more than 1 year and others like mevinphos and TEPP that disappear within 1 day. Some, like parathion, malathion, and many more are broad spectrum, but others, such as temephos, ethion, and TEPP, are restricted in toxicity to a narrow range of insect groups. Some are soil insecticides, some plant systemics, some effective against foliar pests, and others act as fumigants. Some are almost equally toxic to all forms of animal life, while others are sufficiently selective to permit their use in and on livestock for the control of ectoparasites and endoparasites. In no other group of insecticides can one find such a variety of biological performance and adaptability. All act, however, on the cholinesterase system, and many are toxic to birds, a subject for discussion later (Chapter 18) in relation to the role of some insecticides from this class as environmental contaminants.

Carbamate Insecticides

The developments in new insecticides are proceeding so rapidly that one hesitates to refer to the status of any particular group of compounds with respect to their stage in development. In a sense the carbamates can be thought of as the most recent major group of insecticides, but new developments in the pyrethroids (previously discussed), in juvenile hormone mimics, in sex and other pheromones, in antimetabolites, and in microbial insecticides may soon render such an assessment inaccurate. The carbamates do follow the organophosphorous insecticides in a historical sense and are well established as a major group of highly effective and versatile compounds.

The carbamates have been reviewed in a recent book by Kuhr and Dorough (1976), and the reader is referred to that text for a comprehensive treatment. Much of what follows draws heavily on that review.

The carbamates act as inhibitors of acetylcholinesterase in a manner similar to that of the organophosphorous insecticides. In part they owe their activity to a configurational similarity (Figure 10.4) with acetylcholine and the attraction of the $C = O$ group to the OH site on acetylcholinesterase. Thus the carbamates serve as substrate for the enzyme, tying it up and rendering it unavailable for its designated

$$CH_3-\overset{\underset{|}{CH_3}}{\overset{|}{N^+}}-CH_2CH_2O\overset{\overset{O}{\|}}{C}CH_3 \qquad CH_3S-\overset{\underset{|}{CH_3}}{\overset{|}{C}}-CH=N-O\overset{\overset{O}{\|}}{C}NHCH_3$$

acetylcholine   aldicarb

Figure 10.4   Representation of acetylcholine and the carbamate aldicarb, indicating structural similarities.

function of cleaving acetylcholine in the neural junction. In addition, as pointed out by Metcalf (1971), the structure of the insecticidal carbamates includes a bulky side chain capable of interacting with the anionic site of cholinesterase and located at a proper distance from the $C = O$ group to accommodate the enzyme. The union of carbamate and enzyme is less stable than that of the enzyme with organophosphorous insecticides. Thus the inactivation is reversible at a higher rate and recovery in systems poisoned with carbamates is much more rapid.

Carbaryl (Sevin®) (1-naphthyl-N-methylcarbamate) was the first of the truly commercially successful carbamate insecticides. Carbaryl is a white crystalline solid soluble in polar organic solvents but with low solubility in water (40 ppm). It has low mammalian toxicity with an extremely low toxicity via dermal exposure. It is a broad spectrum insecticide, used extensively in agriculture for foliar pests and as a homeowner product for garden pests and chinch bugs in lawns. It is used also for ectoparasite control on livestock and pets. Carbaryl is ineffective against some species of aphids (but highly effective on others) and is weak against a number of species of lepidopterous larvae (e.g., cabbage looper, some cutworms). It is highly toxic to the honeybee.

Carbaryl is stable in storage but short lived in the field with a half-life of 3 to 4 days on plant foliage, 7 to 9 days in soil, and 1 to 5 days in water (Kuhr and Dorough, 1976).

Carbaryl and other carbamates are metabolized rapidly by both the plant and animal systems. The routes of metabolism include hydrolysis at the ester linkage, oxidation (by mixed function oxidases in animals, questionably in plants), and by the formation of conjugates (Kuhr and Dorough, 1976). The result is detoxification and excretion in animals, detoxification followed by dissipation or storage in plants.

The literature on carbaryl metabolism is extensive and covered in detail by Kuhr and Dorough (1976). In vertebrates at least 13 metabolites have been identified. Degradation is rapid, most animals excreting a high percentage of an ingested dose within 24 hours. Carbaryl is nonphytotoxic to most plants at insecticidal dosage rates, but it may affect blossom set on tomatoes under certain weather conditions and is used commercially as a fruit-thinning agent on apples.

Carbaryl is used extensively, with production in the United States reported as 45 million lb in 1971, about 40% of which was used in American agriculture (NAS, 1975).

Carbofuran (Furadan®) is a white crystalline solid with some solubility in most organic solvents (e.g., 4% in benzene; 15% in acetone) and moderately soluble in water (700 ppm). It is a most versatile compound, with insecticidal, acaricidal, and nematicidal properties, effective as a contact or stomach poison and as a plant systemic. It is used on a wide range of agricultural pests in foliar applications; for soil treatments in the furrow at planting time for root maggot, wireworm, and rootworm control; and as a broadcast application to soil at 6 to 10 lb/acre to control some species of nematodes. It is toxic to most insect groups, though weak in activity against some lepidopterous larvae. Normally used in foliar treatment at 0.5 lb/acre, it is effective against alfalfa weevil and grasshoppers at a dosage of 2 to 4 oz. At high dosages in soil it is likely to cause extensive mortality to some species of earthworms. Its mammalian toxicity is high by ingestion but low by dermal exposure. It is highly toxic to birds, kills in the field having been reported (Chapter 18). It is nonphytotoxic at insecticidal dosage rates, and although it is generally recognized as having some stimulatory effect on plant growth, the mechanism for this effect has not been defined.

Carbofuran is not long lasting when applied to plant foliage but, when applied in the soil, can be taken up by plants over a period of several weeks and provides control of sucking and many plant-feeding arthropods. The half-life of carbofuran in soil is related to soil type, being shorter in sand (30 days), intermediate in loam (40 days), and most persistent in muck (80 days). Low residues (in terms of percentage of initial deposit) are likely to persist for more than 1 year. Persistence in soil is related to pH, with lower pH values associated with greater persistence (Getzin, 1973).

Carbofuran is metabolized by plants, insects, and higher animals. When ^{14}C-labeled carbofuran was administered orally to rats, 30% was eliminated in 6 hours and more than 90% in 3 days (Lucier et al., 1972). At least six metabolites are produced. When fed to the lactating cow, most of the metabolites (94%) were excreted in the urine. Among metabolites identified in the urine were 3-hydroxycarbofuran, 3-hydroxy-N-hydroxycarbofuran, 3-hydroxycarbofuran phenol, 3-keto-carbofuran phenol, and carbofuran phenol (Figure 10.5). Secreted in the milk was 0.2% of the dosage. Alfalfa treated with carbofuran yielded four metabolities, the dominant one (45%) being 3-hydroxycarbofuran (Menzie, 1969). Production figures indicate that 8 million lb of carbofuran were produced in the United States in 1971 (NAS, 1975). The rapid increase in its use since it first became available commercially (1967) would suggest that this figure grossly underestimates current usage. Its greatest use is in vegetable, forage, and field crops production.

Methiocarb (Mesurol®) is a recent introduction to the American

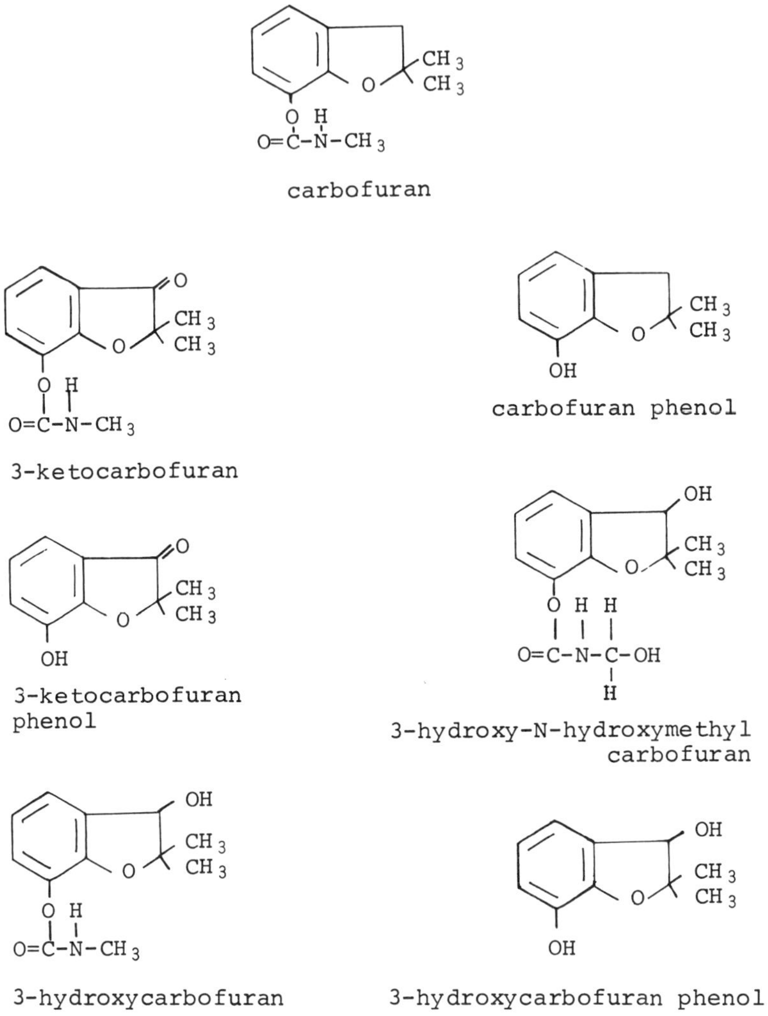

carbofuran

3-ketocarbofuran

carbofuran phenol

3-ketocarbofuran
phenol

3-hydroxy-N-hydroxymethyl
carbofuran

3-hydroxycarbofuran

3-hydroxycarbofuran phenol

Figure 10.5 Carbofuran and its metabolites in the dairy cow (Menzie, 1969).

market but has been used in Europe for a number of years. It is a white
crystalline solid soluble in most organic solvents and practically insoluble in water. It is a general purpose insecticide with some acaricidal
properties. Many of the carbamates have some activity against slugs
and snails, but methiocarb is perhaps the most effective. In this connection it may be useful in world health programs where important
human parasites (e.g., schistosomes) have snails as intermediate hosts.

Methiocarb also has potential as a bird repellent and has been used for this purpose as a seed coating on forest and agricultural seeds. It also shows promise as a bird repellent on susceptible fruit (e.g., blueberries, cherries, grapes) and on field corn.

In plants and animals methiocarb is oxidized to its sulfoxide and sulfone in addition to other metabolites (see Kuhr and Dorough, 1976).

Metalkamate (Bux®) is a moderately toxic soil insecticide used in North America on corn and rice. In Europe it is used also against a number of foliar pests. Metalkamate is a mixture of two compounds, the technical material being a yellowish solid soluble in varying amounts in organic solvents and slightly soluble (50 ppm) in water. Not much is known about the metabolism of this compound. About 6 million lb of metalkamate were produced in the United States in 1971 (NAS, 1975).

Methomyl (Lannate®) is a white crystalline solid with variable solubility in organic solvents and moderate solubility (5.8%) in water. When applied to plant foliage, its residue is short lived (Braun, H. E., personal communication); from a practical point of view effective insect control is provided for less than 7 days. In soil, persistence is of short duration, with 71% degraded in 1 month (Harvey and Pease, 1973). Metabolic decomposition in plants and animals is rapid, with carbon dioxide and acetonitrile produced (Harvey and Reiser, 1973).

Methomyl is a broad spectrum insecticide and acaricide with some nematicidal properties. It is useful in vegetable and field crops production and is especially effective against lepidopterous larvae and aphids.

In addition to its contact action methomyl has some systemic activity, a fact that improves its effectiveness against aphids and other sucking insects on plants. Slightly more than 1 million lb were produced in the United States in 1971 (NAS, 1975).

Pirimicarb (Pirimor®) is a colorless solid soluble in most organic solvents and slightly soluble (0.27%) in water. It has moderate oral and dermal toxicity and acts as a contact, plant systemic, and fumigant with selective toxicity to aphids.

Propoxur (Baygon®) is a white crystalline solid soluble in most organic solvents and slightly soluble (0.2%) in water. It is moderately toxic by ingestion but of low dermal toxicity to mammals. Propoxur has some systemic action in plants, being moderately effective against aphids and some Hemiptera. Propoxur is an important household insecticide, being used as sprays or baits for control of a wide range of pests, including flies, mosquitoes, ants, and cockroaches. It is used extensively as a residual spray for fly control in livestock housing and

Table 10.2   Name, Structure, Acute Oral Toxicity to the Rat, and
Dermal Toxicity of Some Carbamate Insecticides

Name	Structure	Acute Toxicity* (LD_{50} to Mammal)	
		Oral	Dermal
carbaryl: 1-naphthyl methyl carbamate		307 -986	>500 ->1000
carbofuran: 2,3-dihydro-2, 2-dimethyl-7- benzofuranyl methyl carbamate		8 -14	10200
methiocarb: 4-(methylthio)3, 5-xylyl methyl- carbamate		130	>12000
metalkamate: m-(1-ethyl- propyl)phenyl methylcarbamate mixture(1:4) with m-1-methyl- butyl) phenyl methylcarbamate		87 -170	400
methomyl: methyl N-(methyl- carbamoyl) oxy thioacetimidate		17 -24	>1000

in fogging equipment or aerial application for control of adult mosquitoes. When applied indoors, it has residual activity for several
weeks but has a short residue when used outdoors. It is highly toxic to
the honeybee.

Aldicarb (Temik®) is the most toxic of the commercial carbamates,
with high toxicity by either ingestion or dermal exposure. It is used
exclusively in the granular formulation as a soil application. It is taken
up by the plant and provides insect and mite control for 4 to 12 weeks.
In the soil aldicarb persists only a short while, with a half-life of 7 to 10

Table 10.2   (*Continued*)

Name	Structure	Acute Toxicity* (LD$_{50}$ to Mammal)	
		Oral	Dermal
pirimicarb: 2-(dimethyl-amino)-5,6-dimethyl-4-pyrimidinyl dimethylcar-bamate	(structure of pirimicarb)	147	>500
propoxur: 0-isopropoxy-phenyl methyl-carbamate	(structure of propoxur)	95 -104	>1000
aldicarb: 2-methyl-2-(methylthio) propionalde-hyde-X-(methyl-carbamoyl) oxime	(structure of aldicarb)	1.0	5

*Toxicity figures are from Kenaga and End, 1974. In many instances, figures for acute dermal toxicity are based on the rabbit.

days. Ten metabolites were recovered from soil, aldicarb sulfoxide (predominantly) and aldicarb sulfone accounting for much of the residue a few weeks after application (Coppedge et al., 1967). A similarly large number of metabolites are found in plants (Andrawes et al., 1971).

Studies on the metabolism of aldicarb in higher animals indicate rapid metabolism and excretion. In the lactating cow 80% of the labeled material (^{35}S-aldicarb) had been excreted in the urine within 24 hours (Dorough and Ivie, 1968).

The major uses of aldicarb are in cotton, sugar beets, sweet potatoes, potatoes, and peanuts. It is used also in greenhouses in the production of flowers and ornamentals.

As can be seen from this brief discussion of the carbamates, much diversity exists within the group (Table 10.2). A wide range of mammalian toxicity is found, and though most have moderate to low toxicity to mammals by dermal exposure, aldicarb is highly toxic by this route. Some, such as carbaryl and carbofuran, are extremely broad spectrum, but pirimicarb is specifically an aphicide. Many of the carbamates are highly toxic to birds, but all are metabolized by plants and animals and do not bioaccumulate. All of the commercial carbamates are toxic to the honeybee. None of the carbamates are likely to persist in aquatic environments.

Juvenile Hormone Mimics

Growth and reproduction in insects is controlled by a series of hormones that must be present in the right place, at the right time, and in the right concentrations if the insect life cycle is to proceed normally. As has been demonstrated in discussions on pesticide degradation, insects share with higher animals many common metabolic reactions. Because of this, many insecticides are toxic to other animal life. Some processes in insects are, however, shared with few other classes of animals. One such process is molting, a procedure unique to insects and crustaceans.

The molting process involves three hormones. One, produced in the brain and referred to as the brain hormone, directs the prothoracic gland to secrete a second hormone, the molting hormone, called ecdysone. This hormone initiates the molts (or ecdyces), insects molting several times before reaching maturity. The form into which the insect molts is determined by a third hormone, the juvenile hormone, produced in a small pair of glands, the corpora alata, located just behind the brain. When the juvenile hormone is secreted in high concentrations, the immature insect molts into another immature form, but if the juvenile hormone is absent, the molting produces a pupa or adult, depending on the stage from which the insect is molting and whether or not a pupal stage is part of the life cycle of that particular insect.

The story of how this pattern of events was unraveled is an exciting chapter in insect physiology and has been described in popular and scientific literature (Menn and Beroza, 1972). It led, however, to the realization that, if the juvenile hormone were present when the insect was supposed to molt to an adult form, this would not occur and the sexual maturity required for the reproduction of the species would be prevented. Hence arose the concept of using these as insecticides, insecticides that would be active only against those animals in which the juvenile hormone determines the outcome of the molting process and that, theoretically, would be innocuous to other forms of life (Williams, 1967).

The "juvenile hormone mimics" ushered in what Williams (1967) called "third-generation pesticides," characterizing the first generation as lead arsenate and the insecticides of the pre-World War II era, and DDT and the rash of new insecticides that followed it as the "second generation." High hopes were held for these third-generation types. Not all of these expectations have materialized, but enough has been learned to suggest that these will be valuable additions to the insecticide field. It has become evident as well that insects can and have developed resistance to them, their efficacy in the field being sometimes less than desired (Cerf and Georghiou, 1972).

Methoprene (Altosid®) (isopropyl 11-methoxy-3,7,11-tri-methyldodeca-2,4-dienoate) has been tested widely in the laboratory and field and has found practical application as a mosquito larvicide. Methoprene is a colorless liquid soluble in organic solvents and practically insoluble (1.4 ppm) in water. Although it is effective, theoretically, and in laboratory tests, against a wide variety of insects, the problem in field control has been to deliver the insecticide to the site of action in the insect at the proper time. Effectiveness as an insecticide depends on methoprene's being present in the insect at a time when juvenile hormone is not supposed to be, that is, when the insect is about to molt to the pupal or adult stage. It has also been shown (Riddiford, 1972) that juvenile hormone prevents normal development in the egg.

Methoprene is of low mammalian toxicity, with an acute oral LD_{50} to the rat of 34,000 mg/kg. Although it is toxic to most insects, there is a marked difference in susceptibility, and since the timing of treatment in relation to insect stage of development is so critical, it can be used selectively with little danger to nontarget species (Miura and Takahashi, 1973).

Methoprene is metabolized by mammals and excreted mostly in the urine and feces. In cattle some methoprene is excreted unchanged in the feces in amounts sufficient to be insecticidal to dipterous larvae that breed in cattle dung. In this respect methoprene shows promise for control of the face fly and some other livestock pests (Zoecon Corp., 1974).

The potential of hormone mimics as insecticides is covered in a recent review (Bowers, 1976). It is apparent that this is an exciting field for further exploration.

Antimetabolites

Just as the "juvenile hormone mimics" are specific to a few classes of animals, so too are the antimetabolites. This is a broad category of chemicals designed to prevent the formation of vital components of the insect system. The surface has barely been scratched in this undertaking, but enough has been learned to suggest it is a path well worth pursuing.

As discussed earlier, one of the essential components of the insect cuticle is chitin, a polymer of N-acetylglucosamine. In nature chitin is relatively restricted to insects and some groups of fungi. Thus chemicals that act specifically to disrupt the formation of chitin or to degrade it might be expected to be narrowly specific in their effects on the environment and thus good candidates for pest control. The production of chitin involves chemical processes mediated by enzymes, and several compounds have been developed that are antagonistic. A number of these compounds have been tested experimentally, and one **Dimilin**® has entered the commercial insecticide market.

Pheromones

Pheromones are chemicals emitted by an organism to affect another organism. They are variously described as allomones or karomones, depending on whether they act within a species or act interspecifically and have been known for many years as important communication systems, especially in social insects. The identification of the sex pheromone of the silkworm (Butenandt et al., 1959) has been followed by extensive studies with sex pheromones and the discovery and synthesis of these compounds for a wide variety of insect pests, especially among the Lepidoptera (Roelofs and Cardé, 1974). In addition pheromones that induce other responses have been investigated (see Birch, 1974).

The sex pheromones have been explored as monitoring tools, as methods for trapping one sex out of the population, and as "confusion" chemicals to prevent mating (Tette, 1974). Significant success has been achieved.

These pheromones are available for a number of important pest species, including the red-banded leaf roller, the codling moth, the oriental fruit moth, the European corn borer, the gypsy moth, and the cabbage looper (Roelofs, 1976).

Microbial Insecticides

Just like man and other animals, insects are subject to diseases caused by bacteria, fungi, viruses, microsporidians, nematodes, and other organisms. In the field these diseases are present in insect populations as endemic pathogens infecting, in most instances, a low percentage of the population. As such they exert a greater or lesser effect in maintaining populations at levels below those that would occur if the diseases were not there. Periodically, many of these diseases reach epizootic proportions with dramatic effects in "wiping out" insect populations. These epizootics may be widespread or local.

Many insect pathogens are known, and a few have been studied with a view toward their manipulation and incorporation in pest control. In North America the first attempts involved the fungus *Beauvaria globulifera* against the chinch bug, but these attempts were largely unsuccessful (Snow, 1895). More recently there has been renewed interest in insect pathogens, owing in large part to the leadership of the late Professor E. H. Steinhaus and his students at the University of California and to the former Institute for Insect Pathology (now Forest Pathology Laboratory) in Sault Ste. Marie. This latter institution concentrates its efforts on forest pests. Thus was born the concept of the use of microbial agents for insect control or microbial insecticides (Steinhaus, 1956). Although the viral and bacterial pathogens have been studied most intensively, recent work indicates that mermithid nematodes may be useful in control of mosquitoes and other biting flies and fungi, especially *Coelomyces* sp., may also be developed for control of these pests. To date the mass production of some insect pathogens has required their growth in the living host. Although it is possible to produce large quantities by this method (especially if tissue culture procedures can be improved), a giant step awaits the development of artificial media. Despite these difficulties two microbial insecticides are commerically available, and an insecticide based on *Bacillus thuringiensis* Berliner has been marketed for several years.

Bacillus popilliae (milky disease) is a pathogen of the Japanese beetle and certain related scarabaeid larvae. It is produced in larvae collected in the field, the disease being incubated in these in the laboratory. Heavily infected larvae are then macerated, mixed with an inert carrier, and marketed under the name Doom®. Doom® can be applied to turf to control susceptible larvae, and because the bacterium persists in the soil for many years and reproduces itself with each larval infection, a single treatment is effective for many years. Diseased larvae develop high concentrations of the bacterium in blood, which becomes milky, hence the common name "milky disease." Warm soil temperatures are required if the pathogen is to prove lethal. Milky disease is pathogenic only to some scarabaeid larvae. It is innocuous to other animal life and to plants (Dutky, 1963).

Heliothis virus (Viron H®) is a polyhedrosis virus of the corn earworm and closely related species. Many viruses are pathogenic to insects and have been described as polyhedroses, granuloses, or viruses without inclusion bodies, based on the configuration in which the virus particles (virons) occur in the host tissues (Steinhaus, 1949). The viruses that occur in polyhedra in host tissue are among the most pathogenic, and one that attacks the cell nuclei in *Heliothis* sp. has been sold for use on tobacco and cotton.

Viruses are rather specific and, to date, those showing greatest promise are pathogens of lepidopterous larvae and certain species (sawflies) of Hymenoptera. They have not been registered yet for food crops (except on cotton, which in part, is a food crop). On plant foliage virons within polyhedra become nonpathogenic in a few weeks but in soil retain infectivity for many years (Jaques, 1974).

Superficially the nature of viruses and other insect pathogens would suggest they are ideal insecticides in that they occur naturally and are rather specific. Extensive testing has failed to detect any effects on nontarget organisms. There is a worry, however, in that microorganisms in general have a high mutation rate, and the possibility exists that, especially under conditions of laboratory culture, mutations might arise that are pathogenic to other organisms and to man. It has been argued that this is unlikely (Steinhaus, 1959).

Bacillus thuringiensis Berliner

A number of commercial products (Biotrol®; Thuricide®; Dipel®) based on toxins produced by the bacterium *B. thuringiensis* are marketed as insecticides. These are referred to sometimes as microbial insecticides but are not, for they do not cause a bacterial infection in the insect (except in a few species). To consider these as chemical insecticides of microbial origin is more correct.

Bacillus thuringiensis was described by Berliner in 1915 from infected larvae of the Mediterranean flour moth from a mill in Thuringia, a town in what is now known as Austria. During the 1940s it was grown in culture and proved lethal to the larvae of the alfalfa caterpillar (Steinhaus, 1951). There followed extensive work with this bacterium in many laboratories, and it was found to produce a toxin generally effective against lepidopterous larvae but with little toxicity to other forms of insect or animal life.

B. thuringiensis grows readily on a wide variety of culture media, and as the cultures mature, sporangia are formed, each of which contains a spore and usually a somewhat diamond-shaped "parasporal

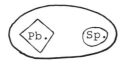

Figure 10.6 Diagrammatic sketch of a sporangium of *Bacillus thuringiensis* containing a spore (Sp.) and parasporal body (Pb.).

body" (Hannay and Fitz-James, 1955) (Figure 10.6). Most of the insecticidal activity of *B. thuringiensis* is due to the δ endotoxin contained in the parasporal body (crystal). Although the precise nature of the toxin has not been determined, it is believed that several polypeptides are involved. Several other toxins are also produced, and though these are insecticidal, they play a lesser role. The chemistry and mode of action of these toxins are discussed in detail by Somerville (1973).

Heimpel and Angus (1959) noted that, in silkworm larvae, feeding ceased soon after ingestion of a mixture of crystals and spores of *B. thuringiensis*. In an ingenious series of experiments in which $BaSO_4$ was incorporated in the food and the insects x-rayed at various intervals after feeding, they demonstrated that within minutes after ingestion of the bacillus culture peristaltic action in the digestive tract ceased. Paralysis in the digestive tract is associated with disruption of the gut lining, a disruption that apparently permits the alkaline juices in the digestive tract to "leak" into the blood. Normally the blood pH is near neutral in lepidopterous larvae and is not highly buffered. Thus leakage from the digestive tract causes a rapid rise in blood pH, an occurence that, whether induced by injected chemicals or leaked digestive juices, brings about gut paralysis (Heimpel and Angus, 1963).

Insecticides based on *B. thuringiensis* (*B.t.*) are effective against a wide variety of leaf-eating caterpillars and are used on vegetable crops. They are nontoxic to nontarget species, including man, and can be used on edible crops right up to harvest. *B.t.* has been used also on the gypsy moth (Kaya et al., 1974) and other lepidopterous larvae on ornamentals and has been tried but with limited success for control of the spruce budworm (Various authors, 1974).

10.4 FUMIGANTS

In the United States the production of fumigants exceeds 180 million lb/year (NAS, 1975). Many of these are highly toxic and nonselective.

They are used extensively by pest control operators for termite control (where buildings are fumigated under vapor-retaining tents) and for rodent and insect control in feed mills, food plants, and food storages. They are used in agriculture in grain storage, in greenhouses, and as soil treatments for some crops. They are used by government and quasi-government agencies to rid plants for export or import of pests, in grain-handling storage and shipping containers, including ocean-going vessels, railway cars, and highway trucks. In addition large quantities of fumigants such as naphthalene and paradichlorobenzene are used in the home to protect clothing and other fabrics from clothes moths and other pests.

Included among the fumigants are HCN, methyl bromide, carbon bisulfide, chloropicrin (often used in combination with others as a "warning gas" since it is highly lacrymal), ethylene dibromide, aluminum phosphide, and others. All are volatile or produce gases at room temperature, and their effectiveness lies in the fact that these gases penetrate throughout a structure or product (e.g., stored grain) and kill pests that are present. Because of high volatility, they are of short persistence, and though they present a hazard to those conducting the fumigation, the nature of the products and their use make environmental contamination of little consequence.

10.5   ACARICIDES

Many of the insecticides are effective also as acaricides. This is especially true in the organophosphorous and carbamate compounds, but this generalization leaves many exceptions. Whereas ethion, demeton, and oxydemeton methyl are excellent acaricides, most of the other organophosphorous insecticides are not sufficiently effective to provide satisfactory field control. Within the carbamates the same situation obtains.

Among the earliest effective acaricides petroleum oils and sulfur have been discussed, as well as the dinitro compounds dinocap and binapacryl. Some of the bis-dithiocarbamates used as fungicides also have acaricidal properties.

Three compounds, dicofol, chlorobenzilate, and chloropropylate, are among the most widely used acaricides, and because they are closely related to DDT, the reader might infer, inappropriately, that they share with DDT the feature of bioaccumulation.

Dicofol (Kelthane®) entered the market in 1955. This compound was discussed briefly earlier. It is a long-lasting acaricide with a rather slow kill of established mite infestations, apparently being more effective against the younger mites. As discussed earlier, dicofol is produced in some insects as a metabolite of DDT, but it has not been shown to convert to DDE. It is nonsystemic and used as a foliar spray or dust in agricultural, nursery, and greenhouse crops; 4 million lb were produced in the United States in 1971 (NAS, 1975).

Chlorobenzilate became available in 1952. It is a yellow solid soluble in most organic solvents but practically insoluble in water. It is moderately persistent on plant foliage. It is hydrolyzed to dichlorobenzilic acid and ethanol. It is used on many fruit and vegetable crops but may be phytotoxic to some cultivars. Chlorobenzilate is used also as a smoke in bee hives for the control of tracheal mites. About 2 million lb were produced in the United States in 1971 (NAS, 1975).

Chloropropylate became available in 1964. It is a white powder soluble in organic solvents but with slight (10 ppm) solubility in water. It is a nonsystemic acaricide, not phytotoxic at normal dosage rates, and used against phytophagous mites on a wide variety of crops. It is moderately persistent and is hydrolyzed, especially under alkaline conditions, to dichlorobenzenilic acid and propanol. Less than 1 million lb were produced in the United States in 1971 (NAS, 1975).

A number of other acaricides are available, and some are included in Table 10.3 as a guide to their structure and generally low mammalian toxicity. Many are esters with chlorinated phenyl-ring structures.

10.6 NEMATICIDES

Nematodes are important agricultural pests, and their control is essential to the production of some crops in some areas. Extensive use is made of halides such as methyl bromide, ethylene dibromide, and chloropicrin; sulfur-containing compounds such as carbon bisulfide, methan, and dazomet; and the organophosphorous compounds zinophos and fensulfothion. These latter compounds are also effective insecticides.

These compounds tend to be nonspecific, and since good and deep penetration of soil is required for control, are used at much higher dosages than insecticides are. As a result soil fumigation frequently results in high mortality, not only of nematodes but also of much of the other animal life in the treated soil. In addition some have fungicidal and herbicidal properties. Thus the life in fumigated soil is likely to be altered drastically.

Table 10.3 Name, Structure, Acute Oral Toxicity to the Rat, and
Dermal Toxicity of Some Acaricides

Name	Structure	Acute Toxicity* (LD_{50} to Mammals)	
		Oral	Dermal
dicofol (Kelthane®): 4,4-dichloro-a-(trichloro-methyl) benzhydrol	Cl-C-Cl structure	87 -500	1931 -3263
chlorobenzilate: ethyl-4,4-dichlorobenzilate	C-O-C₂H₅ structure	700 -3200	>10200
chloropropylate: isopropyl 4,4-dichlorobenzilate	C-O-C-CH₃ structure	5000 -34600	>4000
ovex (Ovotran®): p-chlorophenyl p-chloro-ben-zenesulfonate	S-O structure	2000 -2050	--
fenson: p-chlorophenyl benzenesulfonate	S-O structure	1560 -1740	>2000

The halides used as soil fumigants are highly volatile and disappear rapidly. This is true also for carbon bisulfide. The sulfur-
containing compounds degrade rapidly, and though the organophosphorous nematicides are more persistent, they disappear (with the
exception of fensulfothion) in a few weeks.

Because their activity in soil is transient and their application is
discretionary (i.e., they are placed in the soil in agricultural plantings),

Table 10.3 *(Continued)*

Name	Structure	Acute Toxicity* (LD_{50} to Mammals)	
		Oral	Dermal
tetradifon (Tedion®): p-chlorophenyl 2,4,5-tri-chlorophenyl sulfone		5000 -14700	>1000
tetrasul (Animert V-101®): p-chlorophenyl 2,4,5-tri-chlorophenyl sulfide		3960 -17100	>2000
oxythioquinox (Morestan®): 6-methyl-2,3-quinoxaline-dithiol cyclic S,S-dithio-carbonate		2000 -3000	>2000
dinocap: 2-(1-methyl heptyl)-4,6-dinitrophenyl erotonate		980 -1190	>4700
binapacryl: 2-sec-butyl-4,6,-dinitro-phenyl 3-methyl-1,2-butenoate		136 -225	1010 -1690

*Based on Kenaga and End, 1974 or Martin, 1972; dermal figures may refer to rat or rabbit.

they are unlikely to cause environmental contamination. Recently human health hazards (temporary sterility in males) have been associated with the production of 1,2-dibromo-3-chloropropane (Nemagon®; Fumazone®), and the use of these products is being reviewed.

11

RODENTICIDES AND THEIR USES

Several species of small mammals are pests of the home and in the production of food, fiber, and forest products. While mice and rats are the most destructive and first come to mind, many others are important pests in a variety of circumstances. Squirrels may be undesirable owing to burrowing or entering homes, where they damage a wide variety of items; gophers may interfere with agricultural production; skunks may raid the chicken house or the beehive or dig for grubs in a prize lawn; rabbits and hares may girdle trees in the orchard or forest plantation; the groundhog is destructive in gardens and agricultural plantings, as well as through burrowing activities that interfere with agricultural equipment; the fox may be a predator on poultry, and larger predators such as feral dogs, coyotes, wolves, lynx, and bobcats may cause loss in livestock production.

Whereas some of these "sometimes pests" can be excluded by fencing and good housing and others may be controlled effectively by trapping or shooting, poisoning has been used for many years. Although not all pest animals are rodents, many are, and the term *rodenticide* is used to describe the poisons used for their control.

As discussed earlier, animal systems have many features in common, and chemicals highly toxic to one species are likely to be highly toxic also to many others. As seen earlier, the rat is the test animal used most extensively in toxicological work to predict hazard to human beings. It thus follows that poisons developed to control the rat and its cousins will be toxic to man, his livestock, and the wildlife that share and inhabit his surroundings. As will·be seen in the discussion that follows, some selectivity in toxicity to various animals is found in some compounds, but the indiscriminate toxicity of many makes the use of rodenticides an environmental concern.

11.1 DIRECT POISONS

Strychnine is an alkaloid obtained from the seeds of *Strychnos nux-vomica* and related species of the family Loganiaceae. It can also be synthesized (Woodward et al., 1954). It is a white crystalline powder soluble in some organic solvents and slightly soluble (143 ppm) in water. It is usually available as the hydrochloride, which is water soluble. When used as a rodenticide, it is frequently incorporated in a grain meal or applied to the seed coat of millet or other seed. For predator control it is used in a bait of hamburger or other meat products or sometimes added directly to the carcass of recently killed animals. It is an extremely persistent compound, and baits are toxic over an extended period. Strychnine is highly toxic by ingestion, but its mode of action is poorly understood. In some way it interferes with inhibitory neurons whose function is to counteract certain stimuli in the nervous system. Death comes quickly, preceded by violent convulsions.

Red squill is also of botanical origin, obtained from the bulbs of *Urginea* (=*Scilla*) *maritima*, a member of the family Liliaceae. The most toxic of several glycosides, **scilliroside**, was isolated in 1942 (Stoll and Renz, 1942). It is a yellow crystalline solid soluble in alcohols but almost insoluble in water. Scilliroside is a powerful cardiac glycoside that produces convulsions and respiratory failure in the rat. Rats are especially susceptible (Table 11.1), and though the compound is toxic also to other animals, cats and pigs survived a dosage of 16 mg/kg and fowl survived 400 mg/kg (Barnett et al., 1949). Red squill is reported to have an additional safety feature in that, because of its emetic effect, it is regurgitated by most animals. Mice and rats are unable to regurgitate. It is used as a bait on various seeds but is not very persistent. Chemical methods for its detection are not adequate (Martin and Worthing, 1977).

Sodium fluoroacetate (1080) is the most toxic of the rodenticides. It is a white powder soluble in water but with low solubility in alcohols. The mother compound, fluoroacetic acid, occurs naturally in a number of plants, including *Dichapetalum cymosum* and *Acacia georginae*, plants toxic to animals. It is used as a bait in cereal or in water pans to control ground squirrels, rats, mice, and predators but because of its extreme toxicity is permitted for use only by trained

Table 11.1   Name, Structure, and Acute Oral Toxicity to the Rat of Some Rodenticides

Name	Structure	LD_{50} mg/kg
strychnine		1 -30
scilliroside		0.43 -0.7
sodium fluoroacetate		0.22
antu α-naphthyl-thiourea		6 -8
phosazetim: 0,0-di-(p-chloro-phenyl) N-aceta-midino phosphora-midothioate		3.7♀; 7.5♂

personnel. Sodium fluoroacetate is a powerful inhibitor of respiration, acting on aconitase. Toxicity involves first a lethal synthesis in which the fluoroacetate is converted enzymatically to fluoroacetyl-CoA. This combines with oxaloacetate to form fluorocitrate. Within the tricar-boxylic acid cycle, fluorocitrate inhibits the conversion of citrate to isocitrate, a process catalyzed by aconitase (Peters, 1963).

Table 11.1 (*Continued*)

Name	Structure	LD$_{50}$ mg/kg
norbormide		11 -52
thallium sulfate	Tl$_2$SO$_4$	25
zinc phosphide		45
endrin		10
toxaphene	(CH$_3$)$_2$ ... -C- +Cl$_x$ (various isomers) H$_2$C	100
DDT	Cl—〈 〉—C(H)(Cl-C-Cl with Cl)—〈 〉—Cl	200

Antu is a grayish powder with limited solubility in organic solvents and low solubility (60 ppm) in water. It is especially toxic to the Norway rat but less toxic to other species. It has some emetic properties and is regurgitated by the dog (Martin and Worthing, 1977). Antu causes increased permeability of lung tissues, and death is due to "drowning pulmonary edema" (Metcalf, 1971).

Table 11.1 *(Continued)*

Name	Structure	LD_{50} mg/kg
warfarin: 3-(α-acetonyl- benzyl)-4-hydroxy- coumarin		58♀; 323♂
coumachlor: 3-(d-acetonyl-4- chlorobenzyl)-4- hydroxycoumarin		900 -1200
coumafuryl: 3-(α-acetonyl-4- furyl)-4- hydroxy- coumarin		200
pindone: 2-pivalyl-1,3- indandione		50
diphacinone: 2-diphenyl- acetyl 1,3- indandione		3

Phosazetim (Gophacide®) is a white crystalline powder soluble in chlorinated hydrocarbons but with low solubility in alcohols and water. It acts as an anticholinesterase and is especially toxic to the pocket gopher (LD_{50} 1.0 mg/kg), less toxic to the rat; dogs survived dosages of 20 mg/kg. Because of this it is used as a selective poison bait for gophers.

Norbormide (Raticate®) is a grayish crystalline powder slightly soluble in alcohol with low solubility in most organic solvents and slight solubility (60 ppm) in water. It is a highly selective toxicant to rats, but for most other mammals and birds the acute oral LD_{50} exceeds 1000 mg/kg.

Thallium sulfate, a product used extensively as an ant bait, is moderately toxic to rats and mice and is used as a bait.

Zinc phosphide is especially toxic to mice and is used, usually on cracked corn, as a bait for mice in orchards.

Among the insecticides already discussed, **endrin** and to a lesser extent **toxaphene** have been used as a spray or dust in orchards for mouse control, and DDT is used by pest control operators as a tracking powder for mice and for control of bats.

11.2 ANTICOAGULANTS

Warfarin was the first of a group of rodenticides that owe their effectiveness to their role as anticoagulants. Here they interfere with blood clotting by competing with vitamin K, the prosthetic group in the formation of prothrombin, an essential component in the blood clotting sequence.

Blood clotting is a complicated process depending on the presence of a skeletal structure of fibrinogen. This conversion is accomplished by thrombin, a product dependent on its precursor prothrombin. Thus, when vitamin K is inhibited, the level of prothrombin in the blood is reduced and blood clotting ability impaired. The result is internal hemorrhaging, and it is by this mechanism that the anticoagulants cause death (Metcalf, 1971).

Warfarin is a colorless crystalline solid soluble in alcohols and organic solvents but practically insoluble in water (Martin and Worthing, 1977); 12 million lb of warfarin were produced in the United States in 1971 (NAS, 1975).

Coumachlor (Tomorin®) is a white crystalline solid soluble in alcohols and some organic solvents but practically insoluble in water. The product is marketed as a bait or as a tracking powder. It is highly toxic to dogs and pigs (Martin and Worthing, 1977).

Coumafuryl (Fumarin®) is a related hydroxycoumarin similar to coumachlor.

Pindone (Pival®) is a yellow crystalline solid soluble in most organic solvents and slightly soluble (18 ppm) in water. It was at one time proposed as an insecticide, since it is ovicidal, but the discovery of its anticoagulant activity (Kabat et al., 1944) led to its use as a rodenticide.

Diphacinone is also a yellow crystalline solid closely related to pindone and with much the same clinical and biological properties.

As a group the anticoagulants have moderate to high toxicity to mammals in terms of acute oral LD_{50}'s. Their toxicity lies in repeated exposure, and by daily consumption much greater toxicity is expressed. For example, whereas the acute oral LD_{50} to the rat for coumachlor is 900 to 1200 mg/kg, the LD_{50} for a 2 to 3 week repeated exposure is 0.1 to 1.0 mg/kg/day (Martin and Worthing, 1977). Thus these materials are prepared as baits in foods attractive to mice and rats. In contrast to baits using strychnine or other single dose toxicants, baits used for anticoagulants must remain attractive for several days, and repeated baiting may be required to effect control. Bait "shyness" (avoidance) may develop where quick acting poisons are used and toxic reactions are apparent in the area where the baits are placed. This seldom occurs with the anticoagulants. There have been problems, especially in the Norway rat, because of the development of resistance to the anticoagulants.

12

PISCICIDES AND THEIR USES

Fish farming is becoming increasingly important in North America and throughout the world to augment food supply and to provide recreation to anglers willing to pay for the privilege. Whereas fish farming as such in North America is a relatively recent development, the practice of restocking lakes and streams to augment the natural population of desirable fish species is not. The extent to which this is practiced is indicated by the fact that, in the United States in 1965, 53% (almost 82 million acres) of inland waters were stocked with 1.4 billion fish (Stroud and Martin, 1968).

These massive restocking efforts are in many instances practical only when management practices in lakes, ponds, and streams are conducive to survival of stocks introduced. In many instances nature and some control on harvesting provide acceptable management, but this is not always the case. Some waters may contain such high populations of trash fish that survival and growth of desirable species is poor, and in other areas predatory species may be overly abundant. In the Great Lakes, for example, the sea lamprey has drastically reduced populations of salmonids, and survival of these species in these massive bodies of water requires that the lamprey be controlled.

In addition to these broad problems of competition and predation there are problems with pathogens. In some bodies of water diseases such as kidney disease in trout; infectious pancreatic necrosis, systemic fungal infections, and viral diseases in channel catfish; and tapeworm infections in bass may require that all fish be destroyed to get rid of the disease. Some fish also harbor pathogens of man and may have to be destroyed totally before fish from the infected body of water are safe for human consumption (Lennon et al., 1970).

In North America the application of pesticides to water is controlled rigidly, most of the usage being by government personnel. The compounds that have been used in the United States are discussed by Lennon et al. (1970); these include a number of insecticides highly toxic to fish but not registered for fish control. Included are dichlorvos, azinphosmethyl, endrin, malathion (either by itself or in combination with naled), phosphamidon, endosulfan, rotenone, and toxaphene. These insecticides have been discussed (Chapter 10). With the exception of rotenone none are registered as piscicides. As fish toxicants these insecticides are relatively nonselective and, with the exceptions of endrin, toxaphene, and endosulfan, are nonpersistent. Thus restocking can be undertaken soon (1 month) after the pesticides have been applied. With the persistent insecticides fish survival may be reduced for 1 to several years.

In addition to the insecticides a number of other toxicants have been used and show promise.

Antimycin (Fintrol®) is an antibiotic produced in cultures of *Streptomyces* sp. It is prepared as a liquid concentrate for drip treatment of streams, for spraying on the surface of water, or as a coating on sand particles for distribution. Antimycin is highly toxic to fish, but there is some selectivity among fish species. This selectivity is a function, in part, of inherent toxicity but more importantly is related to the zone of treatment. The Fintrol-5® formulation releases the antimycin within the top 5 feet of water, whereas in the Fintrol-15® preparation, the toxicant is released in the top 15 feet. These sand-coated preparations thus permit a stratification of treatment zone, and since the toxicant is of short persistence, it selectively kills fish that inhabit these zones. The sand formulations are advantageous also where plant growth may hinder effective coverage of the water area with sprays. Under these situations the weight of the sand particle improves penetration to the water below. Antimycin is highly toxic to mammals and acts as an irreversible inhibitor of cellular respiration (Lennon et al., 1970).

TFM (3-trifluormethyl-4-nitrophenol) was developed specifically for control of larvae of the sea lamprey. These larvae develop in the bottom mud in streams where they are free living for 4 to 6 years before moving downstream and entering main water bodies as predators—especially on salmonids. TFM is a crystalline solid formulated as a liquid for controlled application to streams. It is of moderate toxicity to mammals (acute oral LD_{50} to the rabbit 160 mg/kg) and acts on the respiratory system with severe hemorrhaging in the capillaries (Lennon et al., 1970). TFM is used by itself or in combination with niclosamide (Bayluscide®) (5-chloro-N-(2-chloro-4-nitrophenyl) salicylanilide), a molluscicide that synergizes TFM and reduces the cost of treatment. Niclosamide is toxic to fish and zooplankton but of low toxicity to mammals (> 5000 mg/kg).

Squoxin® (1,1-methylenedi-2-naphthol) is a selective piscicide for squaw fishes (Cyprinidae) predaceous on salmonids (MacPhee and Ruelle, 1969). Although it is extremely toxic to other fish species, there is a minimum 10-fold margin of safety between effective dosages for squaw fish and those for other fish. Mammals and birds are reported as surviving low dosages (a few mg/kg/day) for 7 days. It is reported as nonpersistent (Lennon et al., 1970).

In addition to the uses of toxicants for fish control purposes already discussed, there is at times a need to remove fish from reservoirs supplying potable water or for rapid removal of fish in fish hatcheries prior to initiating new stocks. For such purposes Squoxin® is fast and effective and has the added benefits of antibacterial action and short persistence.

In practice piscicides are applied by a wide number of methods. Early attempts using rotenone often involved dragging sacs of the powder through waterways, but more sophistication has evolved. Some applications are made by air, some by motor driven boat either by spraying or by distributing granular material by power- or hand-operated equipment. Sometimes the toxicant is metered into the propeller stream. In stream applications of TFM careful metering devices must be used to ensure that the dosage applied will provide lamprey control but not kill desirable fish species.

13

The need to control some species of birds has always been a problem. Early settlers in America were plagued with "blackbirds" in corn, and the removal by birds of cereal seeds prompted a variety of seed treatments, often with coal tar products, to repel birds but with the effect, in many instances, of reducing germination because of phytotoxicity.

As earlier pointed out (Chapter 4), weaver birds do extensive damage to rice and other crops in Africa and elsewhere, and in North America massive flocks of red-winged blackbirds destroy field corn, ducks damage cereals in western Canada, and a variety of birds, including the robin, sometimes do extensive damage to tender fruits. Birds such as cowbirds, starlings, pigeons, and sparrows feed on grains in storage and may be pests around dwellings and public buildings. A unique problem exists around airports, where birds may collide with aircraft, causing extensive damage and sometimes a crash. Control efforts have varied with the situation involved. Dynamiting congregating sites has been practiced where massive numbers of birds have established roosts, and highly toxic pesticides such as parathion and endrin have been employed in some situations. Spraying large congregations in their roosts with a detergent so that the birds are unable to retain their body warmth has been employed where low night temperatures result in the birds' freezing to death. This practice has not been without public criticism.

To a significant extent bird control has involved baiting with broad spectrum toxicants such as strychnine or zinc phosphide. Although some selectivity can be achieved by careful sizing of the bait (usually cracked corn or weed or cereal seeds) or by bait placement, these methods provide a minimum of discrimination, and desirable as well as pest species are killed.

The search for specific bird toxicants has not been very successful. Methiocarb (Chapter 10) has some repellency and has been used experimentally on corn and on some tender fruits and shows promise for control of robins and other songbirds on blueberries. Birds learn quickly, and methiocarb is distasteful to them. Thus, after sampling a few blueberries treated with the compound, they seek something more pleasing to the palate. To date the compound has not been registered for this purpose, because insufficient evidence has been obtained to demonstrate that the material can be applied in a manner effective for bird repellency without leaving a residue on the fruit at harvest. The problem is that birds are not attracted to blueberries until the fruits are ripe or near ripe, and methiocarb has some degree of persistence. In some countries methiocarb is used successfully for bird repellency as a seed coating.

Fenthion (Baytex®), an insecticide (Chapter 10), has been used for weaver bird control (marketed as Queleton®) in various parts of the world and has been used by pest control operators for pigeon control around public buildings. Fenthion is an effective insecticide of moderate toxicity to mammals but highly toxic to birds. It is a contact and stomach poison; for bird control, use is made of its contact action and its ready absorption through the skin. For bird control it is applied as a paste to roosting areas. This application technique provides a degree of selectivity.

Avitrol® (4-aminopyridine) has been available for many years and is registered for use in the United States and Canada. It is used as a bait on sized-cracked corn; by sizing of the bait a degree of selectivity is obtained. Its use to date has been primarily against red-winged blackbird on corn in the field, for pigeon and sparrow control around public and commercial buildings, and for control of various birds around livestock feeding pens. In this last use some degree of selectivity in birds affected is achieved through placement of baits.

Avitrol® is highly toxic to mammals (acute oral LD_{50} to the rat 29 mg/kg) and more toxic to birds. Birds poisoned with the product become hyperactive, flutter violently, and emit distress calls that disperse other birds, especially those of the same brood, for whom the distress calls have particular significance. This expression of toxicity, in theory at least, is ideal for dispersing birds such as the red-winged blackbird, which move in large flocks and whose sheer numbers mean extensive local damage when they alight to feed in a field of corn in the susceptible milk stage.

In practice Avitrol® is applied to cracked corn at a rate that makes one particle of the treated corn an effective dose. The treated corn is then mixed with untreated cracked corn at about 3% and distributed in the field. Repeated applications are sometimes recommended. Thus in the commercial product, Avitrol®, only 3 bait particles in 100 contain the toxicant, and thus only a small percentage of birds in the flock will be poisoned. Thus the principle in its use is dispersion of flocks rather than their destruction.

There is great argument among research workers and agriculturists concerning the effectiveness of the product, and experimental evidence is not convincing. Dyer (1968) found Avitrol® less effective than some other methods (e.g., patrolling with a jeep and shotgun) in reducing field damage, whereas other workers report more satisfactory results (Goodhue and Baumgartner, 1965; DeGrazio et al., 1971, 1972). The concept is that, if the birds learn that corn fields are unsafe (as evidenced by distress of their siblings), they will abandon corn fields and seek food elsewhere. The contrary argument that seems equally well supported is that the birds will learn only that some corn fields are unsafe (those in which distress has been seen) and will migrate to other fields a few miles down the road. If the former argument is correct, then treatment of selected fields in a corn-growing region is good economics, whereas if the latter argument pertains, treatment of all fields is required and the economics becomes questionable indeed. Surveys of damage indicate millions of dollars lost to birds each year, but as a percentage of the crop in large areas this represents a small fraction (less than 2%) (Dyer, 1968), a loss that would not cover the cost of protection if a majority of the acreage must be treated.

14

PESTICIDES IN SOIL

$\mathbf{A}$ few widely used pesticides are persistent, and some are relatively nonbiodegradable. These reside in our environment for significant periods. Even for the less persistent pesticides some residency time is involved, and questions arise as to where these pesticides persist in our environment and what factors determine the site and duration of such residency.

It is clear that the distribution and fate of pesticides are determined by a host of variables that include the nature of the pesticide and the many factors that determine the environment in which it is found. An attempt is made to depict some of these factors in Figure 14.1 and to define their significance more precisely in the chapters that follow.

14.1 SOURCE OF PESTICIDES

Whether or not pesticides are present in soil is not a question. They are, and in the case of a few such as DDT, copper, and arsenic, they are present in relatively high concentrations in localized areas. They arrived there in a variety of ways only some of which were planned.

Intentional Application

Direct application to the soil surface, incorporation in the top few inches of soil, or application to crops are the routes by which most high concentrations of pesticides reach the soil. These are intentional,

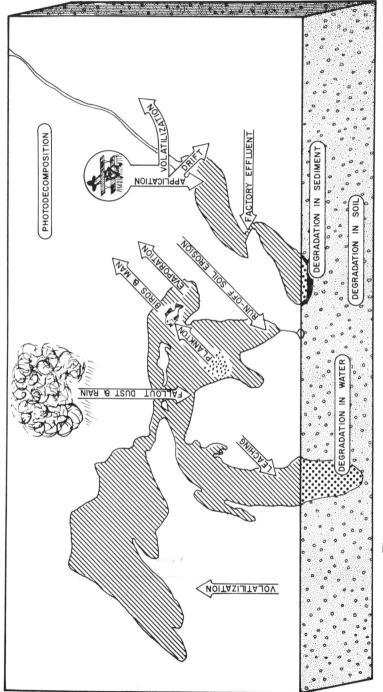

Figure 14.1 Pesticides in the environment and some factors important in their fate.

Table 14.1 Total Acreage of Land in the Province of Ontario, Acreage and Percentage of Acreage Devoted to Agricultural Production, 1969[a]

Land Utilization	Acreage	Percent of Total
Total in province	220,218,880	100
Commercial farms	13,229,561	6.0
Field crops	7,559,000	3.4
Tobacco	120,000	0.05
Vegetables	121,489	0.05
Fruit	77,869	0.03

[a]Harris and Sans, 1971.

designed to achieve control of a pest or pests in a particular setting. These direct applications place relatively high concentrations of pesticides in the soil, but pesticides are applied to only a small percentage of the total land mass. Figures for the United States (NAS, 1971) indicate that, in terms of millions of acres treated for agricultural purposes, 158 are treated with herbicides, 57 with insecticides, and about 8 million with fungicides each year. This represents about 25% of the total land devoted to agricultural production (including pasture) and less than 1% of the land mass. A smaller percentage is treated in Canada.

Harris and Sans (1971) show that fruit and vegetable production occupies 0.08% of the land area of the province of Ontario, and tobacco production, 0.05% (Table 14.1). Since these are the only areas where pesticides are used widely, direct application is thus made to only 0.13% of the land area in the province and to only 2.3% of the land devoted to commercial agriculture. In other areas use is more intensive. In Illinois 84% of the land area is in commercial farms, and pesticides are applied to about 50% of this, or more than 14 million acres (Metcalf and Sanborn, 1975). Figures for forest use of pesticides reflect a low percentage receiving pesticides directly. In the massive aerial spray programs against the spruce budworm, less than 15 million of the 550 million acres of Canadian forests ever received treatment with DDT. In the United States annual use of pesticides in forests is much less than 1% of the forested acreage.

Unintentional Applications

Although the land area receiving direct, intentional application of pesticides is small, large amounts of pesticides reach the soil through drift during pesticide application and through atmospheric fallout.

Drift

Despite the best intentions and the exercise of discretion when pesticides are applied, small to large portions of that being applied are likely to be carried by the wind to nontarget areas. The extent of drift approaches zero where pesticides are being applied in granular form and/or injected into the soil, but as much as 50% or more may be lost where spraying techniques are employed. This is especially true where application is to a high canopy such as an orchard, where the spray stream is directed upward (at least in part) or in aerial applications. The extent of drift to nontarget areas is determined by a host of variables, including air movement, droplet sizes in the spray steam, temperature, and the size of the area being treated. Thus in western Canada, where 2,4-D is used extensively in a vast contiguous acreage of cereals, drift to areas where treatment is not intended is much less hazardous than in eastern areas, where a 20 acre treated orchard may be surrounded by areas where orchard pesticides are contaminants. The drift from ground spraying with a boom a few inches above a plant canopy of alfalfa, potatoes, or cotton and spray nozzles directed downward is much less than from an aircraft flying several feet above the plant canopy or several hundred feet above a forest or city.

The importance of droplet size is indicated by data in Table 14.2. These data illustrate that, even under conditions of a very light breeze (3 mph), small droplets are carried long distances. Many spray applications produce droplets in the 30 to 300 micron range, and in practically all operations some droplets in the 5 micron range are produced. Thus drift is assured. Relative sizes of some droplets are indicated in Figure 14.2 to give some perspective of droplet size emitted by com-

Table 14.2   Drift of Spray Droplets of Differing Sizes in a 3 mph Air Current when Falling 10 Feet[a]

Droplet Size (microns)	Drift (ft)
450	8.5
150	22.0
100	48
50	178
20	11,000
10	44,000
2	21 miles

[a]Akesson and Yates, 1964.

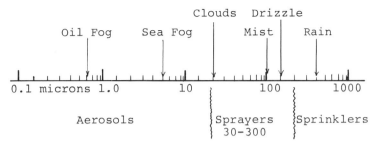

Figure 14.2 Droplet sizes in various types of natural and man-made systems (Fisher and Hikichi, 1971).

mercial sprayers in relation to droplets with which we may be more familiar.

Direct extrapolation from Table 14.2 does not predict accurately the drift pattern from a spray application, for other factors affect droplet size after release from the spray orifice. Once released, each droplet is subject to loss of water, and this loss is a function (in part) of relative humidity. Brann (1965) showed that under conditions of high relative humidity a droplet 125 microns in size lost 12% of its volume by evaporation while traveling 36 ft from its release point and that under low relative humidity 75% of its volume was lost while it was traveling the same distance. Thus water lost by evaporation reduces the size of spray droplets as they are transported downwind from the point of release, and this loss in size means further dispersal for the now smaller droplet.

A number of studies (Akesson and Yates, 1964) have been done in which deposits have been measured at various distances downwind from the point of application. These show a relatively heavy deposit immediately adjacent to the treatment swath with a rapid decline as the distance from treatment increases. This is to be expected, since large droplets drift only a short distance in comparison to those of very small diameter, and the amount of pesticide carried per droplet increases dramatically with the larger droplets. For example, one droplet 1000 microns in diameter carries the same pesticide load as eight 500 micron droplets, sixty-four 250 micron droplets, or one thousand 100 micron droplets. Thus spray drift in biologically significant amounts is confined, usually, to a narrow band immediately adjacent to the treatment zone.

Whereas some studies have concentrated on drift from a specific spray operation, others have looked at soil pesticide levels downwind from a general agricultural area. Tests in Arizona showed that, whereas total DDT residue levels in soil were 6.7 ppm within a few meters of a cotton field, residues declined sharply as the distance from the cotton field increased and were in the 0.1 ppm range at 100 meters and in the 0.01 ppm range at 10,000 meters from the treatment area (Laubscher et al., 1971). In some studies (Decker et al., 1950; Hindin et al., 1966; Akesson and Yates, 1964; and others) attempts have been made to account for the total pesticide released from the sprayer. In general relatively large portions, sometimes exceeding 50% of the released dosage, are unaccounted for in deposits on foliage or that trapped by physical means within a few feet of the point of discharge. With volatile pesticides the unaccountable portion is much higher than with a pesticide of low volatility.

Spray drift of fungicides or insecticides would be largely un-noticed if it were not for programs to monitor pesticides in soil, water, air, and food on a regular basis. Herbicide drift is an entirely different situation. Many high value horticultural crops are highly sensitive to the hormone herbicides. Thus vapor drift and spray drift of 2,4-D and related herbicides have long been regarded as a major problem. More recently the increased use of contact herbicides such as diquat, dinoseb, and paraquat as crop desiccants has also led to damage claims.

Atmospheric Fallout

Beginning in the early 1960s, a number of studies have demonstrated the presence of pesticides in rain water and/or snow. Some of the pesticide is present on small particles contained in the rain droplets and some is dissolved in the droplets. A compilation of reported data (partially reported in Table 14.3) includes wide variations in amounts found in various studies, but when comparable areas are considered (e.g., rural England vs rural North America), values are in surprisingly good agreement. Variations are to be expected. Sampling in England for a 28 month period showed residues of p-p'-DDT ranging from 20 to 130 ng/liter at different seasons in Cornwall and from 25 in Cornwall to 190 ng/liter in Kent for the same season. Similarly large variations were also seen in levels of Gamma-BHC (10 to 230 ng/liter) and dieldrin (1 to 35 ng/liter) (Tarrant and Tatton, 1968). Seasonal use patterns suggested an explanation for some of the variations observed but not all. High levels in March 1966 could not be explained on the basis of field usage of the pesticides at the time of year, and the suggestion was advanced that low rainfall during the period might explain the high levels detected (Wheatley, 1973). Some of these findings were confirmed by a similar study conducted in Ohio (Cohen and Pinkerton, 1966).

Although a number of pesticides may be present in rainfall, it is doubtful that the levels are sufficiently high to constitute a major

Table 14.3 Pesticides in Rainwater—Some Levels Reported[a]

Pesticide	Area	Level ng/l	Reference
DDT[b]	England	3	Wheatley & Hardman, 1965
	England	470	Abbott et al., 1965
	England	46	Tarrant & Tatton, 1968
	California	5	Swift, 1971
	Hawaii	4	Bevenue et al., 1972
	Florida	1,000	NAS, 1971
	Ohio	340	Cohen & Pinkerton, 1966
	Florida	460	Kolipinski et al., 1971
	Antarctic (snow)	4	Peterle, 1969
	Ontario (early winter snow)	43	Frank et al., 1974
	Ontario (late winter snow)	4	Frank et al., 1974
Dieldrin	Ontario (early winter snow)	10	Frank et al., 1974
	Ontario (late winter snow)	0.7	Frank et al., 1974
	England	40	Tarrant & Tatton, 1968
	England	95	Abbott et al., 1965
BHC[c]	England	175	Abbott et al., 1965
	England	260	Tarrant & Tatton, 1968
	USA	70	Cohen & Pinkerton, 1966

[a]Table includes mostly highest levels reported, see text.
[b]Includes metabolites.
[c]Includes several isomers.

source of soil contamination. Edwards (1973) calculated possible deposit of DDT in rainfall on the basis of an annual precipitation of 50 in. and a DDT level of 210 ng/liter in rain. This gave a value of 0.0023 lb/acre, but he suggested this was a maximum value and that the actual value would be less, perhaps about 0.001 (misprinted as 0.01 in Edwards, 1973) lb/acre/year. Ware et al. (1974) showed that the half-life of DDT in soils may approach 20 years, and since the mean levels of DDT in agricultural soils in the United States are about 0.01 lb/acre, it would appear that Edwards' estimate for rainfall deposit is high.

Pesticides may reach soil by direct deposit of atmospheric dust. A number of studies show that much of the pesticides contained in rainwater is adsorbed on particulate matter, and the fact that rainfall in large industrial cities is likely to contain higher than average amounts of pesticides is related to the particulate air pollution in these cities. This particulate matter is settling constantly on the soil surface, carrying with it a small but measurable amount of pesticide. Occasionally such fallout can be dramatic. Cohen and Pinkerton (1966) describe a dust storm that originated in the high plains area of Texas and New Mexico and sent a dust cloud that blanketed much of the eastern part of the United States on January 26, 1965. This dust storm was so severe that, in Lubbock, Texas, a rain gauge was filled with sand to a depth of 3 in. At a collection area of 1090 ft^2 in Cincinnati, Ohio, 175 g of soil were collected and found to contain DDT + DDE (0.8 ppm); chlordane (0.5 ppm); ronnel (0.2 ppm); heptachlor epoxide (0.04 ppm); 2,4,5-T (0.04 ppm); dieldrin (0.003 ppm); sulfur (0.5 ppm); and arsenic (26 ppm). The authors assumed that the pesticides originated with the soil in the high plains area, but the possibility that some of it might have been picked up by the soil as it was blown hundreds of miles through the atmosphere cannot be ruled out.

Normal dust deposits are certainly less dramatic, the average dustfall for Cincinnati being about 180 tons/mi^2/year. On the assumption that such dust carried 1 ppm pesticide, this would represent an annual deposit of approximately 0.3 g/acre. This amount seems insignificant indeed but is brought into perspective when we realize that some seed treatments that add this amount of pesticide per acre have been banned.

14.2 FATE OF PESTICIDES IN SOIL

Many factors are known to influence the behavior and fate of pesticides after contact with soil (Figure 14.3). These include (1) adsorp-

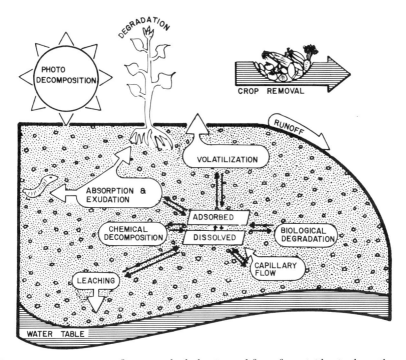

Figure 14.3 Processes influencing the behavior and fate of pesticides in the soil environment (redrawn and modified from Weber et al., 1973).

tion to clay and organic matter, (2) leaching with the downward percolation of water, (3) volatilization to the atmosphere, (4) uptake by soil organisms or plants, (5) movement with runoff water or eroded soil, (6) microbial degradation, (7) chemical degradation, and (8) photolysis. Of primary importance is the chemical nature of the pesticide and the soil type. If we are to predict possible environmental effects or if we are to maximize the effectiveness of soil-active pesticides, the physical and biological processes in soil must be understood. From an environmental standpoint it is most important to understand adsorption and desorption phenomena, for these influence most other processes determining the eventual fate of a pesticide. From a pest control standpoint with soil-active pesticides, the effectiveness, rates, and frequency of application are dependent primarily on sorption phenomena and the tendency of the pesticide to be degraded.

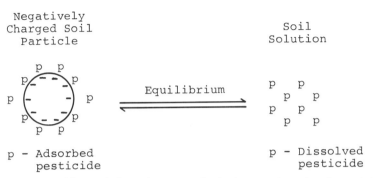

Figure 14.4 Equilibrium balance between adsorbed pesticide molecules and those dissolved in the soil solution.

Adsorption

Pesticides, like other chemical molecules, have varying tendencies to be adsorbed or attracted to clay or organic matter particles or dissolved in the soil solution. For the most part the adsorption sites on clay or organic matter are negatively charged and constitute the "cation exchange capacity" of a particular soil. For each pesticide, for each soil type, and for each set of soil conditions, a different equilibrium is established between the amount of pesticide adsorbed and the amount dissolved in the soil solution (Figure 14.4).

Soil Type

Soil type is one of the most important factors influencing pesticide sorption equilibria. Of special importance are the clay and organic matter, for these are colloidal and have high cation exchange capacities and high surface areas. Clay and organic matter content in soil can vary from less than 1% in sand to well over 50% in heavy clay or peat soils. Adsorption of pesticides to the negatively charged sites on clay or organic matter can occur by dipole-dipole attraction, hydrogen bonding, or by actual ionic binding if cationic pesticides are involved (Bailey and White, 1964). Thus higher rates of pesticide application are required for effective soil pest control if soils are high in either clay or organic matter, since, in these soils, much of the pesticide is adsorbed and not active.

Nature of the Pesticide

The chemical structure of a pesticide determines sorption equilibria by influencing its direct affinity for the clay or organic matter or by influencing its solubility or affinity for the soil solution. Many investigators have noticed a general but not precise inverse correlation between pesticide solubility and adsorption. Ward and Upchurch (1965), studying adsorption of a series of related carbamate and aniline compounds, found that 60% of the variation in adsorption could be attributed to the effects of structure on solubility, while the remaining variation was due to the effects of structure on the direct affinity of the chemical for the adsorbent. This inverse relationship between solubility and adsorption is more likely to be true within groups of chemically similar compounds.

In addition to many other factors the persistence of pesticides in soil is influenced by the way they are formulated for application. Granular formulations are usually the most persistent. Wettable powder and dust formulations are often less persistent than emulsifiable preparations.

There is currently considerable interest in the development of "slow release formulations" for soil-active pesticides. Overall persistence would of course be increased, but the objective is to obtain season long pest control with fewer applications and less total use of pesticide.

Soil Moisture Content

One would expect more pesticide to be adsorbed when soils are dry than when they are moist, since decreases in soil moisture should shift sorption equilibria toward greater adsorption. This is true in moderately light to very light soils but not in heavy soils. In a mathematical study Green and Obien (1969) established that the impact of fluctuations in available soil moisture on the amount of atrazine dissolved in the soil solution was inversely related to the adsorption capacity of soil. Their observations have been confirmed by Harris and Mazurek (1966), who found that the addition of moisture caused the release of pesticides adsorbed on mineral soils but not on muck soils.

Water molecules are themselves polar. Thus, when more water is added to a pesticide-soil complex, the water molecules begin to compete with the pesticide molecules for adsorption sites on the soil colloids and force more of the pesticide into solution.

Soil pH

Most soils have a pH in the 4.5 to 8.0 range. The fate of pesticides in soil varies with pH differences, primarily because of the influence of pH on sorption phenomena. Soil pH also affects chemical degradation. Within a group of similar soils differing mainly in their acidity or alkalinity, pesticide adsorption is usually higher in the more acid soils. The soil colloids are essentially a negatively charged, cation-exchanging environment. Within normal ranges slight increases in soil acidity may convert pesticides from negatively charged anions to uncharged molecules or even to positively charged cations and thus dramatically increase their adsorption. When, however, soils are extremely acid, most of the cation exchange sites are occupied by hydrogen cations, and pesticide adsorption is very low owing to the lack of negative sites. Adsorption is also very low at extremely high pHs, particularly for pesticides that can become anions, because negatively charged anions are not attracted to negatively charged sites in the soil.

Soil Temperature

Pesticide adsorption in soil is an exothermic process (Bailey and White, 1964). When hydrogen bonds or ionic bonds are formed, heat is given off. Thus, when the soil temperature increases, the input of heat can break some of these bonds and cause the desorption of some pesticide molecules. At higher temperatures the greater solubility of pesticides also results in further shifts in the sorption equilibrium toward more pesticide's being available in the soil solution.

Leaching

Solubility is one important aspect related to pesticide persistence in soil; within limits, pesticides highly or moderately soluble in water are leached from soils more quickly than those that are less soluble. Thus the highly insoluble organochlorine insecticides such as DDT and dieldrin tend to persist, are not readily leached, and are found dominantly in the upper few inches of agricultural soils. In studies in six apple orchards in western New York, some containing as much as 259 lb DDT/acre, 70% of the DDT residues were in the upper 2 in. of soil, despite the fact that little DDT had been applied in the 13 years immediately preceding the sampling (Kuhr et al., 1974).

For many soil-active herbicides, some rainfall and some leaching down into the soil are required for effective pest control. In years with excessive amounts of rainfall, highly soluble, poorly adsorbed pesticides may be leached excessively and require reapplication for effective pest control near the soil surface. Under these conditions injury to deep-rooted crops can occur from the excessive leaching of some soil-active herbicides. Whereas the leaching of pesticides down as far as the water table is a possibility, there is currently little evidence that this occurs. In tiled fields, however, movement after leaching into the tiles can be significant.

Movement with Runoff Water

As water moves laterally over a sloping soil surface, pesticides can be desorbed and moved with this runoff water (Bailey and White, 1964). Like the leaching of pesticides, lateral movement with runoff water is likely to be greatest under conditions in which pesticides are the least likely to be adsorbed.

Movement with Eroded Soil

When soil particles are themselves moved by either wind or water erosion, adsorbed pesticide molecules may be transported along with them. No soil-applied pesticide is immune to this type of transport, but the longer a pesticide remains adsorbed at or near the soil surface without degradation, the more likely that movement by soil erosion will be important. As discussed earlier in this chapter, dust storms can be a significant factor in the long distance transport of adsorbed pesticides.

Volatilization

Several factors influence the tendency of pesticides to volatilize and leave soil as a vapor. The structure of the chemical is important because this determines its vapor pressure, as well as its solubility in soil water and its tendency to be adsorbed. Cool, dry conditions in soils with high organic matter or clay content normally result in very little loss of even the most volatile chemicals from the soil, since they are adsorbed tightly. Conversely, warm, moist conditions bring about greater desorption and greater volatilization losses. In a series of studies Harris and coworkers used crickets to assess the volatility of a number of pesticides when incorporated in moist, sandy loam. They reported little volatilization of DDT, γ-chlordane, diazinon, dieldrin, endrin, heptachlor epoxide, carbofuran, or metalkamate but marked evidence of volatility with aldrin, heptachlor, chlorpyrifos, fonofos, phorate, diazinon, and thionazin (Harris and Lichtenstein, 1961; Harris and Mazurek, 1966; Harris, 1970; Harris, 1973). These findings correlate well with reported vapor pressures for the pesticides involved (Martin, 1972). Because of volatility, dinitroaniline and thiocarbamate herbicides (e.g., trifluralin and butylate) have to be applied as preplant soil-incorporated treatments. Spencer and Cliath (1974) have trapped and identified vapors of trifluralin above fields treated with this herbicide. Loss rates were as high as 40% of that applied if the soil was warm and moist. Some investigators (Ashton and Sheets, 1959) have reported that volatile herbicides can be more effective in soils with higher organic matter contents, because the greater adsorption prevents losses by volatility more than it reduces the soil activity of the herbicide.

Microbial Degradation

The primary microorganisms in soil are algae, fungi, actinomycetes, and bacteria. Most of these are dependent on organic compounds for energy and growth. When an organic pesticide is added to soil and reaches an equilibrium between the soil colloids and the soil solution, any molecules in the soil solution are immediately attacked as potential energy sources. If the chemical is readily available in the soil solution, any organisms that can adapt to it as an energy source rapidly increase in numbers until they have completely degraded it. After this they may decrease in number. Soil organisms respond in this manner to 2,4-D (Loos, 1975) and to most short-lived pesticides. Such pesticides are characterized by a logarithmic or sigmoid disappearance curve (Figure 14.5).

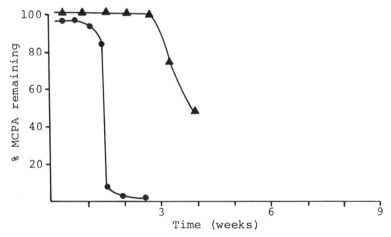

Figure 14.5 Degradation of MCPA in a mineral salts medium inoculated with soil from a field experiment with no previous application of MCPA (▲) and with MCPA applied 1 year previously (●), illustrating the logarithmic disappearnace of MCPA and the shorter persistence in soil with "preadapted" microorganisms. (Redrawn from data by Torstennson et al., 1975.)

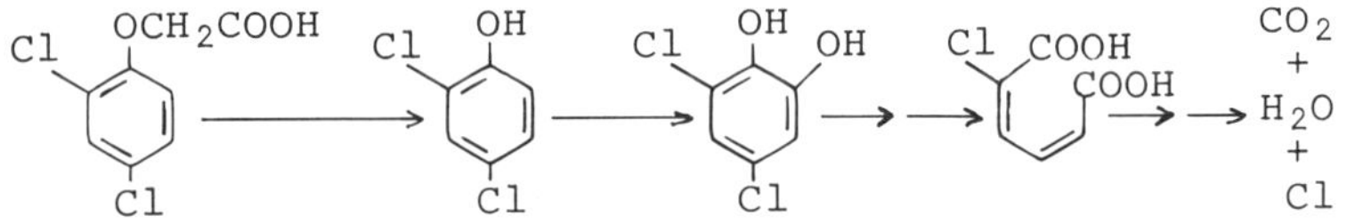

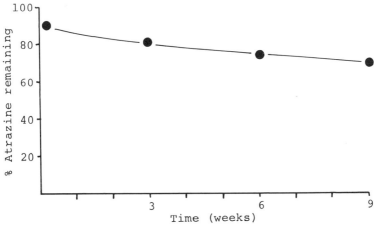

Figure 14.6 Pathway for the microbial degradation of 2,4-D (summarized from Loos, 1969).

Microbial degradation can be rapid and complete. For example, 2,4-D is degraded completely to CO_2, H_2O, and chlorine (Loos, 1969) and has an average half-life in soil of only 2 to 3 weeks (Figure 14.6).

If pesticides are only slightly soluble, tend to be adsorbed in soil, or cannot be adapted to as an energy source for microorganisms, no large population of degrading organisms develops, and the disappearance of the pesticide follows first-order kinetics. Atrazine is an example of a moderately persistent pesticide that has this type disappearance pattern in soil (Figure 14.7) (Kaufman and Kearney, 1970). (See also Chapter 9.)

Any factors that encourage the growth of degrading microorganisms or that increase the availability of pesticides in the soil solution enhance the disappearance of the chemical. Soil organic matter can have slightly opposing effects on pesticide persistence. On the one hand it may decrease the availability of the pesticide, but on the other hand it could improve conditions for the growth of the microor-

Figure 14.7 Rate of atrazine degradation in nonsterile soil at 21°C (redrawn from data by Anderson, 1971).

ganisms. Most other conditions, such as warm temperatures, adequate soil moisture and aeration, unextreme pH, and adequate fertility, encourage microorganisms and increase the desorption and availability of pesticides. The nature of the pesticide itself is important in terms of adsorption, solubility, and microbial degradation. The degree of chlorination often has a large impact on the biodegradability of a pesticide. Among closely related phenoxy herbicides, 2,4,5-T with 3 chlorine substituents is more persistent than 2,4-D with 2 chlorines (Loos, 1975). A similar comparison can be made for the very short residual natural pyrethrins and the chemically similar but more persistent chlorinated pyrethroids (Elliot, 1977). Picloram with three chlorines on a pyridine ring is a highly persistent herbicide (Foy, 1976), while hexachlorobenzene, a fungicide with 6 chlorines (C_6Cl_6), and mirex, an insecticide with 12 chlorines ($C_{10}H_8Cl_{12}$), are both highly resistant to biodegradation (Metcalf and Sanborn, 1975).

Kearney and Kaufman (1975, 1976) have summarized the degradation pathways for most herbicides, and Menzie (1969) and Guenzi (1974) are excellent references, providing data on a wide range of pesticides.

Chemical Degradation

Chemical reactions in soil can destroy the activity of some pesticides and activate others. Whereas adsorption normally decreases the rate of microbial degradation, it may enhance the chemical degradation of some compounds. For example, Armstrong and Chesters (1968) showed that the chemical hydrolysis of atrazine to hydroxy atrazine can be catalysed by adsorption in soil. In chemical degradation pH is an important factor, but the exact influence of high or low pH varies for different pesticides. Among herbicides the chemical degradation of atrazine occurs faster at a low pH (Armstrong and Chesters, 1968). Diazinon, an organophosphorous insecticide, is also broken down more rapidly under acid conditions (Getzin, 1968), but the reverse is true for malathion. Carbofuran, a carbamate insecticide, persisted much longer than expected in acid soil under actual field conditions, a situation that resulted in poisoning of ducks in British Columbia. In this incident granular carbofuran had been applied to rutabagas at the beginning of the season, but ducks picked up toxic granules as late as November.

In the case of organomercury fungicides, some investigators (Booer, 1944) have suggested that chemical conversion from the organo forms to the biologically toxic mercury occurs by a base exchange reaction involving an organomercury-clay complex as an intermediate. Thus in this case a higher clay content might actually result in greater chemical degradation but more pesticidal activity.

Photodegradation

Few organic pesticides are completely resistant to photolysis, but this is probably not a major means of pesticide inactivation or disappearance in soil. Dinitroaniline herbicides such as trifluralin may be broken down in sunlight (Probst et al., 1975), but their application as soil-incorporated treatments prevents excessive exposure to light. When, however, immobile chemicals such as simazine (Comes and Timmons, 1965) are applied and there is little rainfall to leach them into the soil, loss of effectiveness due to photolysis is often suspected.

Uptake of Pesticides by Higher Plants

Within limits, all those soil factors that lead to less adsorption, greater leaching, and greater microbial degradation also result in greater availability for uptake by higher plants and greater pesticide effectiveness. The only exception to this may be the soil-active, volatile herbicides such as the dinitroanilines and thiocarbamates.

When higher plants absorb pesticides from the soil, the absorbed pesticide may be either degraded in the plant or removed when the crop is harvested. This may or may not be significant to the next consumer, but this form of removal does not normally have great significance in the reduction of soil residues. One practical exception to this has been the use of corn to hasten the reduction of atrazine residues in a soil prior to the planting of sensitive ornamental plants (Klingman and Ashton, 1975).

Influence of Cropping Practices

Different cultural practices are likely to influence pesticide persistence, but the way in which this influence will be manifested may be less than predictable. Edwards (1973), citing some specific data, concludes that pesticides persist longer when thoroughly mixed in the soil than when not. This could reflect loss of pesticides by surface erosion and volatilization. When soils containing DDT and aldrin were disced daily for 3 months, both insecticides disappeared more rapidly than when comparable soils were disced only once (Lichtenstein and

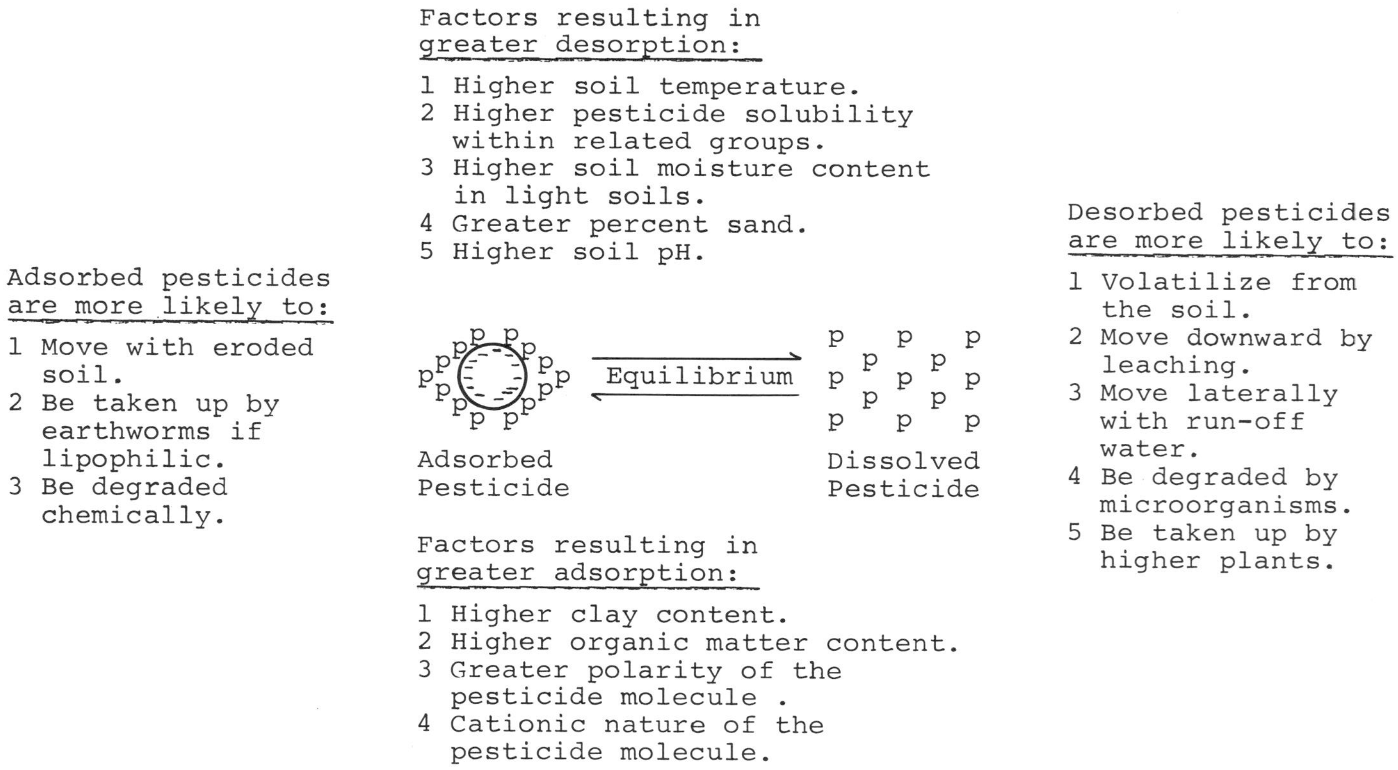

Factors resulting in
greater desorption:

1 Higher soil temperature.
2 Higher pesticide solubility
  within related groups.
3 Higher soil moisture content
  in light soils.
4 Greater percent sand.
5 Higher soil pH.

Desorbed pesticides
are more likely to:

1 Volatilize from
  the soil.
2 Move downward by
  leaching.
3 Move laterally
  with run-off
  water.
4 Be degraded by
  microorganisms.
5 Be taken up by
  higher plants.

```
              p  p  p
pp pp      p  p  p  p
 pp  pp p     p  p  p
pp  pp pp     Equilibrium ⟶
 pp  pp p
pp pp
              Dissolved
              Pesticide
Adsorbed
Pesticide
```

Factors resulting in
greater adsorption:

1 Higher clay content.
2 Higher organic matter content.
3 Greater polarity of the
  pesticide molecule.
4 Cationic nature of the
  pesticide molecule.

Adsorbed pesticides
are more likely to:

1 Move with eroded
  soil.
2 Be taken up by
  earthworms if
  lipophilic.
3 Be degraded
  chemically.

Figure 14.8 Interrelationship of processes influencing the fate of pesticides in soil.

Schulz, 1961). This might appear to contradict the generalization that thorough mixing improves persistence, but against such an interpretation is the obvious fact that the repeated discing exposed new surface residues for volatilization and may have dramatically affected the soil moisture and temperature conditions. It should be remembered, however, that, in the case of DDT, the highest residues are found in orchard soils. In this case incorporation has not been practiced, a high percentage of the total load is in the top 2 in. of soil, and the rate of degradation is indeed slow (Kuhr et al., 1974).

Interrelationship of Soil Processes Influencing Pesticides

From the foregoing discussion it is obvious that there are many different processes that influence the movement, persistence, and activity of pesticides in soil. It is also obvious that there are countless ways in which these processes can interact. Figure 14.8 illustrates some of the many possible interrelationships, with particular emphasis on the factors that influence adsorption-desorption phenomena.

14.3 PESTICIDE RESIDUES IN SOILS

Some of the high values reported for pesticide residues in soil are included in Table 14.4, and with a few obvious exceptions these values reflect direct application to the area from which the soil samples were obtained. Only maximum values are included in this table, and it should not be inferred that these reflect average levels. Quite to the contrary, such high values are unusual. Mean values, although found in many reports, are not very useful, since the number of sites sampled in most studies is quite small and the selection of sites biased frequently to obtain data on a particular use pattern rather than on a random basis. These figures (Table 14.4) give some idea of how high-use patterns are reflected in high residues of the more persistent pesticides. For example, whereas dieldrin and DDT are both highly persistent, DDT levels in vegetable-growing areas are much higher, reflecting the fact that DDT was used as a foliar pesticide with multiple applications per season. By contrast dieldrin was used primarily as a soil insecticide with only one application per season, and, depending on the crop rotation involved, it was applied only every second or third year. A similar application pattern explains the high levels of DDT, arsenic, or copper reported in some orchards and vineyards.

The correlation of pesticide residues and cropping practices can

Table 14.4   Some Maximum Levels of Pesticides Reported from
Various Soils in Canada and the United States in Relation to Crop

Pesticide	Crop	ppm	Reference
Arsenic	Orchard fruits	219	Stevens et al., 1970
	Orchard fruits	124	Miles, 1968
	Orchard fruits	830	Woolson et al., 1971
	Alfalfa	6	Miles, 1968
	Cereals	7	Miles, 1968
	Corn	9	Miles, 1968
	Sugar beet	9	Miles, 1968
	Tobacco	7	Miles, 1968
	Pasture	4	Miles, 1968
	Non-cropland	4.2	Wiersma et al., 1971
	Cropland	5.3	Wiersma et al., 1971
	Vegetables	27	Harris & Miles, 1975
Copper	Grapes	130	Taschenberg et al., 1961
DDT	Orchard fruits	245	Stevens et al., 1970
	Orchard fruits	116	Lichtenstein, 1958
	Grapes	42	Taschenberg et al., 1961
	Vegetables	48	Harris & Mazurek, 1966
	Vegetables	124	Wiersma et al., 1972
	Vegetables	30.4	Brown et al., 1975
	Alfalfa	5	Ware et al., 1971
	Corn	3.7	Harris et al., 1966
	Cereals	4.6	Harris et al., 1966
	Sugar beet	2	Harris & Mazurek, 1966
	Tobacco	5	Harris & Mazurek, 1966
	Turf	87	Fahey et al., 1965
	Pasture	3	Harris & Mazurek, 1966
	Forest	21	Yule & Tomlin, 1971
	Desert	3	Ware et al., 1971
	Tundra	0.15	Brown & Brown, 1970
Dieldrin	Orchard fruit	2.8	Stevens et al., 1970
	Vegetables	0.8	Saha & Sumner, 1971
	Vegetables	1.6	Harris & Mazurek, 1966

be illustrated by a study of residues in selected agricultural soils in
Ontario (Harris et al., 1966; Harris et al., 1977) (Table 14.5).

The range of residues varies widely. For example, in 1966, when
DDT residues were generally at their maximum levels, residues in
the four vegetable sites on muck farms ranged from 10.9 to 96.2 ppm

Table 14.4 (*Continued*)

Pesticide	Crop	ppm	Reference
Dieldrin	Vegetables	1.6	Harris et al., 1966
	Vegetables	17	Wiersma et al., 1972
	Corn	1.2	Decker et al., 1965
	Corn	0.9	Harris & Mazurek, 1966
	Turf	2.2	Fahey et al., 1965
	Pasture	1.1	Harris & Mazurek, 1966
Endrin	Orchards	12.6	Stevens et al., 1970
	Vegetables	0.5	Saha & Sumner, 1971
	Vegetables	3.8	Harris et al., 1966
	Vegetables	2.0	Wiersma et al., 1972
Chlordane	Vegetables	3.9	Saha & Sumner, 1971
	Vegetables	0.6	Harris & Mazurek, 1966
	Vegetables	23.8	Wiersma et al., 1972
	Corn	0.1	Harris & Mazurek, 1966
	Tobacco	0.2	Harris & Mazurek, 1966
	Turf	12.0	Fahey et al., 1965
	Vegetable	3.9	Saha & Sumner, 1971
Endosulfan	Orchards	4.6	Stevens et al., 1970
	Vegetables	0.4	Wiersma et al., 1972
Heptachlor	Vegetables	0.3	Saha & Sumner, 1971
(+ epoxide)	Vegetables	0.2	Harris et al., 1966
	Vegetables	2.2	Wiersma et al., 1972
	Tobacco	0.2	Harris & Mazurek, 1966
	Turf	1.6	Fahey et al., 1965
Diazinon	Vegetable	0.02	Harris et al., 1977
Ethion	Vegetable	7.2	Harris et al., 1977
Parathion	Vegetable	0.03	Harris et al., 1977
Mercury	Wheat	2.05	Sand et al., 1971
	Cropland	1.06	Wiersma & Tai, 1974

for DDT and from 1.3 to 3.3 ppm for dieldrin. Residue levels of organophosphorous insecticides on these same farms were even more variable, ethion residues ranging in 1974 from 0.2 to 7.2 ppm.

Data in Table 14.4 indicate insecticide residues under selected conditions of intense usage and thus reflect maximum levels. To use these data as a base to calculate the load of pesticide in the reservoir of agricultural soils would grossly overestimate the amount present. A better estimate of pesticide levels in agricultural soils is provided in the National Soils Monitoring Program initiated in the United States in 1964 as a part of the National Pesticides Monitoring Program (Bennett, 1967). In the report of this program for 1970 (Crockett et al., 1974) data are presented for 1506 sites in 35 states. In addition the pes-

Table 14.5   Mean Pesticide Residue Levels in Soil in
Selected Agriculture Operations in Southwestern Ontario,
1964–1974[a]

Farm Types	# Farms Sampled	Year	Residue ppm[b]		
			DDT+[c]	Cyclo-dienes[d]	O.P's[e]
Field crops	4	1964	0.7	0.9	ND
		1966	0.3	0.9	ND
		1969	0.4	0.8	ND
		1974	0.4	0.6	ND
Tobacco	4	1964	3.1	0.5	ND
		1966	4.6	0.8	ND
		1969	3.4	0.4	ND
		1974	3.0	0.3	0.008
Vegetable	5	1964	18.4	2.3	0.26
		1966	33.4	3.7	0.30
		1969	20.9	2.7	0.49
		1974	21.5	1.7	2.51
Fruit	2	1964	97.6	ND	ND
		1966	93.0	ND	ND
		1969	62.6	ND	ND
		1974	67.4	ND	ND

[a] Adapted from Harris et al., 1977.
[b] Figures rounded to one decimal except for O.P's.
[c] DDT, DDE, TDE + dicofol (dicofol found only in orchards).
[d] Cyclodienes include aldrin, chlordane, dieldrin, endosulfan, endrin and heptachlor plus their metabolites.
[e] O.P's include chlorpyrifos, diazinon, ethion, malathion, parathion and VC-13.

ticides used and rate per acre are included for each of the crops and states involved in the study. Some of the data from that study are presented in Table 14.6. The extent of these data and the randomness with which sample sites were chosen are the best indication published to date of the residues of persistent pesticides in agricultural soils. These data show that levels are much less than one would have concluded through extrapolation from earlier data. The study showed that dieldrin was the pesticide found most frequently (30.8%) and that DDT (and/or metabolites) was present in almost 23% of the soils sampled.

Residues of aldrin, dieldrin, and heptachlor (as the epoxide) were found more frequently in Iowa and Illinois than in other states,

Table 14.6 Mean Level, Range, and Percentage of Occurrence of Some Chlorinated Hydrocarbon Pesticides in Agriculture Soils, United States, 1970[a]

Pesticide	Mean Concentration ppm[b]		Range ppm	Percent Occurrence[c]
	Arithmetic	Geometric		
Aldrin	.002	.0032	0.01-4.25	13.5
Chlordane	.08	.0044	0.01-13.34	11.0
DDT[d]	.30	.0116	0.01-113.09	22.8
Dieldrin	.04	.0097	0.01-1.85	30.8
Endosulfan[d]	<0.01	--	--	0.3
Endrin	<0.01	--	0.01-0.9	1.8
Heptachlor[d]	0.01	.0016	0.01-0.34	9.8
Isodrin	<0.01	--	0.01-0.18	2.3
Lindane	<0.01	--	--	0.4
Toxaphene	0.06	--	0.79-8.75	1.8
Trifluralin	<0.01	--	--	2.2

[a] Crockett et al., 1974.
[b] Both arithmetic and geometric means are included, the latter (where sufficient occurrence permitted) to indicate better the central tendency of the data.
[c] Of all sites sampled.
[d] Including metabolites.

reflecting the use of these insecticides in corn production, and chlordane residues were detected more frequently in Illinois than in other areas. DDT (and/or metabolites) occurred infrequently in Midwestern states, being most prevalent in the Southeastern states and in California.

14.4 SIGNIFICANCE OF SOIL RESIDUES

Edwards (1973) lists four possible effects on living organisms in soil contaminated with insecticides, and these can be broadened to include effects of pesticides in general.

1. They may be directly toxic to animal and/or plant life in the soil.
2. They may affect these organisms genetically to produce populations resistant to the pesticides.
3. They may have sublethal effects that result in alterations in behavior or changes in metabolic or reproductive activity.
4. They may be taken into the bodies of soil flora and/or fauna and passed on to other organisms.

The extent to which any or all of these effects may apply has not been determined with certainty for any pesticide, but sufficient information is available to demonstrate instances where each type of effect has been noted. As with persistence itself, most of the data on effects relate to DDT and dieldrin. Three groups of organisms are of primary concern, the soil microorganisms, soil invertebrates, and plants.

Effects on Soil Microorganisms

As a living environment the soil owes much of its activity to a host of microorganisms that make it their home. These microorganisms, largely bacteria, fungi, algae, and nematodes, are important to soil nutrition through their role in decay of plant and other organic matter in the soil and as nitrifiers. Anything that disrupts their activity could be expected to affect the nutritional quality of soils and would thus have serious consequences. Because of this, microorganisms have been studied extensively in relation to pesticides, and the volume of evidence indicates clearly that with few exceptions pesticides have no lasting detrimental effects. Any effects noted have been short lived, populations rebounding quickly.

As might be expected, many fungicides temporarily reduce the fungal activity in soils, but with the exception of mercury and benomyl such effects are transient, because the pesticides are not persistent. Dichlone may suppress the activity of the bacteria responsible for nitrogen fixation and because of this is not recommended for use on legume seeds.

Many nematicides temporarily suppress microbial activity in soils, but since these have high vapor pressures, they volatilize rapidly and are soon lost from the soil, permitting microbial populations to rebound.

Microorganisms play a major role in the breakdown of pesticides in soil. Not all these conversions are desirable. Tu et al. (1968) showed that aldrin was converted to dieldrin by five of six genera of fungi and each of seven genera of bacteria tested. When incubated at 28°C for 6 weeks, conversion ranged from almost nil in some species to 9.2% for one culture of *Fusarium* sp.

Many microbes are capable of utilizing pesticides as sources of carbon, and most pesticides studied are attacked at one or more sites by microorganisms. The bacteria *Hydrogenomones* can degrade DDT completely to CO_2, and the list of pesticides known to be degraded by soil microorganisms includes practically all of those currently in use (Edwards, 1973; Khan et al., 1975).

Effects on Soil Invertebrates

Whereas most phyla of invertebrates are represented in the soil fauna, the principal ones belong to the Arthropoda (insects, mites, spiders, millipedes, centipedes), Mollusca (snails, slugs), Annelida (earthworms), and Protozoa (amoeba, paramecium). All of these are important in maintaining soil structure and nutrition, and a number of studies, particularly in England, have determined effects of pesticides on several groups of these animals (Edwards and Thompson, 1973; Edwards, 1970). Unfortunately most of what is available relates to target species of insects, a few beneficial insects, mites, earthworms, and slugs.

As would be expected, the insecticides have been shown to affect arthropod populations. In many instances this was the intended use of the insecticide, and because many of the soil insecticides are not selective (Chapter 10), both beneficial and target species are killed. Much of the data available deal with the persistent organochlorine insecticides, primarily DDT, aldrin, and dieldrin. Edwards (1970) treated field plots with aldrin and DDT at normal dosages and determined the effect on total arthropod weight, predator weight, pest weight, and weight of beneficial arthropods in each plot. As expected, pest weight was drastically reduced with either insecticide, aldrin being more effective than DDT. Total arthropod weight was, however, reduced only slightly by DDT, whereas aldrin treatment caused more than 60% reduction. Weight of predators was about equally reduced by each chemical, approaching 50% of that of the control. Both insecticides caused drastic shifts in the number of species and the numbers of individuals of each species recorded. Somewhat surprisingly, total weight of beneficial arthropods was increased by DDT but drastically reduced following the aldrin treatment. The increase in total weight of beneficial species following DDT use was due to high populations of Collembola made possible by the reduction of predatory mites, which normally feed on these insects and keep populations in check. Other studies have confirmed the lack of toxicity of DDT to Collembola and its high toxicity to predatory mites. The role of DDT in increasing populations of Collembola may not be unique. Edwards (1970) reported that, with the exception of springtails in the family Isotomidae, most of the Collembola are highly resistant to pesticides. Harris et al. (1972) compared DDT and fensulfothion and showed that, while the latter insecticide decreased Collembola populations and held them at low levels for more than 4 months, levels in plots treated with DDT increased above control plots after 2 months and were more than twice control populations at the conclusion of the experiment (136 days).

Effects of insecticides on soil invertebrates are transient, and in most cases, populations return to normal levels within a short time. There are, however, some exceptions. Soil treatments with dieldrin at high dosages have been shown to persist at levels toxic to beetle larvae for several seasons (Gambrell et al., 1968), and the use of dieldrin for termite control depends on the chemicals providing a toxic barrier for many years (Mallis, 1969). Obviously, with such high dosages, the mortality of beneficial soil arthropods is equally high for an extended period. These situations represent the unusual, and there is no evidence that persistent pesticides at levels now found in general agriculture or noncropland are at concentrations likely to affect arthropod populations. The ability of arthropod populations to rebound after drastic population reductions is well recognized (Varty, 1975), and soil applications of nonpersistent insecticides are unlikely to change this pattern. For example, Barrett (1968) showed that, while both the number of arthropods and biomass were reduced by 95% following an application of 2 lb of carbaryl/acre to soil, total biomass was no different from control plots within 7 weeks. This is not always the case, and the extent to which it is may reflect, in part at least, the extent to which an initial population is depleted. Edwards (1970) tested a group of pesticides in relation to direct kill and time required for recolonization of Collembola. Included in the test were the fumigant D-D (dichloropropane-dichloropropene) and the herbicide simazine. D-D caused complete mortality of Collembola within 30 days, and because of its high volatility the chemical had completely disappeared from the soil by this time. Evidence of recolonization by Collembola was not found until 120 days after initial D-D application, and it took almost 2 years for populations to return to normal levels. By contrast, residues of simazine persisted for 6 months, and although 70% of the Collembola population had been killed within 30 days, normal population levels were evident about 6 months later. Of the herbicides tested by Edwards (1970), only DNOC and simazine significantly reduced arthropod populations.

Edwards (1970) showed that the total population of arthorpods was depleted almost as much by cultivation as it was by aldrin, and he concluded that the contribution of arthropods to soil fertility would not be affected by pesticides in an agricultural setting. The possibility for such an effect in forest soils was not excluded.

It is clear from a number of studies that populations of predaceous and parasitic insects in soil have been reduced by the repeated application of pesticides. This has been documented primarily in relation to the organochlorine insecticides and especially where these have been used as broadcast applications (Chapman and Eckenrode, 1973; Read, 1960). Arthropod predators and parasites are highly susceptible to many soil insecticides, but field data suggest these effects are seasonal and populations rebound to normal levels the following year (Tyler, 1978).

There is no question but that genetic changes have taken place in soil arthropods owing to use of insecticide. The investigators were unable to find documentation for this in other than target species, but among the latter resistance to aldrin and dieldrin in cutworms, several species of root maggots, the carrot rust fly, and carrot weevil has been recognized for many years (Harris and Miles, 1975). Corn rootworms have developed resistance also, and there are reports of resistance in the Japanese beetle to chlordane (Niemcyzk and Lawrence, 1973).

Earthworms and slugs are among the largest of soil invertebrates, and though either may achieve pest status, they are important in decomposition of plant residues.

In a few situations such as golf greens and prize lawns, earthworm castings are undesirable, and control measures have been applied. Lead arsenate was used for such control for many years with dosages in excess of 100 lb/acre. More recently aldrin, dieldrin, chlordane, and carbaryl have been used, but again dosages in excess of 20 lb/acre have been required. Some of the newer carbamate insecticides, some organophosphorous compounds, and the fungicides benomyl and thiophanate-methyl are effective at lower dosages.

With the exception of these few insecticides and fungicides, it is unlikely that normal dosages will produce either direct toxicity or residual toxicity to earthworms. Earthworms are highly tolerant of most pesticides, and though they may carry heavy body burdens of the organochlorines, toxicity is not evident. Thompson (1970) found that the small, active surface feeders were more likely to be affected than those feeding below the soil surface. In his tests in a trefoil pasture the largest reductions in earthworm biomass occurred in plots treated with fensulfothion (92%), carbaryl (68%), endrin (67%), and carbofuran (60%). Reductions of less than 40% occurred following treatment with chlorpyrifos, DDT, or metalkamate. The experimental insecticide N-2596 (S-(p-chlorophenyl) O-ethyl ethyl-phosphorodithioate) was the most toxic, causing a biomass reduction of 93%. Edwards and Thompson (1973) include the insecticides chlordane, toxaphene, metalkamate, carbaryl, carbofuran, chlorfenvinphos, fensulfothion, disulfoton, malathion, parathion, phorate, and N-2596 and the herbicides atrazine, chlorpropham, DNOC, monuron, propham, and TCA in their listing of pesticides suppressing earthworm populations.

In orchard soils containing high levels of copper sulfate, earthworm populations are low, and this has been associated with a thick mat of undecomposed organic matter on the soil surface. Low earthworm populations have also been noted where arsenic levels are high. Recently the systemic fungicides benomyl and thiophanate-methyl have been used in apple orchards, and high toxicity to earthworms is reported (Stringer and Wright, 1973; Stringer and Lyons, 1974). In Illinois it was shown that, when benomyl was injected in the soil for control of Dutch elm disease, earthworm populations were severely reduced soon after benomyl treatment but within 1 year had returned to normal (Black and Neely, 1975).

Although a great deal of information is available on the toxicity of pesticides to earthworms, this is not the case for slugs. Few pesticides have proved effective for control, and those that have seem to act as stomach poisons. Evidence suggests that, whereas copper sulfate and lime may reduce slug populations, the effect is largely one or repellency or physical effect on the slime coating characteristic of land slugs. Metaldehyde and methiocarb are toxic when used as baits, and endrin and phorate are reportedly toxic (Edwards and Thompson, 1973). Several of the carbamate insecticides other than methiocarb are somewhat toxic to slugs (Kuhr and Dorough, 1976), but the degree of toxicity reported makes it highly unlikely that residues encountered in the environment pose a hazard to this group of animals.

Edwards (1973) tabulated the available data on soil levels of some pesticides needed to effect 95% mortality of earthworms and slugs. Reported values in ppm (in parentheses) for earthworms were aldrin (7125), endrin (10), toxaphene (715), chlordane (5), DDT (7200), dieldrin (7100), heptachlor (2), and lindane (765). For slugs figures were aldrin (250), chlordane (710), and endrin (720). Residue levels of this magnitude are rarely found.

Residues in Soil Invertebrates

Very little information is available on pesticide levels in arthropods. Since insecticides are generally the most toxic of the pesticides to higher animal life, it is with this group of compounds that we should be most concerned. The fact that insecticides have been developed specifically for arthropod control suggests that these organisms should be generally sensitive and would be killed rather than persist with high body burdens. Many insecticides are toxic to insects in the 1 to 10 mg/kg range, and thus the likelihood of high residues in their bodies is minimal. There are some exceptions. For example, the LD$_{50}$ of DDT to some grasshoppers approaches 10,000 mg/kg, and in such instances relatively high residues might be encountered if the insect were one in which accumulation occurs. Few data are available, but what are indicate that, although residues of organochlorine pesticides can be detected in invertebrates living in contaminated soil, only low levels are found and such levels are of the same order of magnitude as that of the soil (Edwards, 1973). For example, a study of carabid beetles (*Harpalus pennsylvanicus*) in Iowa cornfields showed that the mean residue levels for aldrin, dieldrin, heptachlor (and its epoxide), and DDT (+ DDE) ranged from 7.3 ppb for aldrin to 91 ppb for DDT (+ DDE). These levels were similar to those of the soil from which the beetles were collected (Sellers and Dahm, 1975). There are some exceptions. In one study the DDT residue in beetle larvae was 0.53 ppm, although the soil in which they were found contained only 0.015 ppm of this insecticide (Gish, 1970).

Much more data are available for pesticide residues in earthworms and slugs. Gish (1970) analyzed earthworms from 67 and slugs from 4 sites in an extensive survey of pesticides in soil. He found DDT and DDE in every earthworm sample and DDD in all but 4. More than half of these samples contained a residue greater than 0.1 ppm (dry weight basis). Dieldrin was found in 76% of the samples, and aldrin, endrin, heptachlor (and its epoxide), and gamma chlordane were found in a significant number. In a number of instances endrin was found in earthworm samples, although it was not detected in the soil from which the earthworms were extracted. The levels of residue of total organochlorine pesticides detected averaged 13.77 ppm in earthworms and 1.52 in the soils from which they came, indicating, on a dry weight basis, a concentration factor of 9.0. If this is converted to a living weight of earthworms, the concentration factor from soils to earthworms is approximately 1.6. In this study a few high total organochlorine residues (dry weight basis) were found. Earthworms from 2 cotton fields had residues of 114.6 and 159.4 ppm, from 1 cornfield, 89.4 ppm; from 2 apple orchards, 20.9 and 48.5 ppm; and from a peach orchard, 112.1 ppm. Although these residue levels are among the highest reported (Edwards, 1973), note that these are expressed on a dry weight basis and should be divided by 5.8 for comparison with values in the literature expressed on a wet weight basis.

It has been suggested that the concentration factor of organochlorine insecticides from soils to earthworms may be proportionately higher in soils where residues are lower than in those containing higher levels. This is probably not so. Edwards (1973), plotting the available data, showed good correlation between residue levels in earthworms and those of the soils from which the samples were collected. There is, however, an upper limit to the levels at which pesticides will be stored, for earthworms both excrete and metabolize some pesticides.

Residues in Plants

It has been shown with many pesticides that when these are incorporated in the soil some are taken up by plants growing in the treated soil. There is an extensive literature on this subject (Edwards, 1973) dealing with organochlorine insecticides, and although there are a few reports to the contrary, the vast weight of evidence indicates that, while some uptake occurs with many of these pesticides, aerial portions of plants contain but a small fraction of the residue present in the soils in which they are grown. In general, root crops tend to take up greater residues than others do, and in other crops residues are higher in the roots and lower portions than in the aerial part of the plants. In root crops residues are much higher in the peel than in the pulp.

The uptake of truly systemic pesticides by plants is a different matter, and care must be exercised where these are used to ensure that ample time has elapsed between their soil application and harvest so that excessive residues do not persist. With the possible exception of benomyl and thiophanate-methyl, soil residues of the systemic pesticides disappear within a few months of application provided adequate moisture is present.

One other aspect of the soil residue picture is the possible phytotoxic effects of pesticides on plants. This has been mentioned in connection with the herbicide atrazine, and a number of other herbicides have been shown to persist in sufficient quantity to inhibit growth of sensitive plants the following year. A number of plants, including peas, beans, and rutabagas, are sensitive, and farmers have learned to adjust their rotation practices so that problems of carryover effects are minimized. Perhaps the best known instances of phytotoxicity relate to residues of arsenic. Some soils, especially those supporting orchards for many years, may have high arsenic levels and be quite phytotoxic. In Florida it was found that in some cases new citrus plantings did poorly when planted in old citrus groves, and many vegetables could not be grown in such soils. It was demonstrated that high arsenic residues in the surface soil were responsible.

The established plantings were not affected, because of their deep root system in soil only slightly contaminated. For new plants with roots near the soil surface, exposure to arsenic was at a much higher level and phytotoxic. In more recent work Jacobs et al. (1970) showed that high application rates of the herbicide sodium arsenite to plainfield sand affected the growth of potatoes, sweet corn, peas, and beans, the decrease in growth correlated to higher dosages of arsenic. Peas were more tolerant than beans or sweet corn. Severe growth reduction in peas was still evident when these were planted 3 years after the initial treatment.

Some plants are extremely sensitive to some herbicides, and this presents a problem from the low levels that may drift from the site of application. Allen (1974) showed that 2,4-D at concentrations as low as 0.001 lb/acre and even lower concentrations of picloram or dicamba caused irreversible damage to such sensitive plants as tomatoes.

15

PESTICIDES IN WATER

The widespread use of pesticides makes it inevitable that a portion of these will contaminate surface waters. In addition large amounts of pesticides are applied to waters for control of undesirable insects, plants, or fishes, and those that are persistent reside in the waters to which they are applied or are carried to adjacent areas by water movement. To what extent do pesticides contaminate waters and how significant is this contamination?

15.1 SOURCES OF PESTICIDES IN WATER

Although the relative importance of the different sources of pesticides in water may differ from that in soils, the entrance routes are the same, with both intentional and unintentional sources involved.

Intentional Applications

Pesticides are used to control weeds and algal blooms in ponds and streams; to control insects such as mosquitoes, black flies, and biting midges that breed in water; to control lamprey predaceous on useful fish species; and to remove trash fish or diseased fish from ponds prior to restocking. Each of these procedures may result in contamination of surface waters, the extent of contamination and its duration being a function of conditions in the aquatic environment and the nature of the pesticide involved.

The extent to which ponds, lakes, and streams are treated with herbicides for weed control is not great. Herbicides may be used also in drainage ditches and canals to prevent weed growth from clogging the system and preventing proper operation of pumps. Diquat; 2,4-D (various esters); and atrazine are among the herbicides frequently used for aquatic weed control. Other compounds are used for situations where algal control is the main requirement. Dichlorbenil or copper sulfate may be used, the latter being the material of choice in cranberry bogs, where flooding is a normal production practice and algal growth a frequent problem.

Mosquito control is carried out annually in a large number of communities and recreational areas in North America. Most of the species that bite man breed in shallow water, and control efforts are most effective when directed at this breeding site. In addition aerial application of pesticides or application with fogging machines may be used against the adult mosquitoes in areas where larviciding has not provided adequate control or where communities have elected this approach to relieve the biting problem. Whereas the larviciding programs require direct application to standing waters, programs directed against adults frequently involve treatment in marshy areas where the mosquitoes are breeding and hence direct entry of pesticides into water. Occasionally outbreaks or the threat of outbreak of disease in man (e.g., St. Louis encephalitis, Eastern equine encephalitis) or livestock (e.g., Venezuelan equine encephalitis) when mosquitoes serve as the vector of the pathogen necessitate emergency control, and vast areas may be aerially treated.

In the 1945–1960 period DDT was the insecticide most widely used, because it was highly effective and would persist in water and associated vegetation for several weeks. Although this insecticide may still be used where human health is threatened, it can no longer be used for routine control. Methoxychlor, chlorpyrifos, temephos, or oil are used most frequently as larvicides, methoprene receiving some use in recent years. Each of these materials is short lived, repeated applications being needed. Adulticiding programs rely heavily on malathion and propoxur.

The extent of pesticide use for mosquito control is not known. Mosquito control associations are organized in a number of states, and several American cities have budgets in excess of $1 million for mosquito abatement programs. During the outbreak of Venezuelan equine encephalitis in 1971, 615 million acres were treated aerially with malathion.

Much less insecticide is used for black fly control. In contrast to most man-biting species of mosquitoes, which breed in standing, shallow water, and whose larvae and pupae breathe at the water surface, black flies breed in flowing water with the larvae attached to rocks and other objects in the stream bed. These differences in breeding habits dictate a different approach to control. With mosquitoes an effort is made to treat shallow waters and to concentrate the pesticide at the surface. By contrast, effective control of black fly larvae requires that the pesticide penetrate to the bottom of the stream. As with mosquitoes DDT was used effectively for black fly control for many years but has now been largely replaced by methoxychlor.

The use of lampricides is a relatively recent addition to fish management and has been practiced in North America only in the Great Lakes. The sea lamprey invaded this vast inland water via the Welland Canal some time near 1920. The larval lamprey live in the bottom mud of streams and their estuaries for 4 to 7 years before entering the open waters to become parasitic on lake trout and related species. In the late 1950s it was shown that the pesticide TFM was selectively toxic to lamprey larvae and would provide control at dosages not toxic to desirable species of fish. It has since been found that the addition of Bayluscide® to TFM provides a less expensive treatment for lamprey control (Chapter 12).

Although the sea lamprey is a problem only in the Great Lakes, the potential for pesticide contamination is large. Surveys indicate that more than 10% of the 3000 streams flowing to the Great Lakes contain larval lamprey (Lennon et al., 1971).

The use of piscicides to remove trash fish from ponds, lakes, and streams is not new, and the effectiveness of rotenone for this purpose has been known for many years. Fishing demands have made restocking programs more popular and with them the need to remove competing species so that desirable fish have an optimum chance for survival. The use of fish toxicants in management programs is becoming more common. A review of this subject (Lennon et al., 1971) indicates that, during the 10 year period 1952–1962, biologists in the United States had used fish toxicants on 2500 miles of streams and 225,000 acres of lakes and ponds. Whereas a number of compounds have been tested and used experimentally in a limited way, rotenone and antimycin are employed most frequently.

Unintentional Applications

Waters are contaminated in the same way as soils through pesticide drift during application and as atmospheric fallout on rain and dust.

Additionally, however, water may be contaminated through soil erosion, industrial effluent, sewage, and occasionally by spills into or adjacent to watercourses.

With the exception of atmospheric fallout, most of these contaminations are relatively local. Despite this it must be recognized that some pesticides persist for a long time, and these may be transported rather widely throughout a natural watercourse on soil or other particulate matter.

Atmospheric Fallout

It is established that rain water contains pesticides, and this is an obvious source for contamination of surface waters. Not enough sampling of the atmosphere has been done to establish precise contamination levels, but it is apparent that a low level of atmospheric contamination exists in all parts of the world (Chapter 16). On a basis of a mean value in rain of 210 ppt for DDT, Edwards (1973a) calculated that fallout from rain per year would be approximately 0.0023 lb/acre or about 1 g. He suggested that the 210 ppt figure was probably too high and that 0.001 (misprinted as 0.01 in Edwards [1973], personal communication) would be a more realistic figure. Whether or not this is a realistic value for land masses is questionable. If it is, then a similar annual deposit can be expected from this source for inland waters. Contamination of ocean waters by rainfall may be somewhat less. Atmospheric levels in Barbados and Hawaii suggest that, for DDT (probably the best indicator of pesticide pollution), levels in rainfall may be about 5 ppt. Thus a realistic level of annual input to the oceans from rainfall would be perhaps 100 mg/acre.

As in the case with soil, some pesticide contamination of waters occurs by the settling of atmospheric dust. Over continental areas it would be expected that fallout via this route on inland waters would be comparable to that over soils (Chapter 14). Data for dust fallout over ocean waters are not available.

Soil Erosion

Where pesticides are applied to terrestrial areas, much of the dosage ends up in or on the soil. Those pesticides soluble in water may be carried to nearby waters by surface runoff, and either soluble or insoluble ones may be carried on soil particles in runoff waters or eroded by wind. It is difficult to separate the role of each of these modes of transport, since critical tests have not been done. Leaching tests with the persistent chlorinated hydrocarbons have indicated little movement in soil, and tests of runoff water and ponds in treated areas generally suggest that such contamination is minimal. For example, endosulfan has been detected in irrigation water runoff and runoff after rainfall, the amount in the runoff water decreasing within a short time after application (NRC, 1975).

The quantity of soil eroded each year is immense. In the United States some 4 billion tons of sediment are produced by erosion each year. Half of this is washed into streams and some 1 billion tons reach the ocean. Studies in the Great Lakes area indicate that more than 2 million tons of sediment reach Lakes Erie and Michigan annually and that 37% of this is from sheet erosion on agricultural lands. The extent of sediment load varied drastically from one stream to another but was as high as 270 ppm in the Maumee River subbasin (Environmental Protection Agency [EPA], 1971).

If one assumes that the level of pesticide in the sediment was the same as that reported for soils in the U.S. National Soils Monitoring Program (Crockett et al., 1974), that is, 0.01 ppm, the amount of the chlorinated hydrocarbon insecticides reaching waters by this route would be approximately 40,000 lb. This probably grossly underestimates the amount of pesticides actually transported, since at least a portion of the erosion occurs during the growing season, when the surface soil eroded would carry much higher residues. Research in North Carolina and Georgia on the influence of slope, vegetation, and rainfall on the runoff of herbicides in small watersheds has shown that there is little movement of herbicides except on soil particles (EPA, 1971).

Industrial Effluent

Many industries use pesticides in the manufacture of their products, and effluents may contain high levels. Many instances have been reported where DDT or dieldrin has been discharged from carpet or fabric manufacturing plants, where these pesticides are used in mothproofing. These effluents may result in local contamination and fish kills.

The pesticide industry itself is guilty of a number of instances of release of pesticides in effluents from manufacturing, formulating, and packaging processes. Perhaps the most dramatic is the recent large-scale contamination of the James River with chlordecone (Wilbur, 1976).

Sewage

Many pesticides, along with other contaminants, are contained in municipal sewage. Their source includes discharge from industrial plants, residues from homeowner use, and a variety of fungicides and bactericides used in soaps and cosmetics. The level of contamination may be high; Edwards (1973a) cites instances where dilutions of 1 : 20 are required before the effluent would be safe for fish. The significance of waste water discharge of pesticides in relation to other inputs has seldom been documented. A study in Michigan showed that the Red Cedar River became progressively contaminated with DDT (plus metabolites) in a downstream direction, and this contamination was correlated with the discharge of water treatment plants serving cities with spray programs for Dutch elm disease (Zabik et al., 1971). For some pesticides, for example, hexachlorophene, waste water from urban areas is probably the main source (Sims and Pfaender, 1975).

Spills

The handling of pesticides in the volume now used creates the possibility of large-scale industrial spills during processing, storage, and transportation, as well as the problem of smaller spills into waterways by the applicator. Added to the mishaps brought about by human failings is the problem of flooding in areas where pesticides are stored and fires in manufacturing or storage areas. Although some of these accidents may result in large contaminations of waterways, most are of local significance. Many instances are known. One of the largest involved the accidental discharge of endosulfan into the Rhine River in two separate instances with large fish kills and heavy contamination (Lüssem and Schlimme, 1971).

One final source of local contamination involves the practice of spray applicators' filling their equipment at streams, lakes, and ponds. Where this is practiced, some contamination can occur by the dropping of packages of pesticide in the water, careless disposal of pesticide containers, and in some cases, failure to operate the proper valves on the spray equipment and emptying the sprayer contents into the water.

15.2 PERSISTENCE OF PESTICIDES IN WATER

The persistence of pesticides in water depends on a number of factors, only some of which can be qualified and/or quantified, and these only for some pesticides. Those pesticides that persist in soil also persist in water or in the bottom sediments from which mixing with the overlying water is, to a greater or lesser extent, constantly occurring. Thus any discussion of persistence in water must include the bottom sediments or muds, and the nature and extent of these are perhaps more important than the water itself in terms of persistence and the biological significance of residues.

Nature of the Pesticide

The literature contains a great deal of information on the water solubility, rate of hydrolysis, and many other chemical properties of pesticides. For the most part this is specific information based on laboratory tests in distilled water of specified pH and temperature. Such information is helpful but may not reflect accurately what happens in natural waters.

With the organic pesticides those that are highly water soluble are hydrolyzed rapidly and have a short life in water. This is not true for the inorganics such as the mercurials and arsenicals, whose toxicity is related to the elemental compounds. Thus mercury constantly recycles in the aquatic environment. Its potential as a biocide is not diminished, but its accessibility is, the ocean abyss serving (presumptively) as a massive sink protecting the biosphere from overt toxicity (Lepple, 1973).

Among the organic pesticides, DDT, dieldrin, and endrin are the most persistent. They are each relatively insoluble in water (DDT 0.2 ppb, dieldrin 186 ppb, endrin 100 ppb) and resistant to hydrolysis. These have the potential to persist in surface waters and their associated sediments for long periods, and the consistent reporting of residues in water samples supports this assumption. A close relative to DDT, methoxychlor, shares its chemical characteristics of low water solubility (100 ppb) and resistance to alkali and acid. In laboratory tests using distilled water less than 50% disappeared in 240 days, and yet this insecticide is rarely detected in surface waters except immediately after direct application (NRC, 1975a).

For most pesticides persistence in water is brief irrespective of their solubility. Fenitrothion is only slightly soluble in water (20 ppb) but is readily hydrolyzed, with a half-life of 272 min at 30°C in 0.01 NaOH. Methyl parathion, although much more soluble (60 ppm), has a half-life of 210 min under the same conditions (NRC, 1975b).

Although hydrolysis is an important route of breakdown of many pesticides in water and probably the main route for organophosphorous insecticides, photodecomposition is also important. Photodecomposition can occur at a number of sites in many pesticides, light having been shown to accelerate the rate of hydrolysis in aqueous media. Whether this is directly related to the chemistry of the pesticide or to an indirect effect on microbial degradation has not often been established.

Nature of the Water

Natural waters differ in their composition, pH, temperature, aquatic life, and amount of suspended organic and inorganic material. These factors play an important role in determining the persistence of pesticides.

Chemical Composition

Although it is known that a number of ions act catalytically in the degradation of pesticides, specific studies relating the effect of these in natural waters have not been reported. Some studies have included "hardness" of water (as $CaCO_3$) in characterizing their test medium, but this is not very definitive. Hardness of water is extremely important to the effectiveness of TFM (Kanayama, 1963; Dykstra and Lennon, 1966); for tributaries flowing into the Great Lakes dosages calculated for effective control of lamprey ranged from 1 to 17 ppm, the higher rates required in waters of greater hardness. By contrast, hardness does not influence either the phytotoxicity or persistence of diquat (Yeo, 1967) or endothal (Yeo, 1970).

pH

Since many pesticides degrade by hydrolysis, it might be assumed that pH of natural waters would play an important role in persistence. Carbaryl, for instance, is reported to have a half-life of 1 to 5 days in water, but laboratory tests using distilled water indicated that, although the chemical degraded rapidly at pH values above 8.0, it was most stable at pH 6.3, with a half-life at that pH measured in months (Kuhr and Dorough, 1976). Laboratory studies on fenitrothion indicate that the pesticide is quite stable under slightly acid to slightly alkaline conditions and is stable for 45 days in tap water at pH 7.0, but its persistence in natural waters is only a few days (Sundaram, 1973). Thus, although laboratory findings make it clear that pH is important to the inherent degradation of many pesticides, it may be of only minor significance in natural waters.

Exceptions undoubtedly occur, and interactions of pH and other natural conditions are likely. Greve and Wit (1971) studied both hydrolysis and oxidation rates of endosulfan in water in relation to changes in pH and oxygen content. They found that under anaerobic conditions the half-life of endosulfan was about 5 weeks at pH 7.0 and 5 months at pH 5.5. Although hydrolysis proceeded much more rapidly at the higher pH, the rate of oxidation was similar at either pH level.

Temperature

Increases in temperature increase the rate of chemical reaction and the rate of volatilization of pesticides. Within limits increased temperature increases biological activity and thus would be expected to increase the biological degradation of pesticides in aquatic environments. Nishizawa et al. (1961) found that hydrolysis of fenitrothion was twice as fast at 40°C as it was at 30°C. With carbaryl only 9% was hydrolyzed after 8 days' incubation at 3.5°C, whereas 93% was degraded during the same period when the temperature was 28°C (Karinen et al., 1967). Studies with parathion and paraoxon showed that, when these pesticides were held in aqueous solutions, the rate of hydrolysis doubled for each 10° increase in temperature (Gomaa and Faust, 1972). Tests with the carbamate insecticides carbaryl, propoxur, pyrolan, and dimetilan yielded temperature coefficients for hydrolysis in the same order of magnitude (Aly and El-dib, 1972). This order of temperature effect is probably applicable to many of the organophosphorous insecticides and other pesticides in which much of the decomposition may be chemical.

It is well established that pesticides volatilize from treated surfaces and from surfaces of soil and water in which they may be incorporated. The tendency to volatilize is inherent in the chemical and characterized numerically as vapor pressure. This vapor pressure changes with temperature—the higher the temperature the greater the volatility. Thus higher temperatures tend to reduce the persistence of volatile pesticides.

Khan et al. (1975) have reviewed the literature on the detoxification of pesticides by biota but do not discuss this in relation to temperature. Although it is recognized that most organisms have optimum temperatures and that microorganisms play a major role in the degradation of pesticides, no attention has been given to evaluating how temperature affects the rate of these detoxifications. It has been suggested that, in soils, increased temperatures increase the rate at which pesticides disappear and that this effect is related to the role of higher temperatures in increasing microbial action (Goring et al., 1975).

Aquatic Life

Numerous studies attest to the fact that a wide range of plants and animals detoxify pesticides by a great variety of routes. Any consideration of the biota to which pesticides in the aquatic environment are exposed must include that of the bottom mud, where anaerobic or near anaerobic conditions prevail. In a recent review of this subject Khan et al. (1975) cite 136 references and present (in tabular form) a synopsis of some of the mechanisms involved. They point out that biological degradations include oxidations, dechlorinations, reductions, hydrolyses, and ring cleavage and give examples of each.

Of all the groups involved in direct degradations, microorganisms are probably most important. Goring et al. (1975) identify 48 genera capable of degrading one or more pesticides, many of those listed occurring in natural waters.

It must be recognized that there is a degree of specificity, and this specificity is reflected in the extent to which any pesticide is degraded microbially in a particular water. Many studies bear out this conclusion, one serving to illustrate. Butler et al. (1975) incubated atrazine, carbaryl, and diazinon each at 1 ppm and methoxychlor and 2,4-D each at 0.01 ppm in culture tubes with each of 21 species of planktonic algae and measured degradation by determining the amount of unchanged pesticide in each culture at the end of a 2 week growing period. With carbaryl 80% of the compound disappeared within 1 week from all cultures. The authors concluded that this disappearance was the result of nonbiological breakdown, and carbaryl was dropped from further evaluation. Only 12 of the 21 cultures were able to grow in the atrazine treatments, and none of these brought about significant degradation. All of the cultures reduced the amount of 2,4-D; diazinon; and methoxychlor, although there were major differences in the extent to which this occurred. With 2,4-D, less than 20% was recovered from 7 of the cultures, recovery ranging from 13 to 64%. Methoxychlor was degraded also by all cultures, recovery ranging from 29 to 79%. Less than 50% was recovered from about one-half of the cultures. Diazinon was less affected, and while only 62% was recovered from one culture, 75 to 96% was recovered from the other isolates. As the authors (Butler et al., 1975) point out, this study does not demonstrate conclusively that the pesticides were in fact metabolized, since the products of metabolism were not identified. The loss of pesticide from different cultures at different rates is presumptive evidence; the pesticides with which they worked are generally considered to be biodegradable and, with the possible exception of methoxychlor, ranked as nonpersistent to moderately persistent compounds (Goring et al., 1975).

The biota may be important also by reducing the penetration of light. Heavy mats of algal or other plant growth reduce light, and thus the amount of photodecomposition is reduced.

Sediment

Many studies indicate that bottom sediments in lakes, rivers, and ponds act as reservoirs for persistent pesticides. In early studies in the Columbia River basin, Hindin et al. (1964) reported that DDT residues in the bottom sediment of Entiat River and Wenatchee River were 144 (+ 95 DDD) and 56 mg/kg, respectively, while levels in the raw water at the mouth of these rivers were 4.0 and 0.6 ppt, respectively. Numerous other studies in a wide range of situations confirm this general finding (Edwards, 1973a).

Perhaps the most extensively studied region in North America has been the Great Lakes and their associated streams. Studies showed that persistent pesticides, mostly DDT and dieldrin, could be detected in stream waters and that there appeared to be some correlation between levels in water and rainfall (Miles and Harris, 1971; Harris and Miles, 1975). Sediments contained high residues relative to the waters above them; in the case of DDT the composition of the residues in terms of parent compounds versus metabolites indicated a significant residency time. A summary of pesticide residue data in Ontario for 1968–1969 indicates that for 40 samples of raw river water (unfiltered) the mean DDT (plus metabolites) residue was 19 ppt and, for dieldrin, 6 ppt. By contrast 84 river sediments (on a dry weight basis) had a mean DDT residue of 80 ppb and a dieldrin residue of 4 ppb. Thus residues in sediments were more than 1000 times those of water (EPA, 1971). Frank et al. (1977) reported on an extensive survey of the Lake St. Clair, Detroit River, and Lake Erie portion of the Great Lakes. This area lies in a watershed that includes a large agricultural area on either side of Lake Erie and a number of large industrial cities. Lake Erie receives a high sediment load annually, the estimate for 1968 being 2,775,100 tons (EPA, 1971), most of which derives from sheet erosion. The rate of sedimentation has been studied, and by selecting cores in the sediment and fractioning these into specific segments from the surface downward, it is possible to estimate the age of deposit at each depth.

About 40% of Lake Erie is shallow, and it contains three major basins, smaller ones near the eastern and western ends and a large central basin in the west central portion of the lake. A small projection of this central basin forms the Sandusky basin lying to the west and south of the main basin (Figure 15.1). Frank et al. (1977) did a sampling of sediment in Lake St. Clair in 1970 and again in 1974 on an identical 4 km sample grid that involved 55 sites. They collected samples of the top 2 cm of sediment at each station and found three insecticides, DDT (plus metabolites), dieldrin, and endosulfan. In 1970 DDT (plus metabolites) was present at 6.6 ppb, but this had declined to 2.4 ppb by 1974. In each year about 85% of the total DDT detected was in the form of TDE or DDE. A careful examination of residues at each of the sampling sites suggested a higher level of pesticide near the outflow end of the lake, which suggested a transfer with the flow pattern.

Dieldrin was detected in 24% of the samples in 1970 with a mean residue of 0.1 ppb. This insecticide was not found in the 1974 samples.

Endosulfan was detected in 20% of the samples in 1970, with a mean residue of 0.2 ppb. This insecticide was not found in the 1974 samples.

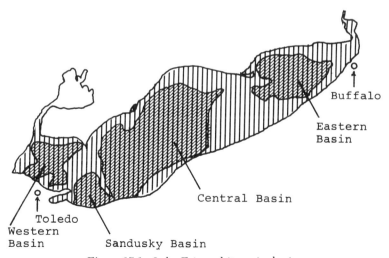

Figure 15.1 Lake Erie and its major basins.

In the Lake Erie sampling, 255 sites were involved. These were on a 10 km grid, except in the western basin, where a 5 km grid was set up and sampling done at alternate grid stations. Only two insecticides (or their metabolites) were found, DDT and dieldrin. PCBs were also detected. DDT (or metabolites) was detected at all sites, with a mean level of 27.9 ppb for the entire lake and made up of 1.3 ppb p,p'-DDT, 18.4 ppb TDE, and 8.2 ppb of DDE. Residues were much higher (+ 2–3) in the basin sediments than in the shallow zones. Residues were higher in the western basin (68.6 ppb) than in the central (25.7 ppb) or eastern basin (26.8 ppb). The ratio of TDE/DDE was similar in each basin at about 2–1.

Dieldrin levels were much lower (about 1.1 ppb) than those of the DDT metabolites and showed a slight increase from the western to the eastern basin.

The study by Frank et al. (1977) documents two important points. The first is that sediments in large lakes are important reservoirs of some persistent pesticides. Legislative actions restricting the use of DDT in the United States and Canada should have eliminated any use of these compounds in the Lake Erie watershed after 1970. Yet significant residues of DDT (plus metabolites) were present, albeit reduced, in Lake St. Clair by about 65% between the 1970 and 1974 samplings. Dieldrin and endosulfan, present in much lower levels than DDT in 1970, could not be detected in 1974. The second point is that sediments, although acting as reservoirs, effect degradation of even the most persistent of pesticides.

One final point is of interest in the Lake Erie study (Frank et al., 1977). They examined one sediment core from the western basin taken to a depth of 112 cm. In this area of the lake sediment rates had been determined at 7.6 mm/year, and it was thus possible by analyzing different core depths to identify when pesticides were deposited in the sediment. In the core examined, the greatest amount of DDT (plus metabolites) was found in the 0 to 2 cm depth, that is, the deposit of 1969–1971. Residues of p,p'DDE were detected in all 2 cm samples to the 8 to 10 cm level, reflecting a deposit of 1958–1960. No DDT per se was found in any of the core samples, suggesting that the original product had been metabolized. DDE was found in all segments of the core to a depth of 12 cm and TDE to a depth of 10 cm. Thus, while the persistence of the parent compound may not be great under anaerobic conditions, persistence of a biologically significant metabolite is. Note also that, in all the sediments analyzed in this study, residues of PCBs were grossly higher than those of DDT and its metabolites.

Suspended Matter

In only a few studies has the suspended matter been isolated from raw waters prior to analysis. These demonstrate the tendency of pesticides to become adsorbed or absorbed to particulate matter whether organic or inorganic, living or dead. As is discussed later, plankton frequently contains pesticides at levels much in excess of the water medium, and it is difficult to separate the role of animate versus inanimate suspended matter in relation to pesticide persistence. Frank et al. (1977) found a correlation between residues of DDT and dieldrin and organic carbon, suggesting perhaps that for these pesticides adsorption or absorption to living organisms is a main route toward the sedimentary sink.

15.3   RESIDUE LEVELS AND PERSISTENCE

The reporting of pesticides in waters is extensive, but the reader should view each report in the context in which it is prepared and with due regard for the method of sampling and sophistication of analytical techniques. Most studies report residues on the basis of unfiltered water, but others have used filters of various capabilities. It is not our purpose to suggest that one method is preferable to another but rather to point out that care must be exercised in comparing values shown in various studies. From an environmental standpoint, whether pesticides are dissolved in waters or carried on filterable particulates may be important, but we do not yet have sufficient data to draw such a conclusion.

Surface Waters

Some reports on pesticide levels in surface waters are summarized in Table 15.1, but the reader should review the text of the reports from which this summary was drawn for the detail of methods and the representativeness of samples. For example, Zabik et al. (1971) and Miles and Harris (1971) report high levels of DDT relative to other reports, but their samples were drawn purposely from areas of high use of this pesticide. By contrast Lichtenberg et al. (1970) surveyed waters from many regions in an attempt to obtain data representative of the natural waters throughout the United States. Glooschenko et al. (1976) found traces of dieldrin and DDT (or metabolites) in some

Table 15.1 Some Mean Concentrations (ppt) of the Four Most Frequently Reported Pesticides in Surface Waters in North America[a]

	Pesticide				
Water Sampled	DDT	Dieldrin	Endrin	Heptachlor	Reference
Rivers & Streams					
U.S.	8.2	6.9	2.4	6.3	Breidenbach et al., 1967
U.S.	10.3	2.3	1.4	2.6	Brown & Nishioka, 1967
U.S.	8.3	5.9	3.6	0.1	Green et al., 1967
Iowa	2.7	6.3	--	--	Johnson & Morris, 1971
Michigan	111.0	--	--	--	Zabik et al., 1971
U.S.	9.3	1.1	0.3	1.4	Manigold & Schulze, 1969
Canada	14.8	1.0	--	--	Frank et al., 1974a
Canada	64.0	7.1	--	--	Miles & Harris, 1971
General					
U.S.	11.5	7.3	0.1	--	Lichtenberg et al., 1970
Florida	17.1	--	--	--	Kolipinski et al., 1971
Coastal					
U.S.	4.02	--	--	--	Cox, 1971
U.S./Canada	2.4	--	--	--	Cox, 1971
Great Lakes	ND	ND	ND	ND	Glooschenko et al., 1976
Farm Ponds					
Surface fed	20.4	1.4	--	--	Frank et al., 1974a
Spring fed	8.5	ND	--	--	Frank et al., 1974a

[a] ppt (parts per trillion) rounded to two decimals.
-- Not included in analysis.
ND Not detected.

samples of water from the upper Great Lakes but were unable to quantify levels, since the sensitivity of their method did not permit this below 5 ppt. Other studies (Gilbertson [cited by Harris and Miles, 1975]; Macek, 1970) indicate that pesticide residues do exist in the Great Lakes with levels in the 2 to 5 ppt range. Kolipinski et al. (1971) obtained higher levels of DDT than many other workers, but their study area, the Florida Everglades, could not be considered typical of North American waters. Thus, with the exceptions noted here, the studies summarized (Table 15.1) indicate that some pesticides persist in American waters and that levels of near 10 ppt for DDT (plus metabolites) and somewhat less for dieldrin are valid. Readers should be cautioned not to infer from summaries of data that the mean levels reported indicate what is normally found in a stream. They do not, as can be seen from an inspection of almost any report. That of Lichtenberg et al. (1970) will serve to illustrate. These workers monitored 110 sites in 1967 and 112 in 1968. For purposes of their survey they divided the United States into nine regions and sampled 7 to 18 water systems in each region. Their results are summarized in Table 15.2 (1967) and Table 15.3 (1968). A perusal of these summaries will show that, while the mean residue level of DDT was higher in 1967 than that of dieldrin, dieldrin occurred more frequently. In 1968 dieldrin similarly occurred more frequently, and in that year its mean residue was higher than that of DDT. Of the 110 sites sampled in 1967, only 13 contained measurable amounts of DDT and 22 contained dieldrin.

In addition to failing to reflect the number of occurrences, mean values frequently do not suggest the range of levels detected. Inspection of Table 15.3 shows that, while no dieldrin was found in water samples from the southwest region, the mean level in the southeast region was 45 ppt. Between regions, residues of DDT ranged from nondetectable to 69 ppt (Table 15.2).

A perusal of mean values for pesticides in surface waters demonstrates that on the whole contamination is at a low level. This feeling is reinforced when residues are expressed in ppb or ppm. Estimates have been made of the "safe" level for pesticides in drinking water, and Ettinger and Mount (1967) have reflected on the potential for fish to accumulate some pesticides within their bodies and have suggested maximum levels in water that would give a reasonable measure of assurance that fish harvested from such waters would be safe to eat (.Table 15.4. More recently maximum permissible levels have been recommended to protect fish and aquatic life (Committee, 1973). These levels are much lower (Table 15.4). A perusal of these figures (note they are presented in ppb, not ppt), along with mean values for

Table 15.2 Residues of Three Organochlorine Insecticides in Surface Waters in the United States by Regions, 1967[a]

Region	No. Samples	Mean Residue ppt			No. of Occurrences		
		Dieldrin	Endrin	DDT[b]	Dieldrin	Endrin	DDT[b]
Northeast	12	13	ND	4	6	0	3
Mid Atlantic	7	3	ND	ND	1	0	0
Southeast	12	12	ND	56	4	0	3
Ohio Basin	9	3	ND	ND	2	0	0
Great Lakes	16	1	5	15	2	1	1
Missouri	14	9	10	69	3	1	3
South Central	18	2	ND	4	3	0	3
Southwest	11	ND	ND	ND	0	0	0
Northwest	11	1	ND	ND	1	0	0
Mean		4.9	1.7	16.4			

[a]Summarized from Lichtenberg et al., 1970.
[b]DDT and/or DDD and DDE.

Table 15.3 Residues of Three Organochlorine Insecticides in Surface Waters in the United States by Region, 1968[a]

Region	No. Samples	Mean Residue ppt			No. of Occurrences		
		Dieldrin	Endrin	DDT[b]	Dieldrin	Endrin	DDT[b]
Northeast	11	8	ND	4	7	0	2
Mid Atlantic	7	3	ND	5	3	0	1
Southeast	12	45	ND	22	6	0	2
Ohio Basin	7	25	ND	7	4	0	1
Great Lakes	17	3	ND	ND	4	0	0
Missouri	15	2	ND	7	3	0	3
South Central	17	1	ND	10	3	0	3
Southwest	14	ND	ND	2	0	0	1
Northwest	12	1	ND	2	2	0	2
Mean		9.8	ND	6.6			

[a]Summarized from Lichtenberg et al., 1970.
[b]DDT and/or DDD and DDE.

Table 15.4   Levels of Some Pesticides Permissible in Potable
Water and Safe Levels Suggested for Surface Waters[a]

		Maximum Suggested	
Pesticide	Permissible Level[ab]	Fish[ac]	Aquatic Life[ab]
Dieldrin	1	0.25	0.005
Endrin	0.5	0.1	0.002
DDT	50	0.5	0.002
Heptachlor	0.1	1.0	0.01
Aldrin	1	0.25	0.01
Lindane	50	5.0	0.02
Chlordane	3	0.25	0.04
Methoxychlor	1000	20.0	0.005
Toxaphene	5	2.5	0.01
2,4-D	20	not given	--
2,4,5-TP(Silvex)	30	not given	--
2,4,5-T	2	not given	--
Organophosphorous plus carbamates	100	not given	--

[a] Levels expressed as parts per billion.
[b] From Committee, 1973.
[c] From Ettinger and Mount, 1967.

pesticide levels in surface waters, indicates a wide margin of safety in terms of known human hazards. Not apparent from such summaries, however, is the fact that pesticides may be at high levels in local situations. In the 1967–1968 survey Lichtenberg et al. (1970) found levels as high as 0.407 ppb for dieldrin, 0.133 for endrin, 0.840 for DDD, and 0.112 for BHC. Although none of these levels approached those considered hazardous in drinking water, note that, for dieldrin, endrin, and DDT, they exceed the levels suggested by Ettinger and Mount (1967) as safe waters for fish for human consumption and are grossly higher than those recommended by the Committee on Water Quality Criteria (Committee, 1973).

Although the persistent chlorinated hydrocarbon insecticides DDT, dieldrin, endrin, and heptachlor dominate the literature in terms of residues in water, they are not the only ones found. Schulze et al. (1973) monitored 20 stations, 1 on each of 20 rivers in the western United States. Streams were monitored from October 1968 to September 1971, mostly on a monthly basis. In this study analyses were made for aldrin; DDT (+ DDE and DDD); dieldrin; endrin; heptachlor (and its epoxide); lindane; chlordane; toxaphene; endosulfan;

2,4-D; silvex; and 2,4,5-T. In addition, samples were analyzed for methyl parathion, parathion, diazinon, and malathion during the latter part of the study. Of all these pesticides only heptachlor (and its epoxide), toxaphene, and malathion were not detected in one or more of the samples. No pesticides were found at levels in excess of those permitted in public water supplies. Approximately one-third of all pesticide occurrences were at one site.

Herbicides were detected in all streams sampled, the highest concentration being 0.99 ppb 2,4-D. 2,4,5-T occurred at all sampling stations at some time during the 3 year period but was found most consistently at two sites, the Arkansas River near Van Buren, Arkansas, and the Canadian River near Whitefield, Oklahoma. At the latter location it was found in 27 of the 33 monthly samples and ranged from 0.01 to 0.05 ppb.

A number of studies have correlated pesticide levels in streams with local use patterns and rainfall. Thus Miles and Harris (1971) found the highest levels of DDT in streams in an agricultural area reflected spring runoff and rainfall patterns throughout the growing season. Richard et al. (1975) found that a 3 in. rainfall immediately after the corn-planting season in Iowa was reflected in 12 ppb atrazine in the South Skunk River the following day. Atrazine levels declined rapidly to 3.9 ppb 2 days later and to 0.4 ppb in 1 week. The herbicide was present in the river throughout the remainder of the growing season, but levels decreased below 0.1 ppb by late September. Similar patterns were observed for DDE and dieldrin. This pattern of pesticide contamination was equally evident in two other rivers monitored.

These studies plus many others support the point made earlier that sheet erosion is a major source of pesticide contamination of streams. It should not be inferred from this that all such erosion derives from agriculture, though this is undoubtedly a major factor. Miles and Harris (1973) calculated pesticide load in streams in agricultural areas versus those in recreational areas of Ontario. They showed that, in terms of contamination of Lakes Erie and Ontario from the Canadian side with DDT, more input came from recreational areas where DDT had been used for mosquito and black fly control than from areas where its use had been primarily in crop production. Along these same lines, data of Brodtmann (1976) on levels of organochlorine pesticides near the mouth of the Mississippi are of interest. He monitored raw water samples on a weekly basis and detected residues of eight pesticides (including heptachlor epoxide and DDD) throughout the season. Residue levels of chlordane, dieldrin, and endrin were highest during the late spring and summer, reflecting what one would expect in terms of seasonal agricultural use. Heptachlor epoxide residues followed a similar trend, but those of heptachlor were reversed. DDT showed peaks in midwinter and again in May but declined from May to October. DDD showed peaks in midwinter, early summer, and late fall. Because agricultural uses of DDT, endrin, and dieldrin were restricted in 1970, the seasonal differences in occurrence cannot be explained by agricultural usage.

Ground Water—Potable Water

The general use of pesticides and the persistent nature of some present the possibility for contamination of ground waters that serve as a drinking water source for much of the rural population of North America. In addition it might be suspected that, because many surface waters are contaminated, albeit usually at low levels, potable waters throughout the country would be contaminated to a similar degree. It appears that this is so, and Westlake and Gunther (1966) cite 0 to 2 ppb as the normal levels. It is probable that zero contamination rarely occurs, though Bevenue (1976) reports that, in terms of pesticide residues, potable water in Hawaii is "almost pristine in character." Brodtmann (1976) analyzed for 10 pesticides in samples from two water districts. In one district no pesticides were detected, but in the other a trace of DDT was found, though it could not be quantified. The author suggested that the sensitivity of the method was not adequate and with refinements in technique identified at low ppt levels eight pesticides (or metabolites) in the Mississippi River water that served as the raw water source for these water districts. Unfortunately he did not report analysis of finished water by using the more sensitive procedure. Richard et al. (1975) found traces of dieldrin, 0 to 28 ppt DDE, and 0 to 483 ppt atrazine in finished water in Iowa cities obtaining their supply from wells and 0.4 to 5 ppt dieldrin, 2 to 5 ppt DDE, and 29 to 405 ppt atrazine in those supplied from surface waters. They analyzed raw and finished water in the Des Moines system and found residues of dieldrin, DDE, and atrazine in each, with little indication that the treatment processes did much to remove these pesticides.

Using methods sensitive in the low ppt level, Burns et al. (1975) concluded that, while artesian wells contained virtually no residues of any of eight organochlorine pesticides used in the area, water from wells depending on surface runoff had levels comparable to those of surface waters. In another agricultural region (southwestern Ontario), Frank et al. (1974) found no dieldrin in any of 14 shallow wells tested and no DDT in 11. Water from 3 wells had DDT (plus metabolites) residues of 4, 40, and 50 ppt, respectively.

It is thus apparent that trace amounts of some pesticides are present in drinking waters of the United States and Canada, but levels are at three or more orders of magnitude below those considered safe (USDI, 1968).

Open Oceans

It has been assumed by a number of authors that the ultimate sink for persistent pesticides is the ocean abyss, but these have been assumptions only and not reinforced by substantial data. Several persistent pesticides, notably DDT, dieldrin, and mirex, have been detected in marine estuaries, and it is possible to estimate the discharge from river systems. Thus Brodtmann (1976) measured the concentration of several pesticides near the mouth of the Mississippi and, knowing the discharge rate in this system, estimated that 9 tons of pesticide would be carried to the Gulf of Mexico during 1970. Actual measurements of pesticide concentrations in the oceans are seldom reported, and those done prior to 1970 may not be accurate. There are two reasons for this. In the first place levels of pesticides in oceans are extremely low and, even in the case of DDT, the residue most likely to be found, the sensitivity of analytical methods is borderline. The second reason is that polychlorinated biphenyls have been falsely identified as organochlorine insecticides in some analyses by electron capture gas chromatography. Schechter (cited by Bevenue, 1976) suggests that the problem of misidentification is more likely with aquatic than terrestrial samples, and Bevenue (1976) concludes that "residues of DDT reported from marine and estuarine conditions deserve cautious acceptance because of their potential mistaken identity with the PCB's."

In studies of ocean waters near the west coast of North America Cox (1971) found DDT residues of 2 to 6 ppt, and a number of authors suggest that the ocean level is near 1 ppt. Recently Harvey et al. (1974) reported on extensive monitoring in the North and South Atlantic, and although DDT (along with PCBs) was found in marine organisms, it could not be found in deep ocean waters, there being no evidence of a gradient of concentration of either DDT or PCBs as the distance from land increased. Contrary to the accepted view, Harvey (1974) cites work purported to show that, in naturally illuminated sea water, the half-life of DDT is about 10 days. It appears that, despite a large number of reports, the fate and role of DDT in open oceans are still very much open to speculation.

A factor that may be important in terms of ocean waters is the tendency for pesticides to concentrate in surface slicks. These are

Table 15.5 Relative Persistence of Some Pesticides in Natural Waters

Non Persistent[a]	Slightly Persistent[b]	Moderately Persistent[c]	Persistent[d]
azinphosmethyl	aldrin	aldicarb	benomyl
captan	amitrole	atrazine	dieldrin
carbaryl	CDAA	ametryne	endrin
chlorpyrifos	CDEC	bromacil	hexachlorobenzene
demeton	chloramben	carbofuran	heptachlor
dichlorvos	chlorpropham	carboxin	isodrin
dicrotophos	CIPC	chlordane	monocrotophos
diquat	dalapon	chlorfenvinphos	
DNOC	diazinon	chloroxuron	
endosulfan	dicamba	dichlorbenil	
endothal	disulfoton	dimethoate	
fenitrothion	DNBP	diphenamid	
IPC	EPTC	diuron	
malathion	fenuron	ethion	
methiocarb	MCPA	fensulfothion	
methoprene	methoxychlor	fonofos	
methyl parathion	monuron	lindane	
mevinphos	phorate	linuron	
parathion	propham	prometone	
naled	Swep	propazine	
phosphamidon	TCA	quintozene	
propoxur	thionazin	simazine	
pyrethrum	vernolate	TBA	
rotenone		terbacil	
temephos		toxaphene	
TFM		trifluralin	
2,4-D			

[a] Half-life less than 2 weeks.
[b] Half-life 2 weeks to 6 weeks.
[c] Half-life 6 weeks to 6 months.
[d] Half-life more than 6 months.

282

calm areas where normal wave and ripple motion has been damped by a film of organic matter. They occur on large bodies of water in which biological activity is high and have been observed widely. Seba and Corcoran (1969) measured pesticide levels in each of seven slicks in the Biscayne Bay area over a period of several weeks. These slicks were relatively persistent within the area, aerial photos showing that about 10% of the bay was covered with slicks during the June to August period of 1968. These workers found that the pesticide level in slicks associated with agricultural drainage areas had up to 137 times as much pesticide as slicks associated with the Florida Current and (in the opinion of Seba and Corcoran [1969]) representative of background pesticide levels in the Caribbean. Aldrin, dieldrin, and DDT were found in all slicks, total residue levels for pesticides ranging from a low mean of 93 ppt for the slick in the Florida Current to a high of 12,750 ppt for one near the outlet of a drainage canal bringing outflow from an agricultural area. Samples of sea water were taken at the same time in the vicinity of these surface slicks and generally had no detectable amount of pesticide (less than 1 ppt).

Persistence of Residues

Many of the factors that determine persistence in water are the same as those operating in soils. Thus persistence in natural waters and their associated suspended matter, biota, and sediments is not always indicated by data determined in laboratory tests using distilled or tap water. Goring et al. (1975) reviewed the literature on persistence of pesticides in soil and listed many of the common chemicals in one of four categories. We have kept the same categories but changed a few pesticides from one category to another, added some, and deleted others (Table 15.5). We make no effort to reference this listing, for it reflects a synthesis of the literature, and the literature is not always in agreement. The reader should consider the listing as a guide to relative persistence but understand that the boundaries between the groups are fuzzy and that, as previously discussed, persistence of any pesticide may vary with water conditions.

15.4 SIGNIFICANCE OF RESIDUES

The aquatic environment is teeming with life. We seldom stop to reflect on the endless chain of events to which each organism is exposed and the dynamics involved in maintaining what we like to call the natural state. In any pond can be found a complex of organisms, some preferring the bottom areas (benthic organisms), others preferring the upper reaches (planktonic organisms), and each of these complexes, though overlapping to some extent and obviously sharing a degree of interdependence, is remarkably distinct. In its simplest form we can picture the detrital matter being decomposed and in the process nurturing a host of bacteria, fungi, and algae that multiply rapidly and release organic matter to enrich the medium. These microorganisms are fed upon by a somewhat smaller number but equally diverse assemblage of grazers that includes microcrustaceans, rotifers, and insects, and these are in turn fed upon by larger crustaceans (e.g., *Cyclops*), by larger rotifers (e.g., *Asplanchna*), and by predaceous insects. To complicate matters even further, some of these predators are predators only during part of their life cycle and may participate as grazers during other stages. For example, *Cyclops* may prey on small rotifers and small crustaceans as an adult but feed on bacteria or algae as young nauplii, and many of the aquatic insects feed on either microflora or microfauna. When pesticides enter such a complex system, the potential for change is apparent, for it is unlikely that the pesticide would affect all organisms equally, and even if it did, this would not guarantee that the equilibrium would not be altered. For example, complete sterilization of a water body would be a transient event and permit new entrants to the system an opportunity to reproduce without competition.

Under natural conditions the events that go on in aquatic communities are difficult to document, and the effects of the intrusion of pesticides into these systems are made more difficult by the lack of good baseline data. We know little about the feeding habits, adaptability in food sources, reproductive potentials, and behavioral traits of many of the aquatic inhabitants. Despite the fact that predaceous insects or some large copepods may represent trophic levels of 3 or 4, our studies with pesticides in water have revolved heavily around effects on fish and birds. Even so, much of the research done on significance of pesticides in the aquatic environment has been questionable both in design and interpretation. Too frequently conclusions have been drawn from a minimum of data where valid replication has not been assured and analytical competence has been questionable. In addition the popular worry about pesticides has prompted a number of workers whose experience and expertise is in descriptive biology to enter experimental biology without due concern for good experimental procedures. Notwithstanding, a significant amount of information is being generated, and a number of recent reviews have attempted to digest these data and translate them into environmental effects. Throughout the text that follows, the reader is referred to such reviews for additional discussion on specific topics.

In those instances where overt toxicity to nontarget organisms occurs, it is easy to draw conclusions on some aspects of significance of pesticide contamination. Thus fish kills can be documented, and at least some aspects of effect (economic, recreational, aesthetic) are readily apparent. Not so apparent, however, are the more subtle effects that may be reflected over time and that we have little competence to analyze. Yet it is the subtle and unrecognized effects that may be most significant in the longer term, and it is certainly speculation concerning these that has led to the great adverse publicity surrounding pesticides in recent years.

The complexity of the aquatic environment suggests the need to discuss two types of effect—direct and indirect. Direct effects are those that alter the numbers, physiology, or behavior of a particular organism (or group of organisms), while indirect effects describe those resulting from the effect a change in one organism (or group of organisms) may have on another.

Direct Effects

Most of the literature deals with direct effects. This is to be expected, for it is comparatively easy to document such effects relative to effects on an ecosystem. In most cases direct effects have been reported in terms of toxicity to an organism or some other measurable criterion occurring under a specified exposure under a given set of circumstances. The dosage of pesticide causing 50% mortality is the most frequent statistic, and this is most helpful. It does not, however, provide a significant depth of understanding about what effects may be expected at exposures below this value or whether, in fact, any significant effects other than death are likely. Some of these data are summarized in Table 15.6 for aquatic organisms, and a quick scanning of these data may give some feeling for the relative toxicity of different pesticides and a perspective on toxic levels in relation to levels commonly found in North American waters. It is clear that levels of contamination (Tables 15.1–15.3) are well below (by several orders of magnitude) those reported toxic to any of the species listed, and thus overt toxicity is not expressed. Does this mean that no problem exists?

A simple yes or no answer would be neither appropriate nor justified. The types of pesticides used range widely in effects, and the large number of compounds makes testing on all species an impossi-

Table 15.6 Acute Toxicity of Selected Pesticides (ppb) to Aquatic Organisms[a]

Pesticide	Effect on Phytoplankton[b] %	Daphnia[c] EC50	Gammarus[d] LC50	Pteronarcys[e] LC50	Culex[f] LC50	Crangon[g] LC50	Bufo[h] LC50	Rana[i] LC50
Aldrin	-85	28	28	8	5	30	2000	
Chlordane	-94	29	160	170				
DDT	-77	.36	4.7	41	70	3	2400	>2000
Dieldrin	-85	250	1400	6	8	68	1100	
Endosulfan	-87	240	9.2	24				
Endrin	-46	20	47	4	15	2.8	570	
Heptachlor	-94	42	150	8	54	110		
Methoxychlor	-81	.78	4.7	30	67	9		
Toxaphene	-91	15	70	7			600	
Abate			960	100	16			>2000
Azinphosmethyl		3.2		8			680	
Chlorpyrifos			.76	50	3			>400
Diazinon		.9	800	60	830			>2000
Dimethoate		2500		140				
Ethion	-69	.01	56	24				
Fenitrothion		.4	12	28	6			
Malathion		1.8	3.8	35	80	246	1900	
Parathion		0.6		8	3	11	1600	
Phosphamidon		8.8	8.4	1400				
Carbaryl	-17	6.4	40	30			7600	>4000

Carbofuran					2700	595
Propoxur	21	25	110			
Allethrin	100	20	28			
Rotenone	23	6000	2900			
Amitrol						
Atrazine	0	3600				
Dalapon		11000		>100000		
Dicamba			1000			
Dichlorbenil		3700	1500	4400		
Diquat	-45				>10000	
Endothall	0	46000	2000			
MCPA		100000				
Monuron	-94	106000				
Paraquat	-53	3700	18000	>100000	54000	
Picloram	0	>380000	48000	120000		
Prometone						
Simazine			21000	50000		
Trifluralin		240	8800	13000		
2,4-D		320000	1800000			
2,4,5-T (acid)		>1500				

Table 15.6 (*Continued*)

Pesticide	Effect on Phyto-plankton[b] %	Daphnia[c] EC50	Gammarus[d] LC50	Ptero-narcys[e] LC50	Culex[f] LC50	Crangon[g] LC50	Bufo[h] LC50	Rana[i] LC50
Copper sulfate								
Dichlone		26	3200					
Nabam								
Benomyl[j]		640						
TFM-2B[j]		7350	26000	15400				

[a] Data included in Table 15.6 has been extracted from numerous reports including Pimentel (1971) who reviewed the literature and cites original work. Where several values are reported in the literature some averaging has occasionally been done.
[b] % decrease in carbon fixation during 4 hours in a mixed natural culture of phytoplankton from a marine estuary when exposed to 1 ppm pesticide (Butler, 1963).
[c] Most figures are from Sanders and Cope (1966) and are the concentrations required to immobilize 50% of the organisms D. pulex after 48 hours at 15.6°C.
[d] Most figures are from Sanders (1969) for Gammarus lacustris.
[e] Mostly Pteronarcys californica, from Sanders and Cope (1968).
[f] Culex fatigans (pipiens), W.H.O. (1968).
[g] The sand shrimp, Crangon septemspinosa, Eisler (1969).
[h] Fowlers toad tadpoles. Data from Sanders (1970).
[i] Bullfrog, Rana catesbeiana. Oral dosage (Tucker and Crabtree, 1970).
[j] Data from Rye and King (1976).

bility. We can only assume that data on one organism are transferable, within limits, to others and that the differences in toxicity represented in available information have not omitted strikingly susceptible species. Thus with a few exceptions (albeit important ones) the general level of pesticide contamination in American waters is of little consequence in terms of direct mortality.

Direct treatment of water for pest control presents a different set of conditions. Here control requires pesticide dosages toxic to living organisms, and in many cases these are in excess of LC_{50} values for many aquatic species. We can thus anticipate effects, direct and indirect.

Effects on Bacteria

Many species of bacteria degrade pesticides. In the process some obtain a suitable carbon source, while in others the benefit to the bacterium is not apparent. The literature on effects of pesticides on bacteria in soil is much more voluminous than that dealing with water and has been reviewed by Tu and Miles (1976) for insecticides. Many of the reports they reviewed were laboratory tests in liquid media. Thus, to the extent that many of the same species of bacteria occur in water as in soil, the results they reviewed should apply equally to water. Tu and Miles (1976) tabulate results of tests with pure cultures of bacteria; 29 species of bacteria are included in their table and 21 insecticides (3 carbamates, 9 organophosphorous compounds, 9 chlorinated hydrocarbons). They conclude that, at recommended dosages, insecticides used in soil are unlikely to have marked effect on bacterial number. In some cases temporary growth suppression was noted, while in others the addition of pesticide was stimulatory. The tabulation of Tu and Miles (1976) shows that, whereas *Escherichia coli, Klebsiella aerogenes,* and three species of *Pseudomonas* common to both soil and water were not affected by any of nine chlorinated hydrocarbon insecticides, growth of *Bacillus megaterium* was inhibited by each of seven tested.

Little work seems to have been done on the toxicity of fungicides to bacteria. Inorganic mercury is converted to methyl mercury by a number of microorganisms (Jensen and Jernelöv, 1969).

The use of herbicides for weed control in ponds is common practice. 2,4-D; diquat; paraquat; and a number of other herbicides are effective. Fry et al. (1973) observed the effect of paraquat on microbial populations after treatment of a reservoir to control two species of angiosperms, *Potamogeton pestinctus* L. and *Myriophyllum spicetum*

L. Two applications of paraquat were employed, the first on June 14 when the growth of the angiosperms was heavy and the second on July 29 when angiosperms were essentially absent. Application rate for the first application was 1 ppm and for the second 0.6 ppm in terms of water in the reservoir.

The design of this experiment, that is, two treatments, permitted two types of observation. The first treatment resulted in death of a standing crop of weeds, and observations after that treatment reflect direct effect of the herbicide plus the effect of a decaying mass of vegetation on which bacteria might thrive. After the second treatment no such increase in decaying matter was present, and the effects could be attributed more directly to paraquat per se.

Results showed that after the first application of paraquat there was an immediate 90% reduction in the number of heterotrophs in water, followed by a sharp rise in numbers to a maximum population 10 times that of pretreatment levels. Paraquat-resistant microorganisms increased by a factor of 90, amylase producers by a factor of 34, and protease producers by a factor of 43.

After the second treatment, only slight changes were noted in any populations. Note, however, that the dosage used in the second treatment was only 0.6 ppm in contrast to 1.0 ppm in the first treatment. No change was noted in the amylase or protease producers.

The limited number of reports of pesticide effects on bacteria deal primarily with total numbers and do not assess the possibility of shifts in species composition or activity. Whereas some studies on microorganisms measured effects in terms of respiration, these studies have dealt with natural flora of mixed species. The fact that morphological and physiological changes can occur is demonstrated by O'Neill and Langlois (1976), who showed that heptachlor caused a thickening in the cell walls and disruption of cell wall structure in *Staphylococcus aureus*.

Effects on Fungi

Little attention has been given to possible effects of pesticides on fungi. The literature contains many reports in which fungi have been included, but only as part of a general category called phytoplankton, and specific effects on fungi have not been isolated. The situation is not much better for fungi in soil except for fungicide evaluations where selectivity has been noted and reported as part of the characterization of these materials (Chapter 8). Tu et al. (1968) isolated 92 pure cultures of microorganisms from soil and grew these with aldrin at 1 ppm. They did not report any toxicity (or comment on it) but demonstrated that several species of *Trichoderma, Fusarium, Penicillium,* and *Aspergillus* were effective converters of aldrin to dieldrin and that *Rhizopus* and *Mucor* were less efficient. *Thrichoderma* has also been shown to degrade malathion and dieldrin (Matsumura and Boush, 1966, 1967), but again these workers did not comment on whether or not toxicity to any of the more than 500 cultures of organisms they tested was observed.

Hodkinson and Dalton (1973) grew 12 species of river fungi in media of varying nutritional levels and added DDT at 2, 10, 20, and 60 ppm. Cultures were incubated at 20°C, and growth was evaluated on the basis of mycelial growth after 100 hours. The addition of DDT increased the growth of all fungi on each of the media tested, but the greatest increases were obtained in low nutrient media. There was little growth effect in 2 or 10 ppm with most species, but at 60 ppm growth was increased more than 50% in 5 of the 12 cultures.

Effects on Algae

A large number of species of algae, unicellular and filamentous, inhabit aquatic environments and form the base of extensive food webs. Some of these dwell near the water surface, while others are adapted to life somewhat below the surface, and together these constitute a major portion of the growth characterized as phytoplankton. In addition some algae are benthic, occupying a niche at the water-mud interface in the bottom of ponds. The first suggestion of a significant effect of pesticides on algae was probably that of Shane (1948), who noted high populations of *Synedra* in a Wilmington reservoir following repeated applications of DDT for fly control. This report was to be followed by many others, such as that of Cook and Connors (1963), who associated blooms of *Anabaena* with use of methyl parathion in Clear Lake, California. Whereas the appearance of such blooms is undesirable from an aesthetic viewpoint, they may be far more important in terms of organic loading of the affected areas, and in the case of *Anabaena* blooms, the alga plus the associated bacteria may render the water toxic to wildlife (Schwimmer and Schwimmer, 1968).

A different sort of concern regarding pesticide effects on algae was raised when it was shown that a number of these were toxic to marine phytoplankton. Ukeles (1962) evaluated the toxicity of 16 pesticides and an organic solvent to each of five species of marine algae believed important as food for oysters and clams. She found that, of the pesticides tested, the herbicides fenuron, monuron, diuron, and neburon were the most toxic and inhibited growth in the low ppb range. Less toxic were the chlorinated hydrocarbon insecticides DDT, lindane, and toxaphene, the last the most toxic of the three. The carbamates carbaryl and nabam were less toxic and the organophosphorous insecticides TEPP and trichlorfon were least toxic, being tolerated in concentrations of several parts per million. There were differences in susceptibility among the algae, the brown pigmented ones (*Monochrysis lutheri* Droop, *Phacodactylum tricornutum* Bohlin) being more sensitive than the green (*Chlorella* sp., *Protococcus* sp., *Dunaliella euchlora* Lerche).

Butler (1963) measured effects of pesticides on algae in terms of the increase or decrease in photosynthesis as determined by the uptake of labeled ^{14}C in a 4 hour period. In one set of tests he used a mixed algal culture (presumably obtained from estuary waters, since zooplankton were present also) and added pesticides to the test culture. Included in these tests were 30 insecticides, 10 herbicides, 4 fungicides, 2 acaricides, and 2 soil fumigants. Of the materials tested, 13 insecticides (9 of 12 chlorinated hydrocarbons, 3 of 14 organophosphorous compounds), 3 herbicides, and 3 fungicides reduced photosynthesis by more than 50%. Some of these results are summarized in Table 15.6. Note that the chlorinated hydrocarbon insecticides caused large reductions. Although giving no supporting data, the author noted that at lower concentrations increased rates of photosynthesis were observed in some instances.

Knowledge that DDT was present in many segments of the aquatic environment focused attention on this compound, and much more is known about the magnitude of its effects on algal productivity than is available for other pesticides. Wurster (1968) studied its effects (in terms of photosynthesis) on four species of marine algae. In addition he tested various concentrations of DDT against a natural phytoplankton community dominated by diatoms collected form Vineyard Sound (Woods Hole). Although there were some differences in reaction among the algal species tested, the general trend was obvious. Increased dosage of DDT decreased the rate of carbon uptake, the effect being especially pronounced in the 10 to 100 ppb range. The same effect was apparent in the natural phytoplankton culture. In some of the cultures there was an indication that low concentrations of DDT were stimulatory, and in the pure cultures there was a leveling out of the effect at higher concentrations. This produced a sigmoid curve that Wurster (1968) interpreted to mean the absence of a threshold dosage. Wurster (1968) suggested that the data showing increased carbon uptake at low levels should not be interpreted as a stimulatory effect. If this is so, then his data indicate clearly a threshold value for DDT with each of the cultures tested.

Few of the fungicides seem to have been tested widely, and those tests that have been done suggest that, with the exception of mercury compounds (most of the input of mercury is other than through pesticides) and copper, most of the fungicides are not likely to present a problem. Three benzimidazole fungicides were tested against four aquatic organisms, one of which was *Chlorella pyrenoidosa* (Canton, 1976). MBC was the most toxic, with an EC_{50} of 0.34 ppm. EC_{50} values for benomyl and thiophanate-methyl were 1.4 ppm and 8.5 ppm, respectively.

The lampricide TFM was tested in the laboratory for toxicity to 10 species of algae in pure culture (Maki et al., 1975). EC_{50} values for growth inhibition measured after 96 hours varied from 4.4 to 15 ppm for four species of green algae and 1.9 to 9.2 ppm for four species of blue-green algae and were 1.2 ppm and 3.2 ppm, respectively, for the diatoms *Nitzschia* sp. and *Navicula pelliculosa*. Exposure to 30 ppm for 96 hours was found to prevent growth but did not kill the cells. Growth was normal when such cultures were placed in "clean" media after such exposure. Field grade TFM was found to be slightly more toxic than analytical material.

Organisms in the aquatic environment are exposed simultaneously to a number of new chemicals, and the question arises whether these might act independently, in a synergistic manner, or whether their actions might be antagonistic. Mosser et al. (1974) tested a DDT-sensitive diatom, *Thalassiosira pseudonana* (= *Cylotella nana*), to determine whether or not PCBs, DDT, and DDE might interact to affect the alga in a manner different from that of the compounds alone. Little growth inhibition resulted when the alga was grown in media containing 10 ppb PCBs or 100 ppb DDE. Growth was, however, inhibited quite markedly when the DDE and PCBs were combined at these dosages. Whereas the DDE + PCB combinations acted synergistically, a quite different effect occurred when DDT and PCBs were combined. PCBs by themselves when added at 50 ppb almost stopped growth, but when DDT was added at 500 ppb, growth was restored to normal levels. This antagonistic effect of DDT on PCBs' inhibition was demonstrated even when the DDT was added 12 to 24 hours after the PCBs. The addition of lower dosages of DDT also counteracted the inhibitory effect of the PCBs but to a lesser extent. There seemed little effect when DDT and DDE were combined.

While the sensitivity of algae to pesticides is a matter for concern, perhaps a concern with greater implications is the high levels of some pesticides taken up by these plants at the bottom of the food chain. Edwards (1973a) tabulates some of the reported data on levels of pesticides measured in algae (and other plants) in relation to residues in the water in which these were growing. He converts these data into concentration factors, that is, amount in plant/amount in water, and shows extremes for DDT in algae of 0.07 and 126,400. The concentration factor for chlordane was 2.0, for endrin 0.67, and for dieldrin 4.091, but only one report was included for each of these three insecticides. As with many other examples, most of the residue data on algae are for DDT and its metabolites. Some of the data available are discussed by Cox (1972), who points out the rapidity with which the pesticide is absorbed by algal cells and the different rates of uptake and absolute uptake by various algal species.

It is apparent that pesticides of low water solubility adsorb readily to particulate matter whether this be organic or inorganic. Thus the high degree of magnification from water to algae is to be expected for materials such as DDT and dieldrin. Similarly for naled, a pesticide of low water solubility, it has been shown that algal cells provide a good adsorbent (Vance and Maki, 1976). Although a few pesticides accumulate in algae, this is not a general phenomenon, especially for pesticides with a significant degree of water solubility. Using a model ecosystem approach (Metcalf et al., 1971), Yu and Sanborn (1975) showed that diatoms (mixed culture of several species) did not pick up parathion. Data of Butler et al. (1975) indicate that, if atrazine or diazinon were absorbed by any of 21 cultures of algae, the amount was small indeed.

It seems reasonable to conclude that algae accumulate nonwater-soluble, persistent pesticides from their environment and that the level of uptake may be several orders of magnitude above that of their surroundings. It is unlikely that significant uptake will occur with nonpersistent or semipersistent pesticides with a significant degree of water solubility.

Effects on Aquatic Invertebrates

Many species of invertebrates frequent marine and fresh water environments and play important roles as grazers, predators, and prey in the complicated food webs of the biosphere. Pesticides have been shown to be toxic to these zooplankton and benthos, the degree of toxicity varying widely among pesticides and among various species and life stages within the animal community. Some values reported for toxicity (immobilization in the case of *Daphnia*) are tabulated in

Table 15.6. The cladoceran *Daphnia pulex* is among the more sensitive. The high degree of sensitivity in this species has been used for many years as a method of detecting and quantifying pesticide residues by bioassay procedures.

Both laboratory and field tests have been used to obtain data relative to pesticide toxicity. The laboratory tests have the advantage of controlled conditions and, provided the test organisms are cultured sufficiently long in "clean" water, should be free of natural pesticide contamination. At best, however, they are artificial and do not adequately reflect the temperature, water quality, light quality, and competitive forces that obtain in the natural setting. In addition much of the experimental work has recorded mortality as the only index of effect and has not considered sublethal effects or their consequences.

In their work with two cladocerans (in part, Table 15.6) Sanders and Cope (1966) found a slight difference between *Daphnia pulex* and *Simocephalus serrulatus*, the former being more sensitive. Organophosphorous insecticides were the most toxic of the various groups of pesticides, and with *D. pulex* EC_{50} values for 48 hr immobilization were less than 1 ppb for 7 of 10 compounds in this class. Dichlorvos was the most toxic, with an EC_{50} value of 0.066 ppb. In other studies (cited by Pimentel, 1971) ethion was even more toxic (Table 15.6). Herbicides were much less toxic to *Daphnia*, and in the Sanders and Cope (1966) study, trifluralin was the only 1 of 11 tested toxic at a concentration less than 1 ppm. This study also showed that, with DDT and malathion, toxicity was greater at lower temperatures and that DDT was more toxic to first and second instar larvae (EC_{50} 2.5 ppb and 2.4 ppb respectively) than to 7 day old (EC_{50} 6.4 ppb) *D. pulex*.

Morgan (1976) investigated the acute toxicity of diazinon to 14 species of stream invertebrates. In some tests exposure was for only 3 and 6 hr, whereas in other tests it continued 1 to 7 days. Results (in part) are given in Tables 15.7 and 15.8. His selection of test animals included larvae from four orders of insects, two species of molluscs, and three classes of crustaceans. Of the animals tested, the chironomid was the most sensitive and the crayfish the least. There were orders of magnitude differences in susceptibility among the insects tested, among the amphipods, and between the two species of molluscs. Exposure time greatly affected LC_{50} values. Increasing exposure time from 24 to 168 hr decreased the LC_{50} value to the mayfly by a factor of 7.6, to the amphipod by 9.6, and to the chironomid by 14.8.

Morgan (1976) extended his study to include sublethal effects. He found that, when the chironomid *Chironomus tentans* was reared in water containing 0.003 ppb diazinon (one-tenth the 96 hr LC_{50}), egg hatch was delayed significantly, the duration of larval life was in-

Table 15.7   LC$_{50}$ Values for Diazinon for 14 Fresh Water
Invertebrates Exposed to the Pesticide for Various Periods
under Static Conditions at 16°C, pH 7.6, and 12-hour
Photoperiod[a]

Test Organism	LC$_{50}$ ppb	Duration of Exposure
Insects		
Trichoptera		
Hydropsycha sparna	220	3 hours
Cheumatopsycha oxa	190	3 hours
Leptocella albida	220	3 hours
Ephemeroptera		
Paraleptophlebia pallipes	134	48 hours
Baetis intermedius	55	48 hours
Plecoptera		
Acromeauria ruralis	294	48 hours
Diptera		
Chironomus tentans	1	48 hours
Molluscs		
Physa gyrina	48	96 hours
Helisoma trivolis trivolis	528	168 hours
Decapod		
Orconectes propinquus	537	48 hours
Amphipods		
Hyalella azteca	22	48 hours
Gammarus lacustris	229	48 hours
Gammarus pseudolimnaeus	3	48 hours
Isopod		
Asellus communis	21	96 hours

[a]From Morgan, 1976.

creased, and the percent pupation and adult emergence was decreased slightly. The total effect on duration of development was an
increase of 33.6% in the time from egg to adult. Investigating other
sublethal effects, Morgan (1976) showed that, when the crayfish *Orconectes propinquus* was reared in water containing 3 ppb diazinon
(one-fifth the 7 day LC$_{50}$), male dominance over the female was
retained, but the number and duration of fights increased and locomo-
tion by both males and females increased also. Exposure to diazinon
at one-third this concentration did not affect these activities in the
crayfish, and the effect of the 3 ppb dosage disappeared within 4 days
after the exposure was terminated.

Table 15.8   LC$_{50}$ Values for Diazinon for Three Fresh Water
Invertebrates Exposed for Various Periods of Time Under
Static Conditions at 16°C, pH 7.6, and 12-hour Photoperiod[a]

| Time of Exposure (hours) | LC$_{50}$ (ppb) for Test Organism | | |
	Paralepto- phlebia pallipes	Gammarus pseudo- limnaeus	Chironomus tentans
24	243	48	0.4
48	134	4	0.1
72	85	3	0.07
96	44	2	0.03
120	--	1	--
144	43	0.7	--
168	32	0.5	0.027

[a]From Morgan, 1976.

Morgan (1976) exposed G. *lacustris* to diazinon at 3 ppb (about one-fiftieth the 7 day LC$_{50}$) and found that activity in the treated tank exceeded that of the control by 1.7. It was shown also that this amphipod did not distinguish between untreated water and that containing concentrations of diazinon toxic to it.

The implications of these findings in terms of long-term effects on aquatic organisms exposed to sublethal concentrations of pesticides are open to speculation and will vary with the organism. Any change that will cause greater exposure (e.g., increased activity) might increase predation, and extension of a life cycle could reduce the number of generations per year and put the organism out of phase with its environment. Quite obviously organisms in nature are constantly evolving, and one could argue that adaptations to the kinds of effects documented by Morgan (1976) would occur. It should be recognized that any measurable effect of a pesticide (or other pollutant) represents a change to which the organism must respond and that whether this results in advantage or disadvantage only the organism and the community in which it lives can demonstrate. Such demonstration takes time and an understanding of ecological interactions that we do not possess. The fact that measurable effects occur at levels of diazinon much below LC$_{50}$ values for toxicity would not be unexpected. Morgan's data suggest that "safe" levels of diazinon in fresh water streams are, for some species at least, less than one-fiftieth the LC$_{50}$

determined on the basis of a 7 day exposure. Although diazinon has seldom been reported in residue surveys, Harris and Miles (1975) found levels in an agricultural drainage canal of 1.4 to 2.0 ppb. Such levels are high enough to induce sublethal effects in chironomids, an important source of food for fish.

In an extensive series of tests Butler (1963) showed high toxicity of DDT to oysters and varying toxicity with several other pesticides. He found that aldrin, BHC, chlordane, dieldrin, endrin, heptachlor, chlordecone, endosulfan, toxaphene, ferbam, naled, disulfoton, and methyl trithion reduced oyster shell growth at levels below 1.0 ppm. Less toxic pesticides included mirex; carbaryl; 2,4-D; 2,4,5-T; eptam, diazinon; phosmet; malathion; and demeton. Other tests with molluscs showed that in the clam *Rangia cuneata* 96 hr LC_{50} values for toxaphene and carbaryl were 460 ppm and 125 ppm, respectively. Comparable values for the herbicides propanil and Molinate® were, respectively, 132 ppm and 197 ppm (Chaiyarach et al., 1975).

Studies on the toxicity to the lobster (*Homarus americanus*) of fenitrothion (McLeese, 1974) and phosphamidon (McLeese, 1974a) showed that, while the 96 hr LC_{50} value at 15°C was about 1 ppb with fenitrothion, phosphamidon was much less toxic, with LC_{50} values of 50 ppb at 12°C and 180 ppb at 4°C. In the latter case, LC_{50} values were the same whether run in static water or continuous flow.

Although ferbam was included in the tests of Butler (1963) and shown to be highly toxic to the oyster, there is a scarcity of data with respect to effects of fungicides on aquatic invertebrates. Canton (1976) included *Daphnia magna* in his tests and showed that, for benomyl, thiophanate-methyl and MBC, the 2 day LC_{50} (ppm) was 0.64, 16, and 0.46, respectively.

Since the dosage of TFM required for lamprey control varies widely with water pH and hardness, dosages vary from one stream to another. Smith (1967) concluded from laboratory tests that, if TFM were applied at the maximum dosage (10 ppm) for the water tested, 100% kill of hydras, turbellarians, and black flies would be expected. In addition high mortality would occur in burrowing mayflies, in some leeches, and in clams. More recently Maki et al. (1975) determined 24 hr and 96 hr LC_{50} values in water where the application rate of TFM would be 9 ppm to effect lamprey control. They confirmed the toxicity of TFM to many invertebrates, finding, as Smith (1967) did, burrowing mayflies and black flies highly sensitive. These workers also reported that sublethal exposure caused some species of insects to emerge as adults sooner and caused hyperactivity in the arthropods. Their data contrast with those of Morgan (1976) for diazinon, where it was shown that one sublethal effect was a delay in emergence.

Many larviciding programs for control of mosquitoes, black flies, and gnats attest to the fact that, when insecticides are introduced to aquatic environments in concentrations adequate for control of target species, many nontarget insects are destroyed (Muirhead-Thompson, 1971). Large numbers of dead insects were noted also in streams and rivers that flowed through eastern forests treated by air with DDT in the early days of the spruce budworm control project. Although the number of dead insects and those that have been affected enough to be dislodged from their normal habitat is large, the significance of this may not be great except in the case of pesticides such as DDT that may persist at toxic levels for some time. Wallace et al. (1973) studied the effects of three insecticides on stream invertebrates when the insecticides were used to control black fly larvae in streams in northern Quebec. Insect populations were estimated on the basis of cone, rock, drift net, and Surber sampling. Temephos and chlorpyrifos were applied at dosages to achieve a 0.1 ppm concentration in the stream for 15 min and methoxychlor to obtain a 0.075 ppm dosage for the same time. Two treatments were applied, the first between June 14 and 26, and the second July 15 to 23. Surber samples were taken before and 24 hr after each treatment. Drift net sampling was for a 24 hr period before and after the application of the pesticides.

Some of the results obtained are summarized in Table 15.9, and while this summary does not do justice to the vast amount of information obtained, it emphasizes two important aspects. The first is that pesticides such as temephos, applied for black fly control, kill many target and nontarget insects. This is apparent from the large increase

Table 15.9   Effect of Temephos Treatment on Stream Insects as Indicated by Relative Number of Insects Taken in Surber Sampling and Drift Nets Before and After Treatment[a]

Insect	Pre-Treatment Drift Net[b]	Pre-Treatment Surber[c]	Post-Treatment Drift Net[b]	Post-Treatment Surber[c]
Simuliidae	5000	87	75000	137
Ephemeroptera	1000	327	15000	234
Plecoptera	800	103	8000	133
Trichoptera	1500	414	25000	305
Chironomidae	200	863	1500	674

[a] From Wallace et al., 1973.
[b] Numbers are extrapolated from a graphical presentation and while indicating trends, may not be accurate.
[c] Figures are totals for 3 sampling sites before and after each of two treatments.

in drift net samples between pretreatment and posttreatment numbers with each of the insect groups. Although the data (Table 15.9) include only those obtained with temephos, effects with methoxychlor and chlorpyrifos were similar. The second and equally important point is that the interpretation of data must be rationalized with respect to the method by which the data are obtained. Thus, on the basis of Surber sampling, one would conclude that none of the pesticides had a great effect on insect populations, while a quite different conclusion would be drawn from the drift net data. In this study additional samplings with polystyrene cones and rocks demonstrated high removal of larvae of the target species with each of the pesticides.

It is obvious from the data presented (Wallace et al., 1973) that large numbers of both target and nontarget insects were killed with the larviciding programs. It is equally obvious (from Surber samples) that a population also survived. The significance of this surviving population is underscored by their data, which show that the pretreatment counts for the second treatment (about 30 days after the first) were about the same for points in the rivers above or below the point where the first treatment had been applied.

The ability of insects to recover their population densities is indicated also in extensive treatment of the Saskatchewan River with methoxychlor (Fredeen, 1974) and with diazinon in a small stream (Morgan, 1976). In the Fredeen study a large river was treated and control of black flies achieved over many miles downstream of the treatment site. By sampling the river bed and establishing recolonization sites for aquatic larvae, he found heavy displacement and/or mortality in Simuliidae, Plecoptera, and Ephemeroptera but less effect on Chironomidae and Trichoptera. The rates of recolonization were rapid, however, with the exceptions of Simuliidae and Plecoptera. Populations of Chironomidae, Trichoptera, and Ephemeroptera had returned to pretreatment levels within 14 days.

Working with diazinon in a small stream at the rate of 3 ppb, Morgan (1976) found that a treatment to give this exposure for 20 min increased drift of chironomid larvae, *Ephemerella* spp., *Chloroterpes* sp., *Hydropsyche* spp., and *Cheumatopsyche* spp. within the first 3 hr after treatment. When the treatment was repeated at 3 day intervals for 11 weeks, differences in sensitivity among stream invertebrates were noted with resulting changes in the benthic community. These changes did not persist, and communities returned to normal within 4 weeks after treatments stopped.

The studies reported here indicate that, whereas large immediate mortality may occur in many species within the aquatic invertebrates following pesticide treatments, permanent changes are unlikely. Fredeen (1974) could find no indication of community shifts, except for black fly populations, after several years of larviciding with methoxychlor in the Saskatchewan River. While this may be so, it must be noted that such large kills, though temporary, may affect the supply of food for fish and also that killed or dying insect larvae might contain pesticide levels toxic to their predators.

Quite a different situation may exist where persistent pesticides are involved. Wallace and Brady (1971) found major differences in invertebrate fauna above and below a point in a small river where a discharge of dieldrin from a textile mill emptied. Sampling sites only a few meters apart indicated that, while colonization of the stream bottom with Ephemeroptera, Trichoptera, and Diptera occurred above the discharge, these were absent below. Populations of Oligochaeta were similar at both sites, but below the discharge the insect fauna comprised only a few chironomids and simuliids.

Since many pesticides are highly toxic to invertebrates, it might be assumed that this sensitivity would result in mortality with little accumulation of residues in the animals. This is not true, at least not for the persistent organochlorine insecticides. Many analyses have been reported on various aquatic organisms, both inland and marine, and these have been tabulated by Edwards (1973a). While most residues in fresh water invertebrates are in fractional parts per million, there are some notable exceptions. Tubificid worms have been reported with 365 ppm dieldrin. The annelids seem less sensitive to the chlorinated hydrocarbons than most other invertebrates, and it will be recalled that, in the soil environment, earthworms survived heavy exposure but contained high residue levels. It has also been shown (Markin et al., 1974) that another group of annelids, the leeches, contained one of the highest residues of mirex following an application of mirex bait for ant control.

Table 15.10 summarizes some of the more recent data on residues of two insecticides in some major groups of aquatic invertebrates under natural conditions. Because most of these data are recent, it might be expected that they would reflect a reduction in DDT levels owing to the withdrawal of most of the uses of this insecticide during the late 1960s. While it is difficult to make comparisons, a perusal of the figures in Table 15.10 versus those of Edwards (1973a) does not indicate any obvious decline.

The St. Margarets Bay area of Nova Scotia is remote from any large-scale use of DDT, and in 1973 Hargrave and Phillips (1976) collected and analyzed a number of benthic invertebrates from the

Table 15.10   Some Pesticide Residues in Aquatic

Pesticide	Organism	Wet Weight Residue, ppb
DDT (R)	Decapods – Crustaceans	12.0-191.0
	Gastropods (Mollusc)	4.2-29.3
	Holotharoidea (Echinoderm)	3.1
	Decapods – Crustaceans	11.0-24.0
	Pelecypods – Mollusca	0.4-10.1
	Polychaeta – Annelida	8.0-24.0
	Euphausiacea – Crustaceans	6.5-64.8
	Branchiopoda – Crustaceans	83.0
	Decapoda – Crustaceans	160.0
	Gastropoda – Mollusc	260.0
	Decapoda – Crustaceans	37.0-92.0
	Pelecypods – Mollusca	27.0
	Pelecypods – Mollusca	ND-70.0
	Arthropoda – Insect Larvae	16.0-24.0
	Decapoda – Crustaceans	52.0-196.0
	Plankton	5.0
	Plankton	40.0
	Plankton – (Seston)	ND-Tr.
	Plankton	Tr.-200.0

deep water (60 to 75 m) and littoral areas (5 to 20 m) of the bay. They found residues of DDT and/or its metabolites in all species; although there was a fair range in residues between different samples within each class or phylum, levels were surprisingly uniform among the different groups (Table 15.10). Most of the residues in the St. Margarets Bay study were of p,p'-DDE, except in the Echinodermata, Pelecypoda, and Polychaeta, where p,p'DDT was also present; $p'p'$DDD was found only in the pelecypods. It is not known whether

Fauna and in Plankton in Natural Environments

Environment	Reference	Remarks
Caribbean (Coast)	Reimold, 1975	
St. Margarets Bay, N.S.	Hargrave & Phillips, 1976	
St. Margarets Bay, N.S.	Hargrave & Phillips, 1976	
St. Margarets Bay, N.S.	Hargrave & Phillips, 1976	
St. Margarets Bay, N.S.	Hargrave & Phillips, 1976	
St. Margarets Bay, N.S.	Hargrave & Phillips, 1976	
Gulf of St. Lawrence	Sameoto et al., 1975	
Carmans River Estuary, L.I.	Woodwell et al., 1967	
Carmans River Estuary, L.I.	Woodwell et al., 1967	
Carmans River Estuary, L.I.	Woodwell et al., 1967	
Everglades Park, Florida		
Everglades, Florida	Kolipinski et al., 1971	
Galveston Bay	Casper, 1967	
Inland water	Moubry et al., 1968	
Maine stream	Dimond et al., 1974	Sampled '71. Treated '60 & '63.
Inland lake, S. Dakota	Hannon et al., 1970	
Carmans River Estuary, L.I.	Woodwell et al., 1967	
Great Lakes	Glooschenko et al., 1976	
Gulf of Mexico, Caribbean	Giam et al., 1973	

this represents a difference in the ability of organisms to degrade DDT within their system or reflects a compartmentalization of residues within aquatic substrates that might be reflected because of different feeding habits. The mussel *Mytilis edulis* (Pelecypoda) sampled in this study is a filter feeder and contained residues of DDE, DDT, and DDD in the order of 22, 40, and 37%, respectively. By contrast the herbivorous snail, *Littorina littorea* (Gastropoda), contained only DDE. The levels of total residue of DDT and metabolites

Table 15.10

Pesticide	Organism	Wet Weight Residue, ppb
Dieldrin	Decapoda - Crustacea	0-<10
	Gastropods - Mollusc	33240.0-62470.0
	Arthropoda - Simulium	14840.0-24530.0
	Arthropoda - Caddisfly Larvae	17280.0-103670.0
	Pelecypoda - Mollusc	ND-10.0
	Plankton	ND-Tr.

was about the same in the mussel and snail. Hargrave and Phillips believed the levels found in their study represent background levels for areas removed from direct exposure to DDT.

A perusal of the data in Table 15.10 with respect to Mollusca indicates some DDT in widely separated waters of North America with varying degrees of local DDT use. The levels found in St. Margarets Bay (4.2 to 29.3), an area of minimum local DDT use, are of the same order as those from Galveston Bay and the Everglades area of Florida. DDT use was certainly extensive in each of the latter two regions. Molluscs from Carmans River contained higher levels.

Laboratory studies make it clear that the uptake of DDT by aquatic invertebrates is directly proportional to the concentration of insecticide in the medium and to the duration of exposure (Bevenue, 1976). Derr and Zabik (1972) demonstrated this with DDE and the Chironomid *Chironomus tentans,* and though they used dosages that they suggested were representative of natural levels of contamination, they found higher levels in their larvae than were reported from field studies. Working with 10 species of invertebrates, Johnson et al. (1971) found that the mean DDT residues were 0.62, 1.25, and 2.00 ppb, respectively, at 1, 2, and 3 days' exposure in a continuous flow system to 0.1 ppb pesticide in the water. Levels of uptake were drastically different. After 3 days *Daphnia* contained the highest residue at 9.17 ppm, while the decapod *Orconectes nais* contained the least at 0.233 ppm. The relationship of increased residue to increased exposure time was demonstrated also with aldrin and three species of invertebrates.

It is clear that many aquatic invertebrates concentrate persistent pesticides in their tissues at levels much in excess of those in the

(*Continued*)

Environment	Reference	Remarks
Caribbean	Reimold, 1975	
Rocky River, N.C.	Wallace & Brady, 1971	In path of discharge
Rocky River, N.C.	Wallace & Brady, 1971	In path of discharge
Rocky River, N.C.	Wallace & Brady, 1971	
Galveston Bay	Casper, 1967	
Great Lakes	Glooschenko et al., 1976	

surrounding environment. Edwards (1973a) tabulates a large number of data reported in the literature and calculates magnification factors, that is, concentration in the organism divided by the concentration in the medium. These show large differences between organisms for a given pesticide and for the range of pesticides, organisms' calculated magnification factors ranging from less than 10 to more than 1 million. Determining these magnification factors in natural waters is difficult and may be imprecise, since the water residues are often near or beyond the sensitivity of the analytical method, and where high levels are found, these are usually transitory. Laboratory experiments in which the level of pesticide can be monitored carefully are a better basis on which to extrapolate. Johnson et al. (1971) exposed a number of species of invertebrates to ^{14}C-labeled DDT and aldrin in a fresh water continuous flow system and measured the uptake after 1, 2, and 3 days. Their data, summarized in part in Table 15.11, show that magnification was greatest in the cladoceran *Daphnia magna* for both DDT and aldrin. Aldrin uptake in decapods was not included in this study, but with DDT, these organisms accumulated the pesticide much less than *Daphnia, Gammarus,* or four of the six insects tested did. Among the insects there were large differences. While larvae of *Culex pipiens* concentrated DDT in 2 days by a factor of 133,600, naiads of the odonate *Libellula* sp. concentrated the pesticide only by a factor of 910 during this same period. Hamelink et al. (1971) found that in an experimental area (farm pond, artificial pools) where an initial application of DDT declined in level in the water as days from treatment increased, this was reflected in DDT levels in the invertebrates. This would be expected, for it has been shown that many invertebrates can metabolize DDT (O'Brien, 1967). By contrast Dar-

Table 15.11 Residues of Two Insecticides in Each of Several Species of Aquatic Invertebrates and Magnification Factor After the Exposure Period Indicated[a]

Pesticide	Period of Exposure	Organism	Concentration		Magnification Factor
			Water (ppb)	Organism (ppm)	
DDT	3 days	Daphnia magna adult	803	9.17	114100
	3 days	Gammarus fasciatus adult	813	1.68	20600
	3 days	Orconectes nais adult	803	.233	2900
	3 days	Palaemonetes kadiakensis adult	1000	.503	5000
	3 days	Hexagenia bilineata nymph	521	1.68	32600
	3 days	Siphlonurus sp. nymph	470	1.08	22900
	2 days	Ischnura verticalis naiad	1013	.375	3500
	2 days	Libellula sp. naiad	793	.072	910
	3 days	Chironomus sp. larva	463	2.2	47800
	2 days	Culex pipiens larva	1046	13.9	133600
Aldrin	3 days	Daphnia magna adult	167	2.4	141000
	3 days	Hexagenia bilineata nymph	213	0.66	31400
	3 days	Chironomus sp. larva	213	0.48	22800

[a]Adapted from Johnson et al., 1971.

row and Harding (1975) concluded from tests with marine copepods (*Calanus* spp.) that these were unable to metabolize DDT. In sea water containing DDT at about 0.2 ppb, prolonged exposure was reflected in increased residues in the organisms for about 2 weeks followed by a leveling off. From these data, on a wet weight basis, the copepods had a magnification factor at 7 days of 40,600; at 14 days, 73,100; at 21 days, 800,000; and at 28 days, 84,100.

While it is obvious that significant levels of persistent pesticides, primarily DDT and its metabolites, occur in aquatic invertebrates, most other pesticides are not found except in a transitory way during or immediately after pesticide treatment. The "transitory" nature varies with the pesticide. Fitzpatrick and Sutherland (1976) found residues of chlorpyrifos in the salt marsh snail (*Melampus bidentatus*) immediately after treatment and up to 5 weeks later when the granular formulation was used but for only 3 weeks after treatment when the emulsion was employed. Fredeen et al. (1975) found residues of methoxychlor in insect larvae only during the actual larviciding in the Saskatchewan River and then only in larvae dying from the treatment.

Effects on Amphibians and Reptiles

Little work has been done on the effects of pesticides on amphibians and reptiles. Sanders (1970) included Fowlers toad and western chorus frog tadpoles in acute toxicity tests with some pesticides, and Tucker and Crabtree (1970) reported some values for the bullfrog (Table 15.6). These seem tolerant to pesticides relative to many invertebrates but not as tolerant as mammals (Chapter 19). Hoffman and Surber (1949) observed effects of an aerial application of DDT at 1 lb/acre for gypsy moth control in two watersheds (acreage 52,000). In these treatments efforts were made to avoid treating lakes directly, and though a few dead frogs, tadpoles, and salamanders were found, mortality was considered minimal and not nearly so prominent as among fishes. Many frogs were found after the spray had been applied, and these appeared normal. In reviewing the pesticide-wildlife situation, DeWitt and George (1959) report that cold-blooded terrestrial vertebrates "have tolerated" DDT in amounts up to 1 lb/acre. These workers reported observations on the effects of the fire ant control program where dieldrin and heptachlor (mostly the latter) were distributed in granular form. In treated areas ground dwelling species (*Rana* sp.) were reduced severely in numbers, while tree frogs (*Hyla, Acris*) were not. The toxicity of three pesticides to various amphibians was tested by Cooke (1972). He treated frog spawn by immersing it in 0.5 ppm DDT in saline and found that, when freshly laid spawn was treated, tadpoles hatching from it behaved normally but contained 19.4 ppm DDT 1 day after hatching. Four days later, when the external gills had been lost, the DDT level had dropped to 7.6 ppm owing to growth. Tadpoles from treated spawn were hyperactive relative to controls, and growth was retarded to the point where, when hyperactivity ceased (about day 13), treated tadpoles were only 62% as heavy as untreated controls. Treatment of spawn 2 or 3 days before hatching did not result in any DDT in tadpoles or any observable effects.

Frog and toad tadpoles were treated with DDT and dieldrin at various concentrations, and frog tadpoles were treated with 2,4-D. In addition tadpoles of the common newt (*Triturus vulgaris*) were treated with DDT. Both DDT and dieldrin were toxic, but there were differences in sensitivity with different ages of tadpoles and with different species. 2,4-D was not toxic, even when added at 50 ppm and exposure time was extended to 48 hr.

It would appear from these tests that frog tadpoles would not survive in a 0.5 ppm DDT medium and that sublethal effects on behavior and morphology would occur at levels much below this. Abnormalities in the snout and a "kink in tail" were observed, as well as color changes and retarded removal of external gills with DDT or dieldrin exposure at sublethal levels.

Some of the abnormalities associated with DDT and dieldrin in these experiments had been observed previously (Cooke, 1970), and the "kink in the tail" has been reported in tadpoles in a pond believed to be contaminated with atrazine (Hazelwood, 1970).

Meeks (1968) conducted an exhaustive study of uptake of DDT by marsh flora and fauna after treatment of a 4 acre marsh near Lake Erie with chlorine-36-ring-labeled DDT at the rate of 0.2 lb A.I./acre. He found residue levels in the leopard frog ranged from a mean of 1.3 ppm 1 day after treatment to 0.2 ppm after 12 months. Leopard frog tadpoles contained, however, high levels of 13.5 ppm after 1 day, 6.5 ppm after 7 days, and 3.0 ppm after 2 weeks. All residue data reported by Meeks (1968) are on a dry weight basis. Conversion to wet weight basis, in the case of tadpoles, places the 13.5 ppm residue at approximately 3.0 ppm. Judged from the data of Cooke (1972), it might be expected that these levels would cause abnormal behavior and probably mortality. It is of interest to note that Meeks (1968) reports that "tadpoles were not encountered the second year."

Snakes seem more susceptible to chlorinated hydrocarbon insecticides than toads and frogs do. Dead snakes were reported in Florida following mosquito control programs using DDT (Mills, 1952), and

George and Stickel (1949) found four dead rough green snakes in a tall grass area that had been treated with 10% DDT dust at 44 lb/acre for control of the lone star tick. In their Pennsylvania study following DDT treatment for gypsy moth control, Hoffman and Surber (1949) found dead water snakes, though frogs and salamanders in the same area were affected little. In areas treated with heptachlor for fire ant control, snake populations were reduced (DeWitt and George, 1959). These authors reported dead snakes associated with many areas of heptachlor treatment for fire ant control and in addition mortality and reduced populations of turtles.

Residues of pesticides in reptiles have been reported in only a few studies. Using labeled DDT, Meeks (1968) found that, after field application of DDT at 0.2 lb/acre, residue levels in snakes and turtles varied with time, being highest about 2 months after application but persisting in measurable amounts for 15 months. Higher residues were found in the northern water snake than in the fox snake (Table 15.12). Fleet et al. (1972) studied the pesticide residue levels in snakes from two flood plains in Texas, one in an area of cotton production and heavy pesticide (DDT, toxaphene, methyl parathion) use, the other a grazing area with little use of pesticides. In this study 10 species of snakes were collected in the low pesticide use area but only 8 in the area of high pesticide use, though total number of snakes was higher (as judged by number captured or seen during a specified search time) in the latter area. Pesticide levels (in part in Table 15.12) were grossly higher in the fat of snakes from the high pesticide use area. Total residue levels were found to be comparable in four species, the blotched water snake, common water snake, ribbon snake, and cottonmouth. Each of these snakes is semiaquatic with similar food prey. Less residue was found in snakes (DeKay's snake, copperhead, speckled king snake, rough green snake) that are primarily terrestrial.

Effects on Fish

Soon after DDT was introduced it was found that it was not only highly effective as an insecticide but also toxic to fish. Tests in Algonquin Park, Ontario, in 1944–1945 showed that small eastern brook trout, as well as other species, succumbed when placed in water containing even minute amounts of the chemical, amounts that for any pesticide used prior to the DDT era would have been considered insignificant (Kerswill, 1958). Two particular areas where DDT was extremely effective was in mosquito and biting fly control and in the

Table 15.12 Some Pesticide Residues Reported in Amphibians and Reptiles in Nature

Pesticide	Animal	Residue (ppm)[a]	Reference
DDT	Frogs	0.13-1.58	Dimond et al., 1975[b]
DDT	Tadpoles	1.23-1.67	Dimond et al., 1975[b]
DDT	Toad (Bufo americana)	2.39	Dimond et al., 1975[b]
DDT	Snake (Thamnophis sirtalis)	0.91-3.20	Dimond et al., 1975[b]
DDT	Salamander (Plethodon cinereus)	0.17-2.06	Dimond et al., 1968[c]
DDD	Bullfrogs	5.0	Hunt & Bischoff, 1960[d]
DDT-R	Common water snake	0.6-1.2	Fleet et al., 1972
DDT-R	Copperhead	1.1	Fleet et al., 1972[e]
DDT-R	Speckled king snake	2.3-5.4	Fleet et al., 1972[e]
DDT-R	Common water snake	454.5-696.9	Fleet et al., 1972[f]
DDT-R	Ribbon snake	272.7-913.7	Fleet et al., 1972[f]
DDT-R	Cottonmouth	369.6-1014.1	Fleet et al., 1972[f]
Dieldrin	Copperhead	0.1	Fleet et al., 1972[e]
Dieldrin	Cottonmouth	3.4-9.7	Fleet et al., 1972[f]
DDT	Leopard frog	0.1-2.3	Meeks, 1968[g]
DDT	Northern water snake	30.3	Meeks, 1968[h]
DDT	Fox snake	5.1	Meeks, 1968[h]
DDT	Blanding's turtle	5.8	Meeks, 1968[h]
DDT	Painted turtle	7.4	Meeks, 1968[h]

[a] Parts per million on wet weight basis; total animal unless indicated otherwise in footnote.

[b] Collected from a forest area one year after DDT treatment at 1.0 lb/acre.

[c] Frogs mostly Rana clamitans, some R. pipiens, R. palustris and R. catebiana.

[d] Samples from DDT treated forest - year of treatment.

[e] Collected 8 months after treatment at Clear Lake at 0.020 ppm.

[f] Residue in fat from a low pesticide use area.

[g] Residue in fat body from a high pesticide use area.

[h] Two weeks after area was treated with DDT, 0.2 lb A.I./acre; ppm given on a dry weight basis.

[i] Residue in fat, 2 months after area was treated with DDT, 0.2 lb A.I./acre; ppm given on a dry weight basis.

control of major forest defoliators such as the spruce budworm, black-headed budworm, and gypsy moth. These uses placed DDT in the aquatic environment where large numbers of fish were exposed. Despite the early work showing that DDT was highly toxic to fish and the observed death of large numbers of all kinds of fresh water fishes following its early use, the benefits apparently seemed, to those planning control programs, to outweigh the risks (Mills, 1952). Writing in *Audobon* in 1952, Landuska wrote, "from some standpoints the DDT picture is not so ominous as it appeared a few years ago. Many costly forest pests are effectively controlled with a one-half pound per acre dosage. This amount appears to have little effect on most forest vertebrates. Even these programs, however, require considerable planning and careful guidance if damage to the highly susceptible inhabitants of productive waters is to be avoided. Of course, a forest that is defoliated or killed outright by an insect outbreak is neither a healthy watershed nor a productive wildlife habitat. And in some circumstances the nominal loss of fish and wildlife accompanying such insect abatement may be heavily outweighed in favour of the ultimate advantages in the over-all wildlife picture." The rationale advanced in this quotation is still valid, but the concept of "nominal loss" in the case of fish was to be shattered by evidence of excessive mortality in several large-scale DDT spray operations.

The first major demonstration of massive fish mortality was associated with a large-scale spray operation to control the spruce budworm in New Brunswick forests (Chapter 5). The program began in 1952 using 1.0 lb DDT/acre and was continued in 1953 and 1954 using the insecticide at one-half this dosage. The area sprayed was so large that avoidance of trout streams was impossible and drift contamination of major rivers equally unavoidable. In fact no great effort was made during the early spray years to avoid spraying over water areas, since the potential for damage to the fishery was not appreciated. In 1954, more than 1 million acres were treated, the treated area including one-third of the watershed of the Miramichi River, a major spawning ground of the Atlantic salmon (Kerswill and Elson, 1955). Atlantic salmon spend 3 years in these tributaries. The young salmon are born from eggs laid in the gravelly base of a cold stream. These eggs hatch in 60 to 100 days, and the young alevins or fry remain in the gravel until the yolk sac attached to their bellies is absorbed. They then move to the water above and feed on small organisms. They grow slowly, becoming small parr the second year and larger parr the third year, leave the river as silvery smolts about the fourth year, and run to sea. After spending 1 to several years at sea, ranging as far as the Newfoundland, Labrador, and Greenland waters (Netboy, 1969), they return as grilse or larger salmon to the rivers in which they were produced to spawn and repeat the cycle. Information on the life history of the salmon permitted fisheries personnel to assess the direct kill on salmon and to predict the significance of this kill on future salmon supplies (Kerswill, 1958).

When the area containing major tributaries of the Miramichi were first treated with DDT at 0.5 lb/acre, counts were made of fish survival and compared with records from earlier years and with data from tributaries not in the treated area. The 1954 spraying of part of the Miramichi watershed included treating the area of the Northwest Miramichi for a distance of about 40 miles from its headwaters, and efforts were concentrated on that tributary plus the Dungravon River, whose headwaters area was treated also. In addition to observations on natural fish populations, cages containing $2\frac{1}{2}$ to $4\frac{1}{2}$ in. salmon that had been seined from natural waters prior to the spray application were placed in streams. Some cages were placed in streams where treatment was proposed, others in areas not to be treated. Of 75 parr placed in two cages in the Northwest Miramichi where spraying occurred, 68 were dead within 21 days of treatment. Only 1 of 50 parr caged in an untreated area died in a similar period. Patrolling of streams inside and outside the spray area showed that many fish were killed in the treated area, while no abnormal mortality was observed elsewhere. During several days after spraying, many small parr (up to $2\frac{1}{2}$ in., i.e., 1953 fry) were found dead and dying. In addition dead trout up to 13 in. long and suckers of various sizes were observed (Kerswill and Elson, 1955).

Detailed observations on young salmon showed that none of the fish that hatched in 1954 (i.e., fingerlings) survived in the treated area of the Northwest Miramichi. In the Dungravon River a few survived but only in areas downstream from the treated portion. Among small parr, about 15% survived in the Northwest Miramichi, and only 33% in the Dungravon. Survival of large parr that would have gone to sea in 1955 was slightly better, but less than half survived in the Northwest Miramichi and about 35% in the Dungravon. Although the actual spraying was done in June of 1954, large numbers of dead parr were still being observed in the Northwest Miramichi in late August and September (Kerswill and Elson, 1955).

The attack by the spruce budworm on New Brunswick forests continued, and the same area of the Northwest Miramichi was treated again with DDT in 1956 and 1957 with a repetition of fish kill. Figure 15.2 indicates the abundance of salmon during this period of treatment in relation to earlier years. The run of spawning salmon was not affected by the 1954 sprays, and this was reflected in an abundance of

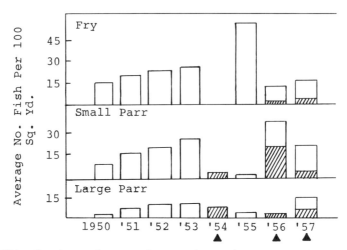

Figure 15.2 Abundance of young salmon in the Northwest Miramichi River. Solid triangles indicate years when DDT was sprayed in the area. Cross-hatching indicates fish found in sprayed parts of rivers. [Redrawn (in part) from Keenleyside, 1959.]

fry in 1955. The low numbers of small and large parr in 1955 and large parr in 1956 reflect the dramatic kill of the 1954 fry and small parr.

In nature the onset of the winter season brings with it falling temperatures in New Brunswick rivers. To simulate this natural occurrence, live parr were collected in October from streams treated at 0.5 lb, 0.25 lb, and zero dosages of DDT. Some were placed in tanks, and the water temperature was lowered between October 5 and November 10 from 57°F to 53°F. This resulted in 40% mortality of parr from the 0.5 lb treatment, about 3 times the death rate observed in the fish from the 0.25 or untreated waters. When the water temperature was dropped rapidly to 33°F, 80% of the fish from the high treatment area died in 8 days, but all fish from the low dosage area and from the untreated waters were living after 40 days (Kerswill, 1961).

The demonstrated high toxicity of DDT to fish focused attention on these as possible sensitive organisms to pesticides. Many excellent laboratory studies have been done to assess this point, and some of the results reported are summarized in Table 15.13. Additional data on other pesticides and other fish species can be found in many reports, including those of Butler (1963), Cope (1963), Edwards (1973a), Eisler (1970), Henderson et al. (1959), Hiltibran (1967), Holden (1973), Muncy and Oliver (1963), and Pimentel (1971).

The reader should be cautioned, however, that a summary such as Table 15.13 is a guide only. Acute toxicity is influenced by a host of variables in testing methods and conditions of test materials, and these are not detailed in a summation of this sort.

It is apparent from a perusal of the accumulated data (Table 15.13) that the chlorinated hydrocarbon insecticides are in a class by themselves in terms of lethality to fish. Their extreme toxicity is approached only by rotenone, a compound known historically as a fish poison. It is thus not surprising that the chlorinated hydrocarbons as a class have been incriminated repeatedly in poisoning by both accidental and deliberate placement in fish-inhabiting waters.

With the exception of DDT most of the fish poisonings with chlorinated hydrocarbon insecticides have been small, local incidents due to contamination during spray operations adjacent to streams or to effluent from pesticide-producing or pesticide-handling plants. A massive fish kill in the Mississippi River was believed due to discharge of endrin in effluent from a plant manufacturing this chemical (Mount and Putnicki, 1966). Runoff following application to agricultural soils can be significant, but only in a few cases has the amount been sufficient to produce fish kills. Lauer et al. (1966) reported fish kills in Louisiana bayous and associated these with runoff from sugar cane areas where endrin was used for pest control. Other instances could be cited where more extensive fish kills have occurred. Endosulfan released in the Rhine River resulted in a fish kill for many miles of that large waterway (Lüssem and Schlimme, 1971). In 1958 a sprayplane carrying endrin crashed in a Louisiana bayou, and fish were killed for 35 miles downstream (Wade, 1969).

The high toxicity of organochlorine insecticides to fish resulted in an intensive search for alternate effective insecticides for use in aquatic areas. In the spruce budworm control projects in Canada and the United States, dimethoate, fenitrothion, and phosphamidon have been used in recent years. These are organophosphorus insecticides, and as shown in Table 15.13, this class of compounds in general, and these three in particular, have low acute toxicity to fish. In a similar way chlorpyrifos, malathion, temephos, and propoxur have, to a large extent, replaced DDT in the control of mosquitoes and black flies. In these programs one chlorinated hydrocarbon, methoxychlor, is still used. This insecticide is about as toxic to fish as DDT, and though some reports suggest sublethal and indirect effects on fish from its use, most of the evidence indicates little problem. For example, the dosage required for black fly control (about 0.3 ppm for 15 min in running water), though well above the lethal level for a 48 hr exposure, seems to be tolerated.

The main substitutes for DDT in forest spraying, dimethoate, phosphamidon, and fenitrothion, have similarly not been shown to cause direct mortality at the dosages required. For example, phosphamidon was used to spray a creek over a distance of 1300 ft at the

Table 15.13 Acute Toxicity of Selected Pesticides to Some Fresh Water Fish. LC$_{50}$ in ppb. Values Are for 48 hours Unless Indicated Otherwise[a]

Pesticide	Striped Bass[b]	Fathead Minnow	Rainbow Trout	Bluegill	Goldfish	Channel Catfish	Carp
Aldrin	7.2	28	3.0	13†	28†		
Chlordane	11.8		10*	77†	81†		
DDT	0.53	19†	7†	8†	21†	16†	10†
Dieldrin	19.7		19†		31†		
Endosulfan	0.1		1.2	3.4	8*		
Endrin	0.094	1.8†	1.8*	0.35*			
Heptachlor	3.0		9.0	19	230	140	
Methoxychlor	3.3	25*	7.2	30	56†		
Toxaphene	4.4			6.8†			
Temephos	1000.0						
Azinphosmethyl		235†	130†	22†	4200†	3250†	695†
Chlorpyrifos	0.58		20				
Diazinon			380*	52*			
Dimethoate			19000*	9600			
Ethion				230			
Fenitrothion	14.0	8650†	170†	103†	10700†	8910†	8600
Malathion	17.8	2050†	2000	47			6540†
Ethyl Parathion		8900†	2750	5720†	9000†	5710†	7130†
Methyl Parathion			8000				
Phosphamidon							
Carbaryl	1000.0	13000†	4380†	6760†	13200†	15800†	5280†
Carbofuran			280	240		210	
Propoxur		250					
Allethrin		19					

Rotenone					
Amitrole			22	>50000	
Atrazine	440000*	12600			
Dalapon		>500000	115000		
Dicamba		35000	130000		175000
Dichlorbenil		20000	20000		
Diquat	140000†	20000	145000	35000†	
Endothal	480000		280000	175000	
MCPA			10000†		
Monuron				75900*	
Paraquat	64000*	34000*	400000*		
Picloram			26500*	32500	
Prometone		56000	118000		
Simazine		152	210	100	
Trifluralin					
2,4-D			960	1300	
2,4,5-T (acid)		1300	500		
Copper Sulfate			150	>500	
Diclone		48	340*	>400	>330
Nabam				21100	140*
Benomyl		480			
TFM-2B		3850			

^{a}Data from numerous reports with extensive use of that collected by Pimentel (1971).
Water temperatures not constant. LC_{50}. Korn and Earnest, 1974.
^{b}Juvenile fish in saline 96 hours.
^{c}Rye and King, 1976.
*24 hour exposure.
†96 hour exposure.

dosage of 1 lb active/acre. Coho salmon fry were placed in the area prior to treatment, but no mortality was observed and the fry appeared to feed and act in a normal way. Laboratory-type assays conducted in aquaria in the stream using stream water and maintaining it at stream temperature showed that no mortality of juvenile salmon occurred with phosphamidon at 1 ppm for 24 hr. For either 24 or 48 hr exposures, mortality did not exceed 10% until concentrations of about 5 ppm were reached. Complete mortality occurred at 10 ppm (Schouwenburg and Jackson, 1966). The extensive use of phosphamidon and fenitrothion for spruce budworm control in eastern Canada has not produced direct fish mortality. Some indirect effects due to reduced food supply may occur (Varty, 1975).

Reference to Table 15.13 will suggest that the acute toxicity of herbicides to fish is low and that problems of direct fish kill should not be encountered. In contrast to the comparatively low concentrations of insecticides used for mosquito or black fly control, herbicides are recommended at high rates (diquat 0.5 to 1.0 ppm; 2,4-D and silvex, 2.0 ppm; endothal, 1 to 3 ppm; dichlobenil, 10 lb/acre; fenac, 20 lb/acre). In spite of this the margin of safety seems adequate and fish kills are not observed. Hiltibran (1967) tested 22 herbicides on eggs and small fry of bluegill, green sunfish, smallmouth bass, lake chubsucker, and stoneroller. He found that fertilized eggs hatched even in high concentrations (several ppm) of the herbicides and that small fry were not killed at dosages likely to be encountered in normal application procedures.

Factors Affecting Toxicity to Fish

A number of factors affect the toxicity of pesticides to fish. These include temperature, age of fish, duration of exposure, degree of salinity (estuarine species), formulation of pesticide, and amount and nature of suspended materials. While the last of these may be one of the most important, it is perhaps the most difficult to measure accutately. It is known that many pesticides are adsorbed to particulate matter and that this adsorption reduces the concentration of "free" pesticide in the water. If the suspended matter includes a large proportion of items that will end up in the digestive tract of the fish, then this may be reflected in toxicity. If, however, the suspended matter is largely inorganic and nonfish food items, a high amount of suspended material will probably reduce the toxicity of a given dosage applied in nature. Frank (personal communication) found that, when chlorpyrifos was used as a mosquito larvicide in Ontario wetlands, the chemical disappeared from filtered water very quickly but significant amounts were detected in particulate suspended matter for some days.

In experiments with toxaphene as a fish poison in lakes Stringer and McMynn (1958, 1960) found that "turbidity" of the water influenced the dosage necessary to achieve complete kill. While dosages of about 0.02 effected complete kill in lakes where the water was clear, only partial kill was obtained in three shallow lakes where turbidity was prominent (Secchi disc reading 1 to 3 ft). These authors suggest that turbidity may hasten the breakdown of toxaphene by microorganisms. Other workers have confirmed the reduced toxicity of toxaphene in turbid waters (Lennon et al., 1971).

Effect of Water Temperature

The effect of water temperature on the acute toxicity of pesticides to fish is apparent in a number of studies. In many cases the effect is not dramatic, though, in general, higher temperatures increase toxicity. This can be important in natural situations. Two spills of endosulfan occurred in the Rhine River. The first spillage occurred in June when the water temperature was $19 \pm 1°C$ and endosulfan residues reached 5 ppb in the West German section and 0.7 ppb in the Dutch section. Fish kills occurred in both sections. The second spill occurred in November when the water temperature was $11°C$ and residues as high as 1 ppb endosulfan were recorded. No fish kill accompanied the November spillage (Lüssem and Schlimme, 1971; Greve and Wit, 1971), and this was probably due to the reduced toxicity of endosulfan at the lower temperature. Macek et al. (1969) showed that the sensitivity of rainbow trout to endosulfan was increased more than fourfold by increasing the water temperature from 1.6 to 12.7°C. Although a number of studies could be cited to indicate temperature effects, that of Macek et al. (1969) will serve to illustrate the phenomenon. They found that, with most pesticides, increased temperatures increased the toxicity of the pesticide to either rainbow trout or bluegills. There was considerable difference, however, in the magnitude of the effect (Table 15.14). The increase of toxicity with increased temperature is not without exceptions. Methoxychlor was more toxic at the lower temperatures with both species of fish, the magnitude of the effect being about twofold. This negative temperature coefficient of about the same order of magnitude is known also for DDT (Holden, 1973). Whether or not it applies to fish has not been determined, but with respect to toxicity to insects, pyrethrum has been shown to be more toxic at low temperatures, and this is true also for some of the new synthetic pyrethroids (Harris and Kinoshita, 1977).

Table 15.14   24 and 96 hour LC_{50} Values (ppb) of Various Pesticides to
Rainbow Trout at Three Temperatures[a]

Pesticide	24 Hour LC_{50}			96 Hour LC_{50}		
	$1.6^\circ C$	$7.2^\circ C$	$12.7^\circ C$	$1.6^\circ C$	$7.2^\circ C$	$12.7^\circ C$
Aldrin	24	8.1	6.8	3.2	3.3	2.2
Azinphosmethyl	25	15	13	6.8	6.2	5.5
Chlorpyrifos	550	110	53	51	15	7.1
Dieldrin	13	3.1	3.1	2.4	1.1	1.4
Endosulfan	13	6.1	3.2	2.6	1.7	1.5
Endrin	15	5.3	2.8	2.5	1.4	1.1
Heptachlor	17	12	13	7.7	7.0	7.3
Methoxychlor	55	45	74	30	42	62
Naled	1300	620	240	340	220	160
Trifluralin	328	239	98	210	152	42

[a]Macek et al., 1969.

Effect of Duration of Exposure

Data in Table 15.14 indicate that the longer a fish is exposed to a
pesticide, the greater the mortality. These data are from a static system, and the LC_{50} values are those for the initial concentration.
They do not reflect the fact that pesticide has been taken up by the fish
and some has degraded in the water. Thus the values do not reflect a
full 24 hr or 96 hr exposure to the toxicant at the level stated. Eisler
(1970a) bioassayed the loss of pesticide during a period in standing
water by setting up duplicate tests with mummichogs (*Fundulus
heteroclitus*) and 12 insecticides. In one group he added the fish 45
min after adding the insecticide and in the other the fish were not
added until the pesticide had been in the water for 96 hr. The fish
were then kept in the test tanks for 96 hr and LC_{50} values determined. The ratio of the LD_{50} value after 45 min exposure to that
after 96 hr was calculated. Aldrin, malathion, and heptachlor are
known to degrade and/or volatilize from water quite rapidly, and this
was reflected in the bioassays. Ratios for the three insecticides were
0.13, 0.19, and 0.25, respectively. DDT, dieldrin, and endrin are
known to be persistent in water, and this was reflected also in the
bioassays. Ratios for these three insecticides were 1.00, 1.10, and 1.25,
respectively. With mevinphos and methyl parathion, ratios were, respectively, 3.82 and 6.60. These latter two values indicate an increase
in toxicity during 96 hr with two compounds that would be expected

to degrade during that time. No explanation can be offered.

Eisler (1970a), extending his tests, found that, when the test exposure was continued for 10 days (static exposure, LC_{50} values based on initial concentrations), the LC_{50} values for malathion; methoxychlor; dioxathion; p,p'-DDT; and aldrin were essentially the same at 96 and 240 hr. Toxicity was greater with dieldrin ($\times$ 1.5) and endrin ($\times$ 1.8) at the longer exposure. Much greater toxicity was found with heptachlor, mevinphos, methyl parathion, and lindane at the longer exposures.

In nature exposure of fish to high concentrations of pesticides is likely to be of short duration. Eisler (1970a) prepared concentrations of five pesticides at the LC_{75} levels and immersed mummichogs in these solutions for periods of 15, 120, 360, 720, and 1680 min. The mummichogs were transferred at the end of the exposure period to pesticide-free tanks, and mortality was observed for 21 days. He found that, with heptachlor, endrin, and DDT, there was little mortality when exposure was 120 min or less, significant mortality with exposure between 120 and 360 min, and high mortality with longer exposure. With methyl parathion and dichlorvos, even a 30 min exposure resulted in significant mortality.

Effect of Age of Fish

Not enough work has been done in this area to permit any generalizations. The few laboratory tests reported are conflicting and may indicate differences in response with different pesticides (Holden, 1973). For example, Lee et al. (1975) tested mirex and methoxychlor on young and adult mullet in continuous flow systems. At mirex dosages of 0.01, 0.1, 1.0, and 10.0 ppm they obtained significant mortality at the 0.10 and 1.0 ppm levels with young juveniles (20 to 43 mm) but no mortality at any of these dosages with older juveniles (70 to 150 mm) or adults (260 to 380 mm). When methoxychlor was the test insecticide, mortality at the 96 hr test period was about the same for young juvenile or adult mullet. Perhaps the best evidence comes from observations on fish kills after forest spraying with DDT. These indicate that for Atlantic salmon small fish are most sensitive, but it is impossible to tell whether this is related to the pesticide's exerting its effect through contact (gill or dermal) or in the food supply. In field experiments using toxaphene as a piscicide, Stringer and McMynn (1958, 1960) reported that, in lakes where incomplete kill had been obtained, it was the large fish that survived, suggesting greater susceptibility in smaller fish.

Effect of Salinity

Wade (1969) calculated the TLM (LC_{50}) value for dieldrin for sheepshead minnow (*Cyprinidon variegatus*) and sailfin molly (*Poecilia latipinna*) at salinity values of 0, 15, and 30 ppt, at three water temperatures for each species and under two levels of pH. No consistent effect of salinity could be detected, though with the minnow at 25°C, the LC_{50} value for low salinity was only about one-third of those at the higher salt concentrations.

In tests with mummichogs the toxicity of DDT, endrin, and heptachlor was not influenced by water salinity in the range 0.12 to 0.36%, but altering salinity markedly affected the toxicity of dichlorvos and methyl parathion.

Effects of Water Hardness

Water hardness is a general term and is interpreted in the ease with which soap suds can be generated. Hard water contains more calcium and (usually) magnesium ions than soft water, and hardness is usually expressed numerically in terms of mg/liter (ppm) of calcium carbonate. A number of investigators have examined the influence of water hardness on toxicity and found little general effect, though there have been some exceptions (Holden, 1973).

TFM is an effective lampricide, but the activity of the compound is related closely to water conditions, especially pH and hardness. It is most effective in soft water of low pH. Applegate and King (1962) evaluated TFM in natural water from a bay, a creek, and a river in the upper Great Lakes area and found that the dosage required for effective lamprey control varied from 3 to 10 ppm, the lowest dosage associated with the softest water. A similar range was found also for other fish. For example, the TFM dosage required to kill 25% of a test population of yellow perch was 7.25 ppm in the softest and 20.5 ppm in the hardest of the three waters tested.

Species of Fish

Macek and McAllister (1970) determined the LC_{50} values for nine insecticides on 12 species of fish representing five families. Their results, summarized in Table 15.15, indicate that, while differences among species were not great (within about one order of magnitude) with DDT, toxaphene, methyl parathion, fenthion, and mexacarbate, large differences were found with lindane and malathion, and differences in excess of a thousandfold found with azinphosmethyl. Gold-

Table 15.15 96-hour LD_{50} Values (ppb) for Nine Insecticides Tested Against 12 Species of Fish in Five Families[a]

Family Species	Azin-phos-methyl	Car-baryl	DDT	Fen-thion	Lin-dane	Mala-thion	Methyl Para-thion	Mexa-carbate	Toxa-phene
Ichtaluridae									
Channel catfish	3,290	15,800	16	1,680	44	8,970	5,710	11,400	13
Black bullhead	3,500	20,000	5	1,620	64	12,900	6,640	16,700	5
Cyprinidae									
Goldfish	4,270	13,200	21	3,404	131	10,700	9,000	19,140	14
Fathead minnow	235	14,600	19	2,440	87	8,650	8,900	17,000	14
Carp	695	5,280	10	1,160	90	6,590	7,130	13,400	4
Centrarchidae									
Redear sunfish	52	11,200	5	1,880	83	170	5,170	16,700	13
Bluegill	22	6,760	8	1,380	68	103	5,720	11,200	18
Largemouth bass	5	6,400	2	1,540	32	285	5,220	14,700	2
Salmonidae									
Rainbow trout	14	4,340	7	930	27	170	2,750	10,200	11
Brown trout	4	1,950	3	1,330	2	200	4,740	8,100	3
Coho salmon	17	764	4	1,320	41	101	5,300	1,730	8
Percidae									
Yellow perch	13	745	9	1,650	68	263	3,060	2,480	12

[a] From Macek and McAllister, 1970.

Table 15.16 96-hour LC$_{50}$ Values (ppb) for 12 Insecticides Tested Against Six Species of Estuarine Fish[a]

Insecticide	Atlantic silverside	Bluehead	Striped killifish	American eel	Mummichog	Northern puffer
Organochlorines						
Aldrin	13	12	17	5	8	36
p-p'-DDT	0.4	7	1	4	5	89
Dieldrin	5	6	4	0.9	10.5	34
Endrin	0.05	0.1	0.3	0.6	1.0	3.1
Heptachlor	3	0.8	32	10	50	188
Lindane	9	14	28	56	60	35
Methoxychlor	33	13	30	12	46	150
Organophosphorous						
Dichlorvos	1,250	1,440	2,300	1,800	2,680	2,250
Dioxathion	6	35	15	6	20	75
Malathion	125	27	250	82	240	3,250
Methyl parathion	5,700	12,300	13,800	16,900	58,000	75,800
Mevinphos	320	74	75	65	300	800

[a]From Eisler, 1970.

fish were about the most tolerant of the species tested, and in general the Ichtaluridae and Cyprinidae were less sensitive than the Centrarchidae, Salmonidae, and Percidae. There were some exceptions. Carp and bullheads were among the most sensitive to toxaphene. A similar experiment by Eisler (1970) showed gross differences in toxicity of some pesticides to six species of estuarine fish. Whereas differences in susceptibility among the six species were within about an order of magnitude for aldrin, lindane, methoxychlor, dichlorvos, dioxathion, methyl parathion, and mevinphos, larger differences were found with endrin, dieldrin, heptachlor, DDT, and malathion. The northern puffer was remarkably resistant to all pesticides tested, but among others relative susceptibility differed with different insecticides (Table 15.16).

Attempts to discover chemicals with differing toxicities to fish species for use as piscicides have not been widely successful. The lampricide TFM is selectively toxic to lamprey, but the safety margin is not large. Rye and King (1976) found the LC_{50} value for the lamprey to be 0.78 ppm and that for rainbow trout 3.85 ppm. Applegate and King (1962) determined the ratio of the concentration needed to kill lamprey larvae and that which would kill 25% of each of 11 other fish species. They found that, while the safety margin approached a factor of 10 for smallmouth bass, it was much less for other species. Values obtained (in parentheses) were as follows: largemouth bass (5.1), smallmouth bass (8.5), bluegill (5.1), walleye (1.3), yellow perch (2.2), yellow bullheads (1.6), white suckers (1.5), blacknose shiners (3.2), golden shiners (4.0), fathead minnows (4.1), and rainbow trout (3.8).

While most piscicides tend to be broadly toxic, some degree of selectivity is found in Squoxin® (1,1-'methylene di-2-naphthol) (Sonar 300), which has been shown to kill two species of squawfish at concentrations not toxic to game fish. The safety margin between the toxic dose for squawfish and salmonids would appear to be greater than 10 (MacPhee and Ruelle, 1969).

Combinations of Pesticides

The aquatic environment is exposed to a variety of pesticides and related materials, but few tests have investigated possible interactions in terms of fish toxicity. From a practical standpoint one of the most significant interactions might be with PCBs, since these are widely distributed. Halter and Johnson (1974) found that, with the coho salmon, the addition of Aroclor 1254 (a PCB) did not change the toxicity of DDT to eggs or alevins. While each compound was toxic, the effect of the combination was not additive, presumably because DDT killed more rapidly than the PCB did. A similar lack of additive effect was found when methyl parathion was added to endrin and tested on largemouth bass fingerlings (Fabacher, 1976). Not all tests with mixtures of pesticides have given this result. Marking and Mauck (1975) tested the toxicities of a number of potential forestry insecticides in combinations against rainbow trout. They included in their tests azinphosmethyl, mexacarbate, trichlorfon, Volaton®, a pyrethrum extract, and two synthetic pyrethroids. They found no great divergence from an additive effect when various combinations were used. In a few cases toxicity of mixtures was less than additive and in a few cases it was greater. A similar spectrum of effects was demonstrated by Macek (1975) using various concentrations and combinations of 19 pesticides on bluegills, but some of these effects were more prominent; 11 combinations had greater than additive toxicity, the greatest effect being demonstrated with parathion and malathion. With combinations of these two insecticides, mortality was increased 11.7 times over that of the same dosages used separately. Malathion and fenthion when combined caused mortality in bluegills 4.37 times greater than when the compounds were used separately.

Sublethal Effects

A wide variety of abnormal behavioral and pathological conditions, as well as failures in reproduction, have been reported in fish exposed to sublethal concentrations of pesticides. It has been shown also that some fishes have become resistant to pesticides and others have the ability to detect pesticides in water and to avoid them.

Temperature Effects

Perhaps one of the best documented effects has been the role of DDT in influencing temperature selection in Atlantic salmon. Ogilvie and Anderson (1965) acclimated under-yearlings at 8°C and 17°C and exposed them for 24 hr to various sublethal dosages of DDT. They then determined the effect of these exposures on temperature preference. They found that temperature preference differences between treated and untreated fish were more pronounced in fish acclimated at 17°C than at 8°C. The effect was not, however, consistent. Those fish exposed to low DDT concentrations (5 ppb) selected water temperatures lower than normal, while those exposed to DDT concentrations above 10 ppb selected water temperatures higher than control fish did. More recently, Ogilvie and Miller (1976) exposed under-yearling

Atlantic salmon to 50 ppb DDT for 24 hr and conducted tests for a period of 7 weeks to determine how long the shift in termperature preference would continue. (The dosage selected, 50 ppb, was high, and only seven of the original test fish survived the entire period.) The effect continued for 4 weeks. The greatest selection preference difference occurred on the first day, the treated fish selecting a temperature about 4°C higher than control fish. After 1 week the difference declined to about 2°C. No differences were noted after 6 weeks. Using brook trout and 20 ppb DDT, Gardner (1973) found that treated fish selected a lower temperature and this preference continued for at least 9 days following exposure. Since DDT has a negative temperature coefficient, the tendency for salmon exposed to near lethal dosages to seek higher temperatures would improve their chance for survival, while the opposite tendency in brook trout would increase toxicity. As previously discussed (Kerswill, 1961), when salmon parr were collected in the Miramichi River in New Brunswick from portions treated with DDT and from untreated portions and placed in cold water, a high percentage of those from the treated stream died. The temperature selection phenomena reflect some effect of DDT in the sensory organs of the lateral line and thus an effect on the peripheral nervous system (Anderson, 1968). Anderson and Peterson (1969) showed that DDT affected the cold-blocking temperature of the tail reflex, and because this is controlled in the spinal cord (Roots and Prosser, 1962), an effect on the central nervous system is demonstrated also.

Behavioral Effects

In addition to its thermocline and cold-blocking effects noted earlier, DDT also affects learning ability in fish. This was shown for brook trout by Anderson and Peterson (1969) in an inability to learn to distinguish dark and light sides of a culture tank. Dill and Saunders (1974), exposing eggs of Atlantic salmon to various concentrations of DDT, found that the normal behavioral pattern of alevins in the ground bed and in their move to the waters above were abnormal and retarded in alevins from eggs exposed to high but sublethal concentrations of DDT. They found also that fry from treated eggs had impaired balance.

Behavioral effects have been noted in a number of studies with pesticides other than DDT. Low levels (0.005 of LC_{50}) of TEPP were shown to affect the learning ability of goldfish, and this was demonstrated also with toxaphene at levels well below threshold values for acute toxicity (Warner et al., 1966). Cairns (1972) showed that diazinon at a concentration as low as 1.25 ppm resulted in hyperactivity in rainbow trout and affected their pattern of distribution. Fish exposed to this level of diazinon were more aggressive and fought more frequently. At a diazinon level of 5 ppm fish behaved normally, while at 12.5 ppm fish were sluggish and engaged in few fights. Experiments with the tin-plated barb showed that at sublethal dosages of diazinon schooling behavior was abnormal, both the size of the school and the depth at which it was formed being altered. There was also a reduced rate of swimming.

Some fish seem inherently to have the ability to detect some pesticides and, where possible, avoid them. Hansen (1969) showed that previously unexposed sheepshead minnows (*Cyprinodon variegatus*) could recognize some but not other pesticides and did not distinguish between different concentrations of the same pesticide. He found that, though they did not react to malathion or carbaryl, they could detect and would avoid chlorpyrifos, DDT, endrin, and the butoxyethyl ester of 2,4-D. This avoidance occurred both at lethal concentrations and at levels well below the LC_{50}. Folmar (1976), testing the avoidance reaction of rainbow trout fry to nine herbicides, found that the fish would avoid the higher concentrations of dalapon; the dimethyl amine salt of 2,4-D; xylene; and acrolein but did not avoid the lowest concentrations tested. Fish did not avoid glyphosate, Aquatrol K®, diquat, or TCA. Fish were most sensitive to copper sulfate, avoiding concentrations of 0.01, 0.001, and 0.0001 ppm (expressed as the copper ion).

Physiological Effects

A number of studies have indicated reduced cholinesterase activity in fish exposed to organophosphorous insecticides, and changes in respiration have also been noted. Such changes have not, however, been consistently related to exposure levels, and the data are not sufficient to permit conclusions (Holden, 1973). Several workers have suggested decreased activity in fish with sublethal exposures, but few have attempted to quantify such effects. Post and Leasure (1974) used rainbow trout, brook trout, and coho salmon as experimental animals and exposed them in flowing water systems to dosages of malathion that resulted in brain acetylcholinesterase levels approximately 25%, 50%, and 75% of normal. The fish were then subjected to a water flow of 0.875 ft/sec, and their ability to compete with this water flow was determined.

The concentration of malathion required to effect similar percentage depletions of brain acetylcholinesterase activity was about twice as great with coho salmon as with either species of trout, though brook trout were slightly more sensitive than rainbow. The time required for acetylcholinesterase levels to return to normal also differed among the three species. Time required for replenishment in coho was 42 days; rainbow trout, 35 days; and brook trout, 25 days. Reducing the brain acetylcholinesterase level had a drastic effect on the stamina of fish.

Kruzynski and Leduc (1972) reported that stamina of Atlantic salmon was reduced in fish receiving methoxychlor at the rate of 0.01 to 0.16 mg/kg/day but that fish receiving 1.0 and 2.0 mg/kg/day had markedly increased stamina.

One of the reactions to exposure to pesticides in fish has been the development of resistance. Resistance of fish to pesticides was first demonstrated by Vinson et al. (1963), who showed that mosquitofish from a Mississippi delta were resistant to DDT. This was associated with DDT use in cotton production. Since then resistance in mosquitofish has been shown for most of the organochlorine insecticides and some of the organophosphorous compounds. This resistance has been reported also in Texas (Dziuk and Plapp, 1973). In addition resistance has been shown to a number of pesticides by several other species of fish, including bluegill, golden shiners, and green sunfish (Ferguson et al., 1964). Mosquitofish have developed tolerance to endrin to the point where they carry such high levels of pesticides that they are toxic to a number of their predators (Rosato and Ferguson, 1968).

Pathological Effects

Holden (1973) summarizes a number of reports of pathology in fish tissues following sublethal exposure to various pesticides. He notes that many of the reports deal with dosages in the toxic range, and though the pathology was observed in living fish, conclusions about whether these were sublethal or acute toxicity effects were difficult.

Kennedy and Walsh (1970) examined bluegills and channel catfish from ponds that received four applications of malathion at two dosages (0.02 and 0.002 ppm) during an 11 week period. They could find no abnormal pathology in blood, brain, spinal cord, eye, gill, heart, kidney, liver, gall bladder, pancreas, intestinal tract, spleen, epidermis, dorsal muscle, gonads, or thyroid. Using the bluegill and sublethal dosages of methoxychlor, they obtained different results. They treated two ponds at 0.01 ppm and two at 0.04 ppm methoxychlor and examined fish at intervals thereafter to observe effects. There was no mortality attributable to the methoxychlor treatments, but especially in the higher dosage abnormal pathology occurred in the liver, blood vessels, and capillary beds. Changes in the liver cells were ill defined. Degenerative changes included some shrinkage in the liver parenchyma, cytoplasmic granularity, and a partial loss of the radial orientation of liver chords. Changes were noted from soon after exposure to day 28 but were gone by day 56. The appearance and disappearance of these symptoms correlated closely with the accumulation of methoxychlor in the fish tissue and its subsequent elimination. Changes were noted also in the circulatory system.

Similar liver pathology was reported by Kurzynski and Leduc (1972) following tests in which methoxychlor was given to Atlantic salmon in their food. These workers did not report pathological findings in the circulatory system but found changes in the kidney in cells lining the tubules and shrinkage of the tubules. In this study it should be noted that some fish died. Thus an interpretation of the sublethality of the exposure level in the particular fish in which pathology was exhibited is a bit fuzzy.

Among other pathological conditions might be noted the appearance of hemorrhagic lesions on fish exposed to some chlorinated hydrocarbons (Mount and Putnicki, 1966) and lesions on the olfactory lobe of fish exposed to copper (Gardner and LaRoche, 1973).

An additional interesting finding is that of Weis and Weis (1975) in which it was shown that a number of pesticides retarded fin regeneration in *Fundulus* (killifish). In their tests killifish were maintained in aquaria and the lower half of the caudel fin was removed. The insecticides carbaryl, DDT, malathion, and parathion were then added to different aquaria at 10 ppb, and the effect on fin regeneration was noted. All insecticides retarded regeneration, but DDT had the least effect.

Effects on Reproduction

Severe hatchery losses in lake trout fry led to a study concluding that high levels of DDT in mother fish resulted in high mortality in fry. Spawn for several hatcheries in New York had been collected from Adirondack lakes, and in 1955 all fry from more than 347,000 eggs taken from Lake George trout died after absorption of the yolk sac just prior to feeding. It was found that eggs of Lake George females crossed with males from other areas produced fry that failed to survive, though when males from Lake George were crossed with females from other areas, fry survival was normal. Burdick et al. (1964) found that Lake George fish carried extremely high levels of DDT and DDE in relation to fish from some other lakes. Individusl fish were stripped, and the DDT and DDE content was compared with that of the mother fish. Although there were a number of instances where high levels in eggs were associated with high levels in the mother fish, there were sufficient instances where this did not hold that such a correlation could not be demonstrated. They found that, when the ether extract of the eggs indicated an equivalent of 2.9 ppm DDT in terms of the weight of fry at egg sac absorption, mortality would occur.

Other workers have reported fry mortality at the yolk sac absorption stage and a toxic level to fry very close to that reported by Burdick et al. (1964) (Holden, 1973). In the work of Burdick et al. (1964) no attempt was made to learn whether some fry surviving the yolk sac absorption stage might experience mortality during and soon after the swim up stage. This was shown for rainbow and cutthroat trout, where 30 to 90% mortality was found after the swim up stage in fry from eggs containing a DDT burden of 0.4 ppm (Cuerrier et al., 1967).

Data on DDT content in eggs and fry mortality are few for other than species of trout. In another salmonid Johnson and Pecor (1969) found that in 1968 coho salmon eggs from Lake Michigan carried 1.1 to 2.8 ppm DDT. Fry hatching from these eggs had a mortality of 15 to 73%.

In another study on lake trout Canadian workers captured 10 large females from Lake Muskoka in November of 1967. Eggs were fertilized by males from the same lake and incubated separately. In addition to egg samples, muscle and gonads from parent fish were analyzed for DDT. The eggs hatched in January 1968, and in early May, prior to yolk sac absorption, fry were collected and analyzed. Fry that died and those that survived through July were analyzed also. In this study fry mortality was 45 to 90% when computed at swim up stage. Although the levels of total DDT (wet weight basis) in adult muscle ranged from 53 to 128 ppb and in gonad from 4.8 to 15.7 ppm, high values in the one were not consistently associated with high values in the other. Eggs varied from 7.4 to 14.2 ppb, but again high egg levels were not correlated with high levels of either muscle or gonad. Total DDT in the fry just prior to swim up were surprisingly uniform (9.84 to 16.61 ppb) but not correlated with the egg batches from which they were derived. Fry that died during the swim up period had residues of 6.75 to 15.14 ppb, and 8 weeks after swim up levels were 1.15 to 4.65 ppb (Boelens, 1974). With respect to the data of Burdick et al. (1964) and Cuerrier et al. (1967) the DDT levels in eggs from Lake Muskoka fish were high and should have resulted in extensive fry mortality. The syndrome reported by Burdick et al. (1964) was not seen in the Boelens (1974) study, and though fry mortality by swim up stage was high (45 to 90%), no data were presented to show how this range of mortality compared with "normal" mortality. (Note: presumably the values in Table 3 [Boelens, 1974] are ppm, not ppb, as indicated in the subcaption.)

15.5 RESIDUES IN FISH

When fish are exposed to pesticides either through contact in water or in their food, these are taken up in varying degrees, and residues will be found in their tissues. The extent of residues will be a function of the level of exposure, duration of exposure (for some pesticides), species of fish, route of exposure, nature of pesticide, and the ability of these fish to metabolize and/or excrete the compound. The nature of the pesticide (as reflected in the ability of the fish to metabolize and excrete it) is the most important factor in terms of persistence of residues. The organochlorine compounds are the most persistent of the pesticides in the aquatic environment and are also the most persistent residues in fish. Among the organochlorines DDT is the most ubiquitous, and it is not surprising that it is the most prevalent residue in fish. Fish accumulate DDT from either food or dermal exposure (Cope, 1963); in fish it persists in the body, primarily in fatty tissues, either as the parent compound or as one or more of its metabolites (Frank et al., 1974a). Some very high residues of DDTR have been reported from fish after high exposures in the laboratory or field. Levels in excess of 100 ppm have been found in lake trout, and this level was exceeded by more than tenfold in the fat from these fish (Burdick et al., 1964). Equally high values were reported for residues of DDD in black crappie, brown bullhead, and largemouth bass from Clear Lake after it had been treated with this insecticide for gnat control (Hunt and Bischoff, 1960). A large trout caught in Lake Muskoka, Ontario, was estimated to contain almost 0.6 g of DDTR (Frank et al., 1974a). Such high levels are the exception, but these exceptions, occurring as they did in natural habitats, demonstrate the degree to which residues of DDT and its metabolites can accumulate.

Many studies have been done to monitor the levels of pesticides in fish, and some of these are included in a tabulation by Edwards (1973a). Keith and Hunt (1966) found DDTR the most prevalent residue in 15 species of fish collected in California from 1963 to 1965. Residues often exceeded 100 ppm in fat of channel and white catfish and striped bass, and this was believed to reflect the feeding habits of these species. For example, small carp containing 1.4 ppm DDT were found in the stomach of a striped bass. Most of the 16 species of fish examined contained, however, less than 1 ppm DDTR on a whole fish, wet weight basis. Toxaphene was detected in 7 of the 16 species analyzed, but in practically all cases the level was well below 1 ppm. There were some exceptions. One sample of tui club had 8.0 ppm and one sample of rainbow trout had 2.57 ppm. In this study dieldrin was detected also in 7 species, 5 of these being the same species in which toxaphene was found. Levels of dieldrin were low, below 0.1 ppm in all instances. Traces of endrin, BHC, and heptachlor epoxide were sometimes detected. Keith and Hunt (1966) reported that most of the fish they analyzed had been taken from sites contaminated by waste agricultural water and considered this the main source of pesticides to these fish.

Perhaps the most extensive survey in fish is the nationwide monitoring program of the United States Bureau of Sport Fisheries and Wildlife. In this study (Henderson et al., 1971) fish from 50 sites were analyzed for eight primary organochlorine insecticides (DDT, dieldrin, BHC, heptachlor, chlordane, aldrin, endrin, and toxaphene) and three metabolites (DDE, TDE, and heptachlor epoxide). Most of the sites were rivers, but there were six sampling stations on lakes, one on each of the Great Lakes and one on Utah Lake at Provo. Residue data from the sampling done in the fall of 1969 are summarized in Table 15.17. These data are from 147 composite fish samples, most of which included 5 fish. Samples of 3 fish species were obtained at all but three of the 50 sampling stations. Of the total of 44 species, 16 were collected only once, but carp were collected at 22 sites, largemouth bass at 16, and channel catfish at 13 locations. DDT was found in all fish sampled and ranged from 0.03 ppm in longnose sucker from the Kenai River in Alaska to 57.8 ppm in one 10 lb channel catfish from the St. Lucia Canal in Florida. The median level for DDTR for all stations was about 1 ppm, mean levels exceeding 5 ppm at 4 stations. Dieldrin was the second most common residue, being detected in all but 10 of the 147 samples. Residues ranged from nondetectable to 1.59 ppm in a sample of 5 largemouth bass from the Apalachicola River in Florida. The median dieldrin level for all samples was 0.03 ppm, and mean residues greater than 0.1 ppm were found at 10 sites.

Residues of BHC were found about as frequently as those of dieldrin—in 132 of the 147 samples—and were found at all but 1 of the 50 sampling sites. Residues ranged from nondetectable to a high of 4.37 ppm in a sample of 4 brown bullheads from Kanawha River, West Virginia. At this same location a sample of 5 white crappie carried a

Table 15.17 Organochlorine Residues from Fish

Fish Species	No. Sam-ples[b]	BHC Mean	BHC Range	Chl Mean
			Residue in ppm	
Bass, largemouth	16	0.07	0.0-0.47	0.06
rock	2		0.01-0.14	
smallmouth	2		0.01-0.01	
white	1		0.01	
Bloater	2		0.03-0.08	
Bluegill	6	0.01	0.0-0.03	
Buffalo, bigmouth	2		0.03-0.07	
smallmouth	1		0.08	
Bullhead, black	2		0.01	
brown	6	0.77	0.01-4.37	0.05
Carp	22	0.11	0.0-0.99	0.04
Catfish, blue	1		0.14	
channel	13	0.23	0.01-1.50	0.09
flathead	1		0.02	
white	3		0.0-0.23	
Chisel mouth	2		0.02-0.03	
Crappie, black	3	0.02	0.01-0.03	
white	3	0.82	0.07-2.19	
Drum, freshwater	1		0.02	
Goldeye	3	0.05	0.02-0.08	
Goldfish	1		0.51	
Grayling, Arctic	1		0.12	
Mullet, striped	2		0.28-1.14	
Perch, white	4	0.23	0.18-0.26	0.44
yellow	8	0.12	0.01-0.31	
Pickerel, chain	1		0.45	
Pumpkinseed	1		0.09	
Sauger	1		0.01	

BHC residue of 2.19 ppm, while the residue in carp was 0.31 ppm. The only other site where BHC residues exceeded 1 ppm was in the Mississippi River at Luling, Louisiana. Here residues in carp, striped mullet, and channel catfish were 0.99, 1.14, and 1.50 ppm, respectively. The median level for BHC at the 50 sites was about 0.02 ppm.

from Natural Waters in the United States, 1969[a]

on whole fish basis, wet weight

ordane Range	DDTR Mean	DDTR Range	Heptachlor[c] Mean	Heptachlor Range	Dieldrin Mean	Dieldrin Range
0.0-0.95	1.98	0.22-10.15		ND[d]	0.16	0.01-1.59
ND		0.19-1.68		ND		0.0-0.07
ND		0.60-1.37		ND		0.03-0.04
ND		0.43		ND		0.02
ND		1.61-6.06		ND		0.02-0.37
ND	0.63	0.11-1.50		ND		0.01-0.63
ND		0.43-1.73		ND		0.04-0.42
ND		1.42		0.16		0.12
ND		0.06-0.12		ND		0.01-0.03
0.0-0.31	1.34	0.21-3.71	0.06	0.0-0.34	0.06	0.01-0.25
0.0-0.68	1.01	0.06-4.56	.002	0.0-0.04	0.07	0.01-0.54
1.30		1.99		0.22		ND
0.0-1.01	5.76	0.10-57.77	.008	0.0-0.07	0.06	0.0-0.96
ND		2.22		ND		0.03
ND		1.12-1.39		ND	0.20	0.0-0.50
ND		0.30-1.20		ND		0.01-0.02
ND	1.25	0.97-1.41		ND	0.14	0.02-0.36
ND	0.52	0.37-0.65	0.06	0.0-0.17	0.12	0.02-0.27
ND		0.85		ND		0.04
ND	0.54	0.07-0.91		ND	0.04	0.01-0.08
ND		3.80		ND		0.04
ND		0.62		ND		0.01
0.0-0.09		0.58-7.93		ND		0.02-0.39
0.0-1.75	8.91	1.92-20.27		ND	0.43	0.06-0.88
ND	1.73	0.09-6.44		ND	0.07	0.02-0.20
ND		0.23		ND		0.02
ND		0.85		ND		0.05
ND		0.66		ND		0.01

Heptachlor was found at only 1 site (Rio Grande, McAllen, Texas) but at that site was present in the 3 fish species analyzed. Heptachlor epoxide was found in 5 samples from 3 sites. Chlordane was detected in 16 samples from 6 sampling stations. It ranged from nondetectable in most samples to a high of 13.5 ppm in a sample of 5 gizzard shad

Table 15.17

| | | Residue in ppm | | |
| | | BHC | | Chl |
Fish Species	No. Sam- ples[b]	Mean	Range	Mean
Shad, gizzard	2		0.06-0.10	
Squawfish, northern	3	.003	0.0-0.01	
Sucker, bridgelip	1		0.02	
flannelmouth	1		0.02	
klamath	1		0.00	
largescale	6	0.03	0.01-0.12	
longnose	2		0.01-0.03	
redhorse	3	0.07	0.02-0.18	0.07
spotted	2		0.01-0.03	
white	7	0.13	0.01-0.27	0.08
Sunfish, redbreast	1		0.00	
Trout, lake	2		0.01-0.02	
rainbow	3	0.01	0.01-0.01	
Walleye	2		0.01	
Whitefish, lake	1		0.05	
round	1		0.04	

[a]From Henderson et al., 1971.
[b]Most samples contained 3-5 fish, three contained one fish, eight contained two (see text).
[c]Includes heptachlor epoxide.
[d]ND = not detected.

from the Rio Grande site in Texas. Channel catfish and blue catfish from the same site had chlordane residues of 1.01 and 1.30 ppm, respectively. A sample of white perch from the Delaware River had a chlordane residue of 1.75 ppm. No residues of aldrin, endrin, or toxaphene were detected in any of the 147 samples.

A number of other extensive studies (Reinke et al., 1972; Veith, 1975; Frank et al., 1974a) make it clear that DDTR is found in most fresh water fish in North America.

The level of residue differs in different waters and among different fish, and a number of workers have suggested that higher levels of DDT are found in older fish with high lipid content, especially if these are upper level predators. Among other organochlorine pesticides dieldrin is most frequently encountered, but the magnitude of residues is normally less than DDTR by a factor of at least 10. A

(*Continued*)

on whole fish basis, wet weight

ordane Range	DDTR Mean	DDTR Range	Heptachlor[c] Mean	Heptachlor[c] Range	Dieldrin Mean	Dieldrin Range
0.0-13.5		0.79-2.40		0.0-0.45		0.05-0.50
ND	1.37	0.56-2.42		ND	0.01	0.01-0.02
ND		1.11		ND		0.02
ND		0.60		ND		0.01
ND		0.05		ND		0.0
ND		0.27-1.21		ND		0.0-0.09
ND		0.03-1.16		ND		0.01
0.0-0.20	0.60	0.08-1.05		ND	0.05	0.01-0.12
ND		0.62-0.91		ND		0.0-0.30
0.0-0.44	2.89	0.15-8.86		ND	0.07	0.01-0.35
ND		0.07		ND		0.01
ND		0.09-1.58		ND		0.0-0.02
ND	0.34	0.14-0.73		ND	0.02	0.0-0.04
0.0-0.10		0.12-0.91		ND		0.01-0.03
ND		0.74		ND		0.03
ND		0.92		ND		0.01

tolerance level for DDTR in fish for human consumption has been established for many years at 5 ppm, and a level of 0.3 ppm for dieldrin has been suggested (Johnson and Ball, 1972). It is clear from the data available that commercial fish, especially those from the Great Lakes, may exceed the tolerance for DDT, and fish from this area have been confiscated (Johnson and Ball, 1972).

With the exception of the persistent organochlorine insecticides, residues of pesticides in fish are transient. Kennedy et al. (1970) found that, in ponds treated with methoxychlor at 0.04 ppm, residue in bluegills reached a high of 20.85 ppm in 3 days, declined to less than 1 ppm by the 28th day, and could not be detected after 56 days. With this same insecticide Kruzynski and Leduc (1972) fed brook trout 0.5 to 100 ppm in their food for an extended period. They reported methoxychlor residues of 2 to 44 ppm at the end of the exposure period (presumably 32 days). Note that these authors used feeding levels to approximate levels that might be present in insect drift immediately (within 24 hr) following larviciding with methoxychlor for black fly control. This drift of high residue dead and dying insects is a transient occurrence, and though fish might be exposed to such a diet for a few hours in a treated stream, continuous exposure would not occur. Thus this experiment should not be interpreted as a simulation of natural conditions.

Kanazawa (1975) exposed Motsugo to 0.96 ppm diazinon in water and within a few days measured residues of 211 ppm in the fish. As the diazinon degraded in the water, the residues in fish became reduced, so that by day 30, residue in the water was 0.27 ppm and in fish 17 ppm. Similar results were obtained with fenitrothion, malathion, and carbaryl.

Herbicides have been detected in fish but only in low concentrations. Schultz and Whitney (1974) monitored fish from a canal in Florida where 7000 acres had been sprayed with the dodecyl-tetradecyl amino salts of 2,4-D at the rate of 4.48 kg (acid equivalent)/ha. Of 60 samples of fish analyzed, 3 had residues greater than 0.10 ppm (0.162 was the highest), and 8 had residues of 0.01 to 0.10 ppm; 49 of the samples had residues either not detectable or less than 0.010 ppm. Similarly low residues were found in 307 fish analyzed from nine ponds treated with the dimethylamine salt of 2,4-D at 2.24, 4.48, or 8.96 kg/ha (Schultz and Harman, 1974).

The herbicide DCPA (dimethyl tetrachloroterephthalate) is used in Texas, primarily for weed control in onions. It has a half-life in soil of about 100 days and has been detected in small amounts in air and water in the lower Rio Grande Valley. Residues in fish ranged from zero to a high of 8.15 ppm in one sample of menhaden taken at a time when the residue in water was 0.96 ppb. Most other samples of menhaden showed much lower residues, and residues were highest during the agricultural growing season. Residues were detected also in several tissues of three other species of fish (Miller and Gomes, 1974).

One of the contaminants detected in fish in recent years is mercury. Although the source of this is probably natural deposits or through escape of mercury used as a catalyst in chlor-alkali plants, it is reported frequently in literature dealing with pesticide residues in fish. Although there are many reports, only two recent ones will be cited. Benson et al. (1976) measured residues in catfish and smallmouth bass from the Snake River and its impoundment areas. Mercury was found in all fish sampled, ranging in catfish from 0.15 to 0.67 ppm and in bass from 0.20 to 1.30 ppm.

Kelso and Frank (1974) included mercury in their analyses of residues in yellow perch, white bass, and smallmouth bass from Lake Erie. In yellow perch residues ranged from 0.060 to 0.075 ppm in three samples of 10 fish, each collected in May, July, and November.

For white bass levels ranged from 0.101 to 0.339 ppm, and for smallmouth bass two samples gave residues of 0.105 and 0.225 ppm. In this study residues in excess of 0.50 ppm occurred in only two fish, a white bass 4 years old (0.63 ppm) and a white bass 5 years old (1.00 ppm).

The levels and nature of pesticide residues in fresh water fish have been well documented. Mercury, being excluded since its source does not arise from its use as a pesticide, the only pesticides of widespread and almost ubiquitous occurrence are DDT (and its metabolites) and dieldrin.

Pesticide residues have been detected also in fish of the open oceans and from estuarine waters. Estuarine fish reflect the pesticide contaminants being fed into the estuaries, and again high residues of the persistent DDT metabolites are common. Earnest and Benville (1971) collected eight species of fish monthly (not all fish were taken in all months) from two areas of San Francisco Bay and found DDT levels ranging from 23 to 366 ppb on a whole fish, wet weight basis. The highest levels were found in perch (dwarf, shiner, pile, white), and there was a general correlation between high fat content and high residues of DDT (and metabolites).

Results from two studies of fish in coastal water are summarized in Table 15.18. Markin et al. (1974a) selected nine sites along the Atlantic and Gulf coasts and analyzed for DDT, DDE, TDE, Aroclor 1260 (PCB), and mirex. DDT and/or its metabolites were found in all fish samples ranging from a low of 0.007 ppm in one sample of gray snapper to a high of 1.16 ppm in one speckled sea trout (Table 15.18). Mirex was detected in fish from only one sampling area (Savannah, Georgia), levels ranging from 0.006 to 0.024 ppm. This finding with respect to mirex does not agree with some earlier work that indicated much more widespread pollution from this chemical. Markin et al. (1974a) found that extensive methods of clean-up were necessary prior to gas chromatography to distinguish between PCBs and mirex. In this study PCB was found in many fish and would have been reported as mirex if additional clean-up steps had not been taken. They suggested that, because this had not been done in the earlier studies where extensive mirex contamination had been reported, the possibility for misidentification in the earlier reports may explain the apparent discrepancy.

On the west coast Duke and Wilson (1971) analyzed coastal fish from three collection sites, La Jolla, California; Seattle, Washington; and Auke Bay, Alaska. The La Jolla collections were from bays and fishing banks along the coast of southern California. The Seattle col-

Table 15.18   Total DDT Residues (DDTR) in Some Coastal Fishes

Atlantic Coast Fish[a]		Pacific Coast Fish[b]	
Species	ppm DDTR[a]	Species	ppm DDTR[b]
Anchovy	0.127	Albacore tuna	0.086
Atlantic whiting	0.020-0.070	Bonito	0.1-1.28
Bluefish	0.223	English sole	13.93
Croaker	0.015-0.236	Hake	0.59-6.61
Drum	0.065	Jack mackerel	0.11-2.8
Flounder	0.018-0.814	Lingcod	17.80
Gray snapper	0.007	Lizard fish	2.13
Grouper	0.041	Ocean whitefish	0.088-0.94
Herring	0.018	Pacific mackerel	0.039-0.055
Mullet	0.007-1.10	Rockfish, blue	11.49
Red snapper	0.086	bocaccio	591.0
Shad	0.078-0.744	olive	24.60
Spanish mackerel	0.042-0.161	rosy	29.80
Speckled sea trout	0.20-1.16	starry	18.1-1026
Spot	0.053	treefish	1.84
Striped sea bass	0.043-0.086	vermilion	162.0
Weak fish	0.040-0.798	Sablefish	103.10
		Sandbass	0.038-0.15
		Sardines	0.96
		Cal.scorpion fish	8.13
		Spiny dogfish	228-473
		White croaker	16.63
		Yellowfin tuna	0.16-0.17

[a]Markin et al., 1974a. Residue on a whole fish, wet weight basis.
[b]Duke and Wilson, 1971. Residue in liver, ppm, wet weight.

lections included fish from along the Washington and Oregon coasts, as well as hatchery specimens, while the Auke Bay samples were mainly from Alaskan rivers. In this study livers were analyzed, and data from most of the La Jolla and Seattle collections are summarized in Table 15.18. Except for sockeye salmon all fish collected at these two stations contained DDT. Residue levels in livers varied widely among fish species, ranging from 0.038 to 1026 ppm on a wet weight basis. Within a species, variations were also great, differences of an order of magnitude being common. A composite sample of 13 starry rockfish from Farnsworth Bank contained 18.11 ppm in the liver, but a sample of 5 fish of the same species taken 1 day later in Santa Monica Bay had 1026 ppm (Table 15.18).

In this study DDT was not found in sockeye salmon, though 46 composite samples (mostly 10 fish) were analyzed. These had been collected over a significant territory ranging from 49° to 53°N–160° to 176°22′W. Similarly no DDT and metabolites was found in chum, coho, king, pink or sockeye salmon, or rainbow trout taken from rivers and coastal areas of Alaska.

No comparison can be made between levels on the east and west coasts of North American on the basis of data in Table 15.18, since the

Atlantic coast data are on a basis of whole fish and the Pacific coast data on a basis of residue in the liver. What is apparent is that, along both the west and east coast near areas where DDT has been used extensively, significant and sometimes high residues are present in coastal fish. In view of the widespread distribution of DDT in the atmosphere and its detection in air, rain, and snow from seemingly all areas, it is surprising that the high trophic level fish taken in the Alaska area did not contain measurable residues in their livers. In view of the past use of DDT in forest insect control in western Canada and the northern United States, it is even more surprising that DDT was not present in sockeye salmon taken in a large area of the Pacific off this area.

Reimold (1975) collected fish off the coast of Puerto Rico and the Virgin Islands within 1 km of the shore and analyzed whole fish for DDT and metabolites, dieldrin, PCBs, and mercury. In 34 species of fish sampled dieldrin residues were detected in 2 samples and DDT in 4.

By contrast mercury was detected in 19 of the species analyzed. The highest mercury residue was 968 ppb found in each of two great barracuda. PCBs were found in 13 species with a high residue of 5391 ppb in striped mullet.

Harvey et al. (1974) reported on residues of DDT and metabolites in fish from two of the commercial fishing grounds, namely, Georges Bank (the New England commercial fishing ground) and the Denmark Strait. Some of the data they reported are summarized in Table 15.19; these show that DDT (and metabolites) is present in significant quantities in major commercial ocean fish. The data suggest higher levels in fish from Georges Bank than in those from the Denmark Strait in the three species (cod, haddock, redfin) taken at both sites. Harvey et al. (1974) included a large number of other fish in their sampling of organisms from the Atlantic Ocean. Among pelagic fish they found 7 ppb DDTR in flying fish (whole fish, wet weight basis) and 4800 ppb in the liver of the silky shark. Analyses of mesopelagic fish from six families indicated low levels of residue, mostly in the <10 ppb range (whole fish, wet weight basis) but ranging as high as 47 ppb in one sample of *Stomias boa*.

It is clear that DDT is present in many open ocean fishes, but in most cases levels are low. Levels reported by different researchers suggest some differences from fish from different waters. For example, cod taken from Georges Bank (Table 15.19) had levels in muscle and livers comparable to those from St. Margarets Bay, Nova Scotia (Hargrave and Phillips, 1976), but these levels were much lower than the 3700 to 13,900 ppb (mean 8100) DDTR residues reported in cod livers

Table 15.19   Total DDT Residues (DDTR) in Livers of
Atlantic Ocean Fish[a]

Species	Georges Bank (ppb)	Denmark Strait (ppb)
Argentine	51	
Cod	2700	170
Haddock	390-1600	260
Halibut		330
Pollock	950-3000	
Redfin	670-1300	190
Wolf-fish		3

[a]Harvey et al., 1974.   Residues in ppb on a wet
weight basis.

from six sampling sites off the coast of the Maritime Provinces of
Canada (Sims et al., 1975). DDT levels in cod livers are further confirmed by analyses indicating a mean of 1.9 ppm DDT in six commercial samples of cod liver oil (Sims et al., 1975).

15.6   INDIRECT EFFECTS

While it is possible to measure direct effects of pesticides in terms of
mortality, reproductive abnormalities, altered behavior, effects on
growth rate, rate of respiration, rate of photosynthesis, or some other
parameter, measurement of indirect effects is much more difficult and
frequently is more assumed than proved. Hurlbert (1975)* has presented a comprehensive treatment on this subject, and the reader is
referred to that review. He considers secondary (or indirect) effects to
be changes in the ecosystem that result from the primary effects, and it
is in this context that we approach the subject.

From a theoretical standpoint the introduction of a selectively
toxic agent into an environment will alter the species composition by
selectively removing a higher percentage of susceptible than resistant
members. Again theoretically this will lead to less competition for
survivors and to the potential, given adequate food supply and the
other prerequisites for survival, for dramatic increases in numbers of
nonsusceptible species. As Hurlbert (1975) points out, however, the

* With reference to Hurlbert (1975) the reader is advised that the unit of reporting
residues in Table II is parts per billion, not parts per million. In addition the LC_{50}
values for the fresh water minnow are, for aldrin and BHC, 28 and 2800, respectively
(Hurlbert, personal communication).

multiplicity of interactions among species within an ecosystem and within each species is likely to dampen the assumed effect, but this may not always occur. Thus the appearance of "algal blooms" following insecticide treatment has been a frequent observation reflecting the fact that the pesticide has reduced the population of microcrustaceans and other grazers.

Perhaps the indirect effects of pesticides can be viewed as three main types of effects.

1. Changes in the physical and/or chemical environment.
2. Changes in competition for food supply.
3. Changes in predator-prey relations.

Changes in the Physical and/or Chemical Environment

The aquatic ecosystem is a sensitive one providing a multiplicity of niches for many animal and plant species. Dramatic upsets in these systems occur when floods, extended drought, or a host of other natural phenomena take place. For example, Kushlan (1974) followed the developments in a Florida marsh when an extended dry period reduced the water level in the marsh and forced aquatic organisms into a small pond. Low oxygen availability caused a high fish kill, but fish mortality was not equal among all species. The fish kill was accompanied by an unusual abundance of phytoplankton. Despite a massive short-term change, the limnological characteristics of the pond were not permanently changed. The phytoplankton levels returned to normal in about 1 month, and the water quality in the pond had recovered within 2 months of the fish kill.

The same sort of effect may occur following an herbicide treatment to a pond with dense vegetation. Destruction of the vegetation results in a high mass decomposition and resulting low oxygen level. McCraren et al. (1969) treated ponds with diuron at 0.5, 1.5, and 3.0 ppm for weed control and found that, in contrast to control ponds where the oxygen level was 9.0 ppm, 3 days after treatment oxygen levels in treated ponds were 3.8, 3.3, and 1.6 ppm, respectively; 30 days after treatment oxygen levels were 6.7, 6.1, and 5.4 ppm (for treatments at 0.5, 1.5, and 3.0 ppm, respectively), while in untreated water the level was 10 ppm at this time. Mortality of bluegills was noted in the pond with the highest treatment, but such mortality did not occur until 3 days after application. These workers used diuron wettable posder in their pond treatments, and this formulation is reported to have an LC_{50} value for bluegills of 25 ppm (Walker, 1965).

It is unlikely that the 3 ppm level used in the study was directly toxic, mortality being more probably a result of low oxygen availability. Anaerobic or near anaerobic conditions have been reported after the use of silvex for control of alligator weed (Cope, 1965) and the use of 2,4-D for milfoil control in tidal creeks (Haven, 1963). In the latter instance 2,4-D was used at the rate of 10 lb/acre, and death of amphipods, clams, and benthic invertebrates occurred. Such mortality did not occur until about 2 weeks after treatment. 2,4-D degrades quite rapidly in natural waters and has been shown nontoxic to oysters and clams at 30 lb/acre (Beaven et al., 1962). The explanation for mortality in the Virginia study seems clearly associated with anaerobic conditions caused by excessive weed decay.

Although depletion of oxygen supply would be expected as a normal posttreatment effect where herbicides are applied to a heavy weed growth in ponds, the small number of reports of this occurring (Pimentel, 1971; Hurlbert, 1975) to the extent that fish or other organisms are killed suggests otherwise. Most reports on the use of herbicides in aquatic environments indicate little, if any, mortality in a wide range of animal species.

It is clear that, where heavy weed growth is present and herbicides are applied, an excessive amount of vegetation must be decomposed. Obviously this reduces the oxygen available for other purposes. In some instances it has been shown that, when the macrophyte growth is removed, phytoplankton may become abundant, and since their photosynthetic rate is high, they may play a dominant role in restoring the oxygen supply. After treatment of a reservoir with paraquat, Fry et al. (1973) record that the water pH fell from 9.3 to 8.1 by the third day posttreatment, indicating a reduction in photosynthesis as the weed growth of submerged angiosperms became chlorotic. The weeds died and were in an advanced state of decay by the eleventh day. By the sixteenth day a growth of the filamentous alga *Zygnema* sp. had appeared, and the macrophytic alga *Chara globularis* appeared 4 days later. Concurrent with the appearance of algae, the water pH rose to 9.8, indicating a high level of photosynthetic activity. Walsh et al. (1971) reported that, when dichlobenil was applied to a small pond to control *Chara* and *Potamogeton*, death of these plants was followed by a phytoplankton bloom and that, 1 month after treatment, oxygen production in the treated pond was five times greater than in a similar pond that had not been treated. With this same herbicide Cope et al. (1969) showed that, when it was applied in granular form to a pond with heavy weed growth, the oxygen level at 3 m depth dropped from 7.6 to 1.5 ppm, and though the oxygen level in surface water was relatively unaffected, the level of oxygen at the 3 m depth did not return to the pretreatment level for 6 months.

A number of effects in water chemistry other than oxygen supply and associated pH variance have been suggested, but there is little documentation (Hurlbert, 1975).

Changes in Competition for Food Supply

When herbicides are used in aquatic environments, the dead plants provide a new food source for a host of microorganisms. Fry et al. (1973) observed this in a dramatic increase in the number of heterotrophs, amylase producers, and protease producers immediately after the first application of paraquat to a reservoir. At the time of this application weed growth was heavy, but after a second application 6 weeks later when no significant weed growth was present, no such increases were noted. These workers also observed a general increase in microbial populations at a time to coincide with release of nutrients by the decaying macrophytes.

A large number of reports of herbicide treatments in ponds (some previously cited) have included data on effects on invertebrates, fish, and the like, but few have been presented in terms of possible secondary effects. Most have been content with reporting mortality or lack of it without any critical appraisal. Some reports (see Hurlbert, 1975) indicate increases in detritus feeders, but so little is known of the food habits of many benthic organisms that conclusions about cause-effect relationships are tenuous at best. Harp and Campbell (1964) found that, after treatment with silvex, populations of oligochaete worms and chironomid and chaoborid larvae increased in comparison with those in untreated ponds. Dragonfly nymphs also became more abundant, but this was not apparent until almost 1 year after treatment.

Most studies suggest that the decomposition of plants killed by herbicides provides a nutrient source, reduces competition, and thus favors the development of algal blooms. Other types of growth may also be favored. As previously cited (Fry et al., 1973; Walsh et al., 1971), the macrophytic algae *Chara* and *Nitella* often appear in dense growth following herbicide treatments.

Competition is important among animal and plant species, but in the former it is difficult to separate competition for the same source of food and predator-prey interactions.

Hurlbert et al. (1972) found that the population of planktonic rotifers increased after ponds had been treated with chlorpyrifos. They attributed this, in part, to reduced numbers of *Moina* (because of toxicity of the insecticide), a competitor for their food supply, and, in part, to the reduced number of *Cyclops*, a predator. Some other good examples come from work with piscicides. For example, the gizzard shad is a phytoplankton feeder, but its young feeds on zooplankton. This species is highly susceptible to rotenone, and this pesticide is used frequently to remove shad selectively from game fish areas. Its removal usually leads to much better production of game fish, since the latter benefit from the increased zooplankton supply available to their young (Zeller and Wyatt, 1967).

Changes in Predator-Prey Relations

Perhaps the best known examples of pesticides' upsetting predator-prey relations are found in the agro-ecosystem, where the species abundance and number of interactions are much less than in aquatic environments. In Nova Scotia the European red mite was not a problem in commercial orchards, it being held at low levels by a complex of arthropod predators, mostly phytoseiid mites. The introduction of DDT changed this relationship, for DDT was quite toxic to the predators but relatively harmless to the mites. Thus the European red mite became a major problem requiring extensive use of acaricides, and this situation did not change until DDT was removed from the apple insect control program. The whole concept of pest management is based on the knowledge that predator-prey relationships can be dramatically altered, especially with insecticides.

In the aquatic ecosystem the effects are undoubtedly as pronounced, but the amplitude of the effect may be dampened by the host of interactions that occur. One of the best documented studies is that of Hurlbert et al. (1972). These workers used 12 shallow (24 cm) unlined ponds, 4 of which were treated with chlorpyrifos at 0.028 kg/ha, 4 treated at 0.28 kg/ha, and 4 left untreated to serve as controls. Treated ponds received three applications of pesticide at about 2 week intervals. All life (except microorganisms) was monitored intensively 21 times during a 3 month period.

Adult populations of insects were affected less by the insecticide treatment than immature stages were, the authors attributing this to greater mobility of the adults as reflected as immigration or emigration. In most instances predaceous insects were more reduced and recovery after treatment much slower than in herbivorous species.

Among the crustacean zooplankton, *Cyclops* and *Moina* suffered high mortality but recovered in 1 to 3 weeks in low-dose ponds. In high-dose ponds recovery did not occur until 3 to 6 weeks after the final treatment. *Diaptomas pallidus* did not seem affected by the low-dose treatments and became abundant after *Cyclops* populations declined. Herbivorous rotifers increased, several species 5- to 25-fold after *Moina* and *Cyclops* populations declined. All planktonic rotifers were much more abundant in treated than in control ponds, and the predaceous rotifer *Asplanchna brightwelli* was similarly more abundant. By contrast all benthic-littoral rotifers were less abundant in treated ponds.

Phytoplankton increased rapidly in treated ponds, high populations persisting up to 6 weeks after the last treatment. The magnitude of increase in relation to controls was 2 for the low-dose and 16 for the high-dose ponds. Blooms of blue-green algae developed in 3 high-dose and 1 low-dose pond.

Hurlbert et al. (1972) attributed these changes in populations to three effects, that is, direct toxicity of chlorpyrifos, changes in competition for food, and predation. They visualized the dominant features of the food web in their ponds as consisting of seven components, some with competing and all with interacting activities (Figure 15.3). They offered explanations for population changes on the basis of these interactions. As pointed out, immediately after treatment there was a decrease in population of *Cyclops, Moina,* herbivorous and predace-

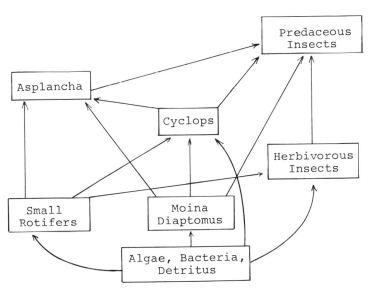

Figure 15.3 A simplified representation of the food web in experimental ponds in the Hurlbert et al. (1972) study [after Hurlbert et al. (1972) but modified to suggest trophic level].

ous insects, and benthic rotifers. These decreases were presumed due to acute toxicity of the insecticide. A decrease was noted also in the delicate phytoplanktonic *Schroederia setigera*, and this was believed due to competition from more vigorous phytoplankton species. Some of these changes were reflected in increases in other organisms. The decreased population of *Cyclops* permitted an increase in *Diaptomas*, since the predation pressure was reduced. Similarly planktonic rotifers increased in numbers, since two of their main predators, *Moina* and *Cyclops*, were less abundant and there was less competition for food from herbivorous insects, *Moina*, and *Cyclops*. *Asplancha* populations increased, because they had an abundant supply of food (small rotifers) and less pressure from predaceous insects. Finally phytoplankton increased because of the reduced population of grazing crustaceans.

It is obvious from this study (Hurlbert et al., 1972) that, in the aquatic environment, indirect effects of pesticides may be expressed in many ways. If one adds the vertebrate species to those included in this study, the ramifications become more complex. It remains to be resolved, however, what the significance of these short-term effects may be.

16

PESTICIDES IN AIR

DDT, the most nearly ubiquitous of all pesticides, is found in virtually every corner of the globe. Although its greatest concentration is in areas of application, it has dispersed widely and can be found from the ice cap of Mount Olympus (Westlake and Gunther, 1966) to the Antarctic (Peterle, 1969) and in all forms of living organisms from tiny microbes to ocean whales (Wolman and Wilson, 1970). This worldwide distribution of DDT is not representative of all pesticides. Quite the contrary. No other pesticide combines the extensive use pattern and properties of persistence that are associated with DDT. Most of the other pesticides have a short life in the environment, and their distribution is confined to the areas near their site of application.

The finding of DDT in areas remote from application prompted a search for methods of transport, and it was soon found that DDT was present in rain and snow collected in various parts of the world (Tarrant and Tatton, 1968; Peterle, 1969) and in atmospheric dust over the land (Cohen and Pinkerton, 1966) and ocean (Risebrough et al., 1968). Thus atmospheric transport of DDT was demonstrated and is now accepted as the major route of long distance movement of pesticides in the environment.

16.1 SOURCES OF PESTICIDES IN AIR

There are many ways by which pesticides reach the atmosphere. Among major routes the following would seem most important.

1. Drift during application.
2. Volatilization during application.
3. Volatilization from treated surfaces.
4. Volatilization following secondary deposit.
5. Escape from pesticide manufacturing and formulating plants.
6. Introduction on dust during dust storms.

Drift During Application

Drift already has been discussed in relation to contamination of soil and water. The amount of drift is a function of atmospheric conditions, the equipment applying the pesticide, the formulation of pesticide being applied, the direction of the discharge relative to the ground, and the height above ground at which the pesticide is released.

The atmospheric conditions important in this connection have several components. Wind speed determines the distance a particle of a given size and weight is carried and its settling velocity. But turbulence may be equally important, especially when pesticides are being applied by air, and updrafts may carry particles of pesticide dust or spray droplets to high elevations and result in pesticide plumes of an unpredictable extent (Orgill et al., 1976).

Temperature and relative humidity are important, especially with sprays. High temperature and low relative humidity combine to hasten the vaporization of the water or solvent used as the spray diluent and leave an increasingly smaller droplet with time. The smaller the droplet, the more likely it is to remain in the atmosphere. For example, at a temperature of 80°F and a low relative humidity a spray droplet 125 microns in diameter loses 75% of its volume and is only 84 microns in diameter after traveling 36 ft from a mist blower (Fisher and Hikichi, 1971). Water droplets of 50 microns survive only 3.5 sec in air at a temperature of 30°C and relative humidity of 50% (Amsden, 1962).

The type of equipment used for pesticide application has a major bearing on the amount of toxicant released into the atmosphere. Less than half of a pesticide applied may actually be deposited on target, and the efficiency of deposit varies markedly with the method of application (Ware et al., 1972). Ware et al. (1969) showed that under the same environmental conditions more drift occurred with a pesticide application by mist blower than by air. The reason for this was the relatively large number of small droplets in the spray stream from the mist blower.

Although many spray procedures use water as the carrier, reduction in drift potential has been achieved through ultra-low-volume spraying in which the pesticide is discharged as technical material or as a high concentrate in a minimum volume of solvent. This has the distinct advantage of reduced vaporization of liquid from the spray droplets. Thus, if a spray nozzle releases droplets in the 100 micron size range, the droplet retains this size until it impinges on the target, and its drift potential does not increase as it travels through the atmosphere.

Direction of discharge affects atmospheric contamination, since this affects the time a spray droplet is in the air between release site and target. Thus orchard spraying with a mist blower in which a portion of the nozzles are directed upward to obtain spray coverage on tall trees is likely to result in a high loss to the atmosphere. Where the foliage canopy is dense, loss is not so great, since a larger percentage of the spray is intercepted by foliage.

The height at which pesticides are discharged affects the amount drifting away from the target area. This is obvious, for the time the droplets require to reach the ground is increased as the altitude of release increases. Other factors are also important. Wind velocity is lower near the ground, and thus the potential for drift due to this factor increases with elevation. Far more important, however, is the residence time of the droplet in the air prior to impact. Amsden (1962) showed that the lifetime of a water droplet of 200 microns diameter in air at 30°C and 50% relative humidity is 56 sec, and during this time it falls 69 ft. A droplet half this diameter has a life of 14 sec and falls only 6 ft during that time.

Volatilization During Application

A number of studies (Spencer et al., 1973; MacKay and Leinonen, 1975) make it clear that evaporation of pesticides of low solubility from water is rapid, and the large surface-to-volume ratio of an aqueous spray would make such evaporation rapid indeed. Three studies give some indication of the magnitude of vaporization during application. The first two of these involve indirect measurement. Decker et al. (1950) applied several insecticides to apples, peaches, and four field crops and measured initial deposit (2 hr post spray) and residues over an extended period thereafter. There were much larger initial deposits with DDT, toxaphene, dieldrin, and lindane (apples and peaches only) than with aldrin, chlordane, or parathion (Table 16.1). Since the application techniques were similar and pesticide formulations all wettable powders (except toxaphene), it is reasonable to assume that the portion of spray lost to drift would be the same for each insecticide.

Table 16.1 Percentage Recovery of Six Pesticides Relative to DDT 2 hours after Spray Application to Various Crops in Relation to Vapor Pressure of the Compounds[a]

	% Recovery[b]			
Pesticide	Apple	Peach	Field Crops[c]	Vapor Pressure[d]
DDT	100	100	100	1.9×10^{-7} at 20°C
Toxaphene	106	79	85	
Dieldrin	192	82	84	1.78×10^{-7} at 20°C
Chlordane	18	30	45	1.0×10^{-5} at 25°C
Aldrin	62	46	56	2.31×10^{-5} at 20°C
Lindane	141	80	--	9.4×10^{-6} at 20°C
Parathion	60	31	--	3.78×10^{-5} at 20°C

[a]Decker et al., 1950.
[b]Based on 100% for DDT.
[c]Alfalfa, sweet clover, red clover, soybeans (mean values).
[d]Martin, 1972.

Thus the difference in deposit might reflect the amount of toxicant volatilized during or immediately (within 2 hr) after the spray had been applied. Although there are some anomalies in the data, they support the inference of Decker et al. (1950) that there is significant volatilization of pesticides during application and that the extent of this is related directly to the vapor pressure of the compounds. Decker et al. (1950) did some laboratory tests in still and moving air using sprays of aldrin and DDT. They found that deposits on a disc from a spray nozzle 4 ft distant were much less with the more volatile aldrin than with DDT. More recently McEwen and Frank (unpublished) applied emulsifiable formulations of four insecticides to head lettuce by using an aqueous spray at 35 gal/acre. Deposits were measured immediately after application (within 15 min) and at 6 hr intervals for 2 days. Results (Table 16.2) show that the deposits from the four insecticides were drastically different immediately after treatment and correlate well with vapor pressure. Although these data provide no indication of the absolute amount of loss from the less volatile compounds, it is obvious that loss of the more volatile compounds was extensive. As in the study by Decker et al. (1950) the difference in deposit must reflect loss due to volatilization during application.

   Some evidence by direct measurement is provided in a recent study of DDT in air after aerial spraying of forest in the Pacific North-

Table 16.2 Deposits (ppm) of Four Insecticides on Lettuce Immediately (15 minutes) After Spraying. Same Dosage Rate and Method of Application[a]

Insecticide	Deposit (ppm)	Vapor Pressure[b]
Mevinphos	7.8	2.9×10^{-3} at 21°C
Leptophos	38.5	
Methamidophos	54.5	3×10^{-4} at 20°C
Methomyl	93.0	5×10^{-5} at 20°C

[a]McEwen and Frank (unpublished data). Mean of three replicated samples at each of three treatment dates.
[b]Martin (1972), Spencer (1973).

west. This spraying was done under special approval after DDT use in the United States had been discontinued; thus background levels of DDT in the atmosphere were low and the levels detected immediately after treatment could be attributed to the recent application. Orgill et al. (1976) measured DDT in the air at several elevations over the treated forest by using a sampler that permitted them to distinguish between particulate and vapor-phase DDT. They found the maximum concentration in a sampling flight 150 m above the forest 2 hr after treatment. In that sample DDT was measured at 19 ng/m³, 3.88 ng of which was in the vapor phase. At other elevations they found the levels were lower but present in vapor and particulate phases in about equal amounts. It would appear from their results that, during forest spraying of DDT, almost half of the DDT not settling on or near the target is lost by volatilization during treatment. If this is true for DDT, then the amount entering the atmosphere as vapor during application of more volatile pesticides must be rather great.

Volatilization from Treated Surfaces

It is a well-known fact that pesticides applied to plants, water, soil, or inanimate objects "disappear" with time, and this disappearance is due to a number of factors, one of which is volatilization. Whereas factors such as the nature of the treated surface, air movement, and temperature affect the rate of volatilization, the most important factor would appear to be the inherent volatility of the pesticide expressed as vapor pressure (Spencer, 1975; Spencer et al., 1973; Wheatley, 1973).

Volatilization has been measured by both direct and indirect methods. Kearney et al. (1964) treated metal planchets with seven s-triazine herbicides and found that the loss of the toxicant was highest in those triazines with the highest vapor pressure and that, consistent with this, loss was more rapid at higher temperatures. They found that the nature of the treated surface was also important, with less rapid loss from soil than from metal planchets and from dry than from moist soils. Studies on volatilization from soil are reviewed by Spencer et al. (1973) and for DDT by Spencer (1975).

Direct measurement of volatilization has been done in the field and laboratory by a number of workers, and gradients of pesticide in air over treated surfaces have been detected readily. Highest levels have been noted soon after treatment (Willis et al., 1969) and are related to dosage rate, soil moisture, soil temperature, air movement, and relative humidity (Harris and Lichtenstein, 1961).

A number of workers have calculated vaporization loss on the basis of known vapor pressure and molecular weight of the toxicants involved. Lloyd-Jones (1971), on the basis of experimental evidence and empirical calculations, concluded that about 50% of the DDT applied to field crops would volatilize in 1 year. His data showed that in England about 2 lb/acre/year (if on the soil surface) would volatilize during the summer and 0.3 lb during the winter. Other estimates indicate the large potential for loss through this route, but it must be remembered that pesticides are normally mixed within the upper few inches of soil and thus do not behave as a surface residue.

Volatilization from treated plant surfaces might be expected to reflect more nearly that from planchets or soil surface. This may, however, be modified by the adsorbency of the surface and the waxy layers to which the pesticides may adhere. Air monitoring over a newly DDT-treated forest indicated that, whereas the volatilization rate was high 26 hr after treatment, DDT levels in the atmosphere had returned to background levels within 86 days (Orgill et al., 1976).

In summarizing their extensive review of the literature on pesticide volatilization, Spencer et al. (1973) conclude that, whereas the vapor pressure of the pesticide is the main criterion in determining the rate of volatilization from a nonadsorbent surface, volatilization under natural conditions is influenced by a variety of other factors associated with the nature of the medium in which the pesticide is distributed. Thus, in soil any factors that restrict movement to the soil surface reduce volatilization, while factors such as increased moisture, which enhances movement, also enhance vaporization.

Note also that, because vapor pressure is the dominant factor, volatilization of pesticides may occur more readily as metabolites than as parent compound or vice versa. In the case of DDT the vapor pressure is much higher for some of the metabolites, and this is true also for lindane. Cliath and Spencer (1972) showed that, in a lindane-treated soil at 30°C, the vapor density of the metabolite PCCH (Y-1,3,4,5,6,pentachlorocyclohexene) was 13.7 times that of lindane and that the vapor density ratio of PCCH to lindane increased with increasing temperature. Thus much of the lindane volatilized from soil would be in the form of PCCH. Similarly it was shown that in soil at 30°C the vapor density of p,p'-DDE was eight times as great as that of p,p'-DDT. Thus in well-aerated soil where DDT is converted to DDE most of the volatilization occurs in the form of the latter.

Volatilization Following Secondary Deposit

As discussed earlier, pesticides in the atmosphere are washed out by rain and return to the earth in water droplets or dust particles to which they become adsorbed. The short-lived pesticides disappear quite rapidly, but those like DDT and, to a lesser extent, dieldrin become surface residues again subject to volatilization. Thus one can visualize for highly persistent compounds a continuing flux of fallout and resuspension, and though definitive data on this cycling are not available, the fact that DDT (plus metabolites) is highest in air over land masses where the pesticide has been used and decreases over oceans supports this assumption. A number of studies confirm long distance transport of pesticides in the atmosphere (Risebrough et al., 1968; Bidleman and Olney, 1974; Harvey and Steinhauer, 1974), and since an air mass moves around the globe in 3 to 4 weeks (Woodwell et al., 1971), it is easy to visualize worldwide distribution through this medium. It is obvious that the scrubbing effect of rain and falling dust dampens this route of transport and tends to retain pesticides near their source of input.

There has been considerable debate concerning the process of codistillation with respect to pesticides. Acree et al. (1963) measured DDT loss from aqueous suspensions in glass jars and concluded from their data that the relative rates of water loss and DDT loss could be explained only by a mass flow of DDT with the water rather than by independent volatility rates. Spencer et al. (1973), reviewing the data of Acree et al. (1963), concluded that these workers had failed to take into account the relative humidity in the jars and the influence this would have on the rate of evaporation of water. Spencer et al. (1973) concluded that codistillation does not apply to pesticides at temperatures below the boiling point of water. Liss and Slater (1974) conclude, however, that, for materials of low solubility, the liquid phase controls the rate of vaporization, and thus for pesticides such as DDT codistillation would be a major factor in pesticide loss. On the basis of principles of mass transfer MacKay and Leinonen (1975) calculate the rate of loss of a number of compounds from a 1 m water column and show that with DDT the half-life due to codistillation (coevaporation) is about 74 hr. With lindane or dieldrin the evaporation is much less rapid, while with various Araclors (PCBs) the half-life is about 10 hr. On the basis of these calculations it might be concluded that in the oceans much of pesticides like DDT is volatilized soon after deposition on the water surface and only that portion incorporated in living matter ever reaches the ocean abyss.

Escape from Pesticide Manufacturing and Formulating Plants

Pesticide plants are likely to be significant point sources of pesticide input. Young et al. (1976) studied the dry aerial fallout of DDT in southern California at 14 coastal stations and 6 inland stations during two 13 week periods in 1973 and 1974. They found that, with one exception, the fallout of DDT increased toward Los Angeles despite the fact that the major agricultural areas did not. Stations established in the Los Angeles basin during a 4 week period in spring of 1974 indicated two areas of high fallout, one around a large DDT-manufacturing plant and the other at a dumping site where waste from this plant had been dumped prior to 1972. In addition it was found that about 85% of the DDT collected on the sampling plates in the plant vicinity consisted of p,p'-DDT and $0,p'$-DDT in about a 5–1 ratio. This indicated a fresh source of DDT. The authors concluded that the escape of DDT from the manufacturing plant and the associated dumping site represented major sources of atmospheric contamination with this pesticide and accounted for a significant fraction of the estimated 1.3 metric tons contributed annually to the California Bight.

Introduction on Dust During Dust Storms

Only one study has documented the significance of this source of atmospheric contamination with pesticides. Cohen and Pinkerton (1966) demonstrated that soil from a dust storm in the Texas-New Mexico area reached Cincinnati some 30 hours later carrying several pesticides, DDT, chlordane, DDE, and ronnel being the main contaminants.

16.2   LEVELS OF PESTICIDES IN THE AIR

Many studies show pesticides in air either as particulate deposits or in the vapor phase. Some of the reported levels are summarized in Table 16.3, but these must be interpreted in the context in which they were

Table 16.3 Some Reported Levels of Pesticide Contamination in the Atmosphere of North America and the Atlantic Region

| Pesticide | Sampling Area | Residue g^{-12}/m^3 | | Reference[a] |
		Range	Mean	
DDT (R)[b]	Rural USA	100-8,500,000		Tabor, 1966
DDT (R)	Urban USA	2,000-8,000,000		Tabor, 1966
p'p'-DDT	Urban USA	100-1,560,000		Stanley et al., 1971
DDTR	Rural Canada	2,300-739,000		Frank et al., 1974
DDTR	Forest, USA	20-18,764		Orgill et al., 1976
p'p'-DDT	Rural USA	100-534,700		Arthur et al., 1976
DDTR	Barbados	0.044-0.629	0.212	Seba and Prospero, 1971
DDTR	Barbados	0.006-0.242	0.078	Risebrough et al., 1968
DDTR	Bermuda	19-57	34	Bidleman and Olney, 1974
DDTR	Bermuda	17-53		Harvey & Steinhauer, 1974
DDTR	Atlantic Ocean	39-170	84	Bidleman and Olney, 1974
DDTR	Atlantic Ocean	<1.0		Harvey & Steinhauer, 1974
DDTR	Providence, R.I.	115		Bidleman and Olney, 1974
Aldrin	Rural USA	100-4,000		Tabor, 1966
Aldrin	Urban USA	ND[c]-8,000		Stanley et al., 1971
Chlordane	Rural USA	ND-31,000		Tabor, 1966
Chlordane	Providence, R.I.	250		Bidleman and Olney, 1974
Chlordane	Bermuda	5-12	8	Bidleman and Olney, 1974
Chlordane	Atlantic Ocean	39-170	84	Bidleman and Oleny, 1974
DEF	Rural & Urban USA	ND-16,000		Arthur et al., 1976; Stanley et al., 1971

357

Table 16.3 *(Continued)*

| | | Residue g^{-12}/m^3 | | |
Pesticide	Sampling Area	Range	Mean	Reference[a]
Dieldrin	Urban USA	ND−29,700		Stanley et al., 1971
Endrin	Urban USA	ND−58,500		Stanley et al., 1971
Lindane	Urban USA	ND−7,000		Stanley et al., 1971
Heptachlor	Urban USA	ND−19,200		Stanley et al., 1971
Malathion	Urban USA	ND−430,000		Stanley et al., 1971
Methyl parathion	Urban USA	ND−129,000		Stanley et al., 1971
Methyl parathion	Rural	ND−2,060,000		Arthur et al., 1976
Parathion	Urban USA	ND−465,000		Stanley et al., 1971
Parathion	Rural USA	ND−4,300		Arthur et al., 1976
Toxaphene	Urban	ND−2,520,000		Stanley et al., 1971
Toxaphene	Rural	ND−1,746,500		Arthur et al., 1976
Diazinon	Rural	ND−77,400		Arthur et al., 1976

[a]See text for additional comments.
[b]DDTR − DDT isomers and metabolites.
[c]ND = Not at detectable level.

obtained. The first extensive air monitoring for pesticides was done by Tabor (1965, 1966). He sampled air in small communities in six agricultural areas and in four cities where pest control programs for mosquito control were being conducted. The sampler used in this work measured particulate pesticide only, and thus the data underestimate the true level, since some of the pesticide would be present in the vapor phase. In most cases total pesticide levels ranged from near the lower limit of detection to about 20 ng/m³. High levels were measured (Table 16.3) in agricultural areas when spraying was actually being done and in cities during actual fogging of DDT for mosquito control.

Stanley et al. (1971) collected 880 composite air samples from nine localities (four urban, five rural) in the United States and analyzed for 19 pesticides. They sampled for a 6 month period at 2 week intervals but staggered sampling to reflect both the summer and winter seasons. DDT was the only pesticide found at all locations, and while it ranged as high as 1560 ng/m³ in the rural area of Orlando, Florida, most samples fell in the 10 to 50 range. Other pesticides found in the sampling program included parathion, methyl parathion, malathion, toxaphene, heptachlor, lindane, BHC, dieldrin, endrin, and the defoliant DEF. Aldrin and 2,4-D were each found in one sample only. The highest pesticide levels were found in the agricultural areas of the south, levels being generally lower in urban than in rural locations. The authors found that low levels of pesticides in the atmosphere were frequently associated with rain but that there was a better correlation between high pesticide levels and reported spraying. They concluded that the effect of rain was more likely to curtail spraying activity than to affect aerial pesticide load directly. The authors reported that most of the pesticide collected in the sampler was deposited on the filter and concluded that pesticides were present mainly as particulates in the atmosphere. A large proportion of the total DDT detected in the samples was p,p'-DDT, and only a small percentage was DDE. This would be expected from a fresh DDT source (note the correlation of high levels with reported spraying). Frank et al. (1974) found similar levels of DDT in air sampled in Ontario and an equally high ratio of DDT to DDE.

Arthur et al. (1976) analyzed air in the Mississippi delta at Stoneville during 1972, 1973, and 1974 using a sampler that permitted detection of both particulate and vapor pesticides; p,p'-DDT and 0,p'-DDT were found in all samples during the 156 week period. By contrast the organophosphorous insecticides, with the exception of methyl parathion, were found only during the growing season. All

Table 16.4 Average Monthly Levels of Four Pesticides in Air, Stoneville, Mississippi, 1974[a]

Month	Total DDT	Toxaphene	Endrin	Methyl parathion
		Level (ng/m^3) of Pesticide		
January	3.0	10.9	0.2	1.0
February	3.6	9.7	0.2	0.3
March	7.6	19.1	0.6	0.3
April	7.7	27.7	0.5	0.6
May	15.6	44.3	0.7	0.6
June	12.8	38.6	0.7	0.9
July	24.3	175.0	9.3	40.9
August	37.9	903.6	27.2	341.1
September	19.4	524.6	18.8	167.9
October	5.1	114.8	4.3	2.0
November	3.3	32.9	1.0	0.0
December	2.1	12.6	0.5	0.0
Average	11.9	159.5	5.3	46.3

[a]Arthur et al., 1976.

pesticide levels were sharply higher in summer than in winter (Table 16.4). Arthur et al. (1976) observed a sharp decline in atmospheric DDT in the highly agricultural Mississippi delta from 1972 to 1974 associated with the discontinued use of DDT. An average DDT level of 99.5 ng/m³ in 1972 had declined to 16.0 in 1973 and 11.9 in 1974.

The high levels of pesticides in air (Table 16.3) over urban and rural North America have all been associated with spraying or fogging applications near where the measurements have been taken. Wood-well et al. (1971) concluded that, for the pesticide DDT, an atmospheric level of 80 ng/m³ would represent a worldwide value, but Stewart (1972) argues that the level is perhaps nearer to 1% of that amount. Data collected over the Atlantic Ocean and in areas such as Bermuda may suggest the general background atmospheric level of DDT, the only pesticide in general atmospheric distribution at quantifiable levels. Risebrough et al. (1968) were the first to measure DDT in the West Indies. They used a nylon mesh screen coated with glycerin and claimed a 50% efficiency in trapping airborne particles more than 1 micron in diameter. By this method they determined that air reaching Barbados carried 0.078 pg DDT/m³ on dust particles. A few years later Seba and Prospero (1971), using the same type of collection procedure but a more nearly complete analysis of collected particles, reported about three times this amount. Bidleman and Olney (1974), using a sampler that collected both particulate and vapor DDT, reported levels considerably higher than in the earlier studies. They found a mean of 34 pg/m^3 and concluded that their higher values reflected the addition of the vaporphase DDT. These workers also found chlordane in their sampling of Bermuda air. Air samples taken at various locations in the western Atlantic (Bidleman and Olney, 1974; Harvey and Steinhauer, 1974) give residue levels for DDT in the same general range (i.e., $\langle 1.0$ to 100×10^{-12} g/m^3).

Orgill et al. (1976), analyzing air over western forests in connection with the 1974 spraying with DDT for tussock moth control, indicated background levels of about 20×10^{-12} g/m^3. In the light of these data it would appear that the level of DDT in air is much lower than suggested in earlier reports (Woodwell at al., 1971; Stewart, 1972). Even Stewart's (1972) estimate of 800×10^{-12} g/m^3 would seem to be one to two orders of magnitude higher than current data suggest.

As we have seen, pesticides other than DDT may occur in air near the site of application, but residues are short lived, probably owing in most cases to photochemical degradation (Wheatley, 1973). Although air samplings in United States communities have found significant and sometimes high levels of malathion, parathion, toxaphene, lindane, and a number of other pesticides, these high levels have been associated with local spray programs, and in few cases have these pesticides been detected during the winter season, when spray operations are minimal.

16.3 SIGNIFICANCE OF PESTICIDE RESIDUES IN AIR

It is recognized that the inhalation of pesticides can be toxic, and for a large number of compounds, LD$_{50}$ values based on exposure by this route have been calculated. Tabor (1965, 1966) calculated the daily intake of DDT for man on the basis of contamination levels he found in each of five communities. He based his calculations on a breathing rate of 1.5 m^3/hr (breathing rate is approximately 0.5 m^3/hr for rest, 1.7 m^3/hr for light work, 3.6 m^3/hr for heavy work) and on the assumption that the lungs would absorb all the DDT inspired. He reported that the intake per day would be 0.2 to 32 μg of DDT. On the basis of an estimated DDT intake of 200 μg/person/day, Tabor (1965, 1966) concluded that the intake from air was minor. Pesticides volatilize from treated surfaces, and this is reflected in locally high pesticide levels in treated fields and a possible danger to field workers.

Some information on inhalation of pesticides has been gathered in studies on safe reentry intervals prior to workers' entering treated crops. Most of the problems in this connection have been with organophosphorous insecticides such as parathion and methyl parathion (McEwen, 1977), and thus most of the data are for these pesticides. Ware and Morgan (1976) calculated the exposure of cotton scouts working for 30 min in a cotton field treated with parathion. They estimated that, if the scout entered immediately after treatment, exposure on hands and forearms would be 3.47 mg, whereas exposure by inhalation would be 1.06 μg. Exposure by each route would be reduced almost 50% if the entry to the treated crop were delayed for 12 hr. It would thus appear that atmospheric levels of pesticides are unlikely to pose a hazard to man in the outdoors. There are exceptions. Oudbier et al. (1974) had people involved in mixing and spraying pesticides wear respirators during the operations and analyzed the amount of pesticide filtered out by these devices. Wettable powders created a far greater inhalation risk during mixing than liquid formulations did. For example, when emulsifiable azinphosmethyl was added to a 500 gal sprayer, no pesticide was picked up in the respirator, but when 2.5 lb of 50% wettable powder was added, more than 3 μg of the chemical was found on the filtering devices.

In a number of studies of atmospheric pollution with pesticides the extent of pollution has been determined by analyses of dust particles caught on oil-coated nylon screens placed vertically (e.g., Risebrough et al., 1968, Seba and Prospero, 1971; Sodergren, 1975) or on coated glass plates (Young et al., 1976) placed horizontally. These sampling devices attest to the significant amount of pesticides, especially DDT, adsorbed on dust particles. Whereas the amount of dust and levels of adsorbed pesticides on outside dust may not be significant in terms of human exposure, a different situation may obtain in enclosed environments, and workers in plants where extensive handling of pesticides is involved must wear respirators. Starr et al. (1974) analyzed the dust in a selected group of American homes. Vacuum cleaner samplings of dust were made in 28 households during a 12 month period. Samples were pooled on a monthly basis, most samples representing at lest two sweepings/week. The households sampled included 16 urban homes (normal exposure to pesticides), 4 farm homes with high agricultural use of pesticides, and 8 homes where at least one member of the family was a pesticide formulator. DDT (and/or metabolites) was found in practically all samples with a mean level (for p,p'-DDT) of 6.90, 8.87, and 30.66 ppm in urban, farm, and formulator homes, respectively. The highest levels found were, respectively, 35.44, 37.80, and 226.15 ppm. Methoxychlor was found in almost 50% of the samples from each group at about the same level as for p,p'-DDT. Chlordane and dieldrin were found in about 25% of the urban and farm homes and in 60 to 80% of the homes of formulators. The levels of dieldrin were less than 50% of those of DDT, while the levels of chlordane approached those for the latter pesticide. Lindane was found in three samples from farm homes and about one-half the samples from those of formulators but was not detected in dust from urban homes. Endrin was found only in homes of formulators, and dacthal was found in a small percentage of dust samples from urban and formulator homes and in 50% of the dust samples from farm homes. Starr et al. (1974) included in their study an analysis of pesticide levels in blood sera of the 28 male householders and 27 spouses. Despite the fact that general levels of pesticides in dust were quite different in the three types of households studied, no correlation could be found between pesticide levels in dust and levels in the blood.

The main significance of pesticides in air is clearly the opportunity this medium provides for global distribution (Woodwell et al., 1971; Harrison et al., 1970). On the basis of estimates provided by Woodwell et al. (1971), Wheatley (1973) calculated that the atmosphere (mass 5×10^{18} kg) would contain 4×10^5 tons of DDT alone, one-sixth of the total production of DDT up to 1974. This estimate was based on a residue level of 84 ng/m³ and as we have discussed is probably an overestimate by at least three orders of magnitude. Nevertheless it is obvious that, with the constant inputs through volatilization and other sources and the fallout in rain and dust, the fact of transportation has been established clearly. The main air mass movement patterns (equatorial easterlies, temperate westerlies, polar easterlies) ensure global distribution for pesticides that are not degraded rapidly by sunlight or other mechanisms. Thus the pesticide DDT will be present throughout the globe for many years regardless of whether its use is discontinued or not. Obviously, continued use will extend its stay in the environment.

Various estimates have been offered for the lifetime of DDT in the atmosphere. Woodwell et al. (1971), using production figures for DDT and assuming rainfall as the method for its removal from the atmosphere, calculated a 3.3 year residence time for DDT, and Bidleman and Olney (1974), using the relative concentrations in air and ocean surface waters, computed it to be 51 days. Peakall (1976) measured DDT residues in snow and rainwater near Ithaca, New York, during the period August 1974 to September 1975 and found highest levels in August–December 1974. Peakall (1976) suggested that this might be related to the spraying of DDT for tussock moth control in the Pacific Northwest during June of that year. He reported that 80% of the total DDT detected was the p,p' isomer, and this is in good agreement with results of air sampling done over the treated forest shortly after application (Orgill et al., 1976). This ratio is, however, also in close agreement with air samples taken in May 1973 at Providence, Rhode Island (Bidleman and Olney, 1974), but does not reflect the relative ratios determined by Bidleman and Olney (1974) in air samples taken in June over the Atlantic. It is thus not possible to identify the source of DDT contamination in the Ithaca rain, but it should be noted that the highest levels coincide with a period of active agricultural operations in the eastern United States, and, as has been discussed, recent cultivation increases the rate of volatilization of DDT resident in soils.

17

PESTICIDES IN FOOD

Prior to the introduction of the modern pesticides significant residues in food involved only two chemicals, mercury and arsenic. Little attention was given to mercury, for its use was limited primarily to soil and seed treatments, and residues in harvested crops were not regarded as significant. With the arsenicals the situation was quite different. The first major incident of arsenic poisoning occurred in beer drinkers in England owing to contamination of hops used in the brewing process. It was soon learned that the use of lead arsenate for insect control, especially on fruits and leafy vegetables, resulted in high levels of arsenic in the edible product, and the first "tolerances" were established for pesticide residues in food. The concept of a "tolerance level" was new and was based on toxicological data indicating a maximum level that could be present in foods without any risk of illness from its consumption.

With the development of the chlorinated hydrocarbons and their widespread use in agriculture it became apparent that residues in food were important, and since then attention to this phase of study has become a major component of pesticide development. Each crop or food product for which a pesticide is registered must be analyzed for residues and a tolerance established prior to the use of the pesticide on the food crop. In the United States this process was formalized in 1954 through an amendment of the Food, Drug and Cosmetic Act. This amendment directed the Food and Drug Administration "to establish safe and legal tolerances for pesticide chemicals on raw agricultural products where the United States Department of Agriculture has established the usefulness of the pesticides." Two criteria are involved.

In the first place it must be established what maximum residue is likely to be present in the raw agricultural commodity when the pesticide is used in a manner effective for the pest control required. Tolerances will not be established much above this level, for it is considered undesirable to have pesticide residues in food, and they will be permitted only if needed. Thus, for many pesticides that degrade rapidly, a zero tolerance is established, since pest control using these chemicals will not result in a residue at harvest. If it is found that the use of a chemical in the manner required for pest control results in a residue, then the second criterion is applied. This involves proof that the residue present is much lower, usually by a factor of 100, than the lowest dosage that can be demonstrated to cause any effect on test animals (Duggan and Dawson, 1967; Hanna, 1974). Note that this is a working guideline and one that may be modified when the nature of the food and quantities consumed are considered. In addition the Delaney amendment to the Food, Drug and Cosmetic Act dictates that in the United States no pesticides shown to be carcinogenic are permitted for use on crops intended for human food.

The problem of residues in food has been addressed at the international level through several committees sponsored by the United Nations. In 1961 a joint meeting of the WHO Expert Committee on Pesticide Residues and the FAO Panel of Experts on the Use of Pesticides in Agriculture directed "that studies be undertaken to evaluate possible hazards to man arising from the occurrence of residues of pesticides in food" (FAO/WHO, 1972). This has been done through these Committees, the Codex Committee, and various working parties of experts. As a result acceptable daily intake (ADI) has been established for a number of pesticides and presented along with suggested tolerances in a series of annual reports of the Joint FAO/WHO meetings.

17.1 SOURCE

Two sources of contamination must be recognized. The first is the result of direct application, while the second is due to contamination or indirect application.

Direct Application

When pesticides are applied to food crops, they degrade through chemical and biological processes at a rate determined by the nature of the chemical and the nature of the plant surface or soil in which the pesticide is placed. With many pesticides residues disappear rapidly during the first few hours or days after application and then more slowly. The literature contains many reports to illustrate this point. McEwen and Frank (unpublished) applied four insecticides to Chinese cabbage at various periods prior to normal harvest date for the crop and analyzed the lettuce on a fresh, wet weight basis. Results (Table 17.1) show differences in the rates of disappearance between the different insecticides and also reduced rates of loss after the first 3 days.

Note also that all of the residue may not be present as the parent compound. Many pesticides may form metabolites that are as persistent or in some cases (e.g., DDT) more persistant than the initial chemical. For example, when endosulfan is applied to leafy vegetables, much of the residue after a few days will be in the form of endosulfan sulfate (Table 17.2). This is recognized in the establishment of tolerances and ADIs, the stated tolerances applying to the composite residue.

Table 17.1 Residue of Four Insecticides on Chinese Cabbage at Harvest When the Insecticides Were Applied at Various Intervals Prior to Harvest[a]

Insecticide	Rate of Application lbs/acre	Application Days Prior to Harvest	Residue at Harvest ppm
Parathion	1.0	0	12.1
		3	1.83
		7	0.27
		14	0.062
		21	0.005
Diazinon	1.0	0	13.3
		3	1.23
		7	0.153
		14	0.049
		21	0.011
Methamidophos	1.0	0	59.7
		3	8.53
		7	1.21
		14	1.90
		21	0.54
Endosulfan	1.0	0	25.2
		3	8.71
		7	4.18
		14	3.17
		21	0.67

[a]McEwen and Frank (Unpublished).

Table 17.2   Residues of Endosulfan and Metabolites on
Chinese Cabbage at Harvest Following Application of
Endosulfan at 1.0 lb/acre 7, 14, and 21 Days Earlier[a]

| | Residue in Fresh Tissue (ppm) | | | |
Application (Days Prior to Harvest)	Endo-sulfan I	Endo-sulfan II	Endo-sulfan sulfate	Total Endo-sulfan
7	0.63	1.02	2.2	3.85
14	0.17	0.41	2.6	3.18
21	0.17	0.07	0.43	0.67

[a]McEwen and Frank (Unpublished).

Indirect Application and/or Contamination

In addition to residues that may result from direct, intentional applications, pesticide residues may be present owing to contamination. This can occur when crops receive drift of pesticides applied to adjacent crops or, in low levels, from atmospheric fallout. Low levels of residues in foods may result also from the growing of crops, especially root crops, in soils containing residues of persistent pesticides.

A number of pesticides have an affinity for lipid materials and accumulate in animal systems. Thus animal products may contain residues. This was discovered in the early days of use of DDT and dieldrin when these pesticides showed up in milk and dairy products. No pesticides in milk were permitted, and as a result large quantities of milk were destroyed. The concept of no residue in milk became untenable with improved analytical capability that permitted the detection of DDT and dieldrin in parts per billion and parts per trillion levels. As a result a tolerance has been established for DDT in milk, and the WHO/FAO suggests that, on a fat basis in milk and milk products, practical limits for DDT and dieldrin are 1.25 and 0.15 ppm, respectively (FAO/WHO, 1972).

17.2   LEVELS OF PESTICIDE RESIDUES IN FOODS

Any perusal of the literature on pesticide residues makes it clear that mankind's concern for the safety of the food supply has been reflected in the extensive coverage of this field. Research reports can be found covering individual pesticides, individual foods, geographic areas, and almost any other category one might wish to pursue. Basically three types of data are available. In the first instance there are a large number of research reports dealing with known treatment protocols and the residues that result at various intervals after application. This is the kind of data that is generated as a requirement for registration and is available for all pesticides registered for use on food or feed crops. A second kind of data is that obtained by regulatory agencies in their role in monitoring foods to ensure that they are not in violation of established tolerances. These data are likely to be biased in that sampling procedures are not strictly at random. Such agencies tend to concentrate on "hot spots," where pesticide use patterns or the particular commodity suggests pesticide residues may be a problem. The third source of data is the most representative of the general food supply and results from routine monitoring of food supply to determine daily pesticide intake.

The most extensive program of this latter type in the United States is the federal Program for Monitoring Pesticide Residues in Food and Feed conducted by the Food and Drug Administration of the U.S. Department of Health, Education and Welfare. This study has two main components: (1) a market basket study and (2) a nationwide survey of unprocessed food and feed. From the standpoint of human intake the market basket study is the most revealing. In this study 127 items of food representing the basic diet of a 19 year old male for a 2 week period is taken from food markets throughout the country, prepared for eating, and analyzed as composite samples in each of 12 categories (Duggan and McFarland, 1967). This sampling has been conducted each year since 1964, and in the sampling from August 1972 through July 1973, samples were collected in 30 different grocery markets in 30 different cities (Johnson and Manske, 1976). Results are summarized in Table 17.3 for each of the pesticides detected in more than 10 of the 360 composite samples. Organochlorine residues were detected in 52% of the composite samples, and residues of organophosphorous insecticides occurred in 31%. The organochlorine residues occurred most frequently in dairy products, meat, fish, and poultry, whereas the organophosphorous residues occurred most frequently in leafy vegetables and fruits.

In addition to the pesticides reported in Table 17.3, the market basket sampling found residues of 16 other pesticides in some composite samples; 11 of these were detected in only 5 or less of the 360 composite samples (Johnson and Manske, 1976).

The data summarized in Table 17.3 demonstrate that, while pesticides are present in foods, the levels are extremely low. These levels, in the United States, are generally lower now than they were some

Table 17.3 Pesticide Residues Found in the 1972–1973 Total Diet and Percentage Intake of Pesticide

		ppm Pesticide Detected in Compo			
Pesticide[b]	Frequency[c]	Dairy Products	Meat, Fish Poultry	Grain & Cereals	Potatoes
Percent of total diet[d]		27.7	15.5	10.7	10.7
Dieldrin	107	.002	.004	--	.001
DDE	81	.002	.012	--	.001
DDT	54	Tr	.006	--	Tr
BHC	59	.001	Tr	--	--
TDE	48	Tr	.002	--	Tr
Diazinon	54	Tr	Tr	--	Tr
Malathion	54	--	--	.024	--
Heptachlor epoxide	46	Tr	Tr	--	Tr
Lindane	39	Tr	Tr	--	Tr
Mercury	32	--	.01	--	--
Endosulfan	29	--	--	--	.001
Arsenic	22	Tr	Tr	--	Tr
Parathion	19	--	--	--	--
Ethion	14	--	Tr	--	--
CIPC	13	--	--	--	.143
Carbaryl	12	--	--	--	Tr
Percentage intake[e]		14.7	40.3	8.2	2.0

[a]From Johnson and Manske, 1976.
[b]Includes only those pesticides occurring in more than 10 of the 360 composite samples.
[c]Number of composite samples containing the pesticide.
[d]Percent of food category in normal diet (Smith, 1971).
[e]Percentage pesticide intake from each food category, average 1964-67 (Duggan, 1968).

Tr Trace
-- No residue found.

years ago. Even then, however, only a small number of samples contained high residues. Summarizing the results of 4 years' data (July 1, 1963, to June 30, 1967), Duggan (1968) reported that 3% of the sam-

Survey in 30 American Cities, Frequency of Residue Occurrence, from Each Food Category, 1964–1967

site Samples in each of 12 Food Categories[a]

Leafy Vegetables	Legume Vegetables	Root Vegetables	Garden Fruits	Fruits	Oils, Fats, Shortening	Sugars & Adjuncts	Beverages
2.6	1.8	2.8	4.5	10.7	1.5	8.0	3.3
Tr	Tr	--	.003	Tr	Tr	--	--
.001	--	.001	Tr	--	Tr	--	--
Tr	--	Tr	Tr	--	Tr	Tr	--
--	--	--	Tr	--	Tr	Tr	--
Tr	Tr	--	.001	--	.001	Tr	--
.001	--	Tr	.001	.001	.003	--	--
--	--	--	--	.003	.023	.001	--
--	--	--	--	--	--	--	--
--	--	--	Tr	--	--	Tr	--
--	--	Tr	--	--	Tr	--	--
.019	--	--	Tr	Tr	--	--	--
Tr	Tr	Tr	--	Tr	Tr	Tr	--
.003	Tr	Tr	Tr	Tr	--	--	--
--	--	Tr	--	.004	--	--	--
--	--	--	--	--	--	--	--
--	Tr	.002	Tr	Tr	--	--	--
3.5	1.5	1.1	10.4	14.4	3.1	0.4	--

ples had residues exceeding legal tolerance or, where no tolerance was established, were considered to have excessive residues. He reported no significant differences between domestic and imported foods in regard to pesticide residues. In these studies about 50% of food samples contained residues. More than 75% of such residues were below 0.11 ppm and 95% were below 0.51 ppm. Residues most frequently found were DDT, DDE, dieldrin, TDE, heptachlor epoxide, lindane, BHC, endrin, aldrin, and toxaphene in domestic foods, while in imported foods the list was the same except that toxaphene was replaced by keldhane. Similar total diet studies have been done in Canada, and residues found are of the same order of

Table 17.4 Some Pesticide Residues in Total Diet Studies in Canada, 1969, 1970, 1971, in Three Categories of Foods

Mean Residue (ppm) of Four Sampling Periods for Each Food Category

Pesticide	Dairy Products			Meat, Fish & Poultry			Leafy Vegetables		
	1969[a]	1970[b]	1971[c]	1969	1970	1971	1969	1970	1971
DDT + metabolites	.007	.004	.006	.028	.011	.020	.014	.009	.001
Dieldrin	.002	.001	.002	--	.001	.001	--	--	--
Aldrin	--	--	--	--	--	--	.019	.003	--
Endosulfan	--	--	--	--	--	--	.006	.002	.038
Heptachlor epoxide	--	--	--	--	--	.001	--	.001	.002
BHC	--	--	.001	--	--	--	--	--	--
Lindane	--	.001	--	--	--	--	--	--	--
Diazinon	--	--	--	--	--	--	--	--	.040
Parathion	--	--	--	--	--	--	--	--	.004

[a] Smith (1971). Samples from Ottawa.
[b] Smith, Sandi and Leduc (1972). Samples from Vancouver.
[c] Smith, Leduc and Charbonneau (1973). Samples from Halifax.

magnitude. Data reported on three categories of foods for each of the 3 years 1969–1971 are presented (in part) in Table 17.4. It would appear that the restrictions placed on the use of persistent chlorinated hydrocarbons in Canada were reflected in the lower residue in leafy vegetables in 1971, but residues in meats and dairy products show no change. Note, however, that the foods included in the analyses summarized in Table 17.4 were taken from different cities each year, and some of the foods would have a local origin.

The 12 composite samples that make up the total diet studies include a variety of foods, and residue levels might be expected to

Table 17.5 Residues of DDT and Its Metabolites and of Dieldrin in Food Items Within the Dairy Products and Meats Categories, United States Total Diet Studies, 1972–1973[a]

	Residue (ppb)[b]			
Product	DDE	DDT	TDE	Dieldrin
Dairy Products[c]				
Whole fluid milk	1–3	--	--	Tr
Evaporated milk	13–16	--	--	2–3
Ice cream	Tr–19	Tr	Tr	Tr–7
Processed cheese	Tr–8	Tr	Tr	5–10
Natural cheese	3–8	Tr–6	Tr–5	5–14
Butter	5–154	Tr–11	Tr	14–56
Ice milk	Tr–9	Tr	--	Tr
Cottage cheese	12–15	--	--	Tr
Meats				
Roast beef	8–559	5–22	Tr–13	2–19
Ground beef	6–83	Tr	--	3–9
Pork chops	Tr–12	Tr–30	32	Tr–7
Bacon	3–18	8–30	Tr	2–3
Chicken	Tr–18	Tr	--	Tr–4
Fish fillet	18–95	18–39	14	--
Tuna or salmon	Tr–18	--	6–10	--
Luncheon meat	10–38	Tr–41	Tr–10	3–11
Frankfurters	14–36	Tr–38	5–10	4–9
Beef liver	20	8	--	--
Eggs	5–46	6	--	Tr
Ham	Tr	Tr	Tr	2–9
Round steak	8–10	Tr–91	19	3–10
Veal	14	11	6	Tr
Lamb	Tr–3	Tr	Tr	8

[a]Johnson and Manske, 1976.
[b]Parts per billion = mg x 10^{-6}/gm.
[c]No residues found in buttermilk, skim milk or nonfat dry milk.
Tr Trace
-- No report.

vary within these categories. Johnson and Manske (1976) included analyses of individual food items in four market basket samples for the two categories "dairy products" and "meat, fish, and poultry." A portion of their data are summarized in Table 17.5 for the most common residues, DDT and its metabolites and dieldrin. These data indicate large variations in residues between and within individual commodities.

17.3 REDUCING RESIDUES IN FOODS

The normal methods used in preparing foods for the table go a long way toward reducing pesticide residues (Geisman, 1974). The extent of such effects varies, however, with the pesticide residue and the nature of the food product.

Lamb et al. (1968) evaluated the effect of various preparative procedures on field residues of three pesticides on spinach. They found that, while washing removed a high percentage (66 to 87%) of carbaryl and significant amounts (17 to 48%) of DDT, only a small amount (0 to 9%) of parathion residues was removed. The addition of detergent to the wash water increased the amount removed. Although steam blanching has little effect on pesticide residues, hot water blanching removed 38 to 60% of DDT, 49 to 71% of parathion, and 96 to 97% of carbaryl. Tests with residues of DDT, malathion, and carbaryl in green beans demonstrated that large percentages of residues of DDT and carbaryl were removed during processing or home preparation and that most of the malathion residue (96%) was removed in cold water washing (Elkins et al., 1968).

Although parathion residues on spinach were greatly reduced by commercial processing, processing of broccoli was quite a different matter. No parathion was removed by commercial washing or steam blanching, and only 30% was removed after washing with detergent added and hot water blanching for 6 min. By contrast residues of carbaryl on broccoli processed for freezing in the home or in a factory were reduced about 90%. When broccoli was washed and cooked fresh, carbaryl residues were reduced about 55% (Farrow et al., 1969). Cooking of broccoli containing residues of malathion reduced residues only 7 to 34% (Kilgore and Windham, 1970). Normal commercial processing of fruits also removes pesticide residues. Fahey et al. (1970) showed that more than 95% of stirofos residues on peaches, pears, and cherries were removed during canning.

A comparison of the pesticide levels in raw products with those in the same food categories gives some indication of the role food preparation for consumption plays in reducing pesticide levels. Comparisons for the years 1964–1969 are presented in Table 17.6.

It can be seen that, with four of the five food classes, residues in the ready-to-eat foods are much less than those in the raw products. As would be expected, however, processing has little effect on the level of DDT in dairy products, since the pesticide is lipophilic and the fat content of many dairy products is not altered between the raw and ready-to-eat product.

Although large reductions in pesticide residues in food crops can be achieved through washing, trimming, blanching, and cooking, this is not true to the same extent for meats. It was shown early (Carter et al., 1948) that, when beef containing DDT was cooked by roasting, broiling, or frying, or in a pressure cooker, there was little difference in residue between the cooked and raw product. Some DDT appeared in the drippings but the percentage was not great. In a similar way residues of dieldrin in meat incorporated in meat loaf are reduced slightly, since some dieldrin shows up in the drippings, but the concentration in the cooked meat loaf is about the same as before cooking (Shafer and Zabik, 1975). Yardick et al. (1971) fed dieldrin to three hogs for 13 days immediately prior to slaughtering and measured the effect of two cooking methods, pan frying and baking, on the residues in bacon. They found that cooked bacon contained 20 to 53% of the residue in the uncooked pork.

Ritchey et al. (1967) showed that loss of DDT or lindane was slight in poultry that were baked or fried and that, while the level of DDT was reduced, the changed DDT showed up as DDD. This was confirmed by more extensive tests (Ritchey et al., 1969) involving the two earlier methods of cooking, as well as by pressure cooking and cooking in a closed sauce pan for 30, 60, or 90 min. Less than 40% of

Table 17.6 DDT Residue Levels in Raw and Ready-to-Eat Foods, United States, 1964–1969[a]

Food Class	Raw Samples		Ready-to-eat Samples	
	Mean	Range	Mean	Range
Large fruits	100	20-250	8	6-12
Grains & cereals	20	5-40	5	3-8
Leaf & stem vegetables	140	80-180	13	7-16
Root vegetables	40	30-40	3	1-7
Dairy products	30	10-40	33	23-38

[a]Duggan and Lipscomb, 1971

the pesticide was removed by any of the cooking methods, and some of that removed was present in fat drippings.

McCaskey et al. (1968) included methoxychlor, chlordane, kelthane, telodrin, and ovex in the feed of chickens and analyzed raw and cooked tissues for residues. They found that residues of methoxychlor did not occur in white or dark meat and only a small amount in the fat. Cooking reduced the residue of chlordane about 60% and removed telodrin residues but had little effect on residue levels of kelthane or ovex.

The story on removal or reduction of pesticide residues in fish suggests that, generally, the same situation holds as with other meats, but one notable exception has been reported. Reinert et al. (1972) studied the effect of preparation and cooking on pesticide levels in yellow perch, bloaters, lake trout, and coho salmon. They found that the amount of pesticide residue removed was related to the amount of oil removed during processing and thus varied among species. In yellow perch, 90% of the DDT residue was removed in the process of preparing fillets, and it was concluded that this reflected the concentration of DDT in the oil and the fact that the oil was concentrated in tissues other than those used for fillets. By contrast, cleaning and filleting of the other three species had little effect on residue levels, since in these the fatty tissue is distributed rather uniformly in the edible cuts. Cooking had only minor effect on the DDT residues. Smith et al. (1975) confirmed the fact that DDT residues were not significantly reduced in coho during cooking and found similar results with Chinook salmon.

Because dairy products make up 27.7% of the average diet (Smith, 1971), it is of interest to note what changes have occurred in the chlorinated hydrocarbon residues in these commodities since their use was curtailed. In Ontario surveys have been made of pesticide levels in bulk milk prior to and following the ban on DDT in the province.

In the earlier surveys the highest residues were found in milk from the southern part of the province, where farming was intensive with accompanying high use of pesticides. Frank et al. (1975) compared results obtained by sampling of bulk milk in 1967, 1968–1969, 1970–1971, and 1973. During each of the surveys milk was collected from bulk tankers (about 300 involved in each survey), each representing a 2 day milk production from 5 to 20 herds. These surveys showed that DDT and dieldrin were the only pesticides found frequently in milk. Lindane and heptachlor epoxide were occasionally found, but chlordane, endosulfan, endrin, and methoxychlor were not detected. The amount of DDT and dieldrin declined from the 1967 to the 1973 survey, as show in Table 17.7. These data show a sharp decline in

Table 17.7 Mean Residue of DDT and Dieldrin in Bulk Milk (ppm, Fat Basis) in Four Surveys in Ontario[a]

Year	No. Tankers Sampled	ppm DDTR[b]	Percent DDE & TDE[c]	ppm Dieldrin
1967	284	0.20	67	0.044
1968-69	374	0.19	80	0.041
1970-71	337	0.12	80	0.035
1973	350	0.05	98	0.015

[a] Frank et al., 1975.
[b] Parts per million DDT + metabolites.
[c] Percentage of total DDT residue present as DDE + TDE.

residues of DDT and dieldrin in 1973 compared with earlier samples and a higher percentage of the total DDT residue present as the metabolites DDE and TDE. Frank et al. (1975) point out that there was a decline, not only in the general level of pesticides in milk, but also in the frequency of high residues.

17.4 SIGNIFICANCE OF RESIDUES IN FOOD

Various committees and working parties of the United Nations have reviewed the data on pesticide levels in foods and their toxicological significance. As a result the concept of ADIs has been advanced as a guideline with respect to specific pesticides. This ADI is the "acceptable daily intake" defined as "the daily dosage of a chemical which, during an entire lifetime, appears to be without appreciable risk on the basis of all facts known at the time. Without appreciable risk is taken to mean the practical certainty that injury will not result even after a lifetime of exposure." The ADI is expressed in terms of mg/kg body weight.

Table 17.8 presents estimates of daily intake of pesticides in the United States based on the level of residues found in various food categories and the percentage that category of food represents in the daily diet (Table 17.3). Dieldrin is the only residue in foods approaching the ADI in any year. Despite the fact that DDT residues are the most consistently found, the daily intake in the United States has remained near 10% of the ADI.

Thus it may be concluded that the general level of pesticide intake in North America is well within established safe levels and

Table 17.8 Daily Intake of Some Pesticides in the United
States for the Years 1965, 1968, and 1970 as Calculated from
"Total Diet" Studies on Ready-to-Eat Foods

Pesticide	Daily Intake (mg/kg)[a]			ADI[b]
	1965	1968	1970	
Chlorinated hydrocarbons				
Aldrin	.000014	Tr	Tr	.0001
DDT (+ metabolites)	.00089	.00064	.00041	.005
Dieldrin	.000072	.000057	.000072	.0001
Endosulfan	--	Tr	.000014	.0075
Endrin	Tr	.000014	Tr	.0002
Heptachlor (+ epoxide)	.000028	.000014	.000014	.0005
Lindane	.000057	.000043	.000014	.0125
Methoxychlor	--	.000014	.000014	.10
Toxaphene	--	.000028	.000014	
Organophosphates				
Diazinon	--	Tr	.000014	.002
Ethion	--	.000014	.000057	.00125
Malathion	--	.000043	.00016	.02
Parathion	--	Tr	Tr	.005
Others				
Carbaryl	.0021	--	--	.01
2,4-D	.00007	.000014	Tr	

[a]Duggan and Corneliussen, 1972. Data extrapolated
to mg/kg on the basis of a 70 kg man.
[b]ADI - Established or tentative, FAO/WHO, 1972.
Tr Trace

should be of no significance. It must be accepted, however, that there
are gross differences in eating habits among individuals, and this affects individual intake. With the existing pesticide levels, however, it
is difficult to see how any diet could result in unacceptably high intake
from market basket foods.

18

PESTICIDES AND BIRDS

Perhaps no area of pesticide use and effect has aroused more public concern or been more extensively debated than that dealing with birds. Both acute and chronic toxicity have been reported, the one shading into the other in situations where food is contaminated and continued ingestion of toxicant has occurred. As with most of the environmental problems with pesticides, the organochlorine insecticides have been involved most heavily and have been the cause of deaths in many bird species and reproductive failures in a host of others. These effects in North America were not predictable, since DDT, the insecticide incriminated most frequently, is not highly toxic to birds in acute toxicity tests and the species chosen for chronic toxicity testing did not show the sensitivity in reproduction later demonstrated in predators and a number of other species. Thus early reports of pesticide-related deaths were viewed with some skepticism, and only recently has the full significance to birds of DDT usage been appreciated.

Although the organochlorine compounds, especially DDT, have been the cause of most of the pesticide-related toxicity and reproductive problems, other pesticides have been incriminated as well. Many of the organophosphorous compounds and carbamate insecticides have much lower LD_{50} values for birds, and this has been reflected in field mortality.

In areas other than North America DDT has not necessarily been the main offender. Aldrin, dieldrin, and heptachlor created serious problems in Great Britain, and in Sweden mercury has been responsible for population declines in some predatory species.

Both lethal and sublethal effects have been documented, and while it is recognized that distinguishing between the two may be somewhat arbitrary, the nature of the result and the criteria for detection are so different that the separation in discussion is desirable.

18.1 LETHAL EFFECTS

Not all species of birds are equally sensitive. This may be a function of physiology, habitat, or food preference. Thus we discuss mortality under several (arbitrary) categories of birds.

Songbirds

In the early years after DDT was introduced it found widespread use in urban and suburban areas for control of Dutch elm disease. In this use it was applied at excessive dosages, 1 or more lb often being used on a single large elm tree. Such high dosages resulted in large amounts' entering the soil under treated trees and in high concentrations in earthworms frequenting such soils. Not surprisingly birds that feed on earthworms and insectivorous birds were poisoned, and dead songbirds were conspicuous in such areas. The early reports of high songbird mortality included, in some cases, analyses of dead birds confirming the presence of DDT in their bodies, and reference was found frequently to "tremors" noted in birds prior to death. Much of the information was not, however, very scientific, and cause-effect relationships were claimed but not substantiated. In addition the American robin was the bird most frequently reported killed, and its ability to repopulate an area created the impression that, although dead birds were found, the significance of this was minimal in terms of the population of the species. The continuing abundance of robins attests to the validity of this impression for this particular species.

In the early years conclusions that pesticides in general or DDT in particular had caused bird mortality were based too frequently on analyses demonstrating that the pesticide was present in the dead bird without a knowledge of the level of pesticide indicative of poisoning. Bernard (1963) brought guidance in this field when he demonstrated through feeding studies on captive robins and house sparrows that, while residue levels on a whole carcass basis or in muscle were unreliable indicators, brain levels gave a good indication of lethality. He concluded that 50 ppm in robin brain or 60 ppm in the brain of the house sparrow indicated a lethal dosage of DDT. Later Hill et al.

(1971) with laboratory experiments determined lethal brain levels to be 18 to 40 ppm for house sparrows, 16 to 24 for cardinals, and 12 to 20 for bluejays.

The extent of poisoning of songbirds in local areas with DDT has been significant. Extensive annual treatment of elms on the campus of Michigan State University resulted in the virtual elimination of the robin population by 1958 (Wallace, 1959), and Hunt (1960) showed that, in communities with Dutch elm disease control programs using DDT, populations of many bird species were slightly to drastically reduced in comparison to untreated areas.

Although most of the reports of songbird mortality in Dutch elm disease spray programs referred to the Midwest, mortality was occurring also in the East. Wurster et al. (1965), studying bird mortality in two neighboring communities in the Northeast United States, one treated with DDT, the other not, recovered 117 dead birds in 15 acres in the treated area but only 10 in an equal area of the untreated community. More than half of the dead and tremoring birds were robins, but in addition chipping sparrow, myrtle warbler, yellow-shafted flicker, hairy woodpecker, downy woodpecker, yellow-bellied sapsucker, black-capped chickadee, white-breasted nuthatch, red-breasted nuthatch, and brown creeper suffered some mortality. Analyses of dead and tremoring robins showed that residues of DDT (mostly DDE) in a large percentage exceeded 100 ppm wet weight in the brain, indicating, according to Bernard's (1963) studies, that the pesticide was the cause of death.

In contrast to the Dutch elm disease control program DDT use for gypsy moth or spruce budworm control was at a much lower dosage rate, usually 0.5 to 1.0 lb/acre. Although mortality of forest birds associated with such programs has been reported (Hotchkiss and Pough, 1946; Prebble, 1975), severe depletion of populations through lethal effects has not been shown. DeWitt and George (1960) conclude that birds can tolerate up to 2.0 lb DDT/acre with no apparent effect and report that no widespread immediate dieoffs of forest vertebrates have accompanied more than 20 million acres of forest spraying. This is substantiated by studies of Dimond et al. (1970) indicating that, though DDT persisted in the forest soil for many years (perhaps 30), residues in earthworms inhabiting such soil were much less than 1 ppm, and while residues in robins were higher than in earthworms, they did not approach levels recognized as likely to be lethal.

Johnson et al. (1976) studied robin populations and breeding success in a series of orchards in New York over a 3 year period, during which the orchards received annually 7 to 10 applications of carbaryl and azinphosmethyl, 2 or 3 applications of lead arsenate (about 50 kg/ha), 2 applications of DDT (11.2 kg/ha), and 2 applications of dieldrin (2.8 kg/ha). Robins nested in the orchards each year, and though clutch size was slightly below normal for the northeast region, breeding success was comparable to that reported in other studies. The soils contained high residues of DDT, those in the mature orchards showing 427 to 441 ppm, in the top 0.3 m. It was shown that for robins early spring feeding was predominantly on rotting apples on the ground. The diet changed to include mostly earthworms and lepidopterous larvae during the breeding season. By late June, however, the diet had changed again, 70% of the food being fruit, mostly cherries. In these studies earthworms were absent from the orchard soils and other ground invertebrates were much less numerous than in adjacent areas. Thus the robins foraged for worms and soil invertebrates in nearby territories where DDT levels in soil were 2 to 11 ppm and earthworms carried DDT residues of 2 ppm or less. As a result robins nesting in the orchard had DDT residues in their brains early in the season comparable to those found in robins from outside the orchard area. During the summer brain residues of DDT were slightly higher in the orchard robins but only at levels of about 2 ppm. During the study one robin was found exhibiting the tremors characteristic of organochlorine poisoning, and analyses indicated a brain residue of 2.6 ppm dieldrin and 15 ppm DDT.

Although songbird mortality due to forest spray programs using DDT has been minimal, some of the insecticides used since the curtailment of DDT have been much more toxic. In 1964, 100,000 acres of stream bank and 60,000 acres of upland forest were treated with phosphamidon at the rate of 0.5 lb/acre. A considerable number of birds were killed and populations temporarily reduced. Warblers were among species most seriously affected (Fowle, 1966). Later studies confirmed that dosages in excess of 0.25 lb active/acre applied by air were likely to cause high mortality. Even at the 0.25 lb dosage some dead birds were found, and it was shown that the birds could receive a toxic dose through their food or, with high dosage, by absorption through their feet (Fowle, 1969). The practical significance of this latter observation is difficult to assess, since in one test perches were "painted" with various concentrations of the insecticide and in another birds were placed on sheets of polyethylene "smeared" with various amounts of the chemical. While the concentrations used in these tests were near field concentrations, the amount applied per unit area was obviously much in excess of what would result from operational spraying with less than 1 gal liquid/acre. Fowle (1969) concluded from limited tests that the LD_{50} of phosphamidon to small forest birds was 1 to 3 mg/kg.

In Quebec, Buckner et al. (1973) monitored songbird populations in a control area and in an area treated with fenitrothion followed 10 to 14 days later with aminocarb (rates not given). They monitored only after the aminocarb treatment and reported that, of 28 species monitored, populations of 2, the bay-breasted warbler and the evening grosbeak, were reduced sharply immediately after spraying. While bay-breasted warblers reestablished within 4 days of treatment, the evening grosbeak seemed to abandon the area in favor of unsprayed territory where insects were more abundant. Buckner et al. (1973) discuss briefly the differing exposures to aerially applied pesticides in relation to the forest niche occupied by various birds. They point out that bay-breasted warblers nests in the upper crown near the ends of branches and are thus maximally exposed to the pesticide cloud. In additional studies Buckner et al. (1973a) reported no significant population effects following treatment with fenitrothion at 2 oz active/acre in each of two applications. In most instances 3 oz/acre seemed harmless to birds, but some reduction occurred in plots receiving 4 oz active/acre. In one treatment with 18 oz active/acre severe reductions occurred. Most affected were members of the families Parulidae (warblers), Fringillidae (grosbeaks, juncos, finches, sparrows), Tyrannidae (fly catcher, phoebe), and two species of kinglets. Robins, thrushes, and woodpeckers were not affected. These workers found declines in numbers of songbirds where aminocarb or mexacarbate were used at 0.75 oz active/acre. A recent appraisal suggests that for fenitrothion the dosage mortality curve for some species (chipping sparrows and yellow throats) is very steep and that a dosage of 4 oz/acre is about where lethality will begin (NRC, 1975b). Since this is the case, it is likely that in extensive aerial treatment of forests some bird mortality will occur, especially in species that frequent the upper forest canopy.

Among other pesticides studied methoxychlor has not caused bird mortality when used for Dutch elm disease control (Wurster et al., 1965), and Kurtz and Studholme (1974) found that, when trichlorfon or carbaryl were used for gypsy moth control, only traces of the pesticides could be found in bird tissues of either ground or canopy feeders, even in the peak of the spraying operations.

Note that the assessment of spray effects on bird populations is difficult. Theoretically, at least, insecticides may alter behavioral patterns to make birds more or less easily seen, and, especially in small plots, local movements may soon nullify immediate effects. Pearce

(1968) suggests that field studies tend to underestimate effects. While population studies may be difficult to interpret, mortality estimates based on dead birds give only an indication of mortality, since the percent of dead birds recovered is unlikely to approach actual mortality. Rosene and Lay (1963) found that, when dead bobwhite quail were planted in normal range vegetation, the percentage recovery was low.

With a few exceptions losses of songbirds due to agricultural pest control have not been documented. Despite widespread use of pesticides, birds such as robins, blackbirds, cowbirds, red-winged blackbirds, and house sparrows that frequent agricultural areas seem to be doing well.

Two wide-scale programs associated with agriculture have, however, resulted in bird mortality among songbirds and gallinaceous species. One of these programs was that directed at the fire ant, using dieldrin and/or heptachlor. In Montgomery County, Alabama, pesticide application (presumably heptachlor at 2 lb active/acre) resulted in an overall decline of more than 50% among 49 species of breeding birds. The reduction was 100% in 7 species of ground inhabitants and 50 to 99% in 17 species inhabiting a low to intermediate strata, but no effect was observed on species inhabiting high strata. Low populations of songbirds continued into the second year after treatment in Texas and Louisiana, with populations reported at 11 to 70% of normal. Nesting success did not, however, seem to be affected (DeWitt and George, 1960).

The second agricultural practice that has led to mortality of songbirds, as well as other species, has been the use of seed treatments on cereals. Little documentation is available, and it is probable that, in most cases, the treated seeds are larger than those taken by most species of songbirds. In a study of the effects of aldrin-treated rice on wildlife Flickinger and King (1972) found that, where treated rice was aerially seeded, mortality was high in a number of bird species and analyses revealed relatively high levels of pesticide in dead birds.

Gallinaceous Species and Waterfowl

In Britain and some other parts of Europe grains were treated with aldrin, dieldrin, heptachlor, or BHC to control wireworms and wheat bulb fly. Mercury was included in many seed treatments to prevent loss of stand from seedling diseases. In the spring of 1961 about 80 incidents of bird mortality were reported, as many as 500 dead wood pigeons being counted in some roosts (Turtle et al., 1963). One of the largest incidents involved a 4000 acre estate in Lincolnshire where birds of 18 species were killed. Included were 5668 wood pigeons, 118 stock doves, 59 rooks, and 89 pheasants (Prestt and Ratcliffe, 1970). In the laboratory Turtle et al. (1963), using the domestic pigeon in feeding trials, calculated LD_{50} values of 55 mg/kg for aldrin, 67 mg/kg for dieldrin, and 167 mg/kg for heptachlor. Although BHC was fed at dosages as high as 600 mg/kg, only a few deaths resulted. Turtle et al. (1963) calculated that, if a wood pigeon ate 50 g of cereal seeds treated at the recommended rate of 2 oz of 40% insecticide/bushel, it would ingest 40 mg or about 80 mg/kg. Such a feeding rate is consistent with normal daily intake and would exceed the LD_{50} value for either aldrin or dieldrin.

As a result of these findings a decision was made in 1962 that aldrin, dieldrin, and heptachlor would no longer be used for seed treatment of grain for spring sowing. For this purpose BHC would be used, since at that time of year, it was equally effective against the seed pests and much less toxic to birds. Aldrin and dieldrin were still permitted for seed treatment for fall seeding, since they provided better protection against the seed pests present at that time of year. This decision reduced bird poisoning, but further incidents occurred, and in 1974 the use of aldrin or dieldrin in wheat seed treatments was discontinued (Hamilton et al., 1976).

Efforts to find other insecticides for seed treatments resulted in the use of chlorfenvinphos and carbophenothion. Following this a number of instances of poisoning of geese were reported, and Bailey et al. (1972) found that these had been poisoned from eating winter wheat while foraging in fields planted with carbophenothion-treated seed. In this incident only geese were poisoned, despite the fact that other seed-eating birds were seen in the fields sown with treated seed. Other incidents of poisoning of wild geese followed, and in 1974 several hundred greyleg geese and 200 or more pink-footed geese were similarly poisoned. Analyses of the gizzards showed the presence of carbophenothion-mercury-treated seed, and esterase determinations supported the conclusion of organophosphate poisoning (Hamilton et al., 1976). These incidents suggested that geese might be more sensitive than other species to the insecticide involved. Tests by Jennings et al. (1975) indicated that, while the LD_{50} for Canada geese was slightly less (29 to 35 mg/kg) than for quail (56.8 mg/kg) or pigeon (34.8 mg/kg), the differences are not great.

Flickinger and King (1972) reported bird mortality where aldrin-treated rice was sown in the Gulf Coast area of the United States. During the 5 year period 1967–1971 they collected and analyzed birds found dead in the area soon after rice seeding and in addition analyzed invertebrate and other components of the environment that might be included in bird diets. Soil residues were low, but snails and crayfish contained sufficient aldrin and dieldrin (9.5 ppm) to be toxic to birds feeding heavily upon them. Dead waterfowl, shorebirds, and passerines contained residues of several pesticides, dieldrin being prominent.

In some cases applications of granular pesticides have caused bird toxicity by direct uptake of the granules. This has occurred in rice fields with a number of dabbling ducks, and the use of carbofuran for root maggot control in cole crops in British Columbia proved lethal to several species of ducks that entered these fields and fed in the surface muds during the fall of the year.

Lutz-Ostertag and Lutz (1970) reported a lethal effect of 2,4-D on embryos of three bird species, but this could not be confirmed in extensive tests with DDT; 2,4-D; picloram and 2,4-D; or 2,4,5-T on eggs of white leghorn hens (Somers et al., 1974, 1974a). A number of studies have indicated that, when pesticides are injected into incubating eggs, embryo mortality and/or teratogenic effects may be experienced when dosages are high (Greenberg and LaHam, 1969, 1970; Paul and Vadlamudi, 1976).

Predaceous and Fish-Eating Species

During the period 1949–1957 DDD was applied several times to Clear Lake for control of gnats. This lake supported a breeding population of more than 1000 pairs of western grebes prior to the DDD treatments, and more than 100 were found dead in December 1954. Autopsies did not indicate any disease present in the birds, and after another dieoff in the spring of 1955 autopsies were equally uninformative. An additional 75 dead grebes were found in December of 1957, and finally 2 birds exhibiting tremors were analyzed for pesticide. Results showed a residue of DDD in their fat of 1600 ppm. After this finding analyses were conducted on a number of species of fish that served as food for the grebes in Clear Lake. These analyses indicated DDD levels of 40 to 2500 ppm in the fat. While the cause-effect evidence was circumstantial, there could be little doubt but that the grebes had suffered chronic toxicity from eating fish highly contaminated with DDD. Mortality was so pronounced that during the nesting season of 1958 and 1959 fewer than 25 pairs of grebes were seen on the 42,000 acre lake (Hunt and Bischoff, 1960).

A second example in which a direct link between food source and pesticide contamination would appear to have been established quite early is that of the golden eagle in Scotland. Golden eagles in the inland Highlands of Scotland began to show population declines in the early 1950s, and dead birds consistently carried residues of dieldrin and DDT—sometimes quite high. In this area the main food source of the eagles is sheep carrion. DDT was used as a sheep dip for ectoparasite control beginning about 1946, and this was replaced later with BHC and dieldrin. Again, while the evidence for cause and effect is circumstantial, the restricted feeding habits of the golden eagle in the Highland area support the view that organochlorine sheep dips were responsible for bird mortality (Lockie and Ratcliffe, 1964; Lockie et al., 1969; Ratcliffe, 1970).

Before discussing pesticide levels found in dead predators it may be useful to cite some studies in which attempts have been made to determine what pesticide residues in dead birds are indicative of poisoning. It is clear from a large number of reports that whole carcass residues are not a good indication of toxicity, at least with the organochlorine insecticides. Since these accumulate and are stored in adipose tissue, the relative amount of fat in the carcass may be important in terms of the ability of the bird to store the pesticide. Stickel and Stickel (1969) in studies with cowbirds found that heavier birds survived longer than lighter ones and that the body residue of DDT was higher in birds that survived than in those that died during feeding studies. They suggested that "storage without metabolism" could be a protective mechanism.

Turtle et al. (1963) reported liver residues of 15 to 20 ppm in feral pigeons that died of dieldrin poisoning, and after an extensive review Moore (1965) concluded that liver residues in excess of 10 ppm dieldrin should be considered presumptive evidence for poisoning with this compound. Robinson (1969) suggests that 20 ppm dieldrin or 30 ppm DDT (plus metabolites) in the liver would incriminate these pesticides as the cause of death. Prestt and Ratcliffe (1970) conclude that, if the 20 ppm liver residue of dieldrin is accepted as diagnostic, then predatory birds killed with this pesticide include marsh harrier, kestrel, buzzard, goshawk, and sparrowhawk in the Netherlands and the rough-legged buzzard, sparrowhawk, hen harrier, kestrel, barn-owl, tawny owl, long-eared owl, and short-eared owl in Great Britain. They correlate death of these birds with the use of aldrin and dieldrin in seed dressings and secondary poisoning by eating pigeons and other seed-eating birds highly contaminated with the insecticide.

Although a number of studies support the thesis that liver residues can be used as diagnostic indices, residues in the brain are perhaps a more valid criterion, since the pesticides involved act on the central nervous system. Bernard (1963) showed that, with DDT and robins, brain levels of 50 ppm were lethal and that a brain level of 60 ppm was lethal in the house sparrow. Other values for lethal levels of DDT (plus metabolites) in the brain include, in ppm, cowbirds 16 to 101, robins 44 to 71, coturnix 104 (Stickel and Stickel, 1969), white leghorn cockerels 28 to 120 (Ecobichon and Saschenbrecker, 1968), double-crested cormorants 24 to 85 (Greichus and Hannon, 1973), kestrels 212 to 301 (Porter and Wiemeyer, 1972), and mallards, approximately 20 to 80 (Friend and Trainer, 1974). Stickel et al. (1966) suggested that a brain residue of 30 ppm DDT (plus metabolites) is a realistic guide to toxicity, and the data just cited support this conclusion for a range of birds. Note that, whereas DDT was the feeding toxicant on which most of these observations were based, this insecticide breaks down in animal tissues primarily to DDE. It is known that DDE is less toxic to birds than DDT is (Heath et al., 1970), and Stickel and Stickel (1969) point out that brain residues of DDE may derive from two sources, direct intake or metabolism of DDT within the bird in question. Stickel and Stickel (1969) reported that in cowbirds the level of DDE in the brain at death is higher in birds fed DDE directly than in those fed DDT. Thus it may well be that, with the use of DDT discontinued in many areas and the environmental burden of this insecticide shifting toward the DDE metabolite, brain levels diagnostic of poisoning with this insecticide derivative will have to be revised upward.

Brain levels diagnostic of lethality with other insecticides have been suggested as 4 to 5 ppm for dieldrin (Stickel et al., 1969), 0.34 to 1.84 for endrin (Ludke, 1976), and about 1 ppm for carbophenothion (Jennings et al., 1975).

If one accepts the levels of pesticides in brain just cited as diagnostic, then it would appear that in North America death due to pesticides has been documented for the great blue heron (DDT) (Call et al., 1976); bald eagles (dieldrin) (Cromartie et al., 1975; Mulhern et al., 1970); nestling prairie falcons (heptachlor epoxide, dieldrin level accepted as diagnostic) (Fyfe et al., 1969); peregrine falcons (DDT) (Peakall et al., 1975); ring-billed gulls (DDT, dieldrin) (Sileo et al., 1977); common egret (dieldrin) (Faber et al., 1972); and the common loon (DDT) (Ream, 1976). Since all of these are either predatory or fish-eating species, it is assumed that their exposure to pesticides was through the food chain and reflects secondary poisoning.

Reidinger and Crabtree (1974) reported on analyses of 169 golden eagles found dead between 1964 and 1971. Two had been poisoned with strychnine, but none carried brain residues suggestive of poisoning by other pesticides. Greichus et al. (1973) did not find near lethal levels in cormorants or pelicans in South Dakota. Pesticide brain levels in dieoffs of loons in Minnesota (Ream, 1976) or of the common murre in Oregon (Scott et al., 1975) did not indicate lethal pesticide loads. In this latter instance brain levels of DDE, though well below the level considered critical (8.7 ppm versus about 30 ppm), were about eight times higher in dead birds than in survivors. This probably reflects mobilization of DDT stored in the fat, since dead birds were thin and weighed less than those that survived. Scott et al. (1975) suggested that under conditions of stress toxicity of DDT might be more pronounced.

The bald eagle is one raptor for which pesticide-induced mortality has been documented; 37 bald eagles found dead in 18 states during 1971–1972 were analyzed for pesticides and autopsied to determine the cause of death. Four of these had died of dieldrin poisoning. Included in this group of dead eagles were 11 found along with golden eagles dead near Casper, Wyoming, and suspected of being poisoned by coyote bait. Eight of these were analyzed and found to have been poisoned with thallium. An additional eagle poisoned with thallium was found in Utah during the reporting period (Cromartie et al., 1975).

Acute Pesticide Toxicity

Tucker and Crabtree (1970) tested a wide variety of pesticides administered orally in a capsule to several species of birds; LD_{50} values obtained are included (in part) in Table 18.1. Note that there are gross differences in toxicity among the pesticides, and differences exist also in the toxicity of a single pesticide to various species of birds. With many pesticides the mallard is much less sensitive to a single dose than the pheasant, coturnix, or house sparrow is, but with three of the insecticides highly toxic to birds, diazinon, parathion, and carbofuran, pheasants are less susceptible than mallards. Table 18.1 also includes data of Heath and Stickel (1965) and of Heath et al. (1970) (cited by Pimentel, 1971) on the LC_{50} values for pesticides included in the diet of 2 week old chicks for 5 days followed by 3 days of untreated food. Again, there are gross differences in toxicity among the pesticides and the relative order of toxicity is not always the same as for a single dose. For example, while aldrin and dieldrin are much less toxic to mallards than endosulfan is when each is given in a single dose, the reverse is true when the insecticides are included in a 5 day diet.

Insecticides that act on the nervous system disrupt the normal

Table 18.1 Acute Toxicity of Selected Pesticides to Some Species of Birds

Pesticide	LD$_{50}$, mg/kg[a]				LC$_{50}$, ppm[b]			
	House Sparrow	Coturnix	Mallard	Pheasant	Bobwhite Quail	Coturnix	Mallard	Pheasant
Aldrin			520	16.8	39	35	160	55
Chlordane			1200		320	325	825	450
DDT		841	2240	1296			1025	500
Dieldrin	47.6		381		39	52	200	52
Endosulfan		69.7	33	79	850	2175	1000	1275
Endrin			5.6	1.8	15	16	21	11
Heptachlor			2000		95	88	575	262
Methoxychlor			2000		5000	5000	5000	5000
Toxaphene			70.7	40		625		525
Temephos	50.1	270	90	31.5	100	250	1500	160
Azinphosmethyl			136	75	450	650	1950	1900
Chlorpyrifos	21	17	75	13		282		
Diazinon			3.5	4.3				
Dimethoate			41.7			350	1000	350
Ethion								
Fenitrothion			150					475
Malathion			1485		3500	2150	5000	3500
Parathion	3.4	6.0	2.0	12.4	190	45	262	365
Phosphamidon			3.0		25	105	750	75
Carbaryl		2290	2179	2000	5000	5000	5000	5000

			0.4					
	1.3		11.9	4.2				
Compound	12.8	28.3						
Carbofuran	1.3		0.4					
Propoxur	12.8	28.3	11.9					
Allethrin			2000	4.2				
Rotenone			2000	1414				
Amitrol			2000			5000	5000	5000
Atrazine					750	5000		
Dalapon							5000	5000
Dicamba				740			5000	5000
Dichlorbenil			2000	1189		5000		1750
Diquat			564			1500	5000	3750
Endothal								
MCPA								
Monuron						5000	5000	4500
Paraquat								
Picloram			2000	2000			5000	5000
Prometone								
Simazine						5000	5000	5000
Trifluralin	668		2000	2000	5000			
2,4-D			1000	472		5000	5000	5000
2,4,5-T (acid)						5000	5000	1775

Table 18.1 (Continued)

| Pesticide | LD$_{50}$, mg/kg[a] | | | | LC$_{50}$, ppm[b] | | | |
	House Sparrow	Coturnix	Mallard	Pheasant	Bobwhite Quail	Coturnix	Mallard	Pheasant
Copper sulfate								
Dichlone			2000	2000				
Nabam	2120	2560		707		5000	2400	5000
Benomyl						5000	5000	5000
Captan					3000			
Mercury[c]	668	2262		360		100	45	150

[a] Dosage given orally in a capsule (Tucker and Crabtree, 1970).
[b] ppm in the diet fed to 2-week old chicks for 5 days followed by untreated feed for 3 days (Heath et al., 1970 (cited by Pimentel, 1971); Heath and Stickel, 1965).
[c] Mercury as "ceresan M" (N-(ethylmercuri)-p-toluene sulphonanilide) 3.2% Hg.

transmission of stimuli, and when this is depressed beyond a certain critical level, involuntary processes such as heartbeat and respiration are no longer controlled and death results. The process is thus dose related at the site of action. Thus the degree of toxicity is determined by exposure to the toxicant and the ability of the bird to keep the toxicant away from sensitive sites by processes involving detoxification, excretion, and, in some cases, storage in less sensitive tissues. The degree of metabolism of pesticides by birds varies among pesticides and among birds. Microsomal enzymes are involved (Menzie, 1969) in a major way and the products of metabolism are many. By far the majority of pesticides are metabolized and excreted rather rapidly, and in such cases, provided intake is well below an acute toxic level, the rate of excretion equals the rate of intake and lethality does not result. For other pesticides, however, this does not occur. The organochlorine insecticides (with some exceptions) are the best example of this type. With these compounds a large number of studies have shown that they accumulate in tissues of birds ingesting them and that tissue levels reflect the level of exposure. It has been shown also that eggs from exposed birds contain these pesticides and that egg residues reflect those of the parent bird.

Experiments in which persistent pesticides were included in poultry rations demonstrate the nature of uptake, storage, and depletion of organochlorine insecticides in birds. Cummings et al. (1967) combined DDT, dieldrin, endrin, heptachlor epoxide, and lindane in a poultry ration and fed these at three low dosages, 0.05, 0.15, and 0.45 ppm for 14 weeks. Eggs were monitored during a 2 week period prior to the pesticide feeding, during the feeding period, and for a 4 week period following the exposure when the hens were on the diet without pesticide added (the basal diet contained low levels (< 10 ppb) of pesticides. Results for eggs (Cummings et al., 1966) showed that all pesticides began to appear in 3 days. Residues increased sharply for about 3 weeks and then plateaued. With dieldrin and heptachlor epoxide there was a greater increase in residues in eggs during the plateau period (3 to 14 weeks) than with other compounds. With each of the pesticides the plateau level reflected the level in the feed. For example, dieldrin at 0.05, 0.15, and 0.45 ppm in the feed gave after 7 weeks residues in the egg on a whole weight basis of approximately 0.05, 0.14, and 0.39 ppm, respectively. Lindane showed less tendency for egg residues. When DDT was fed, one metabolite, DDE, appeared in the eggs and showed a gradual increase in level throughout the experiment. A plateau was not evident for this metabolite except during the final few weeks of feeding in the highest dosage of DDT.

During the 4 week withdrawal period residues of each of the pesticides had decreased, but only with lindane did residues return to preexposure levels by the end of the 4 week period. DDE residues did not decrease but continued to rise during the withdrawal period, indicating the continuing conversion of DDT stored in the birds' tissues. Levels of pesticides in the fat of hens (Cummings et al., 1967) paralleled that of the eggs and plateaued at about 10 times the level in the food. If the level of fat in the eggs is approximately 10% (Stickel, 1973), then the levels in the fat tissue and in the eggs, expressed on a fat content basis, are almost identical. As in the eggs, tissue levels declined during the withdrawal period for all compounds, including DDE.

The results of Cummings et al. (1966, 1967) with this group of organochlorine insecticides are supported by many other studies and demonstrate that, even with these persistent compounds after an initial period of exposure, the hen excretes the compounds at about the rate of intake. This is true for other bird species as well, though the length of period of increasing tissue concentrations and ability to get rid of the pesticide differs among species (Stickel, 1973).

By contrast to results with the more persistent organochlorine compounds, studies with methoxychlor showed that, when this insecticide was fed to hens at 2, 4, 8, or 10 ppm, no residues appeared in the eggs during 85 days of feeding. When dosages were increased to 100 and 1000 ppm in the diet, residues in eggs were 0.2 and 3.0 ppm, respectively, after 2 weeks and remained at these levels during the feeding period. Residues dropped rapidly during a withdrawal period and were undetectable within a month (Olney et al., 1962).

The organophosphorous and carbamate insecticides are metabolized by birds and excreted. There was little residue of malathion in the tissues of hens fed this insecticide, most of it (97%) being excreted rapidly as water-soluble metabolites (March et al., 1956). Feeding carbamates to poultry at sublethal levels results in low residues in tissues and eggs, but these do not persist after the exposure is terminated (Kuhr and Dorough, 1976). Studies with the avicide 4-aminopyridine showed no chronic effects unless fed at high dosages (28 day LC_{50} in feeding tests, 447 ppm and 562 ppm for male and female coturnix, respectively) (Schafer et al., 1975), and similar results were obtained with the insecticide and bird repellent methiocarb (Schafer et al., 1975a).

A significant body of information suggests that birds carrying high residues of persistent organochlorine pesticides in the adipose tissues may be poisoned by these when stress conditions require the utilization of this fat reserve and the pesticide is released. This "lethal mobilization" of pesticides has been documented in a number of instances and inferred in others. Johnson (1976) cites references to demonstrate that when fat reserves are utilized the pesticide may be chemically altered and/or detoxified in the liver; excreted via the feces, urine, bile, or uropygial glands; or transferred to other tissues such as skeletal muscles or the central nervous system. When this last route is followed, death may result.

Perhaps the best evidence for "lethal mobilization" of DDT is provided in a study on cowbirds by Van Velzen et al. (1972). Birds were fed DDT in their diet for 13 days. They were then fed untreated food for 2 days, and this was followed in a portion of the birds by a food restriction of 43% of normal. Birds not previously exposed to dietary DDT survived on the restricted diet, but 7 of the 20 birds previously on the DDT diet died within 4 days. In a second experiment birds were fed diets containing 100, 200, or 300 ppm DDT and subjected to two periods of food restriction, one immediately after the feeding period and the other 4 months later. Birds died during each of the periods of food restriction. Birds that died during the period of food restriction 4 months after dietary exposure to DDT had high brain levels of DDT + DDD (38 to 81 ppm) and in addition had DDE residues of 7 to 75 ppm. The increase in brain DDT was correlated with fat loss in the birds. Other studies confirm a correlation between organochlorine toxicity and stress (Greichus and Hannon, 1973; Clark, 1975), but the toxic level in the brain seems similar in either stressed or unstressed conditions (Clark, 1975).

The significance of stress in bird mortality due to pesticides has been discussed by a number of workers. Mahoney (1974) points out that migratory birds deposit fat prior to migration and that this is used as a source of energy during flight. Porter and Wiemeyer (1972) point out that loss of fat during molting and reproductive periods may represent periods of stress and redistribution of pesticide to the central nervous system. Van Velzen et al. (1972) concluded from their work with cowbirds that "stored DDT residues present a hazard to birds, which utilize stored fat during periods of stress due to reproduction, cold weather, disease, injury, limited food supply, or migration."

The age of birds may affect their sensitivity to death by pesticides, but little information is available on this subject. Friend and Trainer (1974) found that, when feed was given ad libitum to 5 day old, 30 day old, and adult mallards, the youngest ducklings died earlier and had a lower LC_{50} than the older birds did. In this test the 30 day old ducklings had the highest LC_{50} value. Because food intake (and hence pesticide) intake in relation to body weight was not determined and comparative brain residues were not examined, it is not possible to conclude whether apparent differences reflect differing sensitivities among the age groups or different rates of pesticide intake. Rudd (1964), in discussing bird mortality from DDT spraying, states that at 2 to 3 lb/acre this pesticide "will kill adults under certain conditions, will kill considerable young. . . ." The statement implies a higher mortality among young, but no data are given. It is probable that in nature dead nestlings are noted more frequently than adult birds are, because the former are more likely to be found in or in the vicinity of nests, whereas the latter may die in less well searched areas and thus be missed.

It is likely that any factor that imposes stress on a bird will increase its susceptibility to lethal effect from a pesticide. Ludke (1974) found that, when dieldrin and DDE were fed simultaneously to the Japanese quail, the DDE level continued to rise in the quail tissues, in contrast to its plateauing when DDE was fed alone. Later Ludke (1976) demonstrated that the toxic level of endrin in the brain of bobwhite quail was reduced when chlordane also was present. Dieter and Ludke (1975) showed that, when Japanese quail were fed methyl mercury, they became more sensitive to poisoning with parathion, and Sileo et al. (1977) concluded that, though a number of ring-billed gulls found dead did not contain enough of any one pesticide to establish this as the cause of death, death was due to a combination of several factors, including organochlorine pesticides and polychlorinated biphenyls.

18.2   SUBLETHAL EFFECTS

Although lethal effects of pesticides on birds may be dramatic, especially in songbirds, and have aroused much concern, it is sublethal effects that have been most important in terms of long-term survival.

Reproductive Effects

Many reports attest to the fact that populations of a number of bird species have declined since the introduction of modern pesticides. Bird declines have been especially dramatic in certain predatory species, the case of the peregrine falcon being studied most extensively. It is accepted by most biologists that these population declines have been due primarily to behavioral and physiological changes induced by chronic exposure to persistent organochlorine pesticides in their food supply. The population declines have been characterized by:

1. Laying of eggs with thin eggshells (most species).
2. Reduced clutch size.
3. High egg loss during incubation.
4. Egg breakage.
5. High death rate of embryos.
6. High death rate in unhatched chicks and at pipping.
7. High mortality of chicks.
8. Late nesting and unusual nesting behavior.
9. Large proportion of adults in the population.

Although all of these characteristics have not been established in all cases where population declines have occurred, the pattern is well documented.

It was not until after a number of years of use of the organochlorine insecticides that population declines in predatory birds became obvious and still later before the cause-effect relationship could be verified. Once this had been done, a correlation between pesticide levels in birds and thin eggshells was shown, and by working backwards it has been possible to date when the problem first appeared.

The first clarification of the effect of pesticides on raptor populations came from work in England initiated because of claims that peregrines were preying on racing pigeons (Robinson, 1969). Although the peregrine had been reduced in numbers during the war years, populations began to increase after 1945 but suffered a rapid decline following 1955. As a result populations in southern England had by 1961 been reduced to about 60% of the pre-1939 level. Egg breakage and infertility were noted. Egg breakage among peregrines is not normal, and Ratcliffe (1967) reported 47 cases of egg breakage in 168 eyries examined in 1951–1966 compared to only 3 instances of breakage reported in 109 eyries during the 1904–1950 era. A similar increase in egg breakage was noted in. eyries of the golden eagle and sparrow hawk. Eggshell thickness was investigated, and it was found that in all three predators a synchronous and significant decrease had occurred in eggshell thickness around 1945–1947 and that the decrease in thickness was in the calcium carbonate portion. Ratcliffe (1967) noted that the predators involved carried significant amounts of organochlorine insecticides, and correlating the onset of thin shells with the time of introduction of these pesticides suggested a cause-effect relationship. Ratcliffe (1967) reported his data in terms of a thickness index reflecting the ratio of eggshell weight to size. Later, Ratcliffe (1970) reported further on studies on British birds and gave thickness indices for 17 species for the period before the introduction of the postwar organochlorine insecticides and for the period since about 1947. In 9 of the species eggshell thickness had been reduced significantly during the recent era. Eggs of 16 of the 17 species were analyzed, and though there was an apparent relationship between high total organochlorine residues and reduced eggshell thickness, there were some exceptions (Table 18.2). Ratcliffe (1970) discussed at length the possible causes of thin eggshells in birds and looked critically at his data in relation to specific and somewhat isolated populations of some species. He concluded that pesticides were the cause of eggshell thinning.

In North America population declines in some species were being noted. The peregrine falcon was the most outstanding example. In the

Table 18.2 Changes in Eggshell Thickness Index After About 1947 and Total Residue of Organochlorine Insecticides in 17 Species of British Birds[a]

Species	Region	Change in Thickness Index[b]	Organochlorine Residue in Eggs[c]
Peregrine	E. & Central Highlands	−4	4.01
Peregrine	Other regions	−19	15.2
Sparrow hawk	S.E. England	−21	37.2
Sparrow hawk	N. & W. areas	−14	14.0
Golden eagle	W. Scotland	−10	1.8
Golden eagle	E. Scotland	−1	0.2
Merlin		−13	16.4
Kestrel		−5	4.1
Hobby		−5	4.8
Buzzard		0	1.8
Raven		−1	2.1
Carrion crow		−5	0.6
Rook		−5	0.4
Guillemot		0	2.5
Razorbill		0	3.2
Kittiwake		−1	0.7
Black-headed gull		+1	1.5
Shag		−12	3.7
Golden plover		−1	1.8
Greenshank		+1	−−

[a] From Ratcliffe, 1970.
[b] Based in most instances on pre-1945 eggs.
[c] ppm wet weight. Includes BHC, heptachlor epoxide, dieldrin, DDE, TDE, DDT and DME.

preface to *Peregrine Falcon Populations: Their Biology and Decline* (Hickey, ed., 1969) the editor writes, "During the years 1950 to 1965, a population crash of nesting peregrine falcons *(Falco peregrinus)* occurred in parts of Europe and North America on a scale that made it one of the most remarkable recent events in environmental biology." The peregrine was not the only species involved. Reproductive declines had been reported in the brown pelican, bald eagle, and osprey, and though pesticides were suspected as the cause, the relationship had not been proved and thin eggshells had not been identified with lack of reproductive success. Following the first report of thin eggshells in Britain (Ratcliffe, 1967), Hickey and Anderson (1968) investigated this aspect and found that, in declining populations of peregrine falcons, bald eagles, and ospreys, eggshell thickness had been reduced more than 19% in comparison to museum eggs of the same species collected prior to the extensive use of modern insecticides. By contrast no significant change in thickness was found in eggs of the red-tailed hawk, golden eagle, or great-horned owl. In each of the latter three species populations were believed to be stationary and reproductive success satisfactory. Hickey and Anderson (1968) analyzed DDE residues in eggs of herring gulls from five areas and found a correlation between high DDE levels and thin shells. The eggshell-thinning phenomenon was explored further, Anderson and Hickey (1970) examining more than 20,000 eggshells taken prior to 1946 and more than 3000 after that period. They found pronounced eggshell thinning (>20%) in 9 of 25 species of predaceous carrion- and fish-eating birds. Reductions in eggshell thickness were not found in whooping cranes, broad-winged hawk, or rough-legged hawk, but in the common crow eggshell thickness had increased.

Other workers have added to the list of birds in which eggshell thinning has occurred during the past 30 years, and an attempt is made to summarize this information in Table 18.3. In this summary the thickness index (Ratcliffe, 1967) is used and the maximum value reported in the reference cited is given. It should be understood that in many species there are large differences in eggshell thinning between different regions; thus this summary overemphasizes the extent of the problem. Moreover, in some species eggshell thinning greater than that included in the table has been reported. Thus it is well established that eggshell thinning has occurred in many bird species. Our count shows this has happened in birds in at least 13 families and, as discussed later, pesticides, at least DDT and dieldrin, are at least in part and perhaps in a principal way involved. It is less clear what level of eggshell thinning is likely to cause population declines. Peakall et al. (1975), after reviewing the published data and their own work on

Table 18.3 Change in Thickness in North American Bird Eggs Since About 1945

Family	Species	Maximum Thickness Index Change Reported[a]	Location	Reference
Accipitridae	Bald eagle	-30	South Texas	Anderson & Hickey, 1970
	Broad-winged hawk	NS	Man., Wisc., Mich.	Anderson & Hickey, 1970
	Cooper's hawk	-20	Alta., B.C.	Anderson & Hickey, 1970
	Golden eagle	-11	California	Anderson & Hickey, 1970
	Goshawk	-8	California	Anderson & Hickey, 1970
	Marsh hawk	-24	Oregon, Alberta	Anderson & Hickey, 1970
	Mississippi kite	-4[b]	Texas	Parker, 1976
	Red-shouldered hawk			
	Red-tailed hawk	-13	S. Texas	Anderson & Hickey, 1970
	Rough legged hawk	-14	Montana	Anderson & Hickey, 1970
	Sharp-shinned hawk	NS	N.W.T.	Anderson & Hickey, 1970
		-13	Alberta, Quebec	Anderson & Hickey, 1970
Alcidae	Guillemot (Murre)	-12	California	Gress et al., 1971
Anatidae	Common merganser	-15	Wisconsin	Faber & Hickey, 1973
	Hooded merganser	-4	Wisconsin	Faber & Hickey, 1973
	Red-breasted merganser	-23	Wisconsin	Faber & Hickey, 1973

Family	Common name	Value	Location	Reference
Ardeidae	American bittern	-3	Wisconsin	Faber & Hickey, 1973
	Black-crowned night heron	-18	New Jersey	Anderson & Hickey, 1970
	Cattle egret	-3	Louisiana	Faber & Hickey, 1973
	Common egret	-7	Wisconsin	Faber & Hickey, 1973
	Great blue heron	-9	Ontario	Anderson & Hickey, 1970
	Great heron	-9	Louisiana	Faber & Hickey, 1973
	Little blue heron	-4	Louisiana	Faber & Hickey, 1973
	Louisiana heron	-4	Louisiana	Faber & Hickey, 1973
	Snowy egret	-5	Louisiana	Faber & Hickey, 1973
	Yellow-crowned night heron	-1	Louisiana	Faber & Hickey, 1973
Corvidae	Common crow	+10	New Mexico	Anderson & Hickey, 1970
Falconidae	Gyrfalcon	-3	Alaska, N.W.T.	Anderson & Hickey, 1970
	Peregrine falcon	-26	S. California	Anderson & Hickey, 1970
	Prairie falcon	-28	New Mexico	Anderson & Hickey, 1970
	Sparrow hawk	-12	N.Y., Indiana	Anderson & Hickey, 1970
Gaviidae	Common loon	-15	Ontario	Anderson & Hickey, 1970
Gruidae	Whooping crane	NS		Anderson & Hickey, 1970

Table 18.3 (*Continued*)

Family	Species	Maximum Thickness Index Change Reported[a]	Location	Reference
Laridae	Black tern	-10	Wisconsin	Faber & Hickey, 1973
	Caspian tern	-5[b]	Ontario	Gilbertson, 1975
	Herring gull	-10	Great Lakes	Anderson & Hickey, 1970
Pandionidae	Osprey	-21	Md., N.J., Conn.	Anderson & Hickey, 1970
Peleconidae	Brown pelican	-25	S. California	Anderson & Hickey, 1970
	White pelican	-14	B.C.	Anderson & Hickey, 1970
Phalacrocora-cidae	Double-crested cormorant	-30	Wisconsin	Anderson & Hickey, 1970
Podicipedidae	Pied-billed grebe	-3	Wisconsin	Faber & Hickey, 1973
	Red-necked grebe	-13	Wisconsin	Faber & Hickey, 1973
Procellarii-dae	Ashy petrel	reduced	California	Risebrough et al., 1970
Strigidae	Great horned owl	-17	Florida	Anderson & Hickey, 1970

[a]   weight (mg)
      length (mm) x breadth (mm)
[b] Decrease in eggshell thickness.

Alaskan peregrines, conclude that DDE residues in eggs of the peregrine falcon of 20 ppm or more (wet weight basis will mean decimation of the population. If we assume all species are equally sensitive (and they are not), this would result in eggshell thinning of about 15%. On this basis some populations of about 14 species would appear to be in serious trouble (Table 18.3). Parker (1976) suggests, however, that reproduction will be impaired if the eggshell thickness is reduced by 8%. On this basis some populations of 26 species may be adversely affected.

Some word of caution must be made with regard to eggshell thickness. Eggshells are not uniform in thickness, and several measurements must be made to obtain a reliable estimate (Anderson and Hickey, 1970a). Vanderstoep and Richards (1969) point out that the chick embryo derives calcium from the eggshell and that during incubation the shell decreases in thickness by 6.4%, most of the decrease in the case of domestic chickens occurring during the 12 to 18 day incubation period. Kreitzer (1972) found a decrease during incubation of 7.3%. Anderson and Hickey (1970a) were aware of this. They believed, however, that most of the museum eggs used in their comparative study on eggshell thickness had been collected before incubation. This is probably true, for the contents of eggs are removed most easily before significant embryonic development occurs (Kreitzer, 1972). Egg thickness data based on measurements on remains of broken eggs are, however, not very reliable. The subject of eggshell thinning in avian eggs has been dealt with by Cooke (1973) in a recent review, and the reader is referred to that excellent discussion for a full treatment of the matter. Interest in the subject is indicated by the fact that in the review by Cooke (1973) 279 references are cited. Only a few are dealt with here.

A significant number of field studies have found positive correlations between population declines and decreased eggshell thickness and negative correlations between organochlorine residues and eggshell thickness. As pointed out by Cooke (1973), not all of these have been statistically valid, but it is clear that high residues of DDE or dieldrin have been associated consistently with significant eggshell thinning. Ratcliffe (1970) found this to be the case for the peregrine falcon, sparrow hawk, and golden eagle in Britain, and Hickey and Anderson (1968) found a good correlation between high DDE levels and thin eggshells in herring gull eggs. This was confirmed in the peregrine falcon by Fyfe et al. (1969), in common terns by Fox (1976), and in the white pelican by Knopf and Street (1974).

Blus et al. (1972) examined 80 eggs from 12 colonies of the brown pelican, 70 collected in 1969, and 10 in 1970. Eggs were from two breeding sites in South Carolina, one site in California, and nine sites in Florida. They found a negative correlation between DDE and eggshell thickness and described this in a regression equation with a logarithmic base. They interpreted their data to indicate that in the brown pelican the egg-thinning effect from DDE was proportionately greater at low DDE levels. This work was criticized by Hazeltine (1972) on statistical and other grounds, but later work by Blus et al. (1974) seems to substantiate the earlier conclusions.

Faber and Hickey (1973) analyzed eggs of 19 species of aquatic-feeding birds and found that, in 9 species of 13 from the Great Lakes, significant thinning had occurred. In 7 species from Louisiana, only 1, the great heron, had significant thinning. A portion of their data is summarized in Table 18.4 for the Great Lakes birds with these residues converted to a ppm wet weight basis to permit comparison with data reported by Ratcliffe (1970) (Table 18.2). Faber and Hickey (1973) concluded that DDE was an important factor in eggshell thinning for

Table 18.4   Eggshell Thickness Changes (pre-DDT vs. DDT Era) and Mean Organochlorine and PCB Residues in Eggs of Great Lakes Birds[ab]

Bird Species	% Change in Thickness Index[c]	DDE	DDT + TDE	Diel-drin	PCB
Great blue heron	−25*	22	2	.5	8
Red-breasted merganser	−23*	44	3	.9	84
Common merganser	−15*	24	1	1.6	56
Double-crested cormorant	−15*	8	.2	0.3	5
Red-necked grebe	−13*	54	3	0.7	62
Herring gull	−14*	166	2.5	1.0	171
Black tern	−10*	5	.3	.3	9
Black-crowned night heron	−10*	5	.4	.6	5
Common egret	−7*	19	1.5	.3	7
White pelican	−6	2	.1	.1	1
Hooded merganser	−4	2	.1	.1	3
American bittern	−3	.5	0	.01	1
Pied-billed grebe	−1	7	.5	.2	7

[a]Faber and Hickey, 1973.
[b]Converted from lipid weight basis and rounded to significant figures.
[c]Rounded to nearest whole number.
*Significant thinning.

most groups, especially herons, and that, though DDE was important, dieldrin also played a role in two groups. Throughout their data, high levels of DDE were, in most cases, correlated with high levels of PCBs and dieldrin. These authors believed that DDT, TDE, and BHC were not significant to egg thinning, but they were uncertain about the role played by mercury in 1 species. They concluded that the relationship between DDE and eggshell thinning was linear on a log-log basis, except that 2 species did not fit the line. They concluded also that some species are more sensitive to DDE than others and reported that extensive eggshell breakage did not occur unless thinning exceeded 10%.

The thinning of eggshells of a wide number of species of wild birds would suggest that this phenomenon should have been observed in the laboratory. The reason it was not derives perhaps from two points. In the first place the gallinaceous birds used for most laboratory experiments are less susceptible to the thinning effect and, secondly, in laboratory tests eggs are mechanically incubated and breakage is less likely to occur than under natural incubation. Organochlorine insecticide feeding tests with hens and Japanese quail are reviewed by Cooke (1973) and are inconsistent. Cecil et al. (1972) fed white leghorns p,p'-DDT, $0,p'$-DDT, and p,p'-DDE at 5, 25, and 50 ppm for 28 weeks and followed this by a 10 week feeding at 50, 150, and 300 ppm. No changes were detected in egg weight, eggshell calcium, or eggshell thickness. Using higher levels of DDT, however, Britton (1975) reported measurable effects. Coturnix seem to be resistant also to effects of DDT, but this effect may be modified by diet. Bitman et al. (1969) fed DDT at 100 ppm to Japanese quail maintained on a low calcium diet and found that the birds receiving the pesticide had a delay in the production of eggs, laid smaller eggs, and had a higher incidence of egg breakage than untreated controls did. Later Cecil et al. (1971) repeated the experiment using DDT and DDE but providing birds with a diet adequate in calcium content. As in the earlier work there was a delay in ovulation, but after 3 weeks production was comparable in all groups. In the experiments with both low and adequate calcium, there was more egg breakage in treated than in untreated quail, but in each case most of the broken eggs were laid by a few birds. Davison et al. (1976) added an additional wrinkle to the coturnix story. They fed DDE at 0, 2, 10, 40, and 200 ppm and found no change in number of eggs layed, shell thickness, or egg weight. Similar results were obtained when DDT was fed at the same levels. When, however, females on the DDT diet were caged with males, fewer eggs were laid and these had thinner shells. The authors point out that male coturnix are sexually aggressive and conclude that their activities provided sufficient "stress" to cause the effects. DDT did not alter fertility or hatchability and DDE was not tested for its effect on these parameters.

The first good laboratory evidence of DDT's affecting eggshell thickness was provided by Heath et al. (1969, 1970) in 2 year studies with DDT, DDE, and DDD on mallards. In this study DDD impaired reproduction over a period of time but not so severely as DDE and did not cause changes in eggshells. DDT had no effect at the 2.5 or 10 ppm level, but at the 25 ppm dosage, eggshells were thinner, 18% were cracked, and duckling survival was reduced by 35%. Duckling survival per hen was reduced 50% from the controls. DDE at 10 or 40 ppm resulted in eggs with shells 13% thinner than controls and 25% cracking. Hatchability of eggs with uncracked shells was reduced, and though embryo survival during the first 3 weeks of incubation was normal, mortality during the final week was increased by 30 to 50%. Hatched ducklings survived at the normal rate for 14 days. The overall effect was that duckling production per hen was reduced as much as 75%. The sensitivity of mallards to DDE was confirmed by Haegele and Tucker (1974), Risebrough and Anderson (1975), and others (see Cooke, 1973). Other species may be even more sensitive than mallards. Longcore and Samson (1973) found that black duck hens fed DDE (10 ppm) produced only 20% as many ducklings as untreated birds did. They found that egg breakage was four times greater during natural incubation than it was when eggs were handled in a laboratory incubator.

Porter and Wiemeyer (1969) showed that feeding DDT plus dieldrin to American kestrels causes eggshell thinning, and McLane and Hall (1972) found that 2.8 ppm DDE in the diet of screech owls caused eggshell thinning of 12%. Haseltine et al. (1974) found that, when mallards were fed DDE at 10 ppm, eggshell thickness was reduced 20% but that, in ring doves fed this chemical at 40 ppm, no consistent effect was apparent.

Chemicals other than DDT may affect reproduction. Brown et al. (1974) found that, when dieldrin was fed to hens at 10 or 20 ppm, chick mortality increased. Risebrough and Anderson (1975) reported that, when PCB was added to DDE in the feed of mallards (40 and 40 ppm), eggshells were thinner than when DDE was fed alone and the incidence of egg eating increased. Haegele and Tucker (1974) reported that short-term egg thinning was induced in coturnix by feeding parathion; Ceresan m®; Aroclor®; tetraethyl lead; 2,4-D; or carbaryl and that in this species eggshell thinning occurred also when only reduced food intake was involved. These workers reported that, in mallards, short-term egg thinning was induced by feeding heptachlor; Ceresan m®; Aroclor®; 2,4-D; or sodium arsenite.

Significance of Reproductive Effects

The significance of the reproductive effects of pesticides on birds is difficult to assess. As pointed out at the Wisconsin conference (Hickey, 1969), effects on several populations of the peregrine falcon have been dramatic and serious. The report of the American Ornithologists' Union Committee on Conservation 1974–1975 (AOU, 1975) expresses the hope that successful raising of peregrines in captivity will permit the restoration of this species but cautions that unless such birds can be kept free from persistent pesticides the outlook is dim. Fyfe et al. (1969) reported occupied territories of prairie falcons in western Canada down 34%, and Faber et al. (1972) found that nesting success in the common egret had decreased from 52 to 28% during the period 1967–1970. The only other species discussed in the AOU report (AOU, 1975) where pesticides have been involved in the decline is the bald eagle. The AOU report indicates that these birds are much more plentiful than has been reported and that, in 1974, 0.78 young was produced/territory. Only in Maine is the population threatened. Young/territory in 1973 was 0.23 in Maine versus a 48 state average of 0.83.

The brown pelican populations in some areas are in trouble, but though pesticides are involved, other factors may be more important. Blus et al. (1975) discuss the Louisiana population and point out that a population on the mud lumps estimated at 50,000 in 1920 had declined to 4550 by 1938. This was prior to the modern pesticide era, and though the population continued to decline to an estimated 200 breeding pairs and 100 nestlings in 1965, the part played by pesticides in that decline is highly speculative. Gilbertson (1975) reported that, in the Great Lakes area where aquatic feeding birds have high pesticide levels in their eggs (Faber and Hickey, 1973; Frank et al., 1975a), only 16% of eggs of herring gulls hatched and fledgling production was reduced 90%. He found that, whereas reproduction in cormorants was down and eggshell thickness in this species reduced 30 to 35%, common terns, caspian terns, and ring-billed gulls were reproducing well in most areas. Johnson et al. (1975) reported that, despite reduced eggshell thickness, the osprey was reproducing successfully in Idaho, though this is not the case in Sweden, where reproductive success has declined since 1947 (Odsjo and Sondell, 1976).

Many factors affect reproduction in birds, and some of the known causes of eggshell thinning have been discussed by Ratcliffe (1970). Included are imperfect diet (especially if deficient in calcium, manganese, or vitamin D), shortage of food, age and genetic constitution (shells from older birds are more likely to be thin), stress of various kinds, disease, and drugs such as sulfanilamides. High temperatures may cause thin eggshells as well (Risebrough et al., 1970). Thus it cannot be assumed that all eggshell thinning or reproductive failure in wild birds is caused by pesticides. Ream (1976) concluded that, despite the fact that loons in northern Minnesota carried significant pesticide residues in their bodies and in their eggs, the main reason for poor breeding successs was human invasion of their nesting areas.

Without doubt the evidence supports the view that DDT and its metabolites and dieldrin (or aldrin since it converts to dieldrin) are major factors in the reproductive decrease of a number of raptorial and fish-eating birds. There are, however, some facts contrary to this conclusion that should be noted. No explanation has yet been offered for the large differences among bird species in relative susceptibility to egg thinning. A precise definition of the mechanism(s) responsible for egg thinning is not likely to be forthcoming. The second point deals with comparisons between addled or eggs that fail to hatch and those that do. In a significant number of instances in which pesticide residues have been compared, no differences have been found in residues between the successful and unsuccessful eggs. Klaas and Swineford (1976), Frank et al. (1975a), and Johnson et al. (1975) found no differences in shell thickness in eggs of clutches that failed to fledge versus those that did.

Other Sublethal Effects

A large number of reports of population declines cite aberrant parental behavior. Destruction of eggs has been observed (Ratcliffe, 1970; many others) and in the laboratory this has been demonstrated. Longcore and Samson (1973) fed DDE to black ducks and found no differences between treated and untreated birds in the number of eggshells cracked on the first day. Beyond the first day cracking was more pronounced in those receiving the DDE. Of note, however, is that, while none of the cracked eggs were removed from the nests of untreated ducks, 27 (42%) were removed from those receiving DDE.

Decreased egg production has been noted in mallards fed DDE (Risebrough and Anderson, 1975), leghorns fed DDT at high levels (Britton, 1975), and coturnix fed DDT or DDE, though in this species egg production became normal later in the laying period (Bitman et al., 1969; Cecil et al., 1971). In contrast, Brown et al. (1974) found that laying hens fed dieldrin at 10 or 20 ppm had an increase in egg production. Many field studies have reported reduced clutch size among raptors and fish-eating birds. These studies did not establish whether this was due to lowered egg production or to parental destruction.

In both field and laboratory studies late nesting has been reported. Laboratory studies relating this to pesticides have confirmed this in a number of species. Bitman et al. (1968) reported a delay in egg production in quail fed DDT, and Cecil et al. (1971) found this delay to be about 3 weeks. Peakall (1970) found that first egg laying was delayed about 5 days in ring doves when these were on a diet containing 10 ppm DDT. Jefferies (1967, 1971) reported delayed ovulation in Bengalese finches fed DDT or DDE. In a related effect Mahoney (1975) found that, when white-throated sparrows were fed DDT at 5 or 25 ppm, the onset of spring migratory conditions as measured by weight increase and nocturnal activity was delayed 1 week. Effects from feeding DDE were not so pronounced.

There have been few reports of physical impairment in birds due to pesticides. Defective birds have been observed in nature, but a cause-effect relationship has not been established. Gilbertson (1975) reported that common tern chicks with crossed bills and a few with small eyes and duplicated feet were seen in Great Lakes colonies and that, of several thousand ring-billed gulls, about 20 had leg deformities. The author presented no data but stated that the incidence of deformity was higher than usual. Eggs of both species contained significant residues of pesticides and other pollutants, and Gilbertson (1975) stated that analyses of eggs of these and other fish-eating species in Lake Ontario show that these eggs are "among the most contaminated with DDT and PCB of those found anywhere in the world." Defects in chicks have resulted when pesticides were injected into incubating eggs (Greenberg and LaHam, 1969, 1970; Paul and Vadlamudi, 1976).

18.3 MECHANISM OF ACTION

Since most of the toxic and sublethal effects associated with pesticides in birds have been due to insecticides that act on the nervous system, the causes of lethal effects are reasonably clear. The data correlating brain levels with toxicity in DDT and dieldrin is voluminous, and some information is available for a number of other compounds. Jennings et al. (1975) raise some question whether toxicity of carbophenothion is related solely to esterase inhibition, since in poisoned geese they found ruptured blood vessels and speculate that high blood pressure may be a contributing factor.

While lethal effects may be understood, there is no agreement on what causes thinning in eggshells.

Eggshell production in birds involves a number of steps, involving a variety of chemicals, chemical reactions, and enzymes (Cooke, 1973). The process can be summarized as follows: The stimulus for ovulation is provided by a secretion of the pituitary, and the egg, formed in the oviduct, is passed to an enlarged portion known as the shell gland, where calcification of the shell takes place. The end result is a hardened shell, composed mostly of calcium carbonate. It is logical to assume that eggshell thinning is a result of disruption of the normal processes of calcification, and analyses of thin eggshells have shown that percent thinning and percent reduction in calcium content are closely related. While all the calcium that makes up the eggshell derives from the diet, a significant portion (25 to 40%) comes by an indirect route, being stored first as medullary bone deposited in cavities in the structural bones in the bird. This calcium storage in medullary bone is controlled by androgens and estrogens. The medullary calcium, as required, must be remobilized, and this is controlled by secretions of the parathyroid and the ultimobranchial glands. Regardless of the route by which the eggshell calcium is obtained, calcium in the diet must be absorbed from the intestinal tract and carried in the blood to its site of deposition. This absorption is dependent on the steroid, vitamin D, and the action of ATPases for transport of calcium across the intestinal wall. ATPases are important also for the transport of calcium from the blood across the wall and into the lumen of the shell gland.

Calcium is deposited in eggshells as calcium carbonate, and thus the other major requirement is the carbonate ion. It is believed that this comes from metabolic CO_2 and that the enzyme carbonic anhydrase acts in the shell gland to maintain a balance between CO_2 and carbonic acid to ensure a supply of carbonate ions for shell deposition.

This summary suggests a number of points where the altering of normal functions might be critical to eggshell production. Among the major sites for such malfunctions the following might be critical.

1. Interferences with ATPases essential for the transport of calcium across cell membranes.
2. Interferences with vitamin D needed for the absorption of calcium from the intestinal tract.

3. Interferences with the hormones controlling the deposition of calcium in medullary bone.
4. Interferences with the parathyroid and/or ultimobranchial glands controlling mobilization of calcium from medullary bone.
5. Interferences with the carbonic anhydrase that controls the supply of carbonate ions in the shell gland.

Some evidence is available that each of these sites could be involved in pesticide effects.

ATPases

The involvement of pesticides with the activity of ATPases is largely inferential. It is known that DDT, at a low level ($0.3 \mu M$), will inhibit an ATPase from rat brain, but when DDE was used, the concentration had to be increased 1000-fold to effect the same degree of inhibition (Matsumura and Patil, 1969). Koch et al. (1969) also demonstrated inhibition of ATPases in insect nerve, muscle, and brain, but we could find no instances in which an effect on ATPase in birds has been investigated.

Vitamin D

It has been suggested that, since the induction of hepatic enzymes by organochlorine insecticides increases the metabolism of sex hormones, this metabolism increase might extend to vitamin D. Limited experiments along these lines have not supported this suggestion (Peakall, 1969; Nowicki et al., 1972, 1972a).

Medullary Bone Deposition

There is ample evidence that organochlorine insecticides increase the activity of microsomal enzymes in the liver, and Peakall (1967) showed in *in vitro* tests that dieldrin, DDT, and DDE increased the metabolism of testosterone, progesterone, and estradiol by pigeon livers. Theoretically, such an effect could affect the deposition of calcium in medullary bone and hence reduce the amount of mobilizable reserve calcium available when needed during the egg-laying period. Several workers reported reduced deposition of medullary bone when DDT or DDE was fed (Risebrough et al., 1970; Oestreicher et al., 1971; Peakall, 1970), and other studies demonstrate increased metabolism of estradiol (the main hormone involved in medullary bone deposition) in birds dosed with organochlorine pesticides (Peakall, 1969, 1970; Britton, 1975a). Britton (1975a) experimented with the hen, a bird in which eggshell thinning can be demonstrated only at very high dosages of DDT. He showed that the effect of DDT on the ability of liver to metabolize the estrogen was dose and time dependent. On a 500 ppm diet no increased activity was found in 10 days or less. Some increase was noted by day 18, and by day 51 liver microsomal metabolism of estradiol to polar metabolites had more than doubled. The maximum increase (about $2\frac{1}{2}$ times control) occurred in the 600 ppm-dosed birds by day 21 and in the 1200 ppm feeding level by day 14. These data demonstrate an increase in estradiol metabolism due to DDT, and if this were the basis of eggshell thinning, experiments with leghorns fed these dosages should have, with time, resulted in equal eggshell thinning at each of the dosages. This did not occur. Although thinning occurred in the 600 and 1200 ppm feeding levels, no thinning occurred when DDT at 300 ppm was fed for three 28 day periods (Britton, 1975).

Other studies (see Cooke, 1973; Stickel, 1973) suggest that the induction of microsomal enzyme activity by organochlorines and the effect of these on estrogens is not an adequate explanation for eggshell thinning. Peakall (1970) reached this same conclusion after studies on ring doves.

Mobilization of Calcium from Medullary Bone

Remobilization of calcium from medullary bone involves estrogens, secretions of the ultimobranchial glands, and the parathyroid hormone. Haseltine et al. (1974) noted that species such as the pheasant and the chicken are resistent to eggshell-thinning effects from DDT and recover readily from removal of the parathyroid gland, whereas species such as the mallard and ring dove, susceptible to eggshell thinning from DDT, are extremely sensitive to parathyroid gland removal. They reasoned that the correlation might be more than coincidental and conducted experiments with mallards, pheasants, and ring doves in which birds were fed a diet containing 10 ppm DDE and injected with parathyroid hormone. The logic for this experiment was that, if DDE was suppressing the activity of the parathyroid hormone, injection of the hormone should compensate for such suppression. In both mallards and ring doves parathyroid injections had only a short-term effect in nullifying the serum calcium depression in DDE-dosed birds. Haseltine et al. (1974) suggest that doses of parathyroid hormone sufficient to elevate serum calcium levels during the entire period that the egg is in the shell gland might eliminate the DDE thinning. Additional experiments along this line would be revealing, but even if the mobilization of calcium from the medullary bone is involved, it could not alone explain the drastic eggshell thinning reported for the brown pelicans on Anacapa Island, where the calcium carbonate thickness had been reduced more than 90% (Risebrough et al., 1970).

Carbonic Anhydrase

Experiments reported in 1970 offered the explanation that eggshell thinning was a result of deprivation, not of calcium, but rather of the supply of carbonate ions in the shell gland. Peakall (1970) found that the injection of DDT into ring doves caused thin eggshells and inhibited the activity of carbonic anhydrase in the oviduct, and Bitman et al. (1970) reported that, when Japanese quail were fed DDT or DDE and sacrificed with an egg in the shell gland, activity of carbonic anhydrase was reduced in both the shell gland and blood. Pocker et al. (1971, 1971a) disputed this finding, reporting that the active sites on carbonic anhydrase were not blocked by DDT, DDE, or dieldrin and that reduced enzyme activity in the presence of these insecticides occurred only when they were present in excess of their solubility. When this occurred, the enzyme coprecipitated with the insecticides and activity was lost. Such precipitation would not occur in the shell gland *in vivo*, since the lipid content of the medium would ensure solubility. Dvorchik et al. (1971) supported this latter view, finding DDT inhibition only at high dosages where enzyme precipitation occurred and suggested no effect at field exposures. MaGuire and Watkin (1975) found, in *in vitro* experiments, that dieldrin and DDT did cause mild inhibition of bovine carbonic anhydrase at dosages below their maximum solubility levels, a result at odds with that reported by Pocker et al. (1971). Haque and Deagen (1975) suggested that the DDT effect might be due to the metabolite DDA and showed evidence that this compound binds carbonic anhydrase. The significance of this finding is uncertain with respect to eggshell thinning. DDA was identified as a metabolite of DDT or DDD in chickens (Abou-Donia and Menzel, 1968), but it was not found in leghorn cockerels fed DDT (Ecobichon and Sashenbrecker, 1968) or in the pigeon, Bengalese finch, or common grackle (Menzie, 1969). Whereas the feeding of DDE has been shown to cause eggshell thinning in quail, Lamberton et al. (1975) reported that in this species DDE is not metabolized. Peakall (1975) argues the case for inhibition of carbonic anhydrase by DDT and related compounds and suggests that the apparent inconsistency between *in vitro* results (Pocker et al., 1971) and *in vivo* situations (Peakall, 1970; Bitman et al., 1970) can be explained. He believes that in *in vivo* tests a weak binding occurs between the pesticide and carbonic anhydrase and that this inhibits activity of the enzyme. By contrast, in the *in vitro* experiments the presence of organic solvents caused the breaking of this bond and the lack of inhibition.

Other Possibilities

The five categories of possible effects just discussed do not exhaust all possible mechanisms to explain thin eggshells. Other possibilities are reviewed by Cooke (1973) and evidence in support of such theories is discussed. Among these are included various effects on shell constituents other than calcium and carbonate and factors that might operate to affect the thyroid or adrenals. Premature extrusion of eggs and the role of decreased food supply are discussed as well. Jefferies (1975) presents an exhaustive review dealing with possible effects on the thyroid and suggests that most of the sublethal effects attributed to organochlorine insecticides may be due to hyperthyroidism or hypothyroidism.

18.4 RESIDUES IN BIRDS

The residues encountered most frequently are DDE and dieldrin, but, in addition, heptachlor epoxide and chlordecone are common in some areas and mercury is present in many fish-eating birds and raptors. Extremely high residue values have been found in some species in some locations, but there would appear to be differences among species in both exposure and ability to detoxify and eliminate pesticides. These differences exist even among birds occupying essentially the same environmental niche. Keith (1970) reported "My data on white pelicans show that only about 20 percent of the population carry high residues. The remaining birds are equally vulnerable, but do not intercept high contamination in the environment." Thus dramatic differences exist among individual birds of the same species. For example, Hunt et al. (1969) fed DDT to pheasants at dosages of 10, 100, 250, and 500 ppm and measured residues in tissues, eggs, and feces. Regardless of the level of exposure, there were gross differences among birds in each of the tissues examined. At the 100 ppm feeding, residues of DDE in females ranged from 0.2 to 9.3 ppm in the brain, 0.01 to 12.4 ppm in whole blood, 17.9 to 717 ppm in the fat, 6.6 to 68.6 ppm in the ovaries, and 24.6 to 205 ppm in the egg. These workers examined the absorption of DDT from the food in three birds by measuring the intake in the food and excretion in the feces during each of two 5 to 7 day periods and found absorption rates of 96.4 to 98.8%.

Although, as just shown with pheasants, wide differences exist between individuals, there is generally good correlation between the level of intake of organochlorine insecticides and their residues in birds. Higher levels of intake result in higher tissue levels, but as shown by Cummings et al. (1967) a plateau is reached, the plateau being higher the higher the rate of intake. When exposure is discontinued, tissue residues decline, the decline being more rapid with some pesticides than others. Dieldrin and heptachlor epoxide are perhaps the most persistent (Cummings et al., 1967; Robinson, 1970). Levels in eggs reflect intake and tissue levels and on a wet weight basis are frequently about 10% of the level in adipose tissue (Keith and Gruchy, 1972). Since tissue levels reflect intake, it is not surprising that among wild birds pesticide levels are correlated with feeding habits.

The pesticide burden in migratory birds is highly variable and may reflect differing feeding grounds. Fox (1976) found that eggshell quality (presumably reflecting pesticide load) of common terns on Buffalo Lake, Alberta, was poor, even though the fish in the lake contained low residues of pesticide. He contrasted these with a colony in another region, where eggshell quality was high but where analyses of their fish food showed levels of pesticide comparable to those in Buffalo Lake. He believed the difference was due to food taken in other than the nesting site. The terns on Buffalo Lake spend their first 2 years along the coast of California, Mexico, El Salvador, and Peru, and at least in a portion of this range (California) pesticide residues in fish likely to be taken as food are high. Gilbertson (1974) reported that, in Hamilton Harbor, DDE residues in common terns reflected a local source. The residues were low in new migrants and increased as the season advanced.

The extent to which birds are contaminated with organochlorine insecticides is indicated in a number of reports. Prestt and Ratcliffe (1970) reviewed the European literature and reported that residues of one or more of these pesticides were found in 154 species of European birds. Stickel (1973) lists residues in eggs from 77 species. Reports from many investigations in North America indicate that some DDE can be detected in most birds, but fortunately the levels are sufficiently low that deleterious effects are unlikely. Johnson (1975) collected small migratory birds, most of which had been killed by colliding with tall television towers in the Southeastern United States during the period 1964–1973. In all, 128 samples were analyzed involving 19 species and 908 individuals. DDT or metabolites was found in all but 1 sample. Dieldrin was detected in 60 of the 128 samples. High levels of either insecticide were seldom found. The mean residue (lipid, wet weight basis) was 4.31 ppm for DDT and metabolites, and 0.17 ppm for dieldrin. There was no consistent difference in residues between northward or southward migrants, male or female birds, or older versus young of the year. There were, however, trends associated with food habits. Insectivorous and granivorous birds had higher residues than others. Organochlorine residue levels were lower in the latter part of the sampling period than in the earlier specimens.

National pesticide-monitoring programs have been established in the United States using starlings and the wings of black ducks and mallards as indicators. In the starling study, birds are taken from 126 sites, and these show declining residues of organochlorine insecticides. The mean (geometric) of DDT residues was 0.442 ppm in 1972, 0.282 ppm in 1974 (Nickerson and Barbehenn, 1975; White, 1976). Analyses on black duck and mallard wings also indicate a decline in residue levels during the 1972–1973 hunting season compared with earlier data. Extensive tabulations on pesticide residues in birds in relation to dietary intake are presented by Edwards (1973), and Edwards (1973) and Stickel (1973) give tables listing residues reported in eggs. The reader is referred to these reports for detail.

One of the highest areas of pesticide contamination in North America is the Great Lakes region. Frank et al. (1975a) analyzed 307 eggs collected in 1971 from 20 species of aquatic and upland birds. Their data are summarized in Table 18.5. They found that eggs from carnivorous species contained the highest residues of both DDT (including metabolites) and mercury. Eggs from aquatic carnivors had higher residues than those from terrestrial carnivors. Herbivorous and insectivorous birds had lower pesticide residues. Dieldrin residues showed a pattern less consistent with food habits except that residues tended to be higher in seed- and grain-eating species.

On occasion, residues of pesticides other than DDT and dieldrin have been found in wild birds. Heptachlor epoxide, toxaphene, telodrin, endrin, lindane, and chlordecone have been found in areas where these pesticides have been used widely (Stickel, 1973). With none of

Table 18.5 Residues of Pesticides in Eggs of Birds from the Great Lakes and Adjoining Upland Areas and Chief Foods of the Species Involved[a]

Bird Species	Mean Egg Residues, ppm[b]			Chief Foods
	DDT	Dieldrin	Mercury	
Catbird	1.11	Tr	--	Fruit, insects
Brown-headed cowbird	1.44	Tr	--	Seeds, insects
Morning dove	0.26	0.36	--	Seeds, grain
Common grackle	1.01	0.04	0.06	Grain, insects
Ruffed grouse	1.54	0.21	--	Seeds, buds, fruit, insects
Ring-necked pheasant	0.15	0.01	0.06	Grain, seeds, fruit, insects
Eastern phoebe	1.82	ND	--	Insects, few seeds and fruit
Robin	2.00	0.06	0.07	Fruit, earthworms, insects
Starling	2.64	0.21	0.18	Insects, fruit and grain
Common crow	1.98	0.03	0.12	Grain, fruit, insects
Red-tailed hawk	2.47	0.12	0.06	Rodents, reptiles, few birds
Red-shouldered hawk	8.90	ND	3.5	Rodents, reptiles, few birds
Great horned owl	3.88	0.01	0.09	Rodents, birds
Red-winged blackbird	2.79	0.11	0.68	Grain, insects
Canada goose	0.33	0.01	0.12	Seed, grain, aquatic plants
Mallard	2.34	0.13	0.15	Grain, plants, insects
Herring gull	10.4	0.03	0.74	Crustaceans, fish, garbage
Black-crowned night heron	7.82	0.03	0.64	Frogs, fish, crustaceans
Black tern	7.6	0.33	--	Insects, fish
Common tern	22.4	0.46	0.83	Fish, some insects

[a]Frank et al., 1975a.
[b]ppm on a whole weight basis.

these, however, have we seen widespread contamination, and in national monitoring programs they are detected much less frequently and at much lower levels. In a monitoring program on black duck eggs in Canada and through the United States in 1971, all samples contained DDE with a range of 0.09 to 5.94 ppm (wet weight basis) and all contained PCBs with levels of <0.05 to 3.3 ppm. No chlordecone was detected, and only trace amounts of dieldrin and heptachlor epoxide were detected. Residues of organochlorines were lower than those in a survey 7 years earlier. Eggshell thickness seemed to respond to the changed organochlorine content of the eggs. Thickness prior to 1940 was 0.348 mm; in 1964, 0.31 mm; and in 1971, 0.343 mm (Longcore and Mulhern, 1973).

18.5 SUMMARY

The widespread presence of DDT and dieldrin in the environment has resulted in lethal and sublethal effects in a wide range of species of birds. The lethal effects have been dramatic on some songbirds in areas of forest spraying and some urban areas where Dutch elm disease control with DDT has been carried out.

With the exception of the raptors, lethal effects have been confined, in most cases, to local areas of pesticide application. With raptors and some other migratory species, evidence suggests that body pesticide burdens that are not normally lethal may cause toxicity when stress conditions of migration, molting, and reproduction bring about a utilization of stored fat, the pesticide load in that fat being mobilized and transferred to the brain.

Although lethal effects are important, these have decreased in incidence and intensity since use of the persistent organochlorine insecticides has been curtailed. Sublethal effects, reflected primarily as eggshell thinning, egg breakage, and poor reproductive success, have been much more important. Involved principally have been carnivorous birds. Eggshell thinning has been reported in at least some populations of 39 species, and in about 27 species the eggshell thickness reductions are considered of sufficient magnitude to cause reproductive impairment. In only a few species have drastic population declines been documented, and it is apparent that some species are much more susceptible than others. Falconiformes are highly sensitive to eggshell thinning, while gallinaceous species are much less susceptible. There are good correlations between eggshell thinning and DDE content in eggs of several species, and there seems little doubt that high levels of

DDT or DDE in a number of species of carnivorous birds have been responsible for excessive eggshell thinning and reproductive failure. Dieldrin has been similarly incriminated, but this insecticide has played a less prominent role in raptor declines in North America than in Great Britain.

The acceptance of the conclusion that organochlorine pesticides were responsible for thin eggshells and population declines in carnivorous birds has been hampered by lack of an adequate explanation of the physiological mechanism responsible. Such an explanation is still lacking. There is some evidence that the thyroid or parathyroid may be involved, that vitamins or enzymes essential to calcium absorption and transport may be affected, and that the supply of carbonate needed in the shell gland for calcium carbonate deposition may be inadequate owing to the inhibition of carbonic anhydrase by the pesticides or their metabolites. The data available are not conclusive, and a definitive answer is yet to be determined.

National monitoring programs and surveys on a species basis show residues of organochlorine insecticides and mercury decreasing in recent years. Such decreases reflect reduced usage of these pesticides with subsequent reduced exposure of birds.

19

PESTICIDES AND MAMMALS

An extensive literature exists on the acute and chronic toxicity of pesticides to a few species of animals. LD_{50} values determined on the rat are the most widely used index of relative toxicity, and all pesticides are toxic to a greater or lesser degree. As we have seen, acute LD_{50} values for the rat range from near 1 mg/kg for highly toxic insecticides such as TEPP, aldicarb, phorate, and disulfoton to several thousand mg/kg for methoxychlor, many fungicides, and many herbicides (Chapters 7–13). Frequently, differences exist between the sexes in the rat and between different animal species in regard to acute toxicity, and this may extend to chronic toxicity as well, though the data in this latter aspect are less extensive. Although such differences exist, they are not predictable on the basis of current knowledge, and what applies to one pesticide may not apply to another. For example, Luckens and Davis (1964, 1965) found that the big brown bat is about five times as sensitive as the rat to DDT, but with dieldrin and endrin, LD_{50} values for both animals are about the same. Differences undoubtedly reflect differing abilities to metabolize specific pesticides or to store them away from potential sites of action.

19.1   LETHAL EFFECTS

Mortality among various species of animals has been reported with several pesticides. With only a few highly toxic compounds has this been great cause for concern. As in the case with birds, differences exist also in the toxicity of pesticides to mammals in their natural setting, and this may, in part at least, be a function of their habitat and consequent exposure.

Domestic Mammals

Accidental poisoning of farm animals through careless handling and storage of pesticides occurs with some frequency. One of the main sources seems to be the storage of pesticides on the farm in the same areas as that used for feeds. Buck and Van Note (1968) record several instances in which sacs of granular formulations of aldrin and aldrin plus parathion were stored in the same bin with bulk animal feeds and were mixed with the feed in preparing animal rations or contaminated the feed through sweepings or spillage. The result was dead and ill cattle with excessive tissue residues in survivors entering commercial slaughter houses. Braun and Lobb (1976) report an incident where 9 of 18 cattle sheltered in an implement shed died from endosulfan poisoning; 5 lb bags of the pesticide powder had been left in the shed, and the cattle broke the sacks and consumed the contents. Many other instances could be cited where hog and cattle feed have been contaminated with a wide variety of pesticides. In addition cattle have been poisoned when they drank from an unguarded bucket of pesticide solution prepared as a spray for ectoparasite control or as a residual treatment for fly control in the dairy (Frank, personal communication). Poisonings of this type reflect gross carelessness and underscore the necessity of treating pesticides with the care appropriate to their toxicity.

Poisoning of livestock has occurred also where farm animals have fed on plants heavily contaminated with pesticides. This was perhaps more common with some of the earlier pesticides such as the arsenicals and also more common in areas of mixed farming, where small farm units were engaged in vegetable production, as well as in the rearing of livestock or the operating of a small dairy herd.

Poisoning of dogs and cats has been reported following pesticide treatments in a number of areas. Turtle et al. (1963) reported death in these as well as in wild mammals in areas in Great Britain where aldrin and heptachlor were used as seed dressing for cereals. Circumstantial evidence supported strongly the conclusion that the animals had been poisoned from eating dead and dying pigeons and other birds that had fed on the treated seed. Dogs and cats were found dead in areas treated with heptachlor for fire ant control (Scott et al., 1959; DeWitt and George, 1960). In these instances it was not determined whether poisoning was direct or due to consumption of the poisoned small animals and birds that resulted from the treatments.

A different sort of toxicity to animals has been experienced where some plants were treated with 2,4-D. It has been known for many years that under adverse growing conditions cereals and other crops may develop high nitrate levels in their tissues (Cook, 1930). For some crops 2,4-D provides such an adverse condition. Stahler and Whitehead (1950) investigated nitrate levels in sugar beet leaves on seven farms where 2,4-D had been applied mistakenly on 335 acres of this crop. Potassium nitrate levels ranged from 1.81 to 8.77% (dry weight of leaves) in contrast to a normal level of 0.22% where no 2,4-D had been applied. The average level of 4.5% is much in excess of the 1.5% maximum established as a safe feeding level in forages (Davidson et al., 1941). Stahler and Whitehead (1950) point out that, whereas weeds such as pigweed, ragweed, and jimson weed are normally not grazed by cattle, they are eaten readily when they have been treated with 2,4-D. Thus the explanation for toxicity to animals grazing such weeds after 2,4-D treatments lies not in the toxicity of 2,4-D but rather in the high levels of nitrate in the treated foliage (Berg and McElroy, 1953; Fertig, 1952). The nitrates are not toxic but become so when reduced to nitrites in the rumen. The nitrites cause the formation of methemoglobin in the blood and result in an interference with oxygen transport. As a result, anoxia occurs and when severe brings about the death of the animal.

Wild Mammals

Although DDT has been implicated in both acute and chronic toxicity to fish and in reproductive failure in birds, there is little evidence for toxicity to wild mammals, with the possible exception of bats (see Sublethal Effects). Adams et al. (1949) observed mammal populations in an Idaho forest sprayed with DDT at 1.0 lb/acre and in experimental plots (several hundred acres) in Wyoming sprayed at 5 and 7.5 lb/acre in a single season in two equal applications. They detected no effect on populations of several species of small mammals. Similarly, George and Stickel (1949) could find no dead mammals or population changes as a result of the treatment of a variable range of habitats in Texas with DDT for tick control. Deer counts were comparable before and after treatment, and no dead or sick were noted among raccoons, skunks, or rabbits. Stickel (1946) found no change in deer mouse populations in 117 acres treated in June with 2 lb DDT/acre and later (Stickel, 1951) reported no effect on mouse populations following 5 years of annual treatment at the same dosage.

In contrast to the results with DDT, mortality, sometimes extensive, has occurred in wild mammals following application of some cyclodiene insecticides.

Scott et al. (1959) monitored dieldrin applications made in Illinois to control the Japanese beetle; 1400 acres were treated in 1954 with an additional 2500 acres in 1955. Most of the insecticide was applied by air at 3 lb/acre, but in limited areas ground equipment was used. The authors reported that populations of ground squirrels, muskrats, and cottontails were virtually eliminated and that mortality was high in short-tailed shrews, fox squirrels, woodchucks, and the meadow mouse. Other mouse species were not drastically reduced. Following the use of heptachlor at 2 lb/acre for fire ant control, rabbits, opossum, raccoon, cotton-rats, and the white-footed mouse were found dead in several areas, and in Hardin County, Texas, opossum, armadillo, and raccoon virtually disappeared following widespread heptachlor treatment (DeWitt and George, 1960).

In the Netherlands Van Klingeren et al. (1966) found that, in sections of a polder treated with DDT and dieldrin, the hare population had been reduced by 31.5%.

In Great Britain high mortality was noted among wild mammals, particularly foxes, in 1956 and again in 1959–1960. The deaths were believed due to the foxes' eating pigeons that had been poisoned by eating seed wheat treated with insecticides (Turtle et al., 1963). Dieldrin and heptachlor were the main insecticides involved, and it was shown that $2\frac{1}{2}$ to 6 poisoned birds proved lethal to the foxes. On the basis of experimental tests and analyses of dead animals found in the wild, Blackmore (1963) concluded that a tissue residue of 1 ppm dieldrin or 4 ppm heptachlor epoxide was diagnostic of poisoning.

With the exception of dieldrin and heptachlor, there is little evidence that pesticides, when properly used, have caused significant mortality in wild mammals. For example, Buckner and Sarrazin (1975) could find no effect following extensive forest spraying with fenitrothion in Quebec, and in experimental treatments in 2500 acre blocks of forest, no effect on mammal populations was attributable to the use of acephate at 8 or 20 oz/acre (Buckner and MacLeod, 1975).

Aquatic Mammals

There is no evidence of acute toxicity resulting in death among sea mammals. As is discussed later, pesticides may be associated with premature pupping in California sea lions, but the evidence for this is weak.

Human Beings

Most of the environmental problems with pesticides have been associated with DDT and a few other persistent organochlorine insecticides. The situation is quite different with regard to human mortality. The low mammalian toxicity of DDT permitted its direct use on people for louse control, and in intensive occupational exposure in manufacturing plants and with malaria eradication programs hundreds of thousands of people have had prolonged exposure to this insecticide without fatality. In reviewing the literature on the subject, Deichmann (1973) reports, "Acute and fatal intoxications have been rare; when they did occur, they were due to the accidental or intentional (suicidal) ingestion of overdoses." At the calculated LD_{50} dosage for the rat of about 200 mg/kg, a 70 kg person would have to consume about 14 g DDT for a lethal dose. It has been reported (see Deichmann, 1973) that 10 mg/kg will produce illness in some people, but even at this level a dosage of almost 1 g would be required for a 70 kg person. Perhaps the saving feature of DDT is its low dermal toxicity, about 2000 mg/kg (Kenaga and End, 1974).

Quite a different situation exists with many of the organophosphorous insecticides. Parathion, mevinphos, phorate, and many others have extremely high mammalian toxicity, and this toxicity pertains whether the exposure is oral or dermal (Chapter 10). Thus it is not surprising that growers, spraymen, and formulators, accustomed to the relative safety in use of DDT, became ill when they failed to take the safety precautions necessary in handling the more toxic compounds. Mortality with some of these compounds requires less than 100 mg, and while oral consumption of such an amount should not occur in handling the compounds, dermal exposure to this extent is not unlikely for workers without protective clothing. Thus organophosphate poisoning occurs with some frequency. Other insecticides are involved also. Some of the pesticide poisonings are due to exposure in the work place, some through contaminated foods, some through carelessness in storage and labeling (especially in child poisonings), and a significant number as suicides.

A number of studies have been conducted on the incidence of pesticide poisoning. David et al. (1969) reviewed a number of these, and one from Florida will illustrate the situation in a high pesticide use area. In Dade County, with a population of about 1 million, 1004 deaths attributable to poisons were confirmed during the period 1956 through 1965. A large percentage of these were suicides, and with these excluded, the remaining 313 are categorized in Table 19.1. Pesticides contributed a significant portion of the poisonings, and this was

Table 19.1 Human Poisonings in Dade County, Florida, 1956–1965[a]

Cause of Poisoning	Total Poisonings No.	Children Under 5 Years No.	Remarks
Pesticides	39	22	(15 organophosphates)
Medications	118	11	(8 aspirin)
Alcohols	109		
Other	47	12	(4 furniture polish)
Total	313	45	

[a]Davis et al., 1969.

especially pronounced among small children. Parathion was the insecticide most frequently involved in the poisonings.

The Dade County figures indicate the significance of pesticide poisonings. As Davis et al. (1969) suggest, these figures must be viewed as an underestimate, for many pesticide poisonings are not diagnosed or reported as such. Simmons (1969) suggests that there are 100,000 poisonings and 150 fatalities from pesticides each year in the United States, and Copplestone (1977) cites data from a California study that suggest a mortality rate of 0.1/1 million population (suicides excluded). On the basis of limited reports Copplestone (1977) calculates the world death rate from accidental pesticide poisoning at about 20,000/annum but adds that this figure may be too high, since some suicides may be included in some of the raw data used for the calculation. Of the poisonings that occur, the ratio of organophosphate to organochlorine insecticide involvement is about 30–1, 75% of all cases being suicides (Rathus, 1973).

Mass poisonings have occurred in a number of instances where pesticides have contaminated human food or where treated seed has been used for home cooking. Some of these instances are included in reviews on the subject (Davis et al., 1969; Deichmann, 1973; Rathus, 1973).

Instances of gross poisonings result for the most part from equally gross carelessness or indifference to instructions. The treated seed that caused poisoning in Iraq and Turkey was properly labeled, and those receiving it had been warned not to use it for consumption. Contamination of flour with endrin and a number of instances of parathion contamination of foodstuffs were due to poor food sanitation and complete disregard for toxic compounds. In Mexico a truck that included in its load both parathion and flour was involved in a highway accident. Bags of parathion and flour were broken, but the foodstuffs were not discarded, the truck was swept, and what appeared to be flour was packaged and sold as such. It is no surprise that mass poisoning resulted and many deaths occurred.

Carelessness with pesticides—placing them in unmarked containers, placing them in areas accessible to small children, and using pop bottles to store small quantities of household pesticides—has resulted in many poisonings and fatalities in small children. Even pesticides of relatively low toxicity can cause death when massive amounts are consumed. That this can occur with relatively nontoxic compounds is underscored by an incident in a Binghampton hospital in 1962. Common salt was mistaken for sugar in the preparation of formula for infants; 14 babies were fed the formula and 6 died. In this instance $1\frac{1}{2}$ tablespoons of salt was added to 21 oz of formula (Finberg et al., 1963).

19.2   SUBLETHAL EFFECTS

The extent and degree of sublethal effects are documented poorly. In part this results from the profusion of symptoms that may be expressed and the difficulty of distinguishing pesticide poisonings from other disorders. It results also, in the human population, from a failure to report suspected cases.

Domestic Mammals

Many of the insecticides now in use act on the nervous system in mammals and other animals, and subacute dosages give symptoms reflecting this mode of action. It is known that organochlorine insecticides induce hepatic microsomal enzymes, and since these metabolize steroids, it has been suggested that one effect of chronic exposure might be an effect on reproduction (Kupfer, 1969). Laboratory tests on domestic animals have failed to demonstrate such effects with any consistency, and in their natural environment there is no evidence that pesticides affect these animals in any way. With many pesticides the amounts required to produce effects are massive. For example, Peters and Cook (1973) demonstrated that, at high levels of intake, atrazine could affect reproduction in the rat but calculated that, if the cow exhibited the same level of sensitivity, a dosage in excess of 1 lb would be required before any effect would be noted. Reynolds et al. (1976) fed DDT at 250 ppm and methoxychlor at 250 and 2500 ppm in the diet of sheep but could not detect any effects, adverse or favorable.

Whether ranch mink should be included under domestic or wild animals is a moot point, but there is evidence that these animals are highly sensitive to DDT and/or its metabolites. This evidence is available, not from experiments in which DDT "per se" was incorporated in the diet, but rather from tests in which fish carrying high DDE residues were included in mink rations. Gilbert (1969) collected fish of various species from the Mirimachi River, where contamination with DDT was high, and included the meal from these fish in mink diets. Two experimental diets were involved, one with a DDE content of 0.58 ppm, the other with 0.42 ppm. A control group of mink was maintained on a diet with no measurable DDE content. Mink on the DDE-containing rations had lower erythrocyte and leucocyte counts and lower hemoglobin and hematocrit levels than those on control diets. These lower levels are suggestive of a stress condition. Spleens were larger (+ 16% in males, + 28% in females) and adrenals were heavier (+ 11% in males, + 16% in females) in the animals with DDE included in their diets. Although comparisons were made for only a few females, there were more implantation sites but fewer kits in the treated animals and the ratio of female to male kits was higher. In addition the gestation period was slightly longer (50.4 versus 47.6 days) in animals on the diet containing DDE.

Mink ranchers have experienced poor reproductive success, with kit mortality as high as 80% in mink with coho salmon at 15% included in their diet. Aulerich et al. (1971) investigated this matter and fed mink on diets containing fish from various sources, some with high levels of DDE, some in which the levels were quite low. When mink were fed a diet containing 30% Lake Michigan coho salmon, only 6 of 53 females showed signs of whelping, and of the 10 kits born, 8 were dead at birth. The remaining 2 died within 24 hours. Other species of fish from Lake Michigan caused poor reproductive success, and when canned by-product (heads, tails, fins, viscera) of coho from Lake Michigan were included in the mink diet at 30%, it proved lethal in about 3 months. The coho salmon had a total DDT residue of 18.23 ppm, a DDE residue of 10.12 ppm.

When mink were fed coho salmon from Lake Erie (total DDT 2.76 ppm, DDE 0.83 ppm), the reproductive failure was much less, but there was a high mortality in kits during the first 4 weeks. When west coast coho salmon (DDT residue not indicated) were fed, reproductive performance was good and kit survival excellent. Reproduction was good also when mink were fed ocean whiting (total DDT 0.64 ppm, DDE 0.15) but was very poor when the females were fed bloater chub (total DDT 8.81 ppm, DDE 3.87 ppm). Aulerich et al. (1971) cite earlier work in which DDT was fed to mink from weaning to whelping at levels as high as 150 ppm without adverse effect on reproduction. They infer that the metabolites of DDT may be responsible for the effects noted when fish with high DDE levels are included in mink rations. They suggest that mink may not metabolize DDT readily, a suggestion not inconsistent with residue data in wild mink showing that residues were still high in their tissues 7 to 9 years after the area in which they ranged had been treated with DDT. Whether this residue represented prolonged storage or continuing intake from contaminated food was not determined (Sherburne and Dimond, 1969).

Wild Mammals

The extensive use of rats, mice, and rabbits in laboratory toxicological work with pesticides has resulted in a massive literature on sublethal effects. These include modifications in enzyme production, growth rate, reproduction, general activity level, production of tumors, and teratogenic effects with a large number of pesticides. In general, most of the effects noted require high dosages (see later discussion), at levels not likely to be encountered in their natural environment. There are few reports of pesticide effects on wild mammals in their natural habitats. After more than 20 years of massive spray programs on millions of acres of Canadian forests for spruce budworm control, Varty (1975) reported no effects on populations of mammals in the treated areas. During these years large-scale spraying with DDT, fenitrothion, phosphamidon, and dimethoate had been involved. Other studies similarly showed no adverse effects on mammals (Adams et al., 1949; Stickel, 1951). That some effects on mouse populations in the field have not been reported is somewhat surprising in view of the report that parent mortality was increased and litter size reduced when mice were fed 5 ppm mirex, and feeding DDT at 7 ppm resulted in fewer litters per pair (Ware and Good, 1967). Wallace (1971) found an extremely high incidence of mutants in a colony of wild mice from an area in Peru treated heavily with DDT. An association between the two was inferred but no evidence given.

Murphy and Korschgen (1970) fed dieldrin at 0, 5, and 25 ppm for 3 years to deer. At these levels no intoxication resulted and animals fed normally. Fawns from dieldrin-fed does were smaller at birth and exhibited a higher postpartum death rate than those of control animals and the growth rate was reduced in immature females. Fertility of male progeny was not affected. In the animals dosed at the 25 ppm level, the liver size was increased, the pituitary glands were larger, and the thyroids smaller. The significance of these changes was not determined.

As discussed previously, resistance to pesticides has developed in many insect species, a few fish, and some pathogenic fungi. Resistance has been demonstrated also in mammals. The Norway rat developed resistance to warfarin (Boyle, 1960; Cuthbert, 1963), and Ozburn and Morrison (1962) developed a strain of laboratory mice with an LD_{50} level to DDT 1.7 times that of the parent stock. Baker and Morrison (1966) later reported that the resistance (tolerance) was associated with increased lipid content and storage of DDT in the tolerant strain. In the field Webb and Horsfall (1967) found that pine mice exposed to endrin treatments for 11 years had developed a 12-fold tolerance to this insecticide-rodenticide.

Pesticides can affect mammal populations indirectly by altering their food supply. Keith et al. (1959) reported that populations of pocket gophers were reduced drastically following treatment with 2,4-D. These workers found that the gophers preferred forbs to grasses and that, in an untreated range where the plant composition was 75% forbs and 25% grasses, the gopher diet was about 82% forbs. After 2,4-D had been applied for 2 years, the plant composition was changed to 91% grasses and 9% forbs and the gopher diet changed to about a 50–50 mixture. As a result of this effect on diet the population was drastically reduced. Although there was no evidence of 2,4-D toxicity to the gophers, the population declined by 87% when compared to that in untreated areas.

A somewhat similar situation occurred among mice and voles in Indiana following treatment of a clover field with dimethoate (Barrett and Darnell, 1967).

As discussed earlier, there is little evidence to suggest that pesticides used at normal dosages are grossly detrimental to wild mammals, but there may be an exception in the case of bats. Luckens and Davis (1964) showed that the acute LD_{50} value for DDT was much lower (about four-fold) for the big brown bat than for other mammals. In a study on British bats Jefferies (1972) found higher levels of DDT and metabolites in bats than in insectivorous or carnivorous birds (the territories from which the bats and birds were taken were not strictly comparable). Working with several species of Pipistrelles, he found that the LD_{50} for acute oral poisoning with DDT was in the 45 to 90 mg/kg range and that bats metabolize DDT more slowly than birds do.

He found that wild bats carried organochlorines in their tissues at about one-third the lethal level and, that immediately after hibernation, pesticide levels approached the lethal limits. Others also have suggested that bat declines are due to organochlorine insecticides (Cockrum, 1969; Stebbings, 1970), but the evidence is somewhat conflicting. Clark et al. (1975) studied pesticide residues in the free-tailed bat colony at Bracken Cave, Texas, a colony showing normal population maintenance. They found that mothers with high DDE residues produced young with high residues and that DDE was secreted in the milk, the first lactation that the female gives to her own young having the highest DDE concentration. (Beyond the first lactation these bats fed young indiscriminately in regard to parentage, and thus correlations between DDE residues in a parent and her offspring are soon obscured.) The transfer of pesticide residues from the mother through the milk was reflected in the body burden of the young, and by the time nursing was complete, the young contained levels of DDE comparable to those of older bats. In females residue levels increased rapidly to about 1 year of age and then leveled off. Although there was a tendency for older females to contain slightly higher pesticide levels than younger ones, levels fluctuated widely, much of the annual increment of residue being removed during the period of lactation. Males had higher residues than females and a greater tendency to accumulate residues with increasing age. Clark et al. (1975) suggest that, because maximum DDE levels are reached by the end of the nursing period, their first migration may be critical, since the stress involved uses fat reserves and mobilizes stored DDE.

Clark et al. (1975) compared their residue data in a successful colony with data from a colony in Eagle Creek Cave that had declined in the period 1963–1969. They found that residues of DDE were similar in pregnant females, embryos, lactating females, and fallen young of the two populations. In addition residues of DDT and DDE were higher from guano samples from Bracken Cave than from Eagle Creek Cave. The authors concluded that, because the residues were comparable or higher in the successful Bracken Cave colony, the decline in bat population in the Eagle Creek Cave could not be attributed to DDT.

Clark and Lamont (1976) investigated the big brown bat in Maryland, where it had been reported that a high percentage of females were producing dead or abnormally small young. They collected 27 bats believed pregnant and 26 gave birth; 21 litters contained only living young, and in the 5 other litters, 7 pups were dead. After parturition parents and young were killed and analyzed for pesticides and polychlorinated biphenyls. DDT, PCB, DDE, and oxychlordane were found in all females and PCB and DDE in all litters. There was no difference in DDE levels between dead and living pups, but levels of PCBs were higher in those born dead. No correlation was found between pesticide levels in the pregnant mothers and the birth of living or dead pups, but the one female with the highest level of DDT (2.1 ppm) failed to reproduce. The authors cited evidence from studies on mink showing that levels of PCB in their diet as low as 0.64 ppm almost totally prevented reproduction and resulted in kits' dying within 1 day of birth (Platonow and Karstad, 1973); they suggested that PCB levels might explain the incidence of dead pups produced in the bat colony.

Geluso et al. (1976) reported that bat populations in Carlsbad Caverns, New Mexico, declined from an estimated 8.7 million in 1936 to 200,000 in 1973 and that the population in Eagle Creek Cave (cited earlier), Arizona, dropped from 25 million in 1964 to 600,000 in 1970. They studied DDT and dieldrin residue levels in the Mexican free-tailed bat from the Carlsbad Caverns. They suspected that high pesticide levels attained by the young during nursing might be mobilized during the stress of migration and become toxic levels in the brain. To assess this possibility, they collected young bats during the evening exit flight and compared brain and carcass pesticide levels after starvation with and without a 30 min exercise period daily. Bats that were exercised died sooner and weight loss was greater than in unexercised bats. In this study a reference group was selected for analysis immediately upon their leaving the cage, bats being aged and divided into two groups. Mean brain levels of DDE were 3.7 ppm and 1.3 ppm for the younger and older reference groups, respectively. Bats in the other groups died or were sacrificed after 9 days. In the unexercised group mean brain DDE levels were 47 ppm and 70 ppm for the younger and older group, respectively, while in the exercised group the mean DDE level was 160 ppm in each group. Two bats that died after exhibiting symptoms of DDT poisoning had brain residues of DDE of 260 and 330 ppm. The authors noted that the DDE levels in brains of stressed bats were in the order of the 250 to 600 ppm reported lethal to birds and concluded that DDE has contributed to the decline in bat populations. Their data show not only that young migrants are susceptible but also that the propensity for DDE accumulation in the brain is accentuated in older bats and that this thereby increases the risk to this segment of the population.

Luckens (1973) showed that, with the big brown bat, its tolerance to DDT was related to the time at which DDT was administered.

When DDT is given in the fall when the bat is about to hibernate and metabolic activity is low, the bat tolerates a much greater dosage than in the spring.

Aquatic Mammals

Evidence that pesticides may have sublethal effects on aquatic mammals has been reported for only one species. The California sea lion has been observed giving premature birth, and DeLong et al. (1973) analyzed a small number of cows and premature pups and cows and normal pups collected on San Miguel Island. About 2 months separated the sampling periods for the two groups of cows and pups. Higher levels of DDT (80 to 93% DDE) and PCBs were found in cows and their premature pups than in cows and full-term pups. Levels of DDT were about five times higher than those of PCB in each of the normal and premature parturient groups. The data of DeLong et al. (1973) suggest that DDT might be involved in premature births, but more information is needed. An earlier study (Simpson and Gilmartin, 1970) failed to identify either disease or environmental pollutants as the cause of death in aborted sea lion pups, and DDT was not detected (sensitivity 1.0 ppm). This study noted that dead sea lion pups are not a new finding and cited a 1921 report in which 100 live and 106 dead or dying pups were counted in one count on Ano Neuvo Island.

Human Beings

Sublethal effects in human beings can be viewed in the context of two different types of effects. The first deals with acute toxicity from over-exposure, and the second is concerned with chronic effects that may be slow to manifest any symptoms.

Acute Toxicity

The number of cases of acute toxicity of pesticides to human beings is documented poorly, but what records are available indicate that it is a major health concern. Simmons (1969) estimates that there are 100,000 cases annually in the United States. Copplestone (1977) cites figures for California workers suggesting that the incidence in agricultural workers is about 4/1000 and in the pesticide-manufacturing industry about four to five times this incidence. Working from reports of poisoning statistics from 19 countries, the WHO Expert Committee on the Safe Use of Pesticides (WHO, 1973) estimated that, on a global basis, pesticide poisoning may approach 500,000/annum. Although it was recognized that this figure may be grossly in error, there is no question that the number of poisonings is large.

By far the majority of poisonings are due to organophosphorous insecticides, and as the use of these has increased with the withdrawal of the organochlorines, so too has the number of poisonings. Other pesticides have, however, been involved also. Dieldrin used in public health programs has caused rather extensive illnesses in spraymen who did not follow good safety practices. Symptoms may vary from rather mild effects such as headache and nausea to complete incapacity and convulsions. The symptoms are related to the intensity and duration of exposure and may persist for several months after exposure has been terminated (Deichmann, 1973). Illness after overexposure to endrin, heptachlor, and other chlorinated hydrocarbon insecticides has been reported in applying, mixing, and formulating these products. In addition numerous cases are on record of poisoning by accidental consumption, this route being especially prominent in children. Contaminated food, contaminated flour used in baking, contaminated clothing, and a host of other careless practices have all contributed to the poisonings data (Deichmann, 1973; Davis et al., 1969; Rathus, 1973; West, 1964). With the organochlorine insecticides diagnosis of poisonings is not easy, and recovery is sometimes protracted.

All of the types of poisonings listed for the organochlorine insecticides have occurred also with the organophosphates. In addition the high dermal toxicity of many of the latter has presented a problem of toxicity to workers in fields in which the plant foliage contains a high residue that may be dislodged. The first major problems in this type of exposure were reported from California among orchard workers, 23 cases of multiple poisonings in field workers being reported in the period 1949–1973 (Spear et al., 1975). Ethyl parathion was the insecticide involved most frequently, but other insecticides, including methyl parathion, azinphosmethyl, ethion, TEPP, dioxathion, and naled have caused worker illness (McEwen, 1977). The problem has become so acute that in many jurisdictions "reentry" intervals have been established for a number of pesticides. In the United States a 48 hr interval between the application of pesticide and workers' entering the field has been established for ethyl parathion, methyl parathion, demeton, monocrotophos, carbophenothion, oxydemeton methyl, and dicrotophos, and a 24 hr interval has been set for azinphosmethyl, phosalone, EPN, and ethion (*Federal Register*, 1974).

In many of the incidents reported in California a significant period (30 days or more) had elapsed between the last application of parathion and the occurrences of illness in field workers. This was difficult to explain until it was discovered that, under the hot, dry conditions in California orchards, the parathion was metabolized on the plant foliage to paraoxon and that further decomposition occurred only very slowly (Spear et al., 1975a; Spencer et al., 1975). The crops in which the greatest problems have occurred include citrus, grapes, orchard crops, tobacco, and cotton, all crops with dense foliage and a lot of foliage contact by workers when picking, pruning, or doing other chores in these dense canopies. Under such circumstances workers are exposed daily, and since acetylcholinesterase inhibition can be cumulative, this may be an additional consideration. Studies showed that most of the exposure is dermal and that most of this is through the hands and arms (Quinby et al., 1958; Ware et al., 1973, 1975).

The carbamate insecticides are the most recent, widely used group of insecticides and exhibit a wide range of mammalian toxicities. Illnesses have occurred in their use but not to the extent noted for the organophosphates. A few years ago, when the Bertha armyworm threatened the rapeseed crop in western Canada, methomyl was applied by air and ground for control. Growers in that area were accustomed to using DDT and various herbicides but had little experience with highly toxic compounds. As a result carelessness in mixing and application was common, and though no fatalities occurred, about 50 people required medical treatment. Fortunately with both the organophosphate and carbamate insecticides recovery is fairly rapid and work loss minimal.

In addition to the more serious types of illness from acute toxicity, a number of pesticides cause problems as skin irritants or other minor discomforts. Naled has been reported to cause skin irritation (Mick et al., 1970), and the authors have personal knowledge of mild to severe irritation caused by a wide number of pesticides, including difolatan, lime sulfur, dinitro compounds, and many others.

Chronic Effects

Chronic effects of pesticides on human beings have been the subject of an immense amount of speculation with little documentation. This is dealt with later in the discussion of the significance of pesticide residues in human beings. The pesticide around which most speculation has centered has been DDT. It is known that this compound is present in all human tissues and that it can induce microsomal enzymes in the liver (O'Brien, 1967). Such induction could hasten the metabolism of steroids and theoretically exert an effect on reproduction. In addition it has been shown that the metabolism of drugs can be affected, but at the levels of pesticides in human tissues any such effect should be minimal (Street, 1969). The possibility that DDT might induce tumors, cancer, or teratogenic effects in man has been advanced, but no evidence is available (Committee on Occupational Toxicology, 1970). Extensive studies on men with 11 to 19 years of occupational exposure to DDT in its manufacture could detect no adverse effects, despite the fact that these people had a daily intake estimated at 450 times that of the normal population (Laws et al., 1967).

Similarly studies on workers with regular exposure to parathion, toxaphene, DDT, and dieldrin failed to identify any abnormalities in heme synthesis or activities of the adrenal medulla, though such changes have been reported with some pesticides, other commercial chemicals, and drugs (Embry et al., 1972).

In recent years serious chronic illness has occurred in man as a result of overexposure to leptophos and chlordecone. The former is an organophosphorous insecticide but differs from most insecticides in that group in that it has a delayed neurotoxic effect. The first indication of this resulted from the symptoms observed in several hundred water buffalo in Egypt shortly after leptophos had been applied for insect control in cotton. Since then experimental studies on a wide number of animals have shown that the insecticide damages motor nerves, and if the dosage is high, the animals lose muscular control and are unable to stand. With large dosages paralysis becomes more extensive, and death may result from asphyxiation if the chest muscles become affected. The insecticide acts in a manner similar to tri-ortho-cresylphosphate (TOCP). In the 1930s this chemical caused massive poisoning in the United States among people who drank ginger extracts contaminated with this compound (Smith and Elvove, 1930). The condition known as "ginger paralysis" developed 8 to 14 days after the ginger "booze" was drunk, and as many as 15,000 people were affected. Some died; 6 years later a survey of 300 victims showed that 60 were still hospitalized (Shea, 1974).

Widespread use of leptophos has not occurred in North America, though it was registered for use on some vegetable crops in Canada. Recently these registrations have been canceled. Shea (1974) reported that illness characteristic of neurotoxic poisons was observed in six people and that residues of leptophos were found in their tissues, but it was not until 1976 that reports of illness among workers in a leptophos-manufacturing plant became available and the seriousness of the hazard from this insecticide was recognized publicly. The number of people involved has not been disclosed. Reports indicate that more than 200 people have worked in the production plant and at least 12 are believed affected (Milius, 1976).

Chlordecone is a chlorinated hydrocarbon insecticide used mainly as a bait formulation and known to be extremely persistent in the environment. Its mammalian toxicity by acute oral dosage standards is not high (95 to 140 mg/kg), and though it accumulates in the tissues of wild animals, little concern had been expressed about human safety. In 1975 it was learned, however, that wastes from the manufacture of this compound had contaminated a large portion of the James River and in addition more than 70 plant workers involved in its manufacture had become ill (Metcalf, 1977). Chlordecone poisoning was so severe that in July 1975 the State Health Department ordered the plant closed (Wilbur, 1976).

19.3 RESIDUE LEVELS

We have seen earlier that residues of some pesticides are distributed widely in the environment and that some of these are highly lipophilic. It is thus to be expected that residues would be found in mammals.

Domestic Mammals

Perhaps the best overall indication of pesticide levels in domestic animals can be found in the market basket surveys conducted in the United States and Canada (Chapter 17). These indicate that residues other than DDT and dieldrin are rarely found and that levels of these two insecticides are in the ppb range and declining slightly in recent years. An extensive literature is available on the metabolism of pesticides in several domestic mammals, and this shows that, whereas DDT, dieldrin, and heptachlor are persistent, most other pesticides are removed from the body quite rapidly (Menzie, 1969). For example, Reynolds et al. (1976) fed DDT at 250 ppm and methoxychlor at 250 and 2500 ppm to sheep. DDT tissue levels increased rapidly during the first 8 weeks and then reached a plateau of about 524 ppm in the fat. When the feeding was stopped after 18 weeks, the residue declined, with half of the DDT residue gone in 90 days, half the DDD residue in 26 days, and the DDE residue showing about a 223 day period for half of it to be removed. By contrast, feeding methoxychlor at the same level gave a deposit in the fat of 6.8 ppm at the end of 8 weeks and then a slight decline during the succeeding 10 week period. When intake of the pesticide was discontinued, half of the tissue residue was lost in 10 days.

Many instances could be cited of cases where dairy and beef animals have been fed contaminated feed with resulting high pesticide levels in tissue and milk. These are the exception, and the nature of pesticides is such that residues decline when exposure is terminated. In reviewing the pharmacodynamics of dieldrin, Robinson (1970) drew four conclusions:

1. The tissue concentration of dieldrin is a function of the daily intake.
2. Although the relationship cannot be described by a simple linear function, the tissue level for a given exposure level is a function of the duration of exposure (as we have noted earlier, the plateau level reached after a period of time requires adjustment in the linear function).
3. Tissue concentrations are related to blood concentrations.
4. The tissue concentration declines after the exposure has been terminated.

Robinson's conclusions apply equally well to other persistent organochlorine pesticides.

Wild Mammals

As with other animals residues of persistent organochlorine insecticides are detected in a wide range of wild mammals. The subject is reviewed by Dustman and Stickel (1969) and by Stickel (1973). In most instances the levels detected in big game are in the ppb range. Benson et al. (1974) found total DDT residues in the fat of black bear in Idaho ranged from 0.3 to 3 ppb. This pesticide was detected in all specimens tested, and a trace of dieldrin was found in 3 of 8 bears sampled. In South Dakota residues of DDT were found in most samples of white-tailed deer, mule deer, pronghorns, and elk. In most cases residues were less than 0.2 ppm in fatty tissue, but 1.35 ppm was detected in 1 mule deer. Dieldrin at less than 0.1 ppm was detected in 18 of the 47 animals examined, and similarly low levels of lindane were found in 7 of the animals. Laubscher et al. (1971) found DDT levels of 29 to 107 ppb in fat of mule and white-tail deer in a cotton-growing area in Arizona and found a correlation between high soil residues of DDT and high residues in the deer. Deer sampled in Mississippi had 0.07 to 1.7 ppm DDT (adipose tissue) and from 0 to 0.3 ppm mirex (Baetcke et al., 1972). The highest residues of DDT seem to have been found in tule elk in the San Joaquin Valley, where Keith and Hunt (1966) report 14 ppm in fat. The same report gave residue of DDT in mule deer at 0.39 ppm. Polar bear, arctic fox, and arctic sheep were included in data reported by Clausen et al. (1974). In all three species, DDE levels were about 0.2 ppm. Heptachlor epoxide (about 0.05 ppm) was detected in the polar bear and Arctic fox, and aldrin was reported for all species. The levels varied widely with 0.41, 0.038, and 3.06 ppm, respectively, for the sheep, fox, and polar bear. No explanation can be offered for the high aldrin residue in the polar bear.

Among small mammals Keith and Hunt (1966) reported DDT residues of 0.20 ppm in the liver of cottontail and jack rabbit, 0.13 ppm in the deer mouse (whole), and 0.20 ppm in the muscle of the antelope ground squirrel. Several studies show higher residues in carnivores than in herbivores. Dimond and Sherburne (1969) investigated residue levels in small mammals in a forest that had received varying DDT treatment over a period of 9 years. In the year of treatment DDT residues were 1.06 ppm in mice and voles versus 15.58 ppm in shrews; 8 to 9 years after the treatment, residues in mice and voles were 0.04 ppm, while those in shrews were 1.18 ppm. The ratio of metabolites of DDT between the herbivores and carnivores suggested that exposure in the carnivores was to a partially metabolized source of DDT. In the shrews DDE represented 57% of the residue, while in mice this metabolite comprised 40% and in voles only 18% of the total residue. In the same study area hares contained 0.08 ppm DDT in the year of treatment, while in mink the residue was 8.5 ppm. After 9 years the residue in hares was about 0.02 ppm, while in mink 1.6 ppm were found (Sherburne and Dimond, 1969). Again, the contrast is striking between low residues in the herbivorous hare and high residues in mink. Mink feed largely on crayfish, fish, small mammals, and birds.

Additional evidence on the role of food type on residue levels in mammals is presented by Wolfe and Norment (1973) for residues of mirex. They analyzed tissues of a large number of species from an area in Mississippi aerially treated with mirex bait. Included among the mammals was the cotton rat, rice rat, house mouse, harvest mouse, wood mouse, pine mouse, and short-tailed shrew. Taken as overall groups mirex residues in herbivores was 0.01 ppm; in omnivores, 0.21 ppm; in the carnivore, the short-tailed shrew, the residue was 1.0 ppm.

With the possible exception of mink, there is no evidence to suggest that, at the levels found, residues of organochlorine pesticides in these wild mammals are likely to cause deleterious effects. In mink, however, some of the residues reported in wild animals are higher than those found in studies in which reproduction was believed to be impaired (Gilbert, 1969). In addition the nature of their diet suggests that, in DDT-sprayed areas, they may be exposed to residues in their food comparable to those that Aulerich et al. (1971) demonstrated seriously reduced reproduction.

As discussed earlier, bats are sensitive to DDT, and residues approaching or exceeding those considered lethal have been detected in nature (Jefferies, 1972; Geluso et al., 1976). These residues may be responsible for declining bat populations either through direct mortality experienced during migration or other stress periods or by effects on reproduction. The evidence available suggests such effects, but additional information is needed to establish a cause-effect relationship (Clark et al., 1975). It is clear that pesticide residues detected in insects (Baetcke et al., 1972) are such that bats, depending heavily on them for their food supply, have a significant pesticide exposure.

Aquatic Mammals

Some of the highest DDE residues found in any mammal have been found in aquatic mammals. This is not surprising, since many have high lipid content and thus provide an attraction for lipophilic pesticides. In addition their diets may include a large proportion of fish. Primarily, the residues found have been of DDT and its metabolites, dieldrin and mercury. PCBs are present in many cases as well. Residues reported in the literature have been tabulated by Stickel (1973), only a few being cited here. Sladen et al. (1966) found a low level of DDT (39 ppb) in the fat of a crabeater seal from the Antarctic, a finding that stimulated speculation about how the pesticide reached that part of the globe. Holden and Marsden (1967) similarly found pesticide residues in seals and porpoises far from the site of application. Residues were predominantly derived from DDT, but dieldrin was found as well. One porpoise had a combined organochlorine residue of 73.3 ppm. In harbor seals from the California coast Shaw (1971) found DDT residues of 18 to 158 ppm in the blubber. Anas (1974) found large differences in residues in harbor seals from different areas along the western coast of North America. In these analyses DDT and PCBs were not separated. Combined residues in fat were 381 to 2350 ppm in those from San Miguel Island, 459 and 1620 ppm in two seals from Puget Sound, 28 to 110 ppm in three Columbia River seals, and 7 to 28 ppm in three seals taken from Pribilof Islands. In one seal in which DDT and PCBs were analyzed separately, the residues of DDT, DDD, DDE, and PCB were respectively 2110, 4, 96, and 572 ppm. Frank et al. (1973) and Jones et al. (1976) analyzed tissues from a large number of harp seals taken off the eastern Canadian coast. Two populations were involved: one taken from the ice east of Labrador and Newfoundland (Front) and the other from the ice fields around the Magdalen Islands in the Gulf of St. Lawrence (Gulf). Lower residues were found in the blubber of seals from the Front than from the Gulf, the difference being most pronounced among the young. Although DDT and its metabolites were the dominant pesticide residues, dieldrin was found also but at much lower levels. Jones et al. (1976) showed that DDT crossed the placental barrier, since unfed pups contained residues. It is apparent that seals carry quite high pesticide levels when taken in coastal areas of North America. By contrast Clausen et al. (1974) found DDE residues in Arctic mammals to be 0.47, 0.15, and 0.29 ppm, respectively, in the bearded seal, ringed seal, and the hooded seal. In each of the three species aldrin was found also at a level about one order of magnitude less than DDE.

In addition pesticide residues have been found in porpoise, sea lions, and whales. Porpoise from the Arctic contained DDE at 0.32 ppm and aldrin and lindane at about one-tenth this amount. Harbor porpoise from the Bay of Fundy area taken in 1969–1970 contained DDT residues averaging 307 ppm in the blubber of males and ranging as high as 520 ppm in one specimen. Dieldrin was found at levels as high as 13.1 ppm. Levels of both DDT and dieldrin were higher in males than in pregnant and lactating females (Gaskin et al., 1971). DeLong et al. (1973) found DDT in both dams and pups on San Miguel Island in 1970 with residues in blubber as high as 1039 ppm (mean 824 ppm) in females giving premature birth to pups and as high as 203 ppm (mean 103 ppm) in females carrying their pups to full term. A possible cause-effect relationship was discussed by these authors.

Wolman and Wilson (1970) found low levels of DDT in 6 of 23 gray whales collected near San Francisco. Levels ranged as high as 0.36 ppm in blubber. These whales feed during summer in the Bering and Chukchi Seas, their principal food being benthic amphipods, isopods, mysids, molluscs, polychaetes, and hydroids. They migrate to winter calving grounds near Baja California and eat little during their migration or stay on the winter grounds. Pesticide residues were detected in the 4 gray whales sampled in the spring migrants (northbound) in 1968 but in only 1 of 9 in the spring of 1969. Among southbound migrants 1 of 10 had DDT in its tissues, none had dieldrin. Dieldrin was found (0.044 to 0.075 ppm) in the 4 spring migrants of 1968. The gray whale with 0.36 ppm

DDE was estimated to contain a total body burden of about 1.5 g or this compound. DDT and metabolites were found in all of 6 sperm whales examined, levels ranging from 0.010 to 6.0 ppm. Dieldrin (0.016 and 0.019 ppm) was found in 2 of the animals. The sperm whale feeds on ocean fishes, and the higher pesticide levels in contrast to those of the gray whale are not unexpected. Another fish eater, the California sea otter, was found to carry DDT residues of 0.41 to 36 ppm in the fat.

19.4 HUMAN BEINGS

The ubiquitous nature of DDT extends to human beings, all surveys indicating a significant deposit in our tissues regardless of age, color, or domicile. A number of reviews (West, 1966; Wassermann et al., 1974; Durham, 1969; Davies, 1973; and others) have tabulated residue levels in adipose tissues of varying populations, and though DDT-derived pesticides are dominant, dieldrin and BHC are also encountered but at much lower levels.

The highest levels reported for a general population are from a study in Texas, 1969–1972, in which 221 adipose tissues from surgery cases had a mean DDT residue of 23.18 ppm, 17.37 ppm of which was DDE. Dieldrin residues had a mean of 0.35 ppm and BHC 1.29 ppm. In this study there was no difference in residue levels between males and females, but Mexican Americans had higher residues than white Americans (Burns, 1974). The mean DDT residue level found in this study is much higher than the level determined by most other studies. The mean residue in the United States as determined by seven studies tabulated by West (1966) and conducted during the period 1950–1963 was approximately 9.0 ppm. Not included in this mean is a study of meat abstainers, with a residue level of 5.5 ppm, and one of Alaskan eskimos, with a mean residue of 2.8 ppm. A study in 1959–1960 indicated a mean residue of DDT (plus DDE) in the Canadian public of 4.9 ppm (Read and McKinley, 1961).

A number of studies indicate differences in residue levels between males and females. In a study of residents of four U.S. cities Zavon et al. (1965) found total DDT residues in adipose tissue of males and females to be 8.12 ppm and 5.90 ppm, respectively. In this same group dieldrin residues in males were 0.31 ppm and in females 0.24. Heptachlor epoxide levels were 0.11 ppm in males and 0.09 in females. Davies et al. (1973) noted that most studies in the United States indicated a higher body burden of pesticides in black than in white Americans. In a study among Dade County residents with no occupational exposure to pesticides they found that the residue level of DDT increased with age, the increase being most pronounced in the first two decades and greatest in blacks. Regardless of social class, blacks contained higher residues than whites, and differences between male and female, though suggesting higher levels in the former, were not significant.

It is known that pesticides cross the placental barrier and are present in the neonate at levels normally lower than those of the parent (Zavon et al., 1969). Curley et al. (1969) found, however, mean total DDT levels of 8.58 ppm in adipose tissue of stillborn infants, a figure comparable to that of adults. In a separate study of umbilical cord blood of 30 normal babies total DDT residues averaged 0.0136 ppm, again a level consistent with most values reported for adults. By contrast Polishuk et al. (1970) found a mean blood level of 0.0183 ppm in pregnant women and 0.0081 in blood of the fetus. In this latter study there was some evidence that pregnant women metabolized DDT more rapidly than nonpregnant women of the same general age. Adipose levels of DDT-derived materials were about the same in each group (slightly higher, 13.74 vs 12.63 ppm, in nonpregnant women), but whereas nonpregnant women had 3.4 ppm of DDT, pregnant women had 1.6. Reflecting this was the fact that DDE levels in pregnant women were 10.6 ppm, versus 8.3 ppm in those who were not pregnant. In a study of 307 adipose tissues from Israelis, Wassermann et al. (1974a) found a mean DDT (and metabolites) residue of 0.7 ppm in 44 stillborn analyzed in a population in which the residue in the 25 to 44 year group was 14.4 ppm. BHC occurred in the stillborn at 0.04 ppm, dieldrin at 0.02 ppm, and traces of heptachlor epoxide were present.

Different levels of exposure to DDT are reflected in population studies from a number of countries and from studies of different occupational groups. Wassermann et al. (1974a) reported residues of DDT-derived pesticides in human tissue in different countries as 2.9 ppm in Uganda, 4.5 ppm in Kenya, 6.5 ppm in Nigeria, 7.8 ppm in Brazil, 8.5 ppm in South Africa (whites), 13.0 ppm in Thailand, and 14.4 ppm in Israel. In the United States residue levels in man are higher in the south (see Durham, 1969), reflecting apparently the greater abundance of insects in the warmer climate and thus greater insecticide use than in more northern areas.

A number of studies have been done on workers occupationally exposed to insecticides. Perhaps the highest levels were reported by Laws et al. (1967), who studied workers involved in the manufacture of DDT; 35 men with 11 to 19 years of working in pesticide production had DDT residues ranging from 38 to 647 ppm in the fat. Levels in fat were correlated with levels in the blood but were higher in the fat by a factor of 338.

No discussion of pesticide residues in human beings would be complete without mention of residues in mothers' milk. The first study on this subject was reported by Laws et al. in 1951. They analyzed milk from women in Washington, D.C., with no known pesticide exposure and found DDT residues in 30 of 32. Levels ranged as high as 0.12 ppm on a whole milk basis. West (1964) reported on a limited study in California. In six individuals DDT ranged from 0 to 0.12 ppm and DDE from 0 to 0.25 ppm. More recently Hagyard et al. (1973) analyzed milk from six lactating mothers in Arizona. Milk was taken at various intervals postpartum and found to contain total DDT residues of 0.093 to 0.510 ppm. The lowest residues were found within a few days postpartum and the highest after 5 to 7 weeks. The number of samples in this study was low, and in many cases differences between samples were very large. Prior to day 22 postpartum the residues were dominantly DDE, and this was true also for samples taken after 60 days. Between these times, however, the pattern reversed, the total DDT residues being more than 50% DDT. Savage et al. (1973) analyzed 40 samples of milk from mothers in rural Colorado in 1971–1972. DDT and DDE were found in all samples. DDT residues ranged from 0.007 to 0.109 ppm and DDE residues from 0.019 to 0.386 ppm. Dieldrin at levels of 0 to 0.011 ppm was present in 63% of the samples. BHC was present in 87.5% of the samples in levels of 0 to 38 ppb. Measurable amounts of heptachlor epoxide were found in a few samples. In a study in Nova Scotia and New Brunswick, Musial et al. (1974) found mean DDE residues of 0.035 ppm in New Brunswick mothers and 0.017 ppm in those from Nova Scotia. DDT levels were 0.013 ppm and 0.021 ppm, respectively. These are much lower than the mean reported in Colorado of 0.089 ppm DDE and 0.021 ppm DDT. In both the Colorado and Canadian studies the ratio of DDT to DDE residues in the milk was approximately 1 to 3. Studies in other countries have indicated DDT and metabolites are regularly found. Data from several studies are tabulated by Musial et al. (1974).

Significance of Residues

Two excellent reviews on this subject were presented at the Eighth Inter-American Conference on Toxicology and Occupational Medicine, Miami, Florida, in 1973. Durham and Williams (1973) discussed the mutagenic, teratogenic, and carcinogenic properties of pesticides, and Deichmann (1973) presented a more general review of chronic toxicity of organochlorine insecticides. The reader is referred to these reviews for a comprehensive treatment.

Medical Problems Reported

Various authors have attempted to give pesticides a causal role in human diseases such as leukemia, Hodgkin's disease, agranulocytosis, aplastic anemia, allergies, and asthma. While Rathus (1973) concludes that "careful analyses have not shown any relationship to the epidemiological problem," it must be accepted that the limited epidemiological studies reported to date neither incriminate nor exonerate pesticides. Leptophos and chlordecone have caused debilitating diseases in workers occupationally exposed, and illness among plant workers was common in the early days of production of aldrin, dieldrin, and endrin. In the last case the problem was resolved by improved plant sanitation (Deichmann, 1973). It is a basic toxicological tenet that effects are dose dependent and thus should be manifest first in those people with high intake levels. Numerous studies on groups occupationally exposed to pesticides are reassuring in that, with the exception of the cases cited, no adverse effects have been demonstrated. It must be pointed out, however, that most of these studies have involved only the organochlorine insecticides. This is understandable, since these are the oldest group among our mass-produced modern pesticides, it is these that persist in human and other tissues, and, because of this persistence, it is these that have aroused the greatest concern.

Only a limited number of studies involving effects of pesticides have been conducted directly on man, and thus we are left, dominantly, with extrapolating from studies on laboratory animals, mostly the mouse and the rat. Whether or not such extrapolations are adequate and valid is a moot question. It can be said that interpretations based on these animals have worked well as guidelines for man in terms of direct toxic effects. With human ailments and chronic diseases, however, this may not be the case. Diseases tend to be host specific, and the significance of findings or lack of findings in laboratory animals with respect to these may have little relevance. Findings in laboratory animals are, however, suggestive, and where pathological conditions are detected, the relevancy of these to man must be explored.

It is known that the organochlorine insecticides increase the production of hepatic microsomal enzymes (see Triolo and Coon, 1966), and since these are involved in steroid metabolism, an effect on reproduction might follow. Evidence on some species of birds suggests this has occurred, but whether the observed effect is due to the induction of these enzymes or not is still debated (Chapter 18). A large number of studies on mice and rats have failed to demonstrate effects except at very high dosages (Lundberg and Kihlström, 1973; Thomas and Lloyd, 1973; Krause et al., 1975), and it is unlikely that, at residue levels now present in man, any effects are taking place.

Enhancement of microsomal enzymes affects the metabolism of a number of xenobiotics, including drugs and pesticides. This phenomenon was discovered accidentally when rats housed in a room sprayed with chlordane were noted to exhibit very short periods of anesthesia when treated with barbiturates. Street (1969) reviews this topic, pointing out that effects have been demonstrated in rats with DDT at dosages as low as 5 ppm. The implications are significant in terms of medical use of drugs and the frequency of administration. There is little evidence to suggest that, at tissue levels found in the general population, any significant effect is likely. The ability of DDT to increase liver enzyme production has been used medicinally where such an increase is required. Thompson et al. (1969) used DDT at 1.5 mg/kg daily to treat a boy suffering from unconjugated jaundice. No side effects were noted during 6 months of treatment, and plasma bilirubin remained low for at least 7 months after the medication had ceased. The authors attributed this to the persistence of the chemical in the system. Hazeltine (1971) noted that excess bilirubin is most common in newborn infants and that an excess of this compound is an important cause of neonatal jaundice. The condition results when the production of liver enzymes that metabolize bilirubin is delayed; the disease can be alleviated by treatments such as phenobarbitol or DDT that induce microsomal enzymes, though these may act competitively if present at the same time. Milby et al. (1969) found that, during the last quarter of the year, when pesticides, especially defoliants, were being used extensively, there was an increase in hyperbilirubinemia among formula-fed infants in an agricultural area of southern California but that this did not apply to breast-fed infants. Hazeltine (1971) suggests that the differences noted might be attributable to a beneficial effect of DDT from mothers' milk, where the concentrations are much higher than those present in the cows' milk and other ingredients of the mixed formulae.

A correlation between the exposure of workers to pesticides, increased blood pressure, cholesterol levels, and hypertension has been reported in several studies. The correlations were found in both white and nonwhite Americans, being more pronounced in the nonwhite group. High systolic and diastolic blood pressures and hypertension were correlated with high serum levels of DDT and DDE, but whether these were causally related could not be determined, since the workers had exposure to a wide range of pesticides other than DDT. Although high exposure to pesticides, especially the organochlorine insecticides, and increased cardiac risk (by criteria now in vogue) have been identified, their significance has not. As pointed out earlier, general health, morbidity, and mortality studies in workers with intense exposure failed to determine differences between these and the population at large (Sandifer et al., 1972; Deichmann, 1973; Keil et al., 1973; Deichmann and Radomski, 1968).

Pesticides and Cancer

Perhaps no area in the pesticide controversy has generated more concern in the public and more disagreement among scientists than that dealing with the role of organochlorine insecticides in human cancer. The controversy derives from three areas.

1. Inconsistencies in laboratory tests with mice, rats, and other animals regarding the production of tumors in the liver.
2. The interpretation of tumors, that is, whether they are benign or malignant, and, in fact, what constitutes a tumor.
3. Whether tumor production in mice, or for that matter, the lack of it, has any significance to man.

It is not our intention to debate any of these points but rather to point out that the issue is extremely cloudy. For example, Tomatis et al. (1972) fed DDT at 2, 10, 50, and 250 ppm to CF mice during their entire life for two generations. They found an increase in liver tumors in males that was dose related, and in females there were a slight increase in those fed 10 and 50 ppm and a pronounced increase in those fed 250 ppm. They reported that tumors appeared in the dosed animals earlier than they did in controls and that death occurred earlier in those mice with tumors. Four liver tumors gave metastases. Continuing their study for six generations (Turusov et al., 1973), these workers found that, while male controls had an incidence of tumors of 29.9%, those fed 2 and 10 ppm DDT had an incidence of 50 to 56% and that, when the dosage rate was 250 ppm, 86% of the males developed liver tumors. Data for females showed 4.7% in controls, 9% at 10 ppm, 13% at 50 ppm, and 65.5% at 250 ppm. Feeding at 2 ppm did not increase tumor incidence. They found the life span reduced in tumorous animals, but there was no progressive increase in tumor production from generation to generation. One tumor (hepatoma) in the control group metastasized compared to 13 among the four groups dosed with DDT. In contrast to these reports Terracini et al. (1973) found that, in two colonies of BALB/c mice, no increase in incidence of liver tumors was obtained by feeding DDT at 2 or 20 ppm, but an increase was reported when the dosage was 250 ppm. Interestingly, when old mice were fed 250 ppm, the incidence of liver tumors was reduced. In reviewing the literature on the subject Tomatis et al. (1972) cite five reports dealing with DDT and tumor production. One indicates increased incidence in rats, three indicate increased incidence in mice, and one indicates increased production of tumors in rainbow trout. They report also that in "a few reports in rats," one in Rhesus monkey and one in Syrian golden hampster, no increase in the incidence of tumors was noted in DDT-feeding studies.

The situation is no more clearly defined with dieldrin. Van Raalte (1973) cites three reports in which dieldrin was found to cause liver tumors in mice, three reports where it did not cause tumors in the rat, and additional reports that dieldrin did not produce tumors in chickens, pigeons, monkey, or dog.

A screening study involving 130 pesticides done by the Bionetics Research Laboratories detected an increase in tumor incidence in a tumor-susceptible hybrid strain of mice (Innes et al., 1969) with several pesticides.

If cancer induction from pesticides in human beings was occurring, this should be detectable in epidemiological studies in areas of high pesticide use. Such is not the case (Durham and Williams, 1973). Deichmann (1972) points out that, in man, cancer of the liver, the only kind induced in mice, is uncommon in industrialized countries, despite the fact that modern pesticides have been used widely for 25 years. In addition he points out that spontaneous liver tumors are uncommon in man and the rat and questions whether, because of this, the rat may not be a better indicator species than the mouse, since in the latter species spontaneous liver tumors are common. Van Raalte (1973) reports that the kind of nodularity seen in the mouse and interpreted as tumors is of a type "something halfway between hyperplasia and neoplasia," something unknown in man.

There is much controversy concerning what constitutes a malignant tumor. Earlier interpretations were based on the concept that tumors where remission results when the tumorgen is withdrawn are not malignant and are of little significance. Recently an official interpretation refers to "oncogenic" effects and states that a distinction between benign and malignant tumors in laboratory animals is not meaningful in determining the hazard of cancer in man (*Federal Register*, 1975b). The subject is discussed by a number of workers, including Weil (1976). It is apparent that the gross distinction that toxicologists and pathologists make between benign tumors and malignant or cancerous ones is not accepted by the legal authorities responsible for the regulation of pesticides.

The inconsistencies in laboratory findings and lack of agreement among "experts" on the interpretation of such findings make conclusions difficult. Even more difficult is determining the significance of findings in laboratory animals in terms of human cancer. The limited epidemiological studies that have been done give no hint of pesticide involvement. Studies on pesticide levels in human cancer patients have shed little light on the subject.

In several studies tissues from autopsy or blood sera have been analyzed, and in some cases (see Deichmann, 1973) levels of DDT-derived residues have been abnormally high in terminal cancer patients. Casarett et al. (1968), for example, found among 44 autopsies that the 5 with highest organochlorine pesticide levels had in common the fact that they were all emaciated, they all died with carcinoma, and they all had abnormalities of the liver. Hesselberg and Scherr (1974) compared blood levels of DDE and PCBs in 9 patients (8 with carcinoma) and in a control group of 15 normally healthy individuals. DDE levels in the patients had a mean of 21 and a range of 10 to 41 ppm, while in the control group the mean level of DDE was 3 ppm with a range of 2 to 7 ppm. Thus high levels of organochlorine pesticides are found in terminal cancer cases, but they are found likewise in other diseases where the body fat stores are drastically reduced. The mobilization of DDT during fat utilization is well documented, the evidence suggesting that high DDT-derived pesticide levels in cachectic patients result from the total body burden of the pesticide's being compressed into a small lipid mass rather than from a causal relationship to disease onset (Deichmann, 1973; Casarett et al., 1968; Dale et al., 1962). The American Medical Association's Committee on Occupational Toxicology concluded that a link between DDT and carcinogenesis in man remained to be established (Committee, 1970).

It is of interest to note that o,p'DDD has been used for the treatment of adrenal cancer. Up to 10 g/day has been administered for extended periods and remission obtained in some cases (Ottoboni, 1969).

Mutagenic Effects of Pesticides

Equally as complicated as the carcinogenic studies are those dealing with mutagenic effects of pesticides. Reviewing this subject in 1973 Durham and Williams (1973) cite more than 500 reports on the subject, and a book on this topic is available (Epstein and Legator, 1971). The testing system is variable and may involve the use of bacterial systems, *Neurospora*, other plants, *Drosophila*, cell tissue cultures, or reproductive studies in mammals. In one or more of these tests a number of pesticides have shown mutagenic activity. Epstein and Shafner (1968) used a "mutagenic index" (M.I.) to express the relative mutagenic potential of various compounds. The index reflects dominant lethal mutations expressed as total early and late deaths in the embryo divided by the total implantations and multiplied by 100. Using this index and mice as the test animal, they reported effects with DDT (M.I 3) but none with captan. Mutagenic indices for the chemosterilants Metepa, Tepa, and Thiotepa were 27, 38, and 18, respectively. Wallace et al. (1976) concluded from feeding DDT to mice at 250 ppm for five generations that there was no case for gross mutagenic effects by DDT. Fahmy and Fahmy (1969) presented a worrisome picture with regards to genetic effects from pesticides, but the significance to man of the studies they reported is unclear. In addition, at the conference at which this paper was presented (see Kraybill, 1969), there was considerable debate, not only on the relevancy of these findings to man, but also on whether or not all mutations are disadvantageous. Needless to say, the controversy has not been resolved.

Teratogenic Effects of Pesticides

Interest in teratogenicity in pesticides developed after gross deformities in infants were associated with the use of thalidomide to reduce morning sickness in pregnant women. Abnormalities included reduction and other deformities in the limbs and internal abnormalities such as aplasia of the gall bladder, appendix, and kidney; defects in the lung and heart; and abnormalities of the digestive system and anus. The nature of the abnormalities was dependent on the time at which thalidomide was taken, the interval from days 34 to 50 after the last menstrual period being critical (Kajii et al., 1973). The Minamata disease caused by mercury poisoning produced malformed children from affected parents, and deformed babies were reported associated with extensive use of defoliants in Vietnam. In the United States and

Canada the issue was raised when research by the Bionetics Research Laboratories indicated teratogenic effects in mice and rats treated with 2,4,5-T (Courtney et al., 1970). This report aroused much concern, since this and other phenoxyherbicides are used widely. A curtailment in use of 2,4,5-T was instituted immediately. The curtailment took the form of use restrictions limiting the application of this herbicide to areas unlikely to be frequented by pregnant women. Subsequent research determined that the 2,4,5-T used in the Bionetics study contained 27 ppm of a contaminant 2,3,7,8-tetrachlorodibenzo-p-dioxin (TCDD). A large number of studies initiated after this information became available demonstrated that TCDD was extremely embryotoxic and teratogenic at dosages as low as 0.001 ppb and that it was this contaminant and not the herbicide per se that was responsible for the effects reported earlier.

The subject of pesticides as teratogens has been reviewed by Clegg and Khera (1973); their report includes the effects of ethylenethiourea, a contaminant and breakdown product of many of the ethylene-bis-dithiocarbamate fungicides. Experiments by Khera (1973) indicated that, when ethylenethiourea was fed at 10 mg/kg or more to the rat either immediately prior to or during gestation, a variety of malformations occurred. At 5 mg/kg ossification of parietal bones of the skull was delayed. In the rabbit no malformations were seen.

In addition to mercury, TCDD, and ethylenethiourea, a number of other pesticides (or their derivatives or contaminants) have been reported teratogenic in some species. These include diazinon, carbaryl, paraquat, captan, folpet, and difolatan, but in most cases the dosage required to produce an effect was very high and beyond the expectation of exposure in normal use.

Mothers' Milk and Infants

One of the concerns expressed about pesticide residues in the human body has been that breast-fed babies would imbibe an excessive amount of residue via the mothers' milk. Deichmann (1973) pointed out that mothers' milk contained more DDT than cows' milk, since, during the period DDT was freely available, mothers had greater exposure through their diet and through household and garden use of the pesticide. Basing his calculations on a residue level in milk of 150 ppb and a consumption during the early period of infancy of 500 ml/day, he determined that a 3.5 kg infant would consume DDT-derived residues at the rate of approximately 20 μg/kg per day. Deichmann (1973) pointed out that such an intake level was twofold the ADI accepted by

FAO/WHO and that, after 1969, it would be fourfold the dosage, since in that year the ADI was reduced to 5 μg/kg. The intake of dieldrin in breast-fed babies in the United States in the mid-1960s was calculated at 10 times the ADI. Whether or not such an intake represents a hazard is unknown. We were unable to find any evidence in the literature of illness in babies due to intake of organochlorine insecticides from mothers' milk. It must be recalled that ADI values are based on safe levels of intake for a life span approaching 100 years. The relevance of this figure to intake for a few weeks or months is probably remote. One would expect that, if the intake of pesticides in infants was disproportionately high, this would be reflected in tissue residues at an early age. In a study on organochlorine residues of children in the Argentine Republic who died accidentally, Astolfi et al. (1973) found total DDT residue in three newborn ranged from 1.34 to 2.44 ppm; in seven aged 1 to 6 months the range was 1.03 to 4.28 and in four 1 year olds the range was 1.69 to 7.34 ppm. Engst et al. (1969) analyzed residues in 34 children 0 to 3 years old and concluded that residues at birth diminished during the first few months. In neither study was it stated whether or not the babies were breast fed, but we could find no indication in the literature that residues were higher in nursing infants.

19.5 SUMMARY

In contrast to the situation with some species of birds and fish, there is little indication that pesticides have had significant adverse effects on mammals. There are two possible exceptions to this. In both the southern United States and Great Britain bat populations have declined. It is known that bats, especially in the spring, are much more sensitive to DDT than other mammals are, and near lethal levels have been detected in dead bats. Although the case has not been documented as thoroughly as one would wish for a definitive conclusion, the circumstantial evidence is strong and suggests that the decline, in at least some populations, is due to high levels of DDE consumed in their insect diet. The second case involves mink, and again this animal shows a low tolerance to DDT. This sensitivity is expressed as direct mortality but, more importantly, as reproductive failure. Only limited studies have been done on this species, and they have dealt with ranch mink and not those in their natural habitat. These studies suggest that DDT per se is not the culprit but rather the metabolite DDE.

Residues of organochlorine pesticides are found in all wild mammals whether they be terrestrial or aquatic. In general these residues reflect feeding habits, with levels in herbivores much lower than those in carnivores inhabiting the same area. Especially high residues are found in a number of sea mammals, especially those that feed on fish in coastal areas, but there is little evidence that the levels found are causing any distress to the animals.

The human population in industrialized countries has been exposed to a wide range of pesticides for more than 25 years. As a result residues of about 8.0 ppm of DDT-derived compounds are found in the fat of Americans. Levels tend to be slightly higher in southern areas, where pesticide use has been more extensive. Thus levels in tissues of Canadians are slightly less–in the order of 5.0 ppm in the fat. Residues of dieldrin, heptachlor epoxide, and lindane may be found also but at much lower levels. DDT and other organochlorine insecticides, PCBs, and mercury occur also in human blood; they cross the placenta and are present in the neonate. Levels of DDT-derived residues are lower (by about 75%) in the fetus and newborn than in the mother, and though DDT is metabolized very slowly in human beings, pregnant mothers appear to be more efficient in this role and have lower residues than nonpregnant women do. DDT (and metabolites) is secreted in mothers' milk at a level of about 100 ppb, and breast-fed babies during early infancy receive a daily intake of this pesticide at levels greater than the accepted ADI. There is no evidence of any detrimental effect to such infants, and possible beneficial effects have been postulated.

There is no evidence that any of the widely used pesticides are carcinogenic, teratogenic, or mutagenic in man. Some have been proved as such in some laboratory animals, especially when administered in large dosages and/or by routes (such as injection) not characteristic of human exposure. Responses differ in different laboratory mammals, and to date there is no way to assess the significance of these findings in terms of human risk factors. Limited epidemiological studies have not found increased incidence of any disease, morbidity, or mortality in human beings associated with increased use of pesticides.

20

PESTICIDES IN THE BIOSPHERE

Throughout the preceding chapters we have attempted to discuss how pesticides move in various parts of the environment and what the significance of this is to specific life forms in these habitats. While this compartmentalization assists us in dealing with this vast subject, we must realize that in nature these compartments overlap and that the pervasive movement of some pesticides transcends any boundaries we may arbitrarily erect. As a summary for our section dealing with pesticides and their significance to life perhaps we can remove the divisions and think in terms of the biosphere, that portion of the land, air, and water in which life exists.

Most pesticides are transient. They are applied locally and are not sufficiently stable to be transported, in an active form, very far from their site of application. For these the effects are local, and while they may drastically alter the species composition in the immediate area, these effects are short term, and nature's ability to repopulate the area by immigration compensates quickly for the temporary upset. From the point of view of the agriculturalist or forester this repopulation is all too rapid. Thus the corn borer, spruce budworm, late blight fungus, pigweed, and the host of other pests for which the pesticides were used are back the following year in about the same numbers and require that the control practices be instituted again.

A few pesticides, and DDT is the best example, persist in the environment, and their local applications result in much wider distribution. As we have seen with DDT, its direct application to only a small percentage of the North American continent has resulted in residues in practically every surface soil and waterway, in the air around us, and in all life forms (Wurster, 1969). In most cases residues are low and could not be detected without extremely sensitive analytical techniques. Moreover, in most cases the levels are of no biological significance to most of the organisms in which they are found, but they move insidiously within the biomass and its physical environs, concentrating in the biomass because of its lipoidal and adsorptive properties. Although its use has been restricted in most industrialized countries and curtailed in others (see Edwards, 1973), DDT and its metabolites will be recycled through the biosphere for many years, degrading somewhat but at a slow rate during each recycling.

DDT and some closely related compounds had an advantage for pest control in that applications were needed less frequently during the season. However, with the exception of soil insects, the effectiveness was for only one generation of the pest, and the next year brought a return of the problem and a need again for treatment. It was not that the pesticide had disappeared as the nonpersistent compounds do. DDT was there, but it was in the soil, the forest litter, the stream bottoms, or the atmosphere and not on the new plant surfaces, where the new insect population would develop. In addition it was present, usually in low amounts, in the animal and plant life. At first its distribution was local, but as it moved on soil particles, in water, in biological tissues, vaporized, dispersed, and redeposited in every imaginable way, its wide distribution was assured. Thirty years and millions of pounds later, its distribution still reflects to a high degree its intensity of use, but it can be found literally anywhere in the world.

Kenaga (1972) presents an excellent review of the literature on the distribution and bioconcentration of pesticides in the environment and summarizes the available data estimating the relative amounts of DDT-derived compounds in various components. Some of that information is presented in Table 20.1.

The student should recognize that these are "ball park" estimates; one should refer to the original publications if one wishes to make critical comparisons. The figures demonstrate, however, the pervasiveness of DDT in the environment, a subject reviewed in detail by Bevenue (1976).

The attractiveness of biological tissues, especially those containing fat, to DDT was discovered early, and it was not surprising to discover that DDT "biomagnified" in the environment, higher residues being found in some life forms than in others. The concept developed that because of its persistence and other characteristics DDT would accumulate in living organisms in relation to the trophic level of the organism involved. This theory was easy to appreciate. It was based on the knowledge that biomass decreased by a factor of about 10 from one trophic level to the next

Table 20.1   Residues of DDT (and Metabolites) in
Various Segments of the Environment[a]

Segment	Estimate of General Level (ppm)
Air	0.000004
Rain	0.0002
Dust	0.04
Fresh water	0.00001
Seawater	0.000001
Plankton	0.0003
Aquatic invertebrates	0.001
Marine fish	0.5
Freshwater fish	2.0
Predatory birds	10.0
Natural soils	?
Agricultural soils	2.0

[a]From Kenaga, 1972.

higher one, but since DDT had an affinity for lipids, it would pass in
large measure along the food chain, increasing by a factor approaching
10 from food to consumer of that food (Street, 1969a). Numerous examples were available to support the theory. In the Dutch elm disease
control program DDT was applied as a 6% suspension in water at a
dosage of about 1.1 lb active/tree. Immediately after treatment elm
leaves had residues of DDT + DDE of 183 to 283 ppm. The soil under
the tree had about 1 to 18 ppm (higher near the surface), and when the
leaves fell in the autumn, these contained 20 to 30 ppm. Earthworms
in the soil under the trees and feeding on the fallen leaves had residues of about 120 ppm, and robins dying from eating these earthworms had residues as high as 342 ppm in the brain (Barker, 1958).
Studies of Clear Lake following the use of DDD showed pesticide
levels in plankton at 5.3 ppm, and levels in the fatty tissue in bluegills
were 125 to 250 ppm; in bullhead, 342 to 2700 ppm; in grebes, as
much as 1600 ppm; in largemouth bass, 1550 to 1700 ppm; and in
whitefish, 80 to 2375 ppm. Whereas levels were lower in flesh, the
data represent higher residues at higher trophic levels (Cottam, 1965).

Woodwell et al. (1967) published what they referred to as an "especially clear example" of "biological magnification" in a Long Island
marsh. Residues ranged from 0.040 ppm in plankton to 75.5 ppm in an
immature ring-billed gull, that is, an increase of three orders of magnitude to the upper trophic level gull. Markin et al. (1974) found that

in aquatic and terrestrial invertebrates levels of mirex following an aerial application of ant bait were generally higher in predators such as spiders than in herbivorous insects. Dimond et al. (1971) found "food chain concentration evident" when they examined DDT levels in organisms of various trophic levels in an aquatic environment. In a continuation of this study they (Dimond et al., 1974) found that, in a stream in which residue levels in mud were 0.051 to 0.633 ppm, residues of DDT-derived materials were 4 to 10 times higher in trout than they were in crayfish.

Numerous other examples could be cited to demonstrate biological magnification along food chains, and it is clear that in most instances where long-term effects of DDT have been demonstrated, the organisms in trouble are species relatively high in trophic level. Thus it was clear that we needed to develop a system to identify which pesticides might accumulate and persist in biological tissues and to use this information as a guide for recommending where such pesticides might be useful and in what circumstances their use might pose a hazard to nontarget organisms. The model ecosystem approach (reviewed recently by Metcalf, 1977) offered such a possibility. For example, Metcalf et al. (1971) developed a seven-element system involving two food chains. In this procedure labeled pesticide is applied to growing sorghum leaves in a sand-water system in which sorghum is grown as a standing crop. Insect larvae are used as a nutrient source for the alga *Oedogonium* and for diatoms. Added to the water phase of the system are plankton, a snail (*Physa*), *Culex* larvae, and *Gambusia*. Thus two food chains are established, namely, *Estigmene* excreta → alga → snail and *Estigmene* excreta → diatom → plankton → *Culex* → *Gambusia*. Thus, counting from the insect excreta, in one series there is a two-step food chain, in the other a four. When DDT was passed through this system, it and its metabolites were shown to biomagnify, whereas methoxychlor did not.

In drawing conclusions about magnification of DDT up food chains, too frequently the anomalies have been dismissed. Thus in the Carmans River estuary studied by Woodwell et al. (1967) and reported as showing a "systematic increase in DDT residues with increase in trophic level," the level in plankton (identified as mostly zooplankton) was 0.040 ppm, while in the hard clam at the same trophic level, the residue was 0.42 ppm. Crickets had 0.23 ppm, while shoots of *Spartina patens* one step lower in trophic level had 0.33 ppm. While one immature ring-billed gull had 75.5 ppm, most of the other fish-eating birds had residues of 3 to 10 ppm, within the same order of magnitude as the levels in fish that formed their main diet. In many other studies (Frank et al., 1974; Markin et al., 1974; Kolipinski et al., 1971; Meeks, 1968; Keith and Hunt, 1966), though the overall picture indicates that organisms higher in the trophic level have higher DDT residues than those at lower levels, a more critical appraisal of the data raises questions. Thus, whereas, for example, Dimond et al. (1971) showed that, in a stream where DDT had been applied several years earlier, residues in trout and chub were usually higher than those in insects, this did not apply to all sampling dates. Tagatz (1976) developed a model ecosystem using a three-step food chain of turtle grass, the grass shrimp, and the pinfish. When he introduced mirex into this system, he found magnification, but the residue of this persistent insecticide was greater in the grass shrimp than in the pinfish one step up in the trophic scale.

The concept of trophic level as an important parameter in biomagnification of pesticides assumes food as the major route by which organisms take pesticides. This is probably valid for terrestrial animals with low exposure by other routes, but even in this case the amount of pesticide stored in relation to intake is modified by numerous factors, and the trophic level association may be overridden. Thus Jefferies and Davis (1968) found that, when earthworms were exposed in soil to dieldrin at 25 ppm, residues in the earthworms in 20 days were 18.4 to 24.9 ppm. When these were fed to the song thrush for 6 weeks in diets (contaminated worms mixed with "clean" worms) at dosages of 0.32 to 5.69 ppm, residues in the thrushes were 0.09 to 4.03 ppm.

Although there are exceptions, there is a strong body of evidence that, for terrestrial species, the food is the major route of pesticide uptake and that, in most cases, on a whole animal basis, there is little biomagnification from food to consumer, even with DDT. In terms of fat residues, concentrations do increase and by large factors, and the presentation of residue data in terms of deposits in fat tends to distort the overall relationship. Edwards (1973) tabulates data on the accumulation of residues in birds from food, and a review of that tabulation supports this conclusion. High accumulations of residues in terrestrial animals seem more a function of metabolism and excretion than of intake. Quite early, Keith and Hunt (1966) noted differences in residues of DDT in animals at the same trophic level and commented that residues in tissues would be influenced by rate of ingestion, assimilation, degradation, and excretion. In lactating animals it is influenced also by rate of secretion in milk.

In the aquatic environment the situation is more complicated. Residues are more nearly uniform than in terrestrial areas, but the distribution of residues within the water and its interfaces varies dramatically. Keith and Hunt (1966), filtering natural waters, found that the particulate matter contained DDT residues at 10,000 to 100,000 times the level in the filtrate and that, as pointed out by Kenaga (1972), this fact should not be overlooked in calculating magnification factors from water to biological tissues. The extent to which suspended matter may adsorb or absorb a pesticide may alter the biomagnification and distribution of the compound. Hamelink and Waybrant (1976) added DDE and lindane to a large limestone quarry. Shortly after the addition of the pesticide a heavy rain washed a lot of sediment into the quarry. In a sediment trap after 21 days, DDE residue was 1427 ppb versus 0.1 ppb for lindane. At that time pesticide concentrations in the top 6 m of water were 3.28 ppt for DDE and 106.33 ppt for lindane. The effect of this attachment to sediment was reflected in the distribution of the pesticides in relation to depth. By day 21 DDE was distributed rather uniformly from 0 to 15 m, while with lindane the concentration in the upper meter was 147.75 ppt, more than 10 times the amount found in water more than 6 m from the surface. Within 173 days 94% of the DDE was in the bottom mud, while 95% of the lindane was carried in the water. Insecticides are concentrated in both bottom muds and surface slicks and the exposure of animals to these environments affects the levels of pesticides organisms carry.

Kerr and Vass (1973) reviewed pesticide residues in invertebrates in relation to trophic level and could find no correlation. They selected primarily data that reflected background levels and avoided those associated with specific pesticide applications. The student should consult the tables of Kerr and Vass (1973) and the original studies for critical comparisons; in Table 20.2 we present some of these data. We have lumped a number of species in reporting the range of DDT levels, and this reduces the validity of the specific figures. Our ranges are the means from separate studies reported by Kerr and Vass (1973), but in many instances the ranges between the studies were no greater (usually not as great as) than the range for a given species within the group as reported in the original studies. Obviously factors other than trophic level and thus diet are dominant in determining residue levels.

Robinson et al. (1967) analyzed a large number of marine species taken off the Northumberland coast during 1965–1966. They assigned whole number trophic levels to the various species, with plants represented as trophic level 1. Their data on residues of dieldrin and DDE are given in Table 20.3, along with the trophic level assigned by these workers.

Robinson et al. (1967) point out that, though there is good agree-

Table 20.2   Some Reported Residue Levels of DDT in
Aquatic Invertebrates in Different Environments in Relation
to Trophic Level[a]

Habitat	Organism	Trophic Level	Residue DDTR (ppb) Range
Marine benthic	algae	1.0	2-3[b]
	limpet	2.1	3[b]
	snail	2.1	260
	mussel	2.1	20-754
	oyster	2.1	15-60
	shrimp	2.5	43-160
	crab	2.5	42-260
	starfish	3.3	20-78
Marine pelagic	algae	1.0	12-83
	zooplankton	2.1	3-40
	shrimp	2.5	2-120
	squid	4.2	28
Freshwater benthic	clam	2.1	5-43
	midge	2.3	52
	amphipods	2.3	30-460

[a]Adapted from Kerr and Vass, 1973 (see text).
[b]DDE only.

ment in most cases among residues in species in the same trophic level and an indication of increasing concentrations with increasing trophic level, there are exceptions. They cite an "appreciable concentration of dieldrin at each stage" in the food chain microplankton → mussel → eider duck," though their data would question such a conclusion. They note, however, that the residue is lower in the cod than in the sand eel on which it feeds. They show that the levels of dieldrin in livers and eggs of the shag are independent of age and hence of duration of exposure of the birds and conclude that it is not the level of intake but rather the equilibrium between intake and excretion that is the important factor. They suggest that observations that organisms in higher trophic levels have higher residues of organochlorine pesticides could be explained on the basis that vertebrates (which occupy the higher trophic levels) are less efficient at excreting these compounds than the invertebrates are, many of which represent lower levels in food chains.

Whether the intake of pesticides by aquatic organisms is active or passive, via the food, the surface, or in respiration has been the subject

Table 20.3   Residues of Dieldrin and DDE in Marine
Organisms Taken off the Northumberland Coast, 1965–1966[a]

| Species | Trophic Level | Residue ppb (wet weight basis) | |
		Dieldrin	DDE
Serrated wrack	1	0.001	0.002
Oar weed	1	0.001	0.003
Microzooplankton	2	0.020	0.030
Sea urchin	2	0.027	0.050
Mussel	2	0.023	0.024
Cockle	2	0.018	0.012
Limpet	2	0.009	0.003
Macrozooplankton			
(crustacea)	3	0.16	0.16
Lobster	3	0.024	0.024
Shore crab	3	0.025	0.037
Edible crab	3	0.015	0.061
Plaice	3	0.038	0.023
Herring	3	0.057	0.080
Sand eel	3	0.016	0.026
Cod	4	0.009	0.012
Whiting	4	0.040	0.021
*Shag	4	1.56	2.87
*Eider duck	4	0.12	0.25
*Herring gull	4	0.31	0.26
*Lesser black-			
backed gull	4	0.26	0.61
*Cormorant	5	0.19	4.14
*Gannet	5	0.35	1.39
*Grey seal	5	0.09	0.34
*Common dolphin	5	0.04	0.13

[a]Robinson et al., 1967.
*Residues in liver.

of significant debate and contradictory findings. Working with invertebrates Derr and Zabik (1972, 1974) found that the uptake of labeled
DDE was dependent on the concentration in the medium and was the
result of an adsorption-diffusion process, since larvae of the aquatic
midge *Chironomus tentans* took up the same amount of the compound
regardless of whether they were alive or dead. By contrast Burnett
(1971) found that in the sand crab the major areas for passive adsorption would be the carapace and telson but that these, when analyzed
separately, accounted for only one-fifth the uptake of DDT by this
species under conditions where residues were 325 times greater in the

animals than in the water in which they were cultured. In Burnett's test dead animals took up only one-fiftieth the amount of DDT that living animals did. Darrow and Harding (1975) found that labeled DDT was taken up by nonfeeding copepods held in sea water for 8 weeks at 6°C. The DDT concentration in the sea water was 450 ppt. An equilibrium was reached at 50 to 100 ppm after 3 to 4 weeks, but there was no evidence that metabolites were formed. After tests with shrimp, Cox (1971a) suggested a two-step uptake, adsorption on the surface being the initial stage followed by transport within the lipid portion of the animals' tissues.

A large number of studies on fish have failed to establish the relative importance of the various possible routes of pesticide intake. Murphy and Murphy (1971) found that the uptake of DDT from water correlated closely with the uptake of oxygen and thus was greater at 20°C than at 5°C. They concluded that, since oxygen consumption increases with increased body weight, uptake of DDT would similarly increase by the respiratory route and thus higher residues would be expected in larger fish, an observation reported by several workers. They suggested that cutaneous uptake would also be greater in larger fish. They found that dead fish adsorbed little DDT.

DDE residues are high in Lake Michigan fish (Chapter 15), while residues in water are barely detectable. Macek and Korn (1970) used labeled DDT in a flowing water system in which they added the pesticide to water and fed it to brook trout at levels approximating those found in Lake Michigan. After 120 days trout fed 3 ppm in pellets (feeding was at a rate so that pellets did not stand in the water to cause contamination) had 1.92 ppm, while those in contaminated water (3 ppt) and fed clean food had only 0.256 ppm. In each case the rate of tissue residue buildup was linear for 60 days and then declined. Analyses indicated that the fish accumulated 35.5% of the DDT in the pellets but only 3.5% of the DDT in the water. They concluded that the rate of uptake from water at the level of DDT encountered in Lake Michigan would give a 1 ppm residue in a trout in 12 years. By contrast the uptake from their food of smelt and alewife carrying about 3 ppm DDT would result in a residue of 6 ppm after 1 year.

Hamelink et al. (1971) propose that the food chain is not the important consideration in determining pesticide buildup in tissues. They suggest that the tissue residue represents the net effect of a balance between uptake and loss and that this is controlled by adsorption and solubility characteristics of the pesticides and tissues involved. In developing this hypothesis they established a series of ponds in which three systems were operative. In one a complete food chain was provided consisting of algae, invertebrates, and fish. In a second a shorter food chain was present consisting of algae and insects, and in the third an incomplete food chain was established by introducing only algae and fish. In each case DDT was added. When they analyzed algae, invertebrates, and fish, they found a stepwise increase in residue at each step in the trophic level regardless of whether or not the food chain had an opportunity to operate. They found a linear correlation between water concentration of DDT and uptake by invertebrates and explained this on the basis of partitioning of DDT between water and insect lipids. With fish they suggested that the same principle applied but in a two-step process. The first involved a partitioning between DDT in water and the blood in the gills followed by a partitioning between the blood and lipids within the fish. They cited Reinert (1970) as demonstrating a relationship between increased DDT residues and increased body fat in fish, a correlation noted also by other workers, but, as mentioned earlier, the consistency with which this has been demonstrated has not been great. Ernst et al. (1976) examined tissues of five species of fish from the English Channel and found that in general pesticide levels increased with lipid content of the tissues in the order musculature, liver/digestive gland, and visceral adipose tissue. There was, however, a higher content in the lipids in muscle than in other tissues. In general there were positive correlations between residue concentrations in adipose tissue and liver, and weight, but this correlation did not exist in scallops. Reinert (1970), in a study of fish in the Great Lakes, found that in lake trout residue levels increased with fish size but that residue concentrations in the lipid tissue reached an apparent equilibrium, whose level did not increase in larger fish. Hamelink and Waybrant (1976) in their study of lindane and DDE in a limestone quarry found that, in bluegills, there was an initial rapid rise in concentration of residue of DDE that tapered off and reached an equilibrium in 60 days. By contrast with lindane an equilibrium was reached in 5 days. In trout the equilibrium was not reached until 108 days, but with DDE in each case the equilibrium was reached at about 10^5 times the concentration in water. They interpreted these findings as supporting their theory that residue accumulation was the result of exchange equilibria. Their findings on relative residues of lindane and DDE support their earlier calculations of what might be expected on the basis of relative solubility of these two compounds in water (Hamelink et al., 1971).

To a large extent the proposal of Hamelink et al. (1971) implies a passive role of the organism in biomagnification with, for aquatic organisms, the rate of accumulation determined by relative exchange equilibria between an organism and the media in which it lives. Thus, with aquatic organisms, the dominant exchange is between the water outside the body and the lipoidal tissue within. This might at first appear to oppose the view that residue intake in food is important, but in fact it does not, since, once the food is within the digestive tract, the exchange equilibria would operate, this time between the contents of the intestinal tract and lipoidal tissues or constituents of tissues adjacent to it.

It is obvious that we still have much to learn about how and why some pesticides accumulate in certain animal species and certain tissues. Norstrum et al. (1975) tried to describe statistically the relationship between pesticide concentration and uptake in four species of fish in which higher residues were correlated with increased size. They concluded that 50 to 75% of the pesticide intake was via the food but suggested that their model needed refinements before it could provide valid predictions. In later work Norstrum et al. (1976) expanded on this approach with PCBs and methylmercury and obtained better agreement between theoretical values and those found in fish. Moriarty (1975) describes in detail the use of "the compartmental model" to describe the movement and biomagnification of pesticides and analyzes data from several experiments with this approach. He concludes that the model is useful in studying intake and loss of pesticides but that additional information is needed to validate key assumptions in the model.

It is clear that, in all organisms, the rate of residue accumulation and the level of accumulation are dependent on exposure. Perhaps nowhere has this been demonstrated more clearly than in aquatic organisms and birds that feed on them along the Pacific Coast of North America. In this area the focus of contamination has been identified as the Los Angeles area; from there toward the south, north, or west residue levels decline. Thus Duke and Wilson (1971) found DDT-related residues in Santa Monica Bay fish at 370 ppm with a gradient to 1.0 ppm north of Seattle and a gradient to 0.14 ppm off southern Baja. In the filter-feeding sand crab residues were 7.2 ppm near Los Angeles but decreased to 0.1 ppm progressively as samples were taken down the coast from this area (Burnett, 1971). Pelicans on Anacapa Island, 35 miles from the heavily polluted Santa Monica Bay fish, had DDT residues in their eggs of 1215 (intact eggs) and 1818 (broken egg) ppm on a lipid weight basis, while those on Los Coronados Islands, 95 miles south of Anacapa, had 810 ppm, and those on St. Martins Island, 250 miles south, had 192 ppm (Keith et al., 1970; Jehl, 1970).

In a scholarly discussion of the bioconcentration of pesticides

Kenaga (1972) defines bioconcentration as "the amount of a pesticide residue accumulated by an organism by adsorption, and by absorption via oral or other route of entry, which results in an increased concentration of the pesticide by the organism or specific tissue." Adsorption is an important aspect of this bioconcentration in aquatic organisms, especially those with large surface-to-volume ratios, that is, the small organisms. Kenaga (1972) demonstrates this dramatically by calculating that a 1.0 lb/acre application of a pesticide would result in residues of 2.3 and 840 ppm on particles of 1 in. diameter and 0.1 mm diameter, respectively, if these particles had similar surface adsorptive capacities. In addition he points out that small animals usually eat a larger amount, in terms of percent of body weight, than larger animals do and emphasizes the need for accurate characterization of the environment and conditions existing when bioconcentration is calculated. Thus figures on bioconcentration of a pollutant from water to an organism have little relevance unless the concentration in the immediate habitat of that organism is given, and a food-chain pattern is probably of little significance unless exact food patterns are known, a situation not likely in natural aquatic systems. The early simplistic view of DDT's increasing in concentration at each step in the trophic level (biomagnification) has been challenged, the data favoring some other hypothesis.

The extent to which our early views on the movement of persistent pesticides in the biosphere were misleading is summed up in comments by Harvey (1974) in connection with a study of residue levels of organochlorines in the Atlantic. "When we began in 1970, we proceeded on the following assumptions: We would find mainly DDT and its derivatives and little, if any, PCB; we would note decreasing concentrations of DDT in fish and water as we moved seaward; we would observe food chain magnification of chlorinated hydrocarbons in marine mammals; and we would detect toxic effects of DDT and PCB." Harvey (1974) continues, "Fortunately, our approaches were flexible because all four assumptions were wrong." The results of the Atlantic surveys (Harvey, 1974; Harvey et al., 1974) have shed considerable light on what happens to DDT at sea, and these data, together with those cited previously (e.g., Duke and Wilson, 1971; Keith et al., 1970; Jehl, 1970), refute the earlier predictions that the ocean abyss would be the ultimate sink for this pesticide. They found that DDT in measurable quantities was found only in coastal waters, where it adsorbed on particulate matter and settled into bottom sediments. This is reflected in pesticide levels in fish and other aquatic organisms that frequent these habitats. DDT is present, in small amounts, in the atmosphere over the open ocean, but here particulate matter is not abundant in the water, and the pesticide that reaches the water has no surface on which to become adsorbed. The half-life of DDT in naturally illuminated sea water is about 10 days. Thus the pesticide is volatilized or codistilled back into the atmosphere or is decomposed. As a result fish from these waters have extremely low DDT residues. In the Atlantic the highest level of DDT is found in the plankton, while flying fish that feed on these plankton contain residues about 100 times less. Only in sharks and barracuda are levels of DDT found comparable to those in plankton, and Harvey (1974) concludes that "food-chain magnification is not a useful concept with water-breathing animals."

The arguments presented in this chapter do not dismiss the fact that many of the animals suffering ill effects from organochlorine pesticide residues are high trophic level organisms and do contain relatively high pesticide residues. It is important to realize, however, that position in the food chain has little to do with it. Far more important seems to be the ability of the species to eliminate pesticides. Thus, while model ecosystems and studies of food chain components in natural ecosystems are helpful, it would appear that the critical data are those dealing with metabolism within each species. This is reassuring in that it suggests that the present procedure of conducting toxicological and metabolic studies with pesticides on animal species is the best method of predicting which compounds may accumulate in living tissues. The problem remains, however, on which species and how many to test.

Note also that, contrary to the inference in much of the early literature on food-chain magnification, man does not occupy a very high trophic level except in areas and among cultures where fish are a dominant portion of the diet. If this had been made clear at the outset and the danger to fish and birds addressed as dangers to fish and birds rather than as portents for man, perhaps the controversy about the organochlorines could have been more informative and decisions concerning them more rational and less protracted.

21

PESTICIDE LEGISLATION

Legislation with regard to pesticides varies widely among different countries with respect to the agency responsible and the restrictions placed upon the sale, distribution, and use of compounds. It is not our intention to review these but rather to provide some perspective on the rationale for such legislation and describe how it is effected in the United States and Canada.

Before the modern pesticide era extensive legislation on pesticides per se was quite limited, and pesticides were treated along with other poisonous substances. When specific legislation was first introduced, it was aimed at protecting the consumer against misrepresentation of products and not addressed to the problems of safety of use to crops, the user, or the environment. As pesticides assumed a more prominent role, so too did the legislation concerning them, until now in both the United States and Canada specific acts have been enacted and rules and regulations developed that place stringent requirements on these products. In the United States authority rests with the Environmental protection Agency through the Federal Insecticide, Fungicide and Rodenticide Act as amended by the Environmental Pesticide Control Act of 1974. In Canada authority rests with Agriculture Canada under the Plant Products Act. The definition of pesticide under these acts is very broad. "The term pesticide means any substance or mixture of substances intended for preventing, destroying, repelling, or mitigating any pest, and any substance or mixture of substances intended for use as a plant regulator, defoliant or desiccant" (*Federal Register*, 1975a). The Canadian definition is even broader and includes any "substance or thing" but then sets out exemptions for requirements for registration of such "things" as fly swatters and cedar chests of no possible health or environmental concern. In addition to the federal pesticide legislation, legislation dealing specifically with pesticides has been enacted in each of the States and Provinces. Again the responsible agency may be health, agriculture, or environment, but regardless of which agency implements the program, the advice of other involved agencies is sought in the decision-making process.

In reviewing the rationale of pesticide legislation, Glasser (1976) points out that the main objectives of legislation are:

"1. To protect people who may be exposed to acute risks during manufacture, formulation, packaging, transport, and storage;

2. to ensure good packaging, which should carry the proper classification as to risks; and to prevent direct contamination of food and animal feed at any point between manufacture or formulation and use in the field;

3. to protect people who may be exposed to risks when opening containers, diluting a concentrated chemical to obtain the correct field dosage, and subsequently applying the pesticide;

4. when appropriate, to warn against unintentional contamination of non-treated crops, animals, soil, and water;

5. to protect the buyer against the sale or purchase of low-quality products or against misleading claims made on the label or in advertising;

6. to protect the consumers of the treated food or animal feed by ensuring that the pesticide is correctly applied and that adequate intervals between application and harvest are clearly established and, where appropriate, stated on the label. This ensures that pesticide residues, if any, remaining on the food or feed stuffs are of acceptable levels so that there is no risk to the consumer; and

7. to ensure that the consequences of the foregoing requirements do not place unwarranted restrictions on the development of new pesticides."

To ensure that these objectives are met, all pesticides must be registered and carry a registration number on the label before they can be offered for sale. The registration process is thus the means by which necessary assurances about safety and efficacy are determined. The company proposing to sell a pesticide must provide extensive information to the regulating body, and these requirements are set out in detail in rules and regulations surrounding the pesticide acts. The extent of such requirements is indicated by the fact that their discussion occupies 128 pages in the *Federal Register* (*Federal Register*, 1975). Included, where appropriate, are the following:

1. Complete chemical and physical description of the pesticide and products included in its formulation.
2. Data on acute, subacute, and chronic toxicity, including oral, dermal, and inhalation exposure. Data from several animal species.
3. Metabolism and degradation studies, including toxicity studies on metabolites.
4. Efficacy for the purported uses.
5. Residues.
6. Proposed label.

The regulatory body may register the product, refuse to register the product, or give a temporary registration while requiring additional information.

This process of federal registration provides the basic assurance that pesticides are useful for the purposes claimed and that when used "according to directions" should be relatively safe to the user, the consumer of products treated, and to the environment. Thus the pesticide label is of utmost importance, since it specifies the acceptable uses and conditions of use for the product, and the law makes it illegal to use pesticides for uses or by methods not set forth on the label. The label also indicates the classification of the pesticide, and this classification, based on degree of hazard, determines who may use the product. Some pesticides are restricted to use only on a specific permit for a specific job. Others are restricted to licensed pesticide applicators, while others may be sold to the general public.

In addition to federal regulations, States and Provinces may further classify pesticides and place restrictions on use, storage, or display. The system used in Ontario, Canada, may be cited as an example to indicate how such a classification system applies. Here pesticide products are placed, through legislation, in one of six schedules (Ontario Pesticides Act, 1973).

Schedule I includes pesticides that are highly toxic and/or restricted because of environmental reasons. It includes products containing DDT, aldrin, dieldrin, and other restricted compounds, plus highly concentrated pesticides in which the oral LD_{50} of the formulated product is generally less than 50 mg/kg. Pesticides in this schedule may be used only by licensed applicators or by agriculturalists (defined in the act and including foresters). In Schedule I some products are designated as permitted for use only on a permit basis. Products in Schedule I may not be openly displayed in selling outlets and may be sold only through licensed outlets with rigid requirements in storage facilities, a condition of the issuance and maintenance of such license. Records of sale are required.

Schedule II includes pesticides that have toxicities as formulated generally in the 50 to 500 mg/kg range and may be used only by licensed applicators and agriculturalists. They may be sold only through licensed outlets, specific requirements regarding storage and display being a condition of the licensing. Records of sale are required.

Schedule III includes pesticides that have toxicities as formulated generally in the range of 500 to 5000 mg/kg and are available to the general public but may be sold only through licensed outlets. Requirements in storage and display are less rigid for these products, but they may not be dispensed in food-handling outlets unless separately dispensed in a section of the premises meeting conditions that ensure that foods will not be contaminated.

Schedule IV includes pesticides that have toxicities as formulated generally in excess of 5000 mg/kg and are packaged in containers of no more than 2 lb or 1 qt size and in which the containers meet specified standards. No restrictions are placed on the sale of such compounds.

Schedule V includes pesticides of such toxicity as to be placed in Schedule I, but they are exempted from the requirement of a permit for each use when used on agricultural land. They are subject to the same requirements of the vendor as products in Schedule I are.

Schedule VI includes formulations of pesticides meeting the requirements of Schedule IV but packaged in large quantities intended for commercial or industrial use. Such pesticides may be sold only through licensed outlets meeting the requirements of those licensed to sell Schedule III compounds.

Despite these rigid requirements for registration of pesticides, not all developments are predictable, and information may develop after years of use that suggest that registration of a pesticide should be withdrawn. Pesticide legislation now provides procedures by which pesticides may be further restricted or banned from use. Under the United States system such decisions may be challenged, and protracted hearing may delay the emplementation of bans. In Canada pesticide registrations may be withdrawn without the requirement for such hearings.

22

The story of pesticides is an exciting chapter in modern civilization. It embraces the discovery of DDT and, for many people, triumph over the misery and suffering of malaria and the reduction of many other insect-borne human diseases. In this role DDT must be credited or charged, depending on one's point of view, as a major contributor to an increased world population and extended life expectancy in many tropical and subtropical countries where insect-transmitted diseases are major causes of early death. Included also is the role of pesticides in food production, where large per acre yield increases have been sustained and labor input reduced through insect, disease, and weed control. The result has been a food supply of volume and quality not realized before the pesticide era and unlikely to be maintained if these were suddenly unavailable. In addition the forests have remained green and not sacrificed to the appetite of insect hordes.

Against these benefits are the tragedies of reproduction declines in many raptors and fish-eating birds, high pesticide residues that threaten the survival of some fresh water fishing industries, and an increasing incidence of pesticide-related illnesses among workers exposed to some highly toxic compounds. To this must be added the probable role of DDT in declining colonies of bats and the fact that, while pest control can still be obtained, the cost of achieving this with chemical pesticides is increasing constantly.

Small wonder that this chapter in American history has generated so much controversy. On the one hand are benefits rivaling those of the wonder drugs or of the new crop cultivars that ushered in the "green revolution," on the other the threat of extinction of species, the ruining of our environment, and increasing cost of food. To this must be added the concern for long-term effects that might be expressed in permanent species' shifts in our environment or in human disease.

Throughout this book we have tried to document what is known and to indicate the grounds for concern in areas where factual evidence is lacking. While, in many cases, conclusions on benefits are obvious, decisions on risks are not. Even where it is possible to define benefits and interpret risks, the relative weight given to each will be different depending on the point of view of the reader. To the farmer a bushel of grain may be much more valuable than a robin, while to an avid bird watcher, such a weighting of values would be inconceivable. It is this intangibility of environmental effects that has bedeviled the issue and inflamed proponents on both sides of what should have been a rational debate. The issue is even more perplexing when evidence is not hard. DDT and cancer or 2,4,5-T and teratogenic effects are issues that probably never will be resolved to the satisfaction of those who view the evidence with differing standards, standards that are each understandable given the bias of the viewer.

It is clear that pesticides are needed in many areas of modern living and that their uses in food and fiber production, public health, and the home will continue. It is equally clear that they cannot continue to be used in the indiscriminate manner that characterized the heyday of DDT. Common sense dictates that economic poisons must be handled with care and that such care requires different levels of restriction on different pesticides. These cannot be blanket restrictions but must reflect the inherent nature of the compound and the benefits and hazards associated with its use in a specific circumstance. Fortunately this is being recognized in newer legislation, but we must be wary lest, in their zeal to protect man and his environment, legislators needlessly and arbitrarily legislate essential and useful products out of existence.

Studies now underway demonstrate that pest control can be achieved effectively with much less pesticide, provided natural control agents are given optimum opportunity to work and selective pesticides are employed on a need only basis. This represents a valid approach. There is no question but that we have placed great pressures on our physical environment and the life forms that give it a degree of stability. To introduce pesticides needlessly is folly, folly to which the past three decades bear testimony. Surely we have learned a lesson and it should not be necessary to repeat our mistakes again.

Mistakes of the past should be avoided also in our public debate on pesticides. This debate has been characterized by great dramatics and little fact. Those reflecting different points of view seemed less concerned with solutions than with promotions. As a result the scientist took his case to an uninformed public whom he tried to impress with sympathy for birds, fear of the unknown, or economic benefits rather than work with his fellow scientist to identify the problem, resolve it, and explain it. The result has been much wasted effort and decisions made on pesticides by politicians or lawyers rather than by informed scientists. Surely the pesticide issue is mature and is one in which scientists should become sufficiently mature to adopt a consultative rather than adversary stance.

It is not prudent to permit pests to waste a large portion of our food supply or inflict man with disease. It is equally imprudent to contaminate our environment so that its productivity is reduced or its aesthetics despoiled. Pest control requires the input of many disciplines and the cooperation of the pest control expert with those who understand more thoroughly the sensitivities of the natural environment. Understandably these people have their special interests, but these cannot be permitted to override sound planning, be it environmental, economic, or, as in the pest control issue, a combination of both. To bring together the needed expertise may not be easy, but it is essential if our resources are to be used to best advantage for man and the friendly inhabitants that surround him in what we call our environment.

REFERENCES

Abbott, D. C., R. B. Harrison, J. O'G. Tatton, and J. Thomson. 1965. Organochlorine pesticides in the atmospheric environment, February-July 1965. *Nature* **208**: 1317–1318.

Abou-Donia, M. B., and D. B. Menzel. 1968. The metabolism *in vivo* of 1,1,1,-trichloro-2,2-bis(*p*-chlorophenyl) ethane (DDT), 1,1-dichloro-2,2-bis(*p*-chlorophenyl) ethane (DDD) and 1,1-dichloro-2,2-bis(*p*-chlorophenyl) ethylene (DDE) in the chick by embryonic injection and dietary ingestion. *Biochem. Pharmacol.* **17**: 2143–2161.

Acree, F., Jr., M. Beroza, and M. C. Bowman. 1963. Codistillation of DDT with water. *J. Agric. Food Chem.* **11**: 278–280.

Adams, L., M. Hanavan, N. Hosley, and D. Johnston. 1949. The effects on fish, birds, and mammals of DDT used in control of forest insects in Idaho and Wyoming. *J. Wildl. Mgmt.* **13**: 245–254.

Ahrens, J. F., O. A. Leonard, and N. R. Townley. 1970. Chemical control of tree roots in sewer lines. *J. Water Pollut. Control.* **42**: 1643–1655.

Akesson, N. B., and W. E. Yates. 1964. Problems relating to application of agricultural chemicals and resulting drift residues. *Ann. Rev. Entomol.* **9**: 285–318.

Alder, E. 1970. The many types of pesticides—similarities and differences. In: Background for decision—a one-day educational symposium on pesticides. The Governors' Five-state Interdisciplinary Council Proceedings. Division of Agronomy, Minnesota Department of Agriculture, St. Paul. Pp. 6–24.

Alder, E. F., W. L. Wright, and Q. F. Soper. 1960. Control of seedling grasses in turf with diphenyl aceto nitrile and a substituted dinitro aniline. *Proc.* North Central Weed Control Conference. Pp. 23–24.

Alex, J. F. 1966. Survey of weeds of cultivated land in the Prairie Provinces. Experimental Farm, Research Branch, Agriculture Canada, Regina. Map 4, p. 18.

Allen, B. M. 1974. An evaluation of various materials for reducing spray drift in herbicide applications. M.Sc. Thesis. University of Guelph, Guelph, Ontario.

Aly, O. M., and M. A. El-dib. 1972. Studies of the persistence of some carbamate insecticides in the aquatic environment. In: Fate of organic pesticides in aquatic environments. Advances in Chemistry Series III. American Chemical Society. Washington, D.C. Pp. 210–243.

American Ornithologists' Union. 1975. Report of the American Ornithologists' Union Committee on Conservation 1974–1975. *Auk* **92**(4 Suppl.): 1B–16B.

Amsden, R. C. 1962. Reducing the evaporation of sprays. *Agr. Aviation* **4**: 88–93.

Anas, R. E. 1974. DDT plus PCBs in blubber of harbour seals. *Pestic. Monit. J.* **8**: 12–14.

Anderson, D. W., and J. J. Hickey. 1970. Eggshell changes in certain North American birds. Proceedings XVth International Ornithological Congress. Pp. 514–540.

Anderson, D. W., and J. J. Hickey. 1970a. Oological data on eggs and breeding characteristics of brown pelicans. *The Wilson Bull.* **82**: 14–28.

Anderson, G. W., and L. W. Smith. 1969. Persistence in soil of picloram. Canadian Weed Commission (Eastern Section). Pp. 237.

Anderson, J. F., and S. W. Gould. 1974. Defoliation in Connecticut 1969–1974. Connecticut Agricultural Experimental Station Bulletin 749. P. 25.

Anderson, J. M. 1968. Effect of sublethal DDT on the lateral line of brook trout, *Salvelinus frontinalis. J. Fish. Res. Bd. Can.* **25**: 2677–2682.

Anderson, J. M., and M. R. Peterson. 1969. DDT: Sublethal effects on brook trout nervous system. *Science* **164**: 440–441.

Anderson, J. R. 1971. The activity of triazine herbicides in Manitoba and Ontario soils. Ph.D. Thesis. University of Guelph, Guelph, Ontario.

Andrawes, N. R., W. P. Bagley, and R. A. Herrett. 1971. Metabolism of 2-methyl-2-(methylthio)propionaldehyde 0-(methylcarbamoyl)oxime (Temik aldicarb pesticide) in potato plants. *J. Agric. Food Chem.* **19**: 731–737.

Anonymous. 1972. The potato connection. *Newsweek.* November 20. P. 58.

Anonymous. 1976. U.S. OKs sale of DDT to Cambodia. *Honolulu Star-Bulletin.* December 23.

Applegate, V. C., and E. L. King, Jr. 1962. Comparative toxicity of 3-trifluormethyl-4-nitrophenol (TFM) to larval lampreys and eleven species of fishes. *Trans. Amer. Fish. Soc.* **91**: 342–345.

Armstrong, D. E., and G. Chesters. 1968. Adsorption catalyzed hydrolysis of atrazine. *Environ. Sci. Technol.* **2**: 683–689.

Arthur, R. D., J. D. Cain, and B. F. Barrentine. 1976. Atmospheric levels of pesticides in the Mississippi delta. *Bull. Environ. Contam. Toxicol.* **15**: 129–139.

Ashton, F. M. 1965. Relationship between light and toxicity symptoms caused by atrazine and monuron. *Weeds* **13**: 164–168.

Ashton, F. M., and A. S. Crafts. 1973. *Mode of action of herbicides.* John Wiley & Sons, Inc., New York. P. 504.

Ashton, F. M., and T. J. Sheets. 1959. The relationship of soil adsorption of EPTC to oats injury in various soil types. *Weeds* **7**: 88–90.

Aslander, A. 1926. Chlorates as plant poisons. *J. Amer. Soc. Agron.* **18**: 1101–1102.

Astolfi, E., J. C. G. Fernandez, M. B. deJuarez, and H. Piacentino. 1973. Chlorinated pesticides found in the fat of children in the Argentine republic. In: *Pesticides and the environment: A continuing controversy.* Deichmann, W. B. (Ed.). Intercontinental Medical Book Corp., New York & London. Pp. 233–243.

Audus, L. J. 1964. Herbicide behaviour in soil. II. Interactions with soil microorganisms. In: *The physiology and biochemistry of herbicides* (1st ed.). Audus, L. J. (Ed.). Academic Press, New York. 5: 163–206.

Audus, L. J. (Ed.). 1976. *Herbicides: Physiology, biochemistry, ecology* (2nd ed.). Vol. 1. Academic Press, London. P. 608.

Aulerich, R. J., R. K. Ringer, H. L. Seagran, and W. G. Youatt. 1971. Effects of feeding coho salmon and other Great Lakes fish on mink reproduction. *Can. J. Zool.* 49: 611–616.

Bader, A. 1976. Recent advances in synthetic pyrethroids. *Aldrichimica Acta.* 9: 49–51.

Baetcke, K. P., J. D. Cain, and W. E. Poe. 1972. Mirex and DDT residues in wildlife and miscellaneous samples in Mississippi—1970. *Pestic. Monit. J.* 6: 14–22.

Bailey, G. W., and J. L. White. 1964. Review of adsorption and desorption of organic pesticides by soil colloids with implications concerning pesticide bioactivity. *J. Agric. Food Chem.* 12: 324–332.

Bailey, S., P. J. Bunyan, G. A. Hamilton, D. M. Jennings, and P. J. Stanley. 1972. Accidental poisoning of wild geese in Perthshire, November 1971. *Wildfowl* 23: 88–91.

Baird, D. D., R. F. Husted, and C. L. Wilson. 1965. Pre- and post-emergence herbicide activity of N-isopropyl-2-chloroacetanilide on *Echinochloa Crusgalli, Zea Mays* and glycine in the greenhouse. *Proc. Southern Weed Conf.* 18: 653–659.

Baird, D. D., R. P. Upchurch, W. B. Homesley, and J. E. Franz. 1971. Introduction of a new broad spectrum postemergence herbicide class with utility for herbaceous perennial weed control. *Proc. North Central Weed Control Conf.* 26: 64–68.

Baker, P. S., and F. O. Morrison. 1966. The basis of DDT tolerance in the laboratory mouse. *Can. J. Zool.* 44: 879–887.

Baker, R. S., and G. F. Warren. 1962. Selective herbicidal action of amiben on cucumber and squash. *Weeds* 10: 219–224.

Bandeen, J. D., and R. D. McLaren. 1976. Resistance of *Chenopodium album* L. to triazines. *Can. J. Plant Sci.* 56: 411–412.

Barker, R. J. 1958. Notes on some ecological effects of DDT sprayed on elms. *J. Wildl. Mgmt.* 22: 269–274.

Barnett, S. A., J. D. Blaxland, F. B. Leech, and M. M. Spencer. 1949. A concentrate of red squill as a rat poison and its toxicity to domestic animals. *J. Hyg.* 47: 431–433.

Barr, B. A., C. S. Koehler, and R. F. Smith. 1975. Crop losses. Rice: Field losses to insects, diseases, weeds and other pests. Report prepared for the U.S. Agency for International Development UC/AID Pest Management and Related Environmental Protection Project. University of California. Pp. 64.

Barrentine, J. L., and G. F. Warren. 1970. Selective action of terbacil on peppermint and ivyleaf morning glory. *Weed Sci.* 18: 373–377.

Barrett, G. W. 1968. The effects of an acute insecticide stress on a semienclosed grassland ecosystem. *Ecology* 49: 1019–1035.

Barrett, G. W., and R. M. Darnell. 1967. Effects of dimethoate on small mammal populations. *Amer. Midland Natur.* 77: 164–175.

Baskerville, G. L. 1975. Spruce budworm: Super silviculturist. *The Forestry Chronicle.* August. Pp. 138–140.

Baskerville, G. L. 1976. Report of the task-force for evaluation of budworm control alternatives. Department of Natural Resources, Fredericton, New Brunswick. Pp. 210.

Beaven, G. F., C. K. Rawls, and G. E. Beckett. 1962. Field observations upon estuarine animals exposed to 2,4-D. *N.E. Weed Control Conf. Proc.* Suppl. **16**: 449–458.

Behrens, R. 1975. Factors affecting dicamba volatility. Weed Science Society of America Abstract No. 159. P. 59.

Behrens, R., and D. C. Lee. 1966. *Weed Control. Advances in corn production.* Iowa State University Press, Iowa City. Pp. 331–352.

Bennett, I. L. 1967. Forward. *Pestic. Monit. J.* **1**.

Benson, W. W., J. Gabica, and J. Beecham. 1974. Pesticide and mercury levels in bear. *Bull. Environ. Contam. Toxicol.* **11**: 1–3.

Benson, W. W., W. Webb., D. W. Brock, and J. Gabica. 1976. Mercury in catfish and bass from the Snake River in Idaho. *Bull. Environ. Contam. Toxicol.* **15**: 564–567.

Berg, R. T., and L. W. McElroy. 1953. Effects of 2,4-D on the nitrate content of forage crops and weeds. *Can. J. Agric. Sci.* **33**: 354–358.

Bernard, R. F. 1963. Studies on the effects of DDT on birds. Michigan State University. Publication of the Museum, Biology Series **2**: 155–192.

Besser, J. F., J. W. DeGrazio, and J. L. Guarino. 1968. Costs of wintering starlings and red-winged blackbirds at feedlots. *J. Wildl. Mgmt.* **32**: 179–180.

Bevenue, A. 1976. The "bioconcentration" aspects of DDT in the environment. *Res. Rev.* **61**: 37–112.

Bevenue, A., J. N. Ogata, and J. W. Hylin. 1972. Organochlorine pesticides in rainwater, Oahu, Hawaii, 1971–1972. *Bull. Environ. Contam. Toxicol.* **8**: 238–241.

Bidleman, T. F., and C. E. Olney. 1974. Chlorinated hydrocarbons in the Sargasso Sea atmosphere and surface water. *Science* **183**: 516–518.

Birch, M. C. (Ed.). 1974. *Pheromones.* American Elsevier Publishing Co. Inc., New York. P. 495.

Bitman, J., H. C. Cecil, and G. F. Fries. 1970. DDT-induced inhibition of avian shell gland carbonic anhydrase: A mechanism for thin eggshells. *Science* **168**: 594–596.

Bitman, J., H. C. Cecil, S. J. Harris, and G. F. Fries. 1968. Estrogenic activity of o,p'-DDT in the mammalian uterus and avian oviduct. *Science* **162**: 371–378.

Bitman, J., H. C. Cecil, S. J. Harris, and G. F. Fries. 1969. DDT induced a decrease in eggshell calcium. *Nature* **224**: 44–46.

Black, W. M., and D. Neely. 1975. Effect of soil-injected benomyl on resident earthworm populations. *Pestic. Sci.* **65**: 543–545.

Blackmore, D. K. 1963. The toxicity of some chlorinated hydrocarbon insecticides to British wild foxes (*Vulpes vulpes*). *J. Comp. Pathol. Ther.* **73**: 391–409.

Blake, H. T., and P. A. Andrilenas. 1975. Farmers' use of pesticides in 1971— Expenditures. U.S. Agricultural Economic Report No. 296.

Blus, L. J., A. A. Belisle, and R. M. Prouty. 1974. Relations of the brown pelican to certain environmental pollutants. *Pestic. Monit. J.* **7**: 181–194.

Blus, L. J., C. D. Gish, A. A. Belisle, and R. M. Prouty. 1972. Logarithmic relationship of DDE residues to eggshall thinning. *Nature* **235**: 376–377.

Blus, L. J., T. Joanen, A. A. Belisle, and R. M. Prouty. 1975. The brown pelican and certain environmental pollutants in Louisiana. *Bull. Environ. Contam. Toxicol.* 13: 646–655.

Boelens, R. G. U. 1974. Pesticides in the aquatic environment. *Agr. Pest. Soc. Report* 1: 29–38.

Boffey, P. M. 1971. Herbicides in Vietnam: AAAS study finds widespread devastation. *Science* 171: 43.

Booer, J. 1944. The behaviour of mercury compounds in soil. *Ann. Appl. Biol.* 31: 340–359.

Borror, D. J., D. M. DeLong, and C. A. Triplehorn. 1976. *An introduction to the study of insects.* Holt, Rinehart and Winston, New York. P. 852.

Bowers, W. S. 1976. Hormone mimics. In: *The future for insecticides—needs and prospects.* Metcalf, R. L., and J. J. McKelvey, Jr. (Eds.). John Wiley & Sons, New York. Pp. 421–441.

Boyd, J. C. 1971. Field study of a chlordane residue problem: Soil and plant relationships. *Bull. Environ. Contam. Toxicol.* 6: 177–182.

Boyle, C. M. 1960. Case of apparent resistance of *Rattus norvegicus* Berkenhout to anticoagulant poisons. *Nature* 188: 517.

Bozarth, G. A., H. H. Funderburk, Jr., E. A. Curl, and D. E. Davis. 1965. Preliminary studies on the degradation of paraquat by soil microorganisms. *Proc. Southern Weed Conf.* 18: 615.

Brann, J. L., Jr. 1965. Factors affecting the thoroughness of spray application. *N.Y. State Agric. Exp. Stat. J.* Paper No. 1429.

Braun, H. E., and B. T. Lobb. 1976. Residues in milk and organs in a dairy herd following acute endosulfan intoxication. *Can. J. Anim. Sci.* 56: 373–376.

Breidenbach, W. W., C. G. Gunnerson, F. K. Kawahara, J. J. Lichtenberg, and R. S. Green. 1967. Chlorinated hydrocarbon pesticides in major river basins 1957–1965. *Publ. Health Rept.*, Washington. 82: 139–156.

Brian, R. C. 1976. The history and classification of herbicides. In: *Herbicides: Physiology, biochemistry, ecology.* Vol. 1. Audus, L. J. (Ed.). Academic Press, London. Pp. 1–54.

Brian, R. C., R. F. Homer, and J. Stubbs. 1958. A new herbicide 1 : 1'-ethylene-2 : 2'-dipyridylium dibromide. *Nature* 181: 446–447.

Britton, W. M. 1975. Toxicity of high dietary levels of DDT in laying hens. *Bull. Environ. Contam. Toxicol.* 13: 703–706.

Britton, W. M. 1975a. Influence of high levels of DDT in the diet on liver microsomal estrogen metabolism in the laying hen. *Bull. Environ. Contam. Toxicol.* 13: 698–702.

Broadhurst, N., M. L. Montgomery, and V. H. Freed. 1966. Metabolism of 2-methoxy-3,6-dichlorobenzoic acid (dicamba) by wheat and bluegrass plants. *J. Agric. Food Chem.* 14: 585–588.

Brodtmann, N. V., Jr. 1976. Continuous analysis of chlorinated hydrocarbon pesticides in the lower Mississippi River. *Bull. Environ. Contam. Toxicol.* 15: 33–39.

Brooks, G. T. 1969. Metabolism of diene organochlorine (cyclodiene) insecticides. *Res. Rev.* 27: 81–138.

Brown, A. W. A. 1971. Pest resistance to pesticides. In: *Pesticides in the Environment.* Vol. I, Part II. White-Stevens, R. (Ed.). Marcel Dekker, Inc., New York. Pp. 457–552.

Brown, E., and Y. A. Nishioka. 1967. Pesticides in selected western streams—a contribution to the National Program. *Pestic. Monit. J.* 1: 38–46.

Brown, J. F., and H. D. Sisler. 1960. Mechanisms of fungitoxic action of n-dodecylguanidine acetate. *Phytopathology* 50: 830–839.

Brown, J. R., L. Y. Chow, and F. C. Chai. 1975. Distribution of organochlorine pesticides in an agricultural environment, Holland Marsh, Ontario, 1970–1972. *Pestic. Monit. J.* 9: 30–33.

Brown, N. J., and A. W. A. Brown. 1970. Biological fate of DDT in a sub-arctic environment. *J. Wildl. Mgmt.* 34: 929–940.

Brown, V. K. H., J. Robinson, E. Thorpe, and J. W. Barrett. 1974. The toxicity of dieldrin (HEOD) to domestic fowl. *Pestic. Sci.* 5: 567–586.

Bucha, H. C., W. E. Cupery, J. E. Harrod, H. M. Loux, and L. M. Ellis. 1962. Substituted uracil herbicides. *Science* 137: 537–538.

Bucha, H. C., and C. W. Todd. 1951. 3-(p-chlorophenyl)-1,1-dimethylurea—a new selective herbicide. *Science* 114: 493–494.

Buck, W. B., and W. Van Note. 1968. Aldrin poisoning resulting in dieldrin residues in meat and milk. *J. Amer. Vet. Med. Assoc.* 153: 1472–1475.

Buckner, C. H., and B. B. MacLeod. 1975. Impact of aerial applications of Orthene® upon non-target organisms. Chemical Control Research Institute (Canada). Report CC-X-104. P. 48.

Buckner, C. H., B. B. MacLeod, and T. A. Gochnauer. 1973. The impact of forest spraying on populations of small forest songbirds, small mammals and honey bees in the Menjou depot area of Quebec, 1973. Chemical Control Research Institute (Canada). Information Report CC-X-84.

Buckner, C. H., B. B. MacLeod, and D. G. H. Ray. 1973a. The effect of operational applications of various insecticides on small forest birds and mammals. Chemical Control Research Institute (Canada). Information Report CC-X-43.

Buckner, C. H., and R. Sarrazin. 1975. Studies of the environmental impact of the 1974 spruce budworm control operation in Quebec. Chemical Control Research Institute (Canada). Report CC-X-93. P. 106.

Bukovac, M. J. 1976. Herbicide entry into plants. In: *Herbicides: Physiology, biochemistry, ecology* (2nd ed.). Vol. 1. Audus, L. J. (Ed.). Academic Press, London. Pp. 335–364.

Burchfield, H. P. 1959. Comparative stabilities of Dyrene, 1-fluoro-2,4-denitrobenzene, dichlone and captan in a silt loam soil. *Contrib. Boyce Thompson Inst.* 20: 205–215.

Burdick, G. E., E. J. Harris, H. J. Dean, T. M. Walker, J. Skea, and D. Coley. 1964. The accumulation of DDT in lake trout and the effect on reproduction. *Trans. Amer. Fish. Soc.* 93: 127–136.

Burnett, R. 1971. DDT residues: Distribution of concentrations in *Emerita analoga* (Stimpson) along coastal California. *Science* 174: 606–608.

Burns, B. G., M. R. Peach, and D. A. Stiles. 1975. Organochlorine pesticide residues in a farming area, Nova Scotia, 1972–1973. *Pestic. Monit. J.* 1: 34–38.

Burns, J. E. 1974. Organochlorine pesticide and polychlorinated biphenyl residues in biopsied human adipose tissue, Texas, 1969–1972. *Pestic. Monit. J.* 7: 122–126.

Butenandt, A., R. Beckmann, D. Stamm, and E. Hecker. 1959. Über den sexual lockstoff des seidenspinners *Bombyx Mori*. Reindarstellung und Konstitution. *Z. Naturforsch.* B14: 283–284.

Butler, G. L., T. R. Deason, and J. C. O'Kelley. 1975. Loss of five pesticides from cultures of twenty-one planktonic algae. *Bull. Environ. Contam. Toxicol.* 13: 149–152.

Butler, P. A. 1963. *Commercial fisheries investigations*. USDI Bureau of Fish & Wildlife Circular 167. Pp. 11–25.

Cairns, V. W. 1972. Some effects of diazinon on the aggressive behaviour of juvenile rainbow trout (*Salmo gairdner* Richardson). M.Sc. Thesis. University of Guelph, Guelph, Ontario.

Calderbank, A. 1968. The bipyridylium herbicides. *Adv. Pest Control Res.* 8: 127–235.

Calderbank, A. C., and P. Slade. 1976. Diquat and paraquat. In: *Herbicides: Chemistry, degradation, and mode of action* (2nd ed.). Vol. 2. Kearney, P. C., and D. D. Kaufman (Eds.). Marcel Dekker, Inc., New York. Pp. 501–540.

Call, D. J., H. J. Shave, H. C. Binger, M. E. Bergeland, B. D. Ammann, and J. J. Worman. 1976. DDE poisoning in wild great blue heron. *Bull. Environ. Contam. Toxicol.* 16: 310–313.

Canton, J. H. 1976. The toxicity of benomyl, thiophanatemethyl, and BCM to four freshwater organisms. *Bull. Environ. Contam. Toxicol.* 16: 214–218.

Canvin, D. T., and G. Friesen. 1959. Cytological effects of CDAA and IPC on germinating barley and peas. *Weeds* 7: 153–156.

Carson, R. 1962. *Silent spring*. Houghton-Mifflin Co., Boston. P. 368.

Carter, H. R. 1931. Yellow fever: An epidemiological and historical study of its place of origin. Williams and Wilkins, Baltimore. P. 308.

Carter, M. C. 1975. Amitrole. In: *Herbicides: Chemistry, degradation and mode of action* (2nd ed.). Vol. 1. Kearney, D. C., and D. D. Kaufman (Eds.). Marcel Dekker, Inc., New York. Pp. 377–398.

Carter, R. H., P. E. Hubanks, H. D. Mann, L. M. Alexander, and G. E. Schopmeyer. 1948. Effect of cooking on the DDT content of beef. *Science* 107: 347.

Cartwright, P. M. 1976. General growth responses of plants. In: *Herbicides: Physiology, biochemistry, ecology* (2nd ed.). Vol. 1. Audus, L. J. (Ed.). Academic Press, London. Pp. 55–82.

Casarett, L. J., G. C. Fryer, W. R. Yauger, and H. W. Klemmer. 1968. Organochlorine pesticide residues in human tissues. *Arch. Environ. Hlth.* 17: 306–311.

Casida, J. E., T. C. Allen, and M. A. Stahmann. 1954. Mammalian conversion of octamethylpyrophosphoramide to a toxic phosphoramide N-oxide. *J. Biol. Chem.* 210: 607–616.

Casper, V. L. 1967. Galveston Bay pesticide study—Water and oyster samples analyzed for pesticide residues following mosquito control program. *Pestic. Monit. J.* 1: 13–21.

Cecil, H. C., J. Bitman, and S. J. Harris. 1971. Effects of dietary p,p'-DDT and p,p'-DDE on egg production and egg shell characteristics of japanese quail receiving an adequate calcium diet. *Poultry Sci.* 50: 657–659.

Cecil, H. C., G. F. Fries, J. Bitman, S. J. Harris, R. J. Lillie, and C. A. Denton. 1972. Dietary p,p'-DDT, $0,p'$-DDT or p,p'-DDE and changes in egg shell characteristics and pesticide accumulation in egg contents and body fat of caged white leghorns. *Poultry Sci.* 51: 130–139.

Cerf, D. C., and G. P. Georghiou. 1972. Evidence of cross-resistance to a juvenile hormone analogue in some insecticide-resistant houseflies. *Nature* 239: 401–402.

Chaiyarach, S., V. Ratananun, and R. C. Harrel. 1975. Acute toxicity of the insecticides toxaphene and carbaryl and the herbicides propanil and molinate to four species of aquatic organisms. *Bull. Environ. Contam. Toxicol.* 14: 281–284.

Chang, F. Y., G. R. Stephenson, and J. D. Bandeen. 1974. Effects of N,N-diallyl-2,2-dichloroacetamide on ethyl N,N-di-n-propylthiocarbamate uptake and metabolism by corn seedlings. *J. Agric. Food. Chem.* 22: 245–248.

Chang, I., and C. L. Foy. 1971. Effects of picloram on mitochondrial swelling and ATPase. *Weed Sci.* 19: 54–58.

Chapman, R. K., and C. J. Eckenrode. 1973. Effect of insecticide placement on predator numbers and cabbage maggot control. *J. Econ. Entomol.* 66: 1153–1158.

Chen, L. G., A. Ali, R. A. Fletcher, C. M. Switzer, and G. R. Stephenson. 1973. Effects of auxin-like herbicides on nucleohistones in cucumber and wheat roots. *Weed Sci.* 21: 181–184.

Chen, L. G., C. M. Switzer, and R. A. Fletcher. 1972. Nucleic acid and protein changes induced by auxin-like herbicides. *Weed Sci.* 20: 53–55.

Chen, T. M., D. E. Seaman, and F. M. Ashton. 1968. Herbicidal action of molinate in barnyard grass and rice. *Weed Sci.* 16: 28–31.

Cherry, J. H. 1976. Actions on nucleic acid and protein metabolism. In: *Herbicides: Physiology, biochemistry, ecology* (2nd ed.). Audus, L. J. (Ed.). Academic Press, London. Pp. 525–546.

Clark, D. R., Jr. 1975. Effect of stress on dieldrin toxicity to male red winged blackbirds (*Agelaius phoeniceus*). *Bull. Environ. Contam. Toxicol.* 14: 250–256.

Clark, D. R., Jr., and T. G. Lamont. 1976. Organochlorine residues and reproduction in the big brown bat. *J. Wildl. Mgmt.* 40: 249–254.

Clark, D. R., Jr., C. O. Martin, and D. M. Swineford. 1975. Organochlorine insecticide-residues in the free-tailed bat (*Tadarida brasiliensis*) at Bracken Cave, Texas. *J. Mammal.* 56: 429–443.

Clausen, J., L. Braestrup, and O. Berg. 1974. The content of polychlorinated hydrocarbons in Arctic mammals. *Bull. Environ. Contam. Toxicol.* 12: 529–534.

Clegg, D. J., and K. S. Khera. 1973. The teratogenicity of pesticides, their metabolites and contaminants. In: *Pesticides and the environment: A continuing controversy.* Deichmann, W. B. (Ed.). Intercontinental Medical Book Corp., New York & London. Pp. 267–276.

Clemons, G. P., and H. D. Sisler. 1969. Formation of a fungitoxic derivative from benlate. *Phytopathology* 59: 705–706.

Cliath, M. M., and W. F. Spencer. 1972. Dissipation of pesticides from soil by volatilization of degradation products. I. Lindane and DDT. *Environ. Sci. Technol.* 6: 910–914.

Cockrum, E. L. 1969. Editorial: Insecticides and Arizona bat populations. *J. Ariz. Acad. Sci.* **5**: 198.

Cohen, J. M., and C. Pinkerton. 1966. Widespread translocation of pesticides by air transport and rain-out. In: *Organic pesticides in the environment.* Gould, R. F. (Ed.). Advances in Chemistry Series No. 60. American Chemical Society, Washington, D.C. Pp. 163–176.

Colby, S. R. 1965. *N*-glycoside of amiben isolated from soybean plants. *Science* **150**: 619–620.

Collins, S. 1961. Benefits to understory from canopy defoliation by gypsy moth larvae. *Ecology* **42**: 836–838.

Comes, R. D., and F. L. Timmons. 1965. Effect of sunlight on the phytotoxicity of some phenyl urea and triazine herbicides on a soil surface. *Weeds* **13**: 81–84.

Committee. 1973. *Water quality criteria 1972.* National Academy of Sciences, National Academy of Engineering, EPA R3.73.033. P. 594.

Committee on Occupational Toxicology. 1970. Evaluation of the present status of DDT with respect to man. *J. Amer. Med. Assoc.* **212**: 1055–1056.

Cook, R. L. 1930. Effect of soil type and fertilizer on the nitrate content of the expressed sap and the total nitrogen content of the tissue of the small grains. *J. Amer. Soc. Agron.* **22**: 393–408.

Cook, S. F., Jr., and J. D. Connors. 1963. The short term side effects of the insecticidal treatment of Clear Lake, Lake County, California, in 1962. *Ann. Entomol. Soc. Amer.* **56**: 819–824.

Cooke, A. R. 1956. A possible mechanism of action of the urea type herbicides. *Weeds* **4**: 397–398.

Cooke, A. S. 1970. The effect of *pp'*-DDT on tadpoles of the common frog. *Environ. Pollut.* **1**: 57–71.

Cooke, A. S. 1972. The effects of DDT, dieldrin and 2,4-D on amphibian spawn and tadpoles. *Environ. Pollut.* **3**: 51–68.

Cooke, A. S. 1973. Shell thinning in avian eggs by environmental pollutants. *Environ. Pollut.* **4**: 85–152.

Cope, O. B. 1963. *Sport fishery investigations.* USDI Bureau of Fish & Wildlife Circular 167. Pp. 26–42.

Cope, O. B. 1965. Agricultural chemicals and freshwater ecological systems. In: *Research in Pesticides.* Chichester, F. (Ed.). Academic Press, New York. Pp. 115–127.

Cope, O. B., J. P. McCraren, and L. L. Eller. 1969. Effects of dichlobenil on two fish pond environments. *Weed Sci.* **17**: 158–165.

Coppedge, J. R., D. A. Lindquist, D. L. Bull, and H. W. Dorough. 1967. Fate of 2-methyl-2-(methylthio) propionaldehyde *0*-(methyl carbamoyl)oxime (Temik) in cotton plants and soil. *J. Agric. Food Chem.* **15**: 902–910.

Copplestone, J. F. 1977. A global view of pesticide safety. In: *Pesticide management and pesticide resistance.* Watson, D. L., and A. W. A. Brown (Eds.). Academic Press, New York. Pp. 147–155.

Corbett, J. R. 1974. *The biochemical mode of action of pesticides*. Academic Press, New York. P. 330.

Corden, M. E. 1969. Aromatic compounds. In: *Fungicides An Advanced Treatise*. Vol. II. Torgeson, D. C. (Ed.). Academic Press, New York. Pp. 477–529.

Cottam, L. 1965. The ecologists' role in problems of pesticide pollution. *Biosci.* **15:** 457–463.

Courtney, K. D., D. W. Gaylor, M. D. Hogan, H. L. Falk, R. R. Bates, and I. Mitchell. 1970. Teratogenic evaluation of 2,4,5-T. *Science* **168:** 864–866.

Cox, J. L. 1971. DDT residues in seawater and particulate matter in the California current system. *Fish. Bull.* **69:** 443–450.

Cox, J. L. 1971a. Uptake, assimilation, and loss of DDT residues by *Euphausia pacifica*, a euphausiid shrimp. *Fish. Bull.* **69:** 627–633.

Cox, J. L. 1972. DDT residues in marine phytoplankton. *Res. Rev.* **44:** 23–38.

Crafts, A. S. 1945. A new herbicide, 2,4-dinitro, secondary butyl phenol. *Science* **101:** 417–418.

Crafts, A. S. 1956. Weed control: Applied botany. *Amer. J. Bot.* **43:** 548–556.

Crafts, A. S., and C. E. Crisp. 1971. *Phloem transport in plants*. W. H. Freeman and Co., San Francisco. P. 481.

Cramer, H. H. 1967. *Plant protection and world crop production*. Farben Fabriken Bayer. Ag. Leverkusen. Pp. 524.

Crockett, A. B., G. B. Wiersma, H. Tai, W. G. Mitchell, P. F. Sand, and A. E. Carey. 1974. Pesticide residue levels in soils and crops, FY-70-National Soils Monitoring Program (II). *Pestic. Monit. J.* **8:** 69–97.

Cromartie, E., W. L. Reichel, L. N. Locke, A. A. Belisle, T. E. Kaiser, T. G. Lamont, B. M. Mulhern, R. M. Prouty, and D. M. Swineford. 1975. Residues of organochlorine pesticide and polychlorinated biphenyls and autopsy data for bald eagles, 1971–1972. *Pestic. Monit. J.* **9:** 11– 14.

Crosby, D. G. 1976. Herbicide photodecomposition. In: *Herbicides: Chemistry, degradation and mode of action*. Vol. 2. Kearney, P. C., and D. D. Kaufman (Eds.). Marcel Dekker, Inc., New York. Pp. 836–891.

Cuerrier, J. P., J. A. Keith, and E. Stone. 1967. Problems with DDT in fish culture operation. *Natur. Can.* **94:** 315–320.

Cummings, J. G., M. Eidelman, V. Turner, D. Reed, and K. T. Zee. 1967. Residues in poultry tissues from low level feeding of five chlorinated hydrocarbon insecticides to hens. *J. Assoc. Offic. Anal. Chem.* **50:** 418–425.

Cummings, J. G., K. T. Zee, V. Turner, and F. Quinn. 1966. Residues in eggs from low level feeding of five chlorinated hydrocarbon insecticides to hens. *J. Assoc. Offic. Anal. Chem.* **49:** 354– 364.

Curley, A., F. Copeland, and R. D. Kimbrough. 1969. Chlorinated hydrocarbon insecticides in organs of stillborn and blood of newborn babies. *Arch. Environ. Hlth.* **19:** 628–632.

Cuthbert, J. H. 1963. Further evidence of resistance to warfarin in the rat. *Nature* **198:** 807– 808.

Czegledi-Janko, G., and A. Hollo. 1967. Determination of the degradation products of ethylene bis (dithiocarbamates) by thin-layer chromatography, and some investigations of their decomposition in vitro. *J. Chromatography* **31:** 89– 95.

Daciw, M. 1976. Herbicides used agriculturally in western Canada for weed control (Unpublished).

Dale, W. E., T. B. Gaines, and W. J. Hayes, Jr. 1962. Storage and excretion of DDT in starved rats. *Toxicol. Appl. Pharmacol.* **4:** 89–106.

Darrow, D. C., and G. C. H. Harding. 1975. Accumulation and apparent absence of DDT metabolism by marine copepods, *Calanus* spp. in culture. *J. Fish. Res. Bd. Canada* **32:** 1845– 1849.

Davidson, W. B., J. L. Doughty, and J. L. Bolton. 1941. Nitrate poisoning of livestock. *Can. J. Comp. Med.* **5:** 303.

Davies, J. E. 1973. Pesticide residues in man. In: *Environmental pollution by pesticides.* Edwards, C. A. (Ed.). Plenum Press, London & New York. Pp. 313–333.

Davies, J. E., W. F. Edmundson, A. Raffonelli, J. C. Cassady, and C. Morgade. 1973. The role of social class in human pesticide pollution. In: *Pesticides and the environment: A continuing controversy.* Deichmann, W. B. (Ed.). Intercontinental Medical Book Corp., New York & London. Pp. 335–346.

Davis, J. H., J. E. Davies, and A. J. Fisk. 1969. Occurrence, diagnosis, and treatment of organophosphate pesticide poisoning in man. In: *Biological effects of pesticides in mammalian systems.* Kraybill, H. F. (Ed.). *Ann. N.Y. Acad. Sci.* **160:** 383–392.

Davison, K. L., K. A. Engebretson, and J. H. Cox. 1976. *p,p'*-DDT and *p,p'*-DDE effects on egg production, eggshell thickness, and reproduction of japanese quail. *Bull. Environ. Contam. Toxicol.* **15:** 265– 270.

Decker, G. C., W. N. Bruce, and J. H. Bigger. 1965. The accumulation and dissipation of residues resulting from the use of aldrin in soil. *J. Econ. Entomol.* **58:** 266–271.

Decker, G. C., C. J. Weinman, and J. M. Bann. 1950. A preliminary report on the rate of insecticide residue loss from treated plants. *J. Econ. Entomol.* **43:** 919–927.

DeGrazio, J. W., J. F. Besser, T. J. DeCino, J. L. Guarino, and E. W. Schafer, Jr. 1972. Protecting ripening corn from blackbirds by broadcasting 4-aminopyridine baits. *J. Wildl. Mgmt.* **36:** 1316–1319.

DeGrazio, J. W., J. F. Besser, T. J. DeCino, J. L. Guarino, and R. I. Starr. 1971. Use of 4-aminopyridine to protect ripening corn from blackbirds. *J. Wildl. Mgmt.* **35:** 565–569.

Deichmann, W. B. 1972. Research, DDT and cancer. *Indust. Med. Surg.* **41:** 15–18.

Deichmann, W. B. 1973. The chronic toxicity of organochlorine pesticides in man. In: *Pesticides and the environment: A continuing controversy.* Deichmann, W. B. (Ed.). Intercontinental Medical Book Corp., New York & London. Pp. 347–420.

Deichmann, W. B., and J. L. Radomski. 1968. Retention of pesticides in human adipose-tissue—Preliminary report. *Indust. Med. Surg.* **37:** 218–219.

Dekker, J. 1977. Effect of fungicides on nucleic acid synthesis and nuclear function. In: *Antifungal compounds.* Vol. 2. *Interactions in biological and ecological systems.* Siegel, M. R., and H. D. Sisler (Eds.). Marcel Dekker, New York. Pp. 365–398.

DeLong, R. L., W. G. Gilmartin, and J. G. Simpson. 1973. Premature births in California sea lions: Association with high organochlorine pollutant residue levels. *Science* **181**: 1168–1170.

Derr, S. K., and M. J. Zabic. 1972. Biologically active compounds in the aquatic environment: The uptake and distribution of [1,1-dichloro-2,2-bis(p-chlorophenyl)ethylene], DDE by *Chironomus tentans* Fabricius (Diptera: Chironomidae). *Trans. Amer. Fish. Soc.* **101**: 323–329.

Derr, S. K., and M. J. Zabik. 1974. Bioactive compounds in the aquatic environment: Studies on the mode of uptake of DDE by the aquatic midge, *Chironomus tentans* (Diptera: Chironomidae). *Arch. Environ. Contam. Toxicol.* **2**: 152–164.

Desai, P. D. 1972. Herbicidal action of pronamide on plant roots. Ph.D. Thesis. University of Guelph, Guelph, Ontario.

Devlin, R. M. 1974. DDT: A renaissance. *Environ. Sci. Technol.* **8**: 322–325.

DeWitt, J. B., and J. L. George. 1959. *Bureau of Sport Fisheries and Wildlife Pesticide—Wildlife Review: 1959.* Fish & Wildlife Service Circular No. 84 (Revised). P. 36.

Dickens, R., and A. E. Hiltbold. 1967. Movement and persistence of methane arsonates in soil. *Weeds* **15**: 299–304.

Dieter, M. P., and J. L. Ludke. 1975. Studies on combined effects of organophosphates and heavy metals in birds. I. Plasma and brain cholinesterase in coturnix quail fed methyl mercury and orally dosed with parathion. *Bull. Environ. Contam. Toxicol.* **13**: 257–262.

Dill, P. A., and R. C. Saunders. 1974. Retarded behavioral development and impaired balance in Atlantic salmon (*Salmo salar*) alevins hatched from gastrulae exposed to DDT. *J. Fish. Res. Bd. Can.* **31**: 1936–1938.

Dimond, J. B., G. Y. Belyea, R. E. Kadunce, A. S. Getchell, and J. A. Blease. 1970. DDT residues in robins and earthworms associated with contaminated soils. *Can. Entomol.* **102**: 1122–1129.

Dimond, J. B., A. S. Getchell, and J. A. Blease. 1971. Accumulation and persistence of DDT in a lotic ecosystem. *J. Fish. Res. Bd. Canada.* **28**: 1877–1882.

Dimond, J. B., R. E. Kadunce, A. S. Getchell, and J. A. Blease. 1968. DDT residue persistence in red-backed salamanders in a natural environment. *Bull. Environ. Contam. Toxicol.* **3**: 194–202.

Dimond, J. B., R. B. Owen, Jr., and A. S. Getchell. 1974. Distribution of DDTR in a uniformly treated stream. *Bull. Environ. Contam. Toxicol.* **12**: 522–528.

Dimond, J. B., R. B., R. B. Owen, Jr., and A. S. Getchell. 1975. DDT residues in forest biota: Further data. *Bull. Environ. Contam. Toxicol.* **13**: 117–122.

Dimond, J. B., and J. A. Sherburne. 1969. Persistence of DDT in wild populations of small mammals. *Nature* **221**: 486–487.

Dorough, H. W., and G. W. Ivie. 1968. Temik-S[35] metabolism in a lactating cow. *J. Agric. Food Chem.* **16**: 460–464.

Duggan, R. E. 1968. Pesticide residue levels in the United States from July 1, 1963, to June 30, 1967. *Pestic. Monit. J.* **2**: 2–46.

Duggan, R. E., and P. E. Corneliussen. 1972. Dietary intake of pesticide chemicals in the United States (III), June 1968–April 1970. *Pestic. Monit. J.* **5**: 331–341.

Duggan, R. E., and K. Dawson. 1967. Pesticides—A report on residues in food. *N.A.C. News & Pestic. Rev.* **25**(6): 3–7.

Duggan, R. E., and G. Q. Lipscomb. 1971. Regulatory control of pesticide residues in food. *J. Dairy Sci.* **6**95–701.

Duggan, R. E., and F. J. McFarland. 1967. Residues in food and feed. *Pestic. Monit. J.* **1**: 1–5.

Duke, T. W., and A. J. Wilson, Jr. 1971. Chlorinated hydrocarbons in livers of fishes from the northeastern Pacific Ocean. *Pestic. Monit. J.* **5**: 228–232.

Duke, W. B., F. W. Slife, and J. B. Hanson. 1967. *Studies on the mode of action of 2-chloro-N-isopropyl-acetanilide.* Weed Society of America P. 50.

Durham, W. F. 1969. Body burden of pesticides in man. In: *Biological effects of pesticides in mammalian systems.* Kraybill, H. F. (Ed.). *Ann. N.Y. Acad. Sci.* **160**: 183–195.

Durham, W. F., and C. H. Williams. 1973. Mutagenic, teratogenic, and carcinogenic properties of pesticides. In: *Pesticides and the environment: A continuing controversy.* Deichmann, W. B. (Ed.). Intercontinental Medical Book Corp., New York & London. Pp. 307–334.

Dustman, E. H., and L. F. Stickel. 1969. The occurrence and significance of pesticide residues in wild animals. In: *Biological effects of pesticides in mammalian systems.* Kraybill, H. F. (Ed.). *Ann. N.Y. Acad. Sci.* **160**: 162–172.

Dutky, S. R. 1963. The milky diseases. In: *Insect pathology, an advanced treatise.* Vol. 2. Steinhaus, E. A. (Ed.). Academic Press, New York. Pp. 75–115.

Dvorchik, B. H., M. Istin, and T. H. March. 1971. Does DDT inhibit carbonic anhydrase. *Science* **172**: 728–729.

Dyer, M. I. 1968. Blackbird and starling research program 1964–1968. Ontario Department of Agriculture and Food, Toronto. P. 29.

Dykstra, W. W., and R. E. Lennon. 1966. The role of chemicals for the control of vertebrate pests. In: *Pest control by chemical, biological, genetic, and physical means. A symposium.* Knipling, E. F. (Chairman). U.S. Department of Agriculture, Agricultural Research Service. Pp. 33–110. 29–34.

Dziuk, L. J., and F. W. Plapp. 1973. Insecticide resistance in mosquitofish from Texas. *Bull. Environ. Contam. Toxicol.* **9**: 15–19.

Earnest, R. D., and P. E. Benville, Jr. 1971. Correlation of DDT and lipid levels for certain San Francisco Bay fish. *Pestic. Monit. J.* **5**: 235–241.

Ecobichon, D. J., and P. W. Saschenbrecker. 1968. Pharmacodynamic study of DDT in cockerels. *Can. J. Physiol. & Pharmacol.* **46**: 785–793.

Edgington, L. V., and G. L. Barron. 1967. Fungitoxic spectrum of oxathiin compounds. *Phytopathology* **57**: 1256–1257.

Edgington, L. V., and L. V. Busch. 1967. Control of Rhizoctonia stem canker in potato. *Can. Plant Dis. Surv.* **47**: 28–29.

Edgington, L. V., and C. Corke. 1967. Biological decomposition of an oxathiin fungicide. *Phytopathology* **57**: 810.

Edgington, L. V., E. Reinbergs, and M. C. Shephard. 1972. Evaluation of ethirimol and benomyl for control of powdery mildew of barley. *Can. J. Plant Sci.* **52**: 693–699.

Edwards, C. A. 1970. *Persistent pesticides in the environment.* CRC Press, Cleveland, Ohio. Pp. 78.

Edwards, C. A. 1973. *Persistent pesticides in the environment* (2nd ed.). CRC Press, Cleveland, Ohio. Pp. 170.

Edwards, C. A. 1973a. Pesticide residues in soil and water. In: *Environmental pollution by pesticides.* Edwards, C. A. (Ed.). Plenum Press, London & New York. Pp. 409–458.

Edwards, C. A., and A. R. Thompson. 1973. Pesticides and the soil fauna. *Res. Rev.* **45**: 1–80.

Eisler, R. 1969. Acute toxicities of insecticides to marine decapod crustaceans. *Crustaceana* **16**: 302–310.

Eisler, R. 1970. Acute toxicities of organochlorine and organophosphorus insecticides to estuarine fishes. USDI Bureau of Sport Fisheries & Wildlife Technical Paper 46. Pp. 3–12.

Eisler, R. 1970a. Factors affecting pesticide-induced toxicity in an estuarine fish. USDI Bureau of Sport Fisheries & Wildlife Technical Paper 45. Pp. 3–19.

Eliasson, L. 1965. Xylem transport of 2,4-D in aspen. *Physiologra Plantarum* **18**: 506–515.

Elkins, E. R., F. C. Lamb, R. P. Farrow, R. W. Cook, M. Kawai, and J. R. Kimball. 1968. Removal of DDT, malathion, and carbaryl from green beans by commercial and home preparative procedures. *J. Agric. Food Chem.* **16**: 962–966.

Elliot, M. 1977. Synthetic pyrethroids. In: *Synthetic pyrethroids.* Elliot, M. (Ed.). American Chemical Society, Washington, D.C. Pp. 1–28.

Embry, T. L., D. P. Morgan, and C. C. Roan. 1972. Search for abnormalities of heme synthesis and sympathoadrenal activity in workers regularly exposed to pesticides. *J. Occ. Med.* **14**: 918–921.

Engst, R., R. Knoll, and B. Nickel. 1969. Occurrence of DDT and DDE in fatty residues and organs of infants. *Pharmacie* **24**: 673–676.

Engst, R., and W. Schnaak. 1967. Metabolism of the fungicidal ethylene bis (dithiocarbamates) maneb and zineb. I. Identification and fungitoxic activity in a model experiment of synthesized metabolites. *Zeitschrift fur Lebensmittel-Untersuchungs-Forschung* **134**: 216–221.

Environmental Protection Agency. 1971. *Agricultural pollution of the Great Lakes basin.* EPA Report 13020–07/71. Pp. 94.

Epstein, S. S., and M. S. Legator (Eds.). 1971. *Mutagenicity of pesticides: Concepts and evaluations.* Cambridge, Mass., Institute of Technology Press. P. 240.

Epstein, S. S., and H. Shafner. 1968. Chemical mutagens in the human environment. *Nature* **219**: 385–387.

Ernst, W., H. Goerke, G. Eder, and R. G. Schaefer. 1976. Residues of chlorinated hydrocarbons in marine organisms in relation to size and ecological parameters. I. PCB, DDT, DDE, and DDD in fishes and molluscs from the English Channel. *Bull. Environ. Contam. Toxicol.* **15**: 55–65.

Esser, H. O., G. Dupuis, E. Ebert, C. Vogel, and G. J. Marco. 1975. S-triazines. In: *Herbicides: Chemistry, degradation, and mode of action* (2nd ed.). Vol. 1. Kearney, P. C., and D. D. Kaufman (Eds.). Marcel Dekker, Inc., New York. Pp. 129–208.

Ettinger, M. B., and D. I. Mount. 1967. A wild fish should be safe to eat. *Environ. Sci. Technol.* 1: 203–205.

Fabacher, D. L. 1976. Toxicity of endrin and an endrin-methyl parathion formulation to largemouth bass fingerlings. *Bull. Environ. Contam. Toxicol.* 16: 376–378.

Faber, R. A., and J. J. Hickey. 1973. Eggshell thinning, chlorinated hydrocarbons, and mercury in inland aquatic bird eggs, 1969 and 1970. *Pestic. Monit. J.* 7: 27–36.

Faber, R. A., R. W. Risebrough, and H. M. Pratt. 1972. Organochlorines and mercury in common egrets and great blue herons. *Environ. Pollut.* 3: 111–122.

Fahey, J. E., J. W. Butcher, and R. T. Murphy. 1965. Chlorinated hydrocarbon insecticide residues in soils of urban areas, Battle Creek, Michigan. *J. Econ. Entomol.* 58: 1026–1027.

Fahey, J. E., P. E. Nelson, and D. L. Ballee. 1970. Removal of Gardona from fruit by commercial preparative methods. *J. Agric. Food Chem.* 18: 866–868.

Fahmy, O. G., and M. J. Fahmy. 1969. The genetic effects of the biological alkylating agents with reference to pesticides. In: *Biological effects of pesticides in mammalian systems*. Kraybill, H. F. (Ed.). *Ann. N.Y. Acad. Sci.* 160: 228–243.

Fang, S. C., M. George, and V. H. Freed. 1964. The metabolism of s-propyl-1-C^{14} n-butylethythiocarbamate (Tillam-C^{14}) in rats. *J. Agric. Food Chem.* 12: 37–40.

Farrow, R. P., F. C. Lamb, E. R. Elkins, R. W. Cook, M. Kawai, and A. Cortes. 1969. Effect of commercial and home preparative procedures on parathion and carbaryl residues in broccoli. *J. Agric. Food Chem.* 17: 75–79.

Fawcett, R. S., and F. W. Slife. 1975. Germination stimulation properties of carbamate herbicides. *Weed Sci.* 23: 419–424.

Federal Register. 1974. Worker protection standards for agricultural workers. 39(92): 16888.

Federal Register. 1975. Pesticide programs: Guidelines for registering pesticides in the United States. 40(123): 26802–26928.

Federal Register. 1975a. Pesticide programs: Registration, re-registration and classification procedures. 40(129): 28242–28286.

Federal Register. 1975b. 40(129): 28244.

Ferguson, D. E., D. D. Culley, W. D. Cotton, and R. P. Dodds. 1964. Resistance to chlorinated hydrocarbon insecticides in three species of freshwater fish. *Biosci.* 14: 43–44.

Fertig, S. N. 1952. Livestock poisoning from herbicide-treated vegetation. *Proc. N.E. Weed Control Conf.* 6: 13–19.

Finberg, L., J. Kiley, and C. Luttrell. 1963. Mass accidental salt poisoning in infancy. *J. Amer. Med. Assoc.* 184: 187–190.

Fisher, R. W., and A. Hikichi. 1971. Orchard sprayers. Ontario Ministry Agriculture & Food Publication 373. P. 44.

Fitzhugh, O. G., A. A. Nelson, E. P. Laug, and F. M. Kunze. 1950. Chronic oral toxicities of mercuri-phenyl and mercuric salts. *Arch. Ind. Hyg. Occuptl. Med.* **2**: 433–442.

Fitzpatrick, G., and D. J. Sutherland. 1976. Uptake of the mosquito larvicide temefos by the salt marsh snail, New Jersey, 1973–1974. *Pestic. Monit. J.* **10**: 4–6.

Fleet, R. R., D. R. Clark, Jr., and F. W. Plapp, Jr. 1972. Residues of DDT and dieldrin in snakes from two Texas agro-systems. *Biosci.* **22**: 664–665.

Flickinger, E. L., and K. A. King. 1972. Some effects of aldrin-treated rice on Gulf Coast wildlife. *J. Wildl. Mgmt.* **36**: 706–727.

Folmar, L. C. 1976. Overt avoidance reaction of rainbow trout fry to nine herbicides. *Bull. Environ. Contam. Toxicol.* **15**: 509–514.

Food & Agriculture Organization/World Health Organization. 1972. *Pesticide residues in food.* Report of the 1971 Joint FAO/WHO Meeting. WHO Techncal Report Series No. 502. P. 46.

Fowle, C. D. 1966. The effects of phosphamidon on birds in New Brunswick forests. *J. Appl. Ecol.* **3** (Suppl.): 169 (Abstract).

Fowle, C. D. 1969. *Effects of phosphamidon on forest birds in New Brunswick.* Canadian Wildlife Service Report Series No. 16. P. 26.

Fox, G. A. 1976. Eggshell quality: Its ecological and physiological significance in a DDE-contaminated common tern population. *The Wilson Bull.* **88**: 454–477.

Foy, C. L. 1976. Picloram and related compounds. In: *Herbicides: Chemistry, degradation, and mode of action.* Kearney, P. C., and D. D. Kaufman (Eds.). Marcel Dekker, Inc., New York. Pp. 777–813.

Foy, C. L., and D. Penner. 1965. Effect of inhibitors and herbicides on tricarboxylic acid cycle substrate oxidation by cucumber mitochondria. *Weed Sci.* **13**: 226–231.

Frank, R., A. E. Armstrong, R. G. Boelens, H. E. Braun, and C. W. Douglas. 1974a. Organochlorine residues in sediment and fish tissues, Ontario, Canada. *Pestic. Monit. J.* **7**: 165–180.

Frank, R., M. V. H. Holdrinet, and W. A. Ripley. 1975a. Residue of organochlorine compounds and mercury in birds' eggs from Niagara Peninsula, Ontario. *Arch. Environ. Contam.* **3**: 205–218.

Frank, R., K. Montgomery, H. E. Braun, A. H. Berst, and K. Loftus. 1974. DDT and dieldrin in watersheds draining the tobacco belt of southern Ontario. *Pestic. Monit. J.* **8**: 184–201.

Frank, R., K. Ronald, and H. E. Braun. 1973. Organochlorine residues in harp seals (*Pagophilus groenlandicus*) caught in eastern Canadian waters. *J. Fish. Res. Bd. Can.* **30**: 1053–1063.

Frank, R., E. H. Smith, H. E. Braun, M. Holdrinet, and J. W. McWade. 1975. Organochlorine insecticides and industrial pollutants in the milk supply of the southern region of Ontario, Canada. *J. Milk Food Technol.* **38**: 65–72.

Frank, R., R. L. Thomas, M. Holdrinet, A. L. W. Kemp, H. E. Braun, and J. M. Jaquat. 1977. Organochlorine insecticides and PCBs in sediments of Lake St. Clair (1970 and 1974) and Lake Erie (1971). The Science of the total Environment, **8**: 205–227.

Frear, D. S. 1976. The benzoic acid herbicides. In: *Herbicides: Chemistry, degradation, and mode of action* (2nd ed.). Vol. 2. Kearney, P. C., and D. D. Kaufman (Eds.). Marcel Dekker, Inc., New York. Pp. 541–607.

Fredeen, F. J. H. 1974. Tests with single injections of methoxychlor black fly (Diptera: Simuliidae) larvicides in large rivers. *Can. Entomol.* **106:** 285–305.

Fredeen, F. J. H., J. G. Saha, and M. H. Balba. 1975. Methoxychlor residues in the Saskatchewan River and selected fauna following injections of methoxychlor black fly larvicide into the river. *Pestic. Monit. J.* **8:** 241–246.

Friend, M., and D. O. Trainer. 1974. Response of different-age mallards to DDT. *Bull. Environ. Contam. Toxicol.* **11:** 49–56.

Friesen, G. 1957. An appraisal of losses caused by weed competition in Manitoba grain fields. *North Central Weed Control Conf.* **14:** 40.

Friesen, H. A. 1973. Identifying wild oats yield losses and assessing cultural control methods. Let's clean up on wild oats. Seminar. Saskatoon, Saskatchewan (Unpublished).

Fry, J. C., M. P. Brooker, and P. L. Thomas. 1973. Changes in the microbial populations of a reservoir treated with the herbicide paraquat. *Water Res.* **7:** 395–407.

Fuller, H. J. 1955. *General botany.* Barnes and Noble, Inc., New York. Pp. 196.

Funckes, A. J., G. R. Hayes, Jr., and W. V. Hartwell. 1963. Urinary excretion of paranitrophenol by volunteers following dermal exposure to parathion at different ambient temperatures. *J. Agric. Food Chem.* **11:** 455–457.

Funderburk, H. H., Jr., and J. M. Lawrence. 1964. Mode of action and metabolism of diquat and paraquat. *Weeds* **12:** 259–264.

Funderburk, H. H., Jr., N. S. Negi, and J. M. Lawrence. 1966. Photochemical decomposition of diquat and paraquat. *Weeds* **14:** 240–243.

Furtick, W. R. 1970. Present and potential contributions of weed control to solution of problem of meeting the world's food needs. In: *FAO International Conference on Weed Control.* Weed Science Society of America. Pp. 1–6.

Fyfe, R. W., J. Campbell, B. Hayson, and K. Hodson. 1969. Regional population declines and organochlorine insecticides in Canadian Prairie falcons. *Can. Field Natur.* **83:** 191–200.

Gambrell, F. L., H. Tashiro, and G. L. Mack. 1968. Residual activity of chlorinated hydrocarbon insecticides in permanent turf for European chafer control. *J. Econ. Entomol.* **61:** 1508–1511.

Gardiner, J. A., R. C. Rhodes, J. B. Adams, Jr., and E. J. Soboezenski. 1969. Synthesis and studies with 2-C^{14}-labeled bromacil and terbacil. *J. Agric. Food Chem.* **17:** 980–986.

Gardner, D. R. 1973. The effect of some DDT and methoxychlor analogs on temperature selection and lethality in brook trout fingerlings. *Pestic. Biochem. Physiol.* **2:** 437–446.

Gardner, G. R., and G. LaRoche. 1973. Copper induced lesions in estuarine teleosts. *J. Fish. Res. Bd. Can.* **30:** 363–368.

Gaskin, D. E., M. Holdrinet, and R. Frank. 1971. Organochlorine pesticide residues in Harbour porpoises from the Bay of Fundy region. *Nature* **233:** 499–500.

Gast, A., E. Knüsli, and H. Gysin. 1955. Über pflanzenwachstums regulatoren. *Experientia* **11:** 107–108.

Gäumann, E. 1958. The mechanisms of fusaric acid injury. *Phytopathology* **48:** 670–686.

Geisman, J. R. 1974. Reduction of pesticide residues in food crops by processing. *Res. Rev.* **54:** 43–53.

Geissbühler, H., H. Martin, and G. Voss. 1975. The substituted ureas. In: *Herbicides: Chemistry, degradation and mode of action* (2nd ed.). Kearney, P. C., and D. D. Kaufman (Eds.). Marcel Dekker, Inc., New York. Pp. 209–291.

Geluso, K. N., J. S. Altenbach, and D. E. Wilson. 1976. Bat mortality: Pesticide poisoning and migratory stress. *Science* **194:** 184–186.

George, J. L., and W. H. Stickel. 1949. Wildlife effects of DDT dust used for tick control on a Texas prairie. *Amer. Midland Natur.* **42:** 228–237.

Getzin, L. W. 1968. Persistence of diazinon and zinophos in soil: Effects of autoclaving, temperature, moisture, and acidity. *J. Econ. Entomol.* **61:** 1560–1565.

Getzin, L. W. 1973. Persistence and degradation of carbofuran in soil. *Environ. Entomol.* **2:** 461–467.

Giam, C. S., M. K. Wong, A. R. Hanks, and W. M. Sackett. 1973. Chlorinated hydrocarbons in plankton from the Gulf of Mexico and Northern Caribbean. *Bull. Environ. Contam. Toxicol.* **9:** 376–382.

Gibney, L. C. 1974. The plight of the tussock moth. *Environ. Sci. Technol.* **8:** 506–507.

Gilbert, F. F. 1969. Physiological effects of natural DDT residues and metabolites on ranch mink. *J. Wildl. Mgmt.* **33:** 933–943.

Gilbertson, M. 1974. Seasonal changes in organochlorine compounds and mercury in common terns of Hamilton Harbour, Ontario. *Bull. Environ. Contam. Toxicol.* **12:** 726–732.

Gilbertson, M. 1975. A Great Lakes tragedy. *Nature Can.* January/March 1975. Pp. 22–25.

Gish, C. D. 1970. Organochlorine insecticide residues in soils and soil invertebrates from agricultural lands. *Pestic. Monit. J.* **4:** 241–252.

Glass, E. H., and S. E. Lienk. 1971. Apple insects and mite populations developing after discontinuance of insecticides: 10 year record. *J. Econ. Entomol.* **64:** 23–26.

Glass, E. H., W. L. Roelofs, H. Arn, and A. Comeau. 1970. Sex pheromone trapping red-banded leafroller moths and development of a long lasting polyethylene wick. *J. Econ. Entomol.* **63:** 370–373.

Glasser, R. F. 1976. Pesticides: The legal environment. In: *Pesticides and human welfare.* (Gunn, D. L., and J. G. R. Stevens (Eds.). Oxford University Press, New York. Pp. 228–239.

Glooschenko, W. A., W. M. J Strachan, and R. C. J. Sampson. 1976. Distribution of pesticides and polychlorinated biphenyls in water, sediments and seston of the upper Great Lakes, 1974. *Pestic. Monit. J.* **10:** 61–67.

Golab, T., R. J. Berberg, S. J. Parka, and J. B. Tepe. 1967. Metabolism of carbon-14 trifluralin in carrots. *J. Agric. Food Chem.* **15:** 638–641.

Gomaa, H. M., and S. D. Faust. 1972. Chemical hydrolysis and oxidation of parathion and paraoxon in aquatic environments. In: *Fate of organic pesticides in aquatic environments.* Advances in Chemistry Series III. Pp. 189–209.

Goodhue, L. D., and F. M. Baumgartner. 1965. The avitrol method of bird control. *Pest Control.* **33:** 16–17; 46–48.

Goring, C. A. I., D. A. Laskowski, J. W. Hamaker, and R. W. Meikle. 1975. Principles of pesticide degradation in soil. In: *Environmental dynamics of pesticides*. Hague, R., and V. H. Freed (Eds.). Plenum Press, New York & London. Pp. 135–172.

Graham, F., Jr. 1974. The return of DDT-Pest control by press release. *Audobon Mag.* September. Pp. 65–71.

Graham, K. 1963. *Concepts of forest entomology*. Reinhold Publishing Corp., New York. P. 388.

Green, R. E., and S. R. Obien. 1969. Herbicide equilibrium in soils in relation to soil water content. *Weed Sci.* 17: 514–519.

Green, R. S., C. G. Gunnerson, and J. J. Lichtenberg. 1967. Pesticides in our national waters. In: *Agriculture and the quality of our environment*. Brady, N. C. (Ed.). American Association for the Advancement of Science Publication 85. Pp. 137–145.

Greenberg, J., and Q. N. LaHam. 1969. Malathion-induced teratisms in the developing chick. *Can. J. Zool.* 47: 539–541.

Greenberg, J., and Q. N. LaHam. 1970. Reversal of malathion-induced teratisms and its biochemical implications in the developing chick. *Can. J. Zool.* 48: 1047–1053.

Greichus, Y. A., A. Greichus, and R. J. Emericka. 1973. Insecticides, polychlorinated biphenyls and mercury in wild cormorants, pelicans, their eggs, food and environment. *Bull. Environ. Contam. Toxicol.* 9: 321–328.

Greichus, Y. A., and M. R. Hannon. 1973. Distribution and biochemical effects of DDT, DDD and DDE in penned double-crested cormorants. *Toxicol. Appl. Pharmacol.* 26: 483–494.

Gress, F., R. W. Risebrough, and F. C. Sibley. 1971. Shell thining in eggs of the common murre, *Uria aalge* from the Farallon Islands, California. *Condor* 73: 368–369.

Greve, P. A., and S. L. Wit. 1971. Endosulfan in the Rhine River. *J. Water Pollut. Control Fed.* 43: 2338–2348.

Grover, R. 1967. Studies on the degradation of 4-amino-3,5,6-trichloropicolinic acid in soil. *Weed Res.* 7: 61–67.

Guenzi, W. D. (Ed.). 1974. *Pesticides in soil and water*. Soil Science Society of America, Inc., Madison, Wisconsin. P. 562.

Gunther, F. A. 1970. The triazine herbicides. *Res. Rev.* Vol. 32. Springer Verlag, New York. P. 419.

Hacskaylo, J., and V. A. Amato. 1968. Effects of trifluralin on roots of corn and cotton. *Weed Sci.* 16: 513–515.

Haegele, M. A., and R. K. Tucker. 1974. Effects of 15 common environmental pollutants on eggshell thickness in mallards and coturnix. *Bull. Environ. Contam. Toxicol.* 11: 98–101.

Hagyard, S. B., W. H. Brown, J. W. Stull, F. M. Whiting, and S. R. Kemberling. 1973. DDT and DDE content of human milk in Arizona. *Bull. Environ. Contam. Toxicol.* 9: 169–172.

Hall, R. C., G. S. Giam, and M. G. Merkle. 1968. The photolytic degradation of picloram. *Weed Res.* 8: 292–297.

Hall, W. C., S. P. Johnson, C. L. Leinweber. 1954. *Amino triazole—a new abcission chemical and growth inhibitor.* Texas Agricultural Experimental Station Bulletin 789. Pp. 1–15.

Halter, M. T., and H. E. Johnson. 1974. Acute toxicities of a polychlorinated biphenyl (PCB) and DDT alone and in combination to early life stages of coho salmon (*Oncorhynchus kisutch*). *J. Fish. Res. Bd. Can.* 31: 1543–1547.

Hamaker, J. W., H. Johnston, R. T. Martin, and C. T. Redemann. 1963. A nicolinic acid derivative: A plant growth regulator. *Science* 141: 363.

Hamelink, J. L., and R. C. Waybrant. 1976. DDE and lindane in a large-scale model lentic ecosystem. *Trans. Amer. Fish. Soc.* 105: 124–133.

Hamelink, J. L., R. C. Waybrant, and R. C. Ball. 1971. A proposal: Exchange equilibria control the degree chlorinated hydrocarbons are biologically magnified in lentic environments. *Trans. Amer. Fish. Soc.* 100: 207–214.

Hamilton, G. A., K. Hunter, A. S. Ritchie, A. D. Ruthuen, P. M. Brown, and P. I. Stanley. 1976. Poisoning of wild geese by carbophenothion-treated winter wheat. *Pestic. Sci.* 7: 175–183.

Hamilton, J. M., and M. Szkolnik. 1958. Standard and new fungicides for the control of apple scab and cherry leaf spot. N.Y. State Horticultural Society Proceedings, 103rd Annual Meeting. Pp. 72–78.

Hamm, P. C., and A. J. Speziale. 1956. Relation of herbicidal activity to the amide moiety of N-substituted alpha-chloroacetamides. *J. Agric. Food Chem.* 4: 518–522.

Hanna, J. G. 1974. *Report on the quality of food sold in Connecticut.* Connecticut Agricultural Experimental Station Bulletin 576. P. 62.

Hannay, C. L., and P. Fitz-James. 1955. The protein crystals of *Bacillus thuringiensis* Berliner. *Can. J. Microbiol.* 1: 694–710.

Hannon, M. R., Y. A. Greichus, R. L. Applegate, and A. C. Fox. 1970. Ecological distribution of pesticides in Lake Poinsett, S. Dakota. *Trans. Amer. Fish Soc.* 99: 496–500.

Hansen, D. J. 1969. Avoidance of pesticides by untrained sheepshead minnows. *Trans. Amer. Fish. Soc.* 98: 426–429.

Hanson, J. G., and F. W. Slife. 1969. Role of RNA metabolism in the action of auxin herbicides. *Res. Rev.* 25: 59–67.

Haque, R., and J. T. Deagen. 1975. Binding of the chlorinated hydrocarbon bis(*p*-chlorophenyl) acetic acid with enzyme carbonic anhydrase. *Bull. Environ. Contam. Toxicol.* 14: 43–46.

Hargrave, B. T., and G. A. Phillips. 1976. DDT residues in benthic invertebrates and demersal fish in St. Margaret's Bay, Nova Scotia. *J. Fish. Res. Bd. Can.* 33: 1692–1698.

Harp, G. L., and R. S. Campbell. 1964. Effects of the herbicide silvex on benthos of a farm pond. *J. Wildl. Mgmt.* 28: 308–317.

Harris, C. I., D. D. Kaufman, T. J. Sheets, R. G. Nash, and P. C. Kearney. 1968. Behavior and fate of *s*-triazines in soils. *Adv. Pest Control Res.* 8: 1–55.

Harris, C. R. 1970. Laboratory evaluation of candidate materials as potential soil insecticides. III. *J. Econ. Entomol.* 63: 782–787.

Harris, C. R. 1973. Laboratory evaluation of candidate materials as potential soil insecticides. IV. *J. Econ. Entomol.* 66: 216–221.

Harris, C. R., R. A. Chapman, and J. R. W. Miles. 1977. Insecticide residues in soils on fifteen farms in southwestern Ontario, 1964–1974. *J. Environ. Sci. Health,* **B12**: 163–177.

Harris, C. R., and G. B. Kinoshita. 1977. Influence of posttreatment temperature on the toxicity of pyrethroid insecticides. *J. Econ. Entomol.* **70**: 215–218.

Harris, C. R., and E. P. Lichtenstein. 1961. Factors affecting the volatilization of insecticidal residues from soils. *J. Econ. Entomol.* **54**: 1038–1045.

Harris, C. R., and J. H. Mazurek. 1966. Laboratory evaluation of candidate materials as potential soil insecticides. *J. Econ. Entomol.* **59**: 1215–1221.

Harris, C. R., and J. R. W. Miles. 1975. Pesticide residues in the Great Lakes region of Canada. *Res. Rev.* **57**: 27–79.

Harris, C. R., and W. W. Sans. 1971. Insecticide residues in soils on 16 farms in southwestern Ontario—1964, 1966 and 1969. *Pestic. Monit. J.* **5**: 259–267.

Harris, C. R., W. W. Sans, and J. R. W. Miles. 1966. Exploratory studies on occurrence of organochlorine insecticide residues in agricultural soils in southwestern Ontario. *J. Agric. Food Chem.* **14**: 398–403.

Harris, C. R., A. R. Thompson, and C. M. Tu. 1972. Insecticides and the soil environment. *Proc. Entomol. Soc. Ont.* **102**: 156–168.

Harris, N., and A. Dodge. 1972. The effect of paraquat on flax cotyledon leaves: Physiological and biochemical changes. *Planta* **104**: 210–219.

Harrison, H. L., O. L. Loucks, J. W. Mitchell, D. F. Parkhurst, C. R. Tracy, D. G. Watts, and V. J. Yannacone, Jr. 1970. Systems studies of DDT transport. *Science* **170**: 503–508.

Harvey, G. R. 1974. DDT and PCB in the Atlantic. *Oceanis* **18**: 18–23.

Harvey, G. R., H. P. Miklas, V. T. Bowen, and W. G. Steinhauer. 1974. Observations on the distribution of chlorinated hydrocarbons in Atlantic Ocean organisms. *J. Mar. Res.* **32**: 103–118.

Harvey, G. R., and W. G. Steinhauer. 1974. Atmospheric transport of polychlorobiphenyls to the North Atlantic. *Atmos. Environ.* **8**: 777–782.

Harvey, J., Jr., and H. L. Pease. 1973. Decomposition of methomyl in soil. *J. Agric. Food Chem.* **21**: 784–786.

Harvey, J., Jr., and R. W. Reiser. 1973. Metabolism of methomyl in tobacco, corn and cabbage. *J. Agric. Food Chem.* **21**: 775–783.

Haseltine, S., K. Uebelhart, T. Peterle, and S. Lustick. 1974. DDE, PTH and eggshell thinning in mallard, pheasant and ring dove. *Bull. Environ. Contam. Toxicol.* **11**: 139–145.

Haven, D. 1963. Mass treatment with 2,4-D of milfoil in tidal creeks in Virginia. *Southern Weed Conf. Proc.* **16**: 345–350.

Hawkins, D. E., and F. W. Slife. 1977. Economic analysis of herbicide use in various crop sequences. *Ill. Agric. Econ.* **17**: 8–13.

Hay, J. R. 1970. Weed control in wheat, oats and barley. In: *FAO International Conference on Weed Control.* Weed Science Society of America. Pp. 38–47.

Hay, J. R. 1976. Herbicide transport in plants. In: *Herbicides: Physiology, biochemistry, ecology* (2nd ed.). Vol. 1. Audus, L. J. (Ed.). Academic Press, London. Pp. 365–396.

Hazeltine, W. 1971. DDT and juvenile jaundice. *Clin. Toxicol.* **4**: 55–61.

Hazeltine, W. 1972. Disagreements on why brown pelican eggs are thin. *Nature* **239**: 410–411.

Hazelwood, E. 1970. Frog pond contaminated. *Br. J. Herpet.* **4**: 177–185.

Headley, J. C. 1968. Estimating the productivity of agricultural pesticides. *Amer. J. Agric. Econ.* **50**: 13.

Heath, R. G., J. W. Spann, and J. F. Kreitzer. 1969. Marked DDE impairment of mallard reproduction in controlled studies. *Nature* **244**: 47–48.

Heath, R. G., J. W. Spann, J. F. Kreitzer, and C. Vance. 1970. Effects of polychlorinated biphenyls on birds. Proceedings of the XVth International Ornithological Congress. Pp. 475–485.

Heath, R. G., and L. F. Stickel. 1965. Protocol for testing the acute and relative toxicity of pesticides to penned birds. In: *The effects of pesticides on fish and wildlife.* U.S. Fisheries & Wildlife Service Circular 226. Pp. 18–24.

Heimpel, A. M., and T. A. Angus. 1959. The site of action of crystalliferous bacteria in Lepidoptera larvae. *J. Insect Pathol.* **1**: 152–170.

Heimpel, A. M., and T. A. Angus. 1963. Diseases caused by certain spore forming bacteria. In: *Insect pathology, an advanced treatise.* Vol. 2. Steinhaus, E. A. (Ed.). Academic Press, New York. Pp. 21–73.

Hellenbrand, K., and R. M. Krupta. 1970. Kinetic studies on the mechanism of insect acetylcholinesterase. *Biochem.* **9**: 4665–4672.

Helling, C. S. 1971. Pesticide mobility in soils. II. Applications of soil thin layer chromatography. *Soil Sci. Soc. Amer. Proc.* **35**: 737–743.

Henderson, C., A. Inglis, and W. L. Johnson. 1971. Organochlorine insecticide residues in fish—fall 1969. *Pestic. Monit. J.* **5**: 1–11.

Henderson, C., Q. H. Pickering, and C. M. Tarzwell. 1959. Relative toxicity of ten chlorinated hydrocarbon insecticides to four species of fish. *Trans. Amer. Fish. Soc.* **88**: 23–32.

Henderson, J. L. 1963. Comparison of laboratory techniques for the determination of pesticide residues in milk. *J. Assoc. Official Agr. Chemists.* **46**: 209–215.

Hesselberg, R. J., and D. D. Scherr. 1974. PCBs and p,p'-DDE in the blood of cachectic patients. *Bull. Environ. Contam. Toxicol.* **11**: 202–205.

Hickey, J. J. (Ed.). 1969. *Peregrine falcon populations: Their biology and decline.* University of Wisconsin Press, Madison. Pp. 596.

Hickey, J. J., and D. W. Anderson. 1968. Chlorinated hydrocarbons and eggshell changes in raptorial and fish-eating birds. *Science* **162**: 271–273.

Hill, E. F., W. E. Dale, and J. W. Miles. 1971. DDT intoxication in birds: Subchronic effects and brain residues. *Toxicol. Appl. Pharmacol.* **20**: 502–514.

Hill, G. D., J. W. McGahen, H. M. Baker, D. W. Finnerty, and C. W. Bingeman. 1955. The fate of substituted urea herbicides in agricultural soils. *Agron. J.* **47**: 93–104.

Hill, R. 1937. Oxygen evolved by isolated chloroplasts. *Nature* (London) **139**: 881–882.

Hiltibran, R. C. 1963. Effect of endothal on aquatic plants. *Weeds* **111**: 256–257.

Hiltibran, R. C. 1967. Effects of some herbicides on fertilized fish eggs and fry. *Trans. Amer. Fish. Soc.* **96**: 414–416.

Hilton, J. L., T. J. Monaco, D. E. Moreland, and W. A. Gentner. 1964. Mode of action of substituted uracil herbicides. *Weeds* **12:** 129–131.

Hindin, E., D. S. May, and G. H. Dunstan. 1964. Collection and analysis of synthetic organic pesticides from surface and ground water. *Res. Rev.* **7:** 130–156.

Hindin, E., D. S. May, and G. H. Dunstan. 1966. Distribution of insecticide sprayed by airplane on an irrigated corn plot. In: *Organic pesticides in the environment.* Gould, R. F. (Ed.). Advances in Chemistry Series No. 60. American Chemical Society, Washington, D.C. Pp. 132–145.

Hodkinson, M., and S. A. Dalton. 1973. Interactions between DDT and river fungi. II. Influence of culture conditions on the compatibility of fungi and *p,p'*-DDT. *Bull. Environ. Contam. Toxicol.* **10:** 356–359.

Hoffmann, C. H., and E. W. Surber. 1949. Effects of an aerial application of DDT on fish and fish-food organisms in two Pennsylvania watersheds. *Prog. Fish. Culturist* **11:** 203–211.

Holan, G. 1969. New halocyclopropane insecticides and the mode of action of DDT. *Nature* **221:** 1025–1029.

Holan, G. 1971. Rational design of insecticides. *Bull. Wld. Hlth. Org.* **44:** 355–362.

Holden, A. V. 1973. Effects of pesticides on fish. In: *Environmental pollution by pesticides.* Edwards, C. A. (Ed.). Plenum Press, London & New York. Pp. 213–253.

Holden, A. V., and K. Marsden. 1967. Organochlorine pesticides in seals and porpoises. *Nature* **216:** 1274–1276.

Hotchkiss, N., and R. H. Pough. 1946. Effect on forest birds of DDT used for gypsy moth control in Pennsylvania. *J. Wildl. Mgmt.* **10:** 202–207.

Hunt, E. G., J. A. Azevedo, Jr., L. A. Woods, Jr., and W. T. Castle. 1969. The significance of residues in pheasant tissues resulting from chronic exposures to DDT. In: *Chemical fallout.* Miller, M. W., and G. G. Berg (Eds.). C. C Thomas, Springfield, Ill. Pp. 335–358.

Hunt, E. G., and A. I. Bischoff. 1960. Inimical effects on wildlife of periodic DDD applications to Clear Lake. *Calif. Fish & Game.* **46:** 91–105.

Hunt, L. B. 1960. Songbird breeding populations in DDT-sprayed Dutch elm disease communities. *J. Wildl. Mgmt.* **24:** 139–146.

Hurlbert, S. H. 1975. Secondary effects of pesticides on aquatic ecosystems. *Res. Rev.* **57:** 82–148.

Hurlbert, S. H., M. S. Mulla, and H. R. Willson. 1972. Effects of an organophosphorus insecticide on the phytoplankton, zooplankton and insect populations of freshwater ponds. *Ecol. Monogr.* **42:** 269–299.

Innes, J. R. M., B. M. Ulland, M. G. Valerio, L. Petrucelli, L. Fishbein, E. R. Hart, A. J. Pallotta, R. R. Bates, H. L. Falk, J. J. Cart, M. Klein, I. Mitchell, and J. Peters. 1969. Bioassay of pesticides and industrial chemicals for tumorigenicity in mice—a preliminary note. *J. Nat. Cancer Inst.* **42:** 1101–1114.

Jacobs, L. W., D. R. Keeney, and L. M. Walsh. 1970. Arsenic residue toxicity to vegetable crops grown on plainfield sand. *Agron. J.* **62:** 588–591.

James, M. T., and R. F. Harwood. 1969. *Herm's Medical Entomology* (6th ed.). The MacMillan Co., London. P. 484.

Jaques, R. P. 1974. Occurrence and accumulation of viruses of *Trichopulsia ni* in treated field plots. *J. Invert. Pathol.* **23**: 140–152.

Jaworski, E. G. 1956. Biochemical action of CDAA, a new herbicide. *Science* **123**: 847–848.

Jaworski, E. G. 1969. Chloroacetamides. In: *Degradation of Herbicides*. Kearney, P. C., and D. D. Kaufman (Eds.). Marcel Dekker, Inc., New York. Pp. 165–185.

Jaworski, E. G. 1972. Mode of action of N-phosphonomethyl glycine: Inhibition of aromatic amino acid biosynthesis. *J. Agric. Food Chem.* **20**: 1195–1198.

Jaworski, E. G. 1975. Chloroacetamides. In: *Herbicides: Chemistry, degradation and mode of action* (2nd ed.). Vol. 1. Kearney, P. C., and D. D. Kaufman (Eds.). Marcel Dekker, Inc., New York. Pp. 349–376.

Jefferies, D. J. 1967. The delay in ovulation produced by *p,p'*-DDT and its possible significance in the field. *Ibis* **109**: 266–272.

Jefferies, D. J. 1971. Some sublethal effects of *p,p'*-DDT and its metabolite *p,p'*-DDE on breeding passerine birds. *Meded. Fakult. Landbouw. Gent.* **36**: 34–43.

Jefferies, D. J. 1972. Organochlorine insecticide residues in British bats and their significance. *J. Zool.* **166**: 245–263.

Jefferies, D. J. 1975. The role of the thyroid in the production of sublethal effects by organochlorine insecticides and polychlorinated biphenyls. In: *Organochlorine insecticides: Persistent organic pollutants*. Moriarty, F. (Ed.). Academic Press, London, New York. Pp. 131–230.

Jefferies, D. J., and B. N. K. Davis. 1968. Dynamics of dieldrin in soil, earthworms, and song thrushes. *J. Wildl. Mgmt.* **32**: 441–456.

Jehl, J. R., Jr. 1970. Is thirty million years long enough? *Pac. Discovery* **23**: 16–23.

Jennings, D. M., P. J. Bunyan, P. M. Brown, P. J. Stanley, and F. J. S. Jones. 1975. Organophosphorus poisoning: A comparative study of the toxicity of carbophenothion to the Canada goose, the pigeon and the Japanese quail. *Pestic. Sci.* **6**, 245–257.

Jensen, K. I. N., J. D. Bandeen, and V. Souza-Machado. 1977. Studies on the differential tolerance of two lambsquarters selections to *s*-triazine herbicides. *Can. J. Plant Sci.* **57**: 1169–1177.

Jensen, K. I. N., G. R. Stephenson, and L. A. Hunt. 1977. Detoxification of atrazine in three gramineae subfamilies. *Weed Sci.* **25**: 212–220.

Jensen, S., and A. Jernelöv. 1969. Biological methylation of mercury in aquatic organisms. *Nature* **223**: 753–754.

Johnson, B. T., C. R. Saunders, H. O. Sanders, and R. S. Campbell. 1971. Biological magnification and degradation of DDT and aldrin by freshwater invertebrates. *J. Fish. Res. Bd. Can.* **28**: 705–709.

Johnson, D. R., W. E. Melquist, and G. J. Schroeder. 1975. DDT and PCB levels in Lake Coeur d'Alene, Idaho, osprey eggs. *Bull. Environ. Contam. Toxicol.* **13**: 401–405.

Johnson, D. W. 1975. Organochlorine pesticide residues in small migratory birds, 1964–1973. *Pestic. Monit. J.* **9**: 79–88.

Johnson, D. W. 1976. Organochlorine pesticide residues in uropygial glands and adipose tissue of wild birds. *Bull. Environ. Contam. Toxicol.* **16**: 149–155.

Johnson, E. V., G. L. Mack, and D. Q. Thompson. 1976. The effects of orchard pesticide applications on breeding robins. *The Wilson Bull.* **88:** 16–35.

Johnson, H. E., and R. C. Ball. 1972. Organic pesticide pollution in an aquatic environment. In: *Fate of organic pesticides in the aquatic environment.* Faust, S. D. (Ed.). Advances in Chemistry Series III. American Chemical Society. Pp. 1–10.

Johnson, H. E., and C. Pecor. 1969. Coho salmon mortality and DDT in Lake Michigan. *North Amer. Wildl. Natur. Resources Conf., Trans.* **34:** 159–166.

Johnson, L. G., and R. L. Morris. 1971. Chlorinated hydrocarbon pesticides in Iowa rivers. *Pestic. Monit. J.* **4:** 216–219.

Johnson, R. D., and D. D. Manske. 1976. Pesticide residues in total diet samples (IX). *Pestic. Monit. J.* **9:** 157–169.

Johnson, R. E., and P. A. Dahm. 1966. Activation and degradation efficiencies of liver microsomes from eight -vertebrate species, using organophosphates as substrates. *J. Econ. Entomol.* **59:** 1437–1442.

Jones, D., K. Ronald, D. M. Lavigne, R. Frank, M. Holdrinet, and J. F. Uthe. 1976. Organochlorine and mercury residues in the harp seal (*Pagophilus groenlandicus*). *The Science of the Total Environment* **5:** 181–195.

Jones, G. E., and G. W. Anderson. 1964. Atrazine as a foliar application in an oil-water emulsion. Proceedings of the North Central Weed Control Conference. Pp. 27–29.

Jones, R. K., J. W. Heuberger, and J. D. Bates. 1963. Apple scab. III. Effects of serial applications of fungicides on leaf lesions on previously unsprayed trees. Inhibition of conidial germination, removal (suppression) of the organism and subsequent development of late terminal infection. *Plant Disease Rept.* **17:** 420–424.

Jordan, L. S., B. E. Day, and W. A. Clerx. 1964. Photodecomposition of triazines. *Weeds* **12:** 5–6.

Kabat, H., E. F. Stohlman, and M. I. Smith. 1944. Hypoprothrombinemia induced by administration of indandione derivatives. *J. Pharmacol. & Exp. Therap.* **80:** 160–170.

Kajii, T., M. Kida, and K. Takahashi. 1973. The effect of thalidomide intake during 113 human pregnancies. *Teratology* **8:** 163–166.

Kanayama, R. K. 1963. The use of alkalinity and conductivity measurements to estimate concentrations of 3-trifluormethyl-4-nitrophenol required for treating lamprey streams. *Gr. Lakes Fish. Common. Tech. Rep.* **7:** 1–10.

Kanazawa, J. 1975. Uptake and excretion of organophosphorus and carbamate insecticides by fresh water fish, Motsugo, *Pseudoasbora parva. Bull. Environ. Contam. Toxicol.* **14:** 346–352.

Kapoor, J. P., R. L. Metcalf, R. F. Nystrom, and G. K. J. Sangha. 1970. Comparative metabolism of methoxychlor, methiochlor and DDT in mouse, insects and in a model ecosystem. *J. Agric. Food Chem.* **18:** 1145–1152.

Kapusta, G., and E. C. Varsa. 1972. Nitrification inhibitors—do they work? *Down to Earth* **28:** 21–23.

Karinen, J. F., J. G. Lamberton, N. E. Stewart, and L. C. Terriere. 1967. Persistence of carbaryl in the marine estuarine environment. Chemical and biological stability in aquarium systems. *J. Agric. Food Chem.* **15:** 148–156.

Kates, A. H. 1965. A note on damage to tobacco by lateral movement of picloram. *Proc. N.E. Weed Control Conf.* **19**: 393–396.

Kaufman, D. D. 1976. The phenols. In: *Herbicides: Chemistry, degradation and mode of action* (2nd ed.). Kearney, P. C., and D. D. Kaufman (Eds.). Marcel Dekker, Inc., New York. Pp. 665–707.

Kaufman, D. D., and P. C. Kearney. 1970. Microbial degradation of triazine herbicides. *Res. Rev.* **32**: 235–265.

Kaufman, D. D., J. R. Plimmer, P. C. Kearney, J. Blake, and F. S. Guardia. 1968. Chemical versus microbial decomposition of amitrole in soil. *Weed Sci.* **16**: 266–272.

Kaya, H., D. Dumbar, C. Doane, R. Weseloh, and J. Anderson. 1974. *Gypsy moth: Aerial tests with Bacillus thuringiensis and pyrethroids.* Connecticut Agricultural Experimental Station Bulletin 744. P. 22.

Kearney, P. C., and D. D. Kaufman (Eds.). 1975. *Herbicides: Chemistry, degradation, and mode of action* (2nd ed.). Vol. 1. Marcel Dekker, Inc., New York. Pp. 500.

Kearney, P. C., and D. D. Kaufman (Eds.). 1976. *Herbicides: Chemistry, degradation, and mode of action* (2nd ed.). Vol. 2. Marcel Dekker, Inc., New York. Pp. 501–1036.

Kearney, P. C., T. J. Sheets, and J. W. Smith. 1964. Volatility of seven *s*-triazines. *Weeds* **12**: 83–87.

Keeley, P. E., C. H. Carter, and J. H. Miller. 1972. Evaluation of the relative toxicity phytotoxicity of herbicides to cotton and nutsedge. *Weed Sci.* **20**: 71–74.

Keenleyside, M. H. A. 1959. Effects of spruce budworm control on salmon and other fishes in New Brunswick. *Can. Fish. Culturist.* **24**: 17–22.

Keil, J. E., C. B. Loadholt, S. H. Sandifer, W. Weston III, R. H. Gadsden, and C. G. Hames. 1973. Sera DDT elevation in black components of two southwestern communities: Genetics or environment. In: *Pesticides and the environment: A continuing controversy.* Deichmann, W. B. (Ed.). Intercontinental Medical Book Corp., New York & London. Pp. 203–213.

Keith, J. A., and I. M. Gruchy. 1972. Residue levels of chemical pollutants in North American birdlife. Proceedings of the XVth International Ornithological Congress, 1970. Pp. 437–454.

Keith, J. O. 1970. Variations in the biological vulnerability of birds to insecticides. In: *The biological impact of pesticides in the environment.* Gillett, J. W. (Ed.). Oregon State University, Environmental Health Science Series No. 1. Pp. 36–39.

Keith, J. O., R. M. Hause, and A. L. Ward. 1959. Effect of 2,4-D on abundance and foods of pocket gophers. *J. Wildl. Mgmt.* **23**: 137–145.

Keith, J. O., and E. G. Hunt. 1966. Levels of insecticide residues in fish and wildlife in California. Proceedings of the 31st North American Wildlife Conference. Pp. 150–177.

Keith, J. O., L. A. Wood, Jr., and E. G. Hunt. 1970. Reproductive failure in brown pelicans on the Pacific Coast. Transactions of the 35th North American Wildlife Natural Resource Conference. Pp. 56–63.

Kelso, J. R. M., and R. Frank. 1974. Organochlorine residues, mercury, copper, and cadmium in yellow perch, white bass and smallmouth bass, Long Point Bay, Lake Erie. *Trans. Amer. Fish. Soc.* **103**: 577–581.

Kenaga, E. E. 1972. Guidelines for environmental study of pesticides: Determination of bioconcentration potential. *Res. Rev.* **44**: 73–113.

Kenaga, E. E. 1977. Evaluation of the hazard of pesticide residues in the environment. In: *Pesticide management and insecticide resistance.* Watson, D. L., and A. W. A. Brown (Eds.). Academic Press, New York. Pp. 51–95.

Kenaga, E. E., and C. S. End. 1974. *Commercial and experimental organic insecticides.* Entomological Society of America Special Publication 74-1. P. 77.

Kennedy, H. D., L. L. Eller, and D. F. Walsh. 1970. Chronic effects of methoxychlor on bluegills and aquatic invertebrates. USDI Bureau of Sport Fisheries & Wildlife Technical Paper 53. Pp. 3–17.

Kennedy, H. D., and D. F. Walsh. 1970. Effects of malathion on two warm water fishes and aquatic invertebrates in ponds. USDI Bureau of Sport Fisheries & Wildlife Technical Paper 55. Pp. 3–16.

Kerr, S. R., and W. P. Vass. 1973. Pesticide residues in aquatic invertebrates. In: *Environmental pollution by pesticides.* Edwards, C. A. (Ed.). Plenum Press, London & New York. Pp. 134–180.

Kerswill, C. J. 1958. Effects of DDT spraying in New Brunswick on future runs of adult salmon. *Fish. Res. Bd. Can. Studies* **525**: 3–8.

Kerswill, C. J. 1961. Investigation and management of Atlantic salmon. *Fish. Res. Bd. Can. Studies* **676**: 3–9.

Kerswill, C. J., and P. F. Elson. 1955. Preliminary observations on effects of 1954 DDT spraying on Miramichi salmon stocks. *Fish. Res. Bd. Can. Prog. Report* **62**: 17–23.

Kettela, E. G. 1975. Aerial spraying for protection of forests infested by spruce budworm. *The Forestry Chronicle.* August: 141–142.

Key, J. L., and J. B. Hanson. 1961. Effects of 2,4-dichlorophenoxy-acetic acid on soluble nucleotides and nucleic acid of soybean seedlings. *Plant Physiol.* **36**: 145–152.

Khan, M. A. Q., M. L. Gassman, and S. H. Ashrafi. 1975. Detoxification of pesticides by biota. In: *Environmental dynamics of pesticides.* Haque, R., and V. H. Freed (Eds.). Plenum Press, New York & London. Pp. 289–329.

Khera, K. S. 1973. Ethylenethiourea: Teratogenicity study in rats and rabbits. *Teratology* **7**: 243–252.

Kilgore, L., and F. Windham. 1970. Disappearance of malathion residue in broccoli during cooking and freezing. *J. Agric. Food Chem.* **18**: 162–163.

Kirk, B. T., and J. B. Sinclair. 1968. Radioautographic comparison of C^{14} labeled Vitavax and Demosan. *Phytopathology* **58**: 1055.

Kirkwood, R. C. 1976. Action on respiration and intermediary metabolism. In: *Herbicides: Physiology, biochemistry, ecology* (2nd ed.). Vol. 1. Audus, L. J. (Ed.). Academic Press, London. Pp. 444–492.

Klaas, E. E., and D. M. Swineford. 1976. Chemical residue content and hatchability of several owl eggs. *The Wilson Bull.* **88**: 421–426.

Klingman, D. L. 1970. Brush and weed control on forage and grazing lands. In: *FAO International Conference on Weed Control.* Weed Science Society of America. Pp. 401–424.

Klingman, G. C., and F. M. Ashton. 1975. *Weed science principles and practices.* John Wiley and Sons, New York. P. 431.

Knake, E. L., and L. M. Wax. 1968. The importance of the shoot of giant foxtail for uptake of preemergence herbicides. *Weed Sci.* **16:** 393–395.

Knopf, F. L., and J. C. Street. 1974. Insecticide residues in white pelican eggs from Utah. The Wilson Bull. **86:** 428–433.

Knüsli, E. 1970. History of the development of triazine herbicides. *Res. Rev.* **32:** 1–9.

Koch, R. B., L. K. Cutkomp, and F. M. Do. 1969. Chlorinated hydrocarbon insecticide inhibition of cockroach and honey bee ATPases. *Life Sci.* **8:** 289–297.

Kolipinski, M. C., A. L. Higer, and M. L. Yates. 1971. Organochlorine insecticide residues in Everglades National Park and Loxahatchee National Wildlife Refuge, Florida. *Pestic. Monit. J.* **5:** 281–288.

Korn, S., and R. Earnest. 1974. Acute toxicity of twenty insecticides to striped bass, *Morone saxatilis. Calif. Fish & Game* **60:** 128–131.

Kottman, R. M. 1967. *Irresistible force—immovable object.* Proceedings of the North American Conference on Blackbird Depredation in Agriculture. Ohio State Univ. Pp. 5–9.

Kovacicova, J., V. Batora, and S. Truchlik. 1973. Hydrolysis rate and *in vitro* anticholinesterase activity of fenitrothion and S-methyl fenitrothion. *Pestic. Sci.* **4:** 759–763.

Krause, W., K. Hamm, and J. Weissmuller. 1975. The effect of DDT on spermatogenesis of the juvenile rat. *Bull. Environ. Contam. Toxicol.* **14:** 171–179.

Kraybill, H. F. (Ed.). 1969. Biological effects of pesticides in mammalian systems. *Ann. N.Y. Acad. Sci.* **160:** 406–417.

Kreitzer, J. F. 1972. The effect of embryonic development on the thickness of the eggshells of coturnix quail. *Poultry Sci.* **51:** 1764–1765.

Kruzynski, G. M., and G. Leduc. 1972. Methoxychlor, a new threat to the Atlantic salmon. *Atlantic Salmon Jour.* **1:** 5 pp.

Kuhr, R. J., A. C. Davis, and J. B. Bourke. 1974. DDT residues in soil, water, and fauna from New York apple orchards. *Pestic. Monit. J.* **7:** 200–204.

Kuhr, R. J., and H. W. Dorough. 1976. *Carbamate insecticides: Chemistry, biochemistry and toxicology.* CRC Press, Cleveland, Ohio. Pp. 300.

Kupfer, D. 1969. Influence of chlorinated hydrocarbons and organophosphate insecticides on metabolism of steroids. In: *Biological effects of pesticides in mammalian systems.* Kraybill, H. F. (Ed.). *Ann. N.Y. Acad. Sci.* **160:** 244–253.

Kurtz, D. A., and C. R. Studholme. 1974. Recovery of trichlorfon (Dylox®) and carbaryl (Sevin®) in songbirds following spraying of forest for gypsy moth. *Bull. Environ. Contam. Toxicol.* **11:** 78–84.

Kushlan, J. A. 1974. Effects of a natural fish kill in the water quality, plankton, and fish population of a pond in the Big Cypress Swamp, Florida. *Trans. Amer. Fish Soc.* **103:** 235–243.

Kutschinski, A. H., and V. Riley. 1969. Residues in various tissues of steers fed 4-amino-3,5,6-trichloro-picolinic acid. *J. Agric. Food Chem.* **17:** 283–287.

Laidlaw, C. W. J. 1976. Re 1-(3-pyridinyl methyl)-3-(4-nitro phenyl) urea—rodenticide. RNR-76-10. Control Products Section, Products and Marketing Branch, Plant Products Division, Ottawa. P. 3.

Lamb, F. C., R. P. Farrow, E. R. Elkins, J. R. Kimball, and R. W. Cook. 1968. Removal of DDT, parathion and carbaryl from spinach by commercial and home preparative methods. *J. Agric. Food Chem.* **16**: 957–973.

Lamberton, J. G., R. D. Inman, and R. R. Claeys. 1975. The metabolism of p,p'-DDE in laying Japanese quail and their incubated eggs. *Bull. Environ. Contam. Toxicol.* **14**: 657–664.

Laubscher, J. A., G. R. Dutt, and C. C. Roan. 1971. Chlorinated insecticide residues in wildlife and soil as a function of distance from application. *Pestic. Monit. J.* **5**: 251–258.

Lauer, G. J., H. P. Nicholson, W. S. Cox, and J. J. Teasley. 1966. Pesticide contamination of surface waters by sugar cane farming in Louisiana. *Trans. Amer. Fish. Soc.* **95**: 310–316.

Lawless, E. W., A. F. Meiners, K. A. Lawrence, G. L. Kelso, and R. Von Rumker. 1974. *Production, distribution, use and environmental impact potential of selected pesticides.* Report of the Midwest Research Institute to the Council on Environmental Quality. P. 404.

Lawless, E. W., R. Von Rumker, and T. L. Ferguson. 1972. *The pollution potential in pesticide manufacturing.* Technical Studies Report 15-00-72-04. Environmental Protection Agency, Washington, D.C. P. 249.

Laws, E. R., A. Curley, and F. J. Biros. 1967. Men with intensive occupational exposure to DDT. *Arch. Environ. Hlth.* **15**: 766–775.

Lay, M. M., and J. E. Casida. 1976. Dichloroacetamide antidotes enhance thiocarbamate sulfoxide detoxification by elevating corn root glutathione content and glutathione s-transferase activity. *Pestic. Biochem. Physiol.* **6**: 442–456.

Lay, M. M., J. P. Hubbell, and J. E. Casida. 1975. Dichloroacetamide antidotes for thiocarbamate herbicides: Mode of action. *Science* **189**: 287–289.

Lee, J. H., J. R. Sylvester, and C. E. Nash. 1975. Effects of mirex and methoxychlor on juvenile and adult striped mullet, *Mugil cephalus L. Bull. Environ. Contam. Toxicol.* **14**: 180–186.

Leitis, E., and D. G. Crosby. 1974. Photodecomposition of trifluralin. *J. Agric. Food Chem.* **22**: 842–848.

Lennon, R. E., J. B. Hunn, R. A. Schnick, and R. M. Burress. 1971. Reclamation of ponds, lakes, and streams with fish toxicants: A review. FAO Technical Paper 100. FIRI/T100. P. 99.

Lepple, F. K. 1973. *Mercury in the environment.* College of Marine Studies, University of Delaware, Newark, Delaware. P. 75.

Lichtenberg, J. J., J. W. Eichelberger, R. C. Dressman, and J. E. Longbottom. 1970. Pesticides in surface waters of the United States—a 5-year summary, 1964–1968. *Pestic. Monit. J.* **4**: 71–86.

Lichtenstein, E. P. 1958. Movement of insecticides in soils under leaching and non-leaching conditions. *J. Econ. Entomol.* **51**: 380–383.

Lichtenstein, E. P., and K. R. Schulz. 1961. Effect of soil cultivation, soil surface and water on the persistence of insecticide residues in soils. *J. Econ. Entomol.* **54**: 517–522.

Lichtenstein, E. P., and K. R. Schulz. 1964. The effects of moisture and micro-organisms on the persistence and metabolism of some organophosphorous insecticides in soils, with special emphasis on parathion. *J. Econ. Entomol.* **57**: 618–627.

Liss, P. S., and P. G. Slater. 1974. Flux of gases across the air-sea interface. *Nature* **247**: 181–184.

Lloyd-Jones, C. P. 1971. Evaporation of DDT. *Nature* **229**: 65–66.

Lockie, J. D., and D. A. Ratcliffe. 1964. Insecticides and Scottish golden eagles. *Br. Birds* **57**: 89–102.

Lockie, J. D., D. A. Ratcliffe, and R. Balharry. 1969. Breeding success and dieldrin contamination of golden eagles in west Scotland. *J. Appl. Ecol.* **6**: 381–389.

Longcore, J. R., and B. M. Mulhern. 1973. Organochlorine pesticides and polychlorinated biphenyls in black duck eggs from the United States and Canada—1971. *Pestic. Monit. J.* **7**: 62–66.

Longcore, J. R., and F. B. Samson. 1973. Eggshell breakage by incubating black ducks fed DDE. *J. Wildl. Mgmt.* **37**: 390–394.

Loos, M. A. 1969. Phenoxyalkanoic herbicides. In: *Degradation of herbicides.* Kearney, P. C., and D. D. Kaufman (Eds.). Marcel Dekker, Inc., New York. Pp. 1–50.

Loos, M. A. 1975. Phenoxy alkanoic acids. In: *Herbicides: Chemistry, degradation, and mode of action* (2nd ed.). Vol. 1. Kearney, P. C., and D. D. Kaufman (Eds.). Marcel Dekker, Inc., New York. Pp. 1–128.

Lord, F. T. 1949. The influence of spray programs on the fauna of apple orchards in Nova Scotia. III: Mites and their predators. *Can. Entomol.* **81**: 202–214; 217–230.

Lucier, G. W., O. S. McDaniel, C. Williams, and R. Klein. 1972. Effects of chlordane and methyl mercury on the metabolism of carbaryl and carbofuran in rats. *Pestic. Biochem. Physiol.* **2**: 244–255.

Luckens, M. M. 1973. Seasonal changes in the sensitivity of bats to DDT. In: *Pesticides and the environment: A continuing controversy.* Deichmann, W. B. (Ed.). Intercontinental Medical Book Corp., New York & London. Pp. 63–75.

Luckens, M. M., and W. H. Davis. 1964. Bats: Sensitivity to DDT. *Science* **146**: 948.

Luckens, M. M., and W. H. Davis. 1965. Toxicity of dieldrin and endrin to bats. *Nature* **207**: 879–880.

Ludke, J. L. 1974. Interaction of dieldrin and DDE residues in Japanese quail (*Coturnix coturnix japonica*). *Bull. Environ. Contam. Toxicol.* **11**: 297–302.

Ludke, J. L. 1976. Organochlorine pesticide residues associated with mortality: Additivity of chlordane and endrin. *Bull. Environ. Contam. Toxicol.* **16**: 253–260.

Ludwig, R. A., and G. D. Thorn. 1960. Chemistry and mode of action of dithiocarbamate fungicides. *Adv. Pest Control Res.* **3**: 219–252.

Lukens, R. J. 1969. Heterocyclic nitrogen compounds. In: *Fungicides, an advanced treatise.* Vol. II. Torgeson, D. C. (Ed.). Academic Press, New York. Pp. 395–445.

Lundberg, C., and J. E. Kihlström. 1973. DDT and the frequency of implanted ova in the mouse. *Bull. Environ. Contam. Toxicol.* **9**: 267–270.

Lüssem, H., and Schlimme. 1971. Lokalisierung des pflanzen schutzmittels endosulfan im Rhein und dessen Wirkung auf fische. *Gas-Wasserfach, Wasser-Abwasser.* **112**: 18–21.

Lutz-Ostertag, Y., and M. H. Lutz. 1970. Action néfaste de l'herbicide 2,4-D sur le développement embryonnaire et la fécondité du gibier à plumes. *Comptes Rendus. Acad. Sci. Paris,* Ser. D. **271**: 2418–2421.

Macek, K. J. 1970. Biological magnification of pesticide residue in food chains. In: *The biological impact of pesticides in the environment.* Gillet, J. W. (Ed.). Oregon State University, Corvallis. Pp. 17–21.

Macek, K. J. 1975. Acute toxicity of pesticide mixtures to bluegills. *Bull. Environ. Contam. Toxicol.* **14**: 648–652.

Macek, K. J., C. Hutchinson, and O. B. Cope. 1969. The effects of temperature on the susceptibility of bluegills and rainbow trout to selected pesticides. *Bull. Environ. Contam. Toxicol.* **4**: 174–184.

Macek, K. J., and S. Korn. 1970. Significance of the food chain in DDT accumulation by fish. *J. Fish. Res. Bd. Can.* **27**: 1496–1498.

Macek, K. J., and W. A. McAllister. 1970. Insecticide susceptibility of some common fish family representatives. *Trans. Amer. Fish. Soc.* **99**: 20–27.

MacKay, D., and P. J. Leinonen. 1975. Rate of evaporation of low solubility contaminants from water bodies to atmosphere. *Environ. Sci. Technol.* **9**: 1178–1180.

MacPhee, C., and R. Ruelle. 1969. A chemical selectively lethal to squawfish (*Ptychocheilus oregonensis* and *P. umpquae*). *Trans. Amer. Fish. Soc.* **98**: 676–684.

MacRae, I. C., and M. Alexander. 1965. Microbial degradation of selected herbicides in soil. *J. Agric. Food Chem.* **13**: 72–76.

MaGuire, J., and N. Watkin. 1975. Carbonic anhydrase inhibition. *Bull. Environ. Contam. Toxicol.* **11**: 625–629.

Mahoney, J. J., Jr. 1974. Residue accumulation in white-throated sparrows fed DDT for 5 and 11 weeks. *Bull. Environ. Contam. Toxicol.* **12**: 677–681.

Mahoney, J. J., Jr. 1975. DDT and DDE effects on migratory condition in white-throated sparrows. *J. Wildl. Mgmt.* **39**: 520–527.

Maki, A. W., L. D. Geissel, and H. E. Johnson. 1975. *Toxicity of the lampricide 3-trifluoromethyl-4-nitrophenol (TFM) to 10 species of algae.* USDI Fisheries & Wildlife Service Investigations in Fish Control 56. P. 17.

Mallis, A. 1969. *Handbook of pest control.* MacNair-Dorland Co., New York. P. 1158.

Manigold, D. B., and J. A. Schulze. 1969. Pesticides in selected western streams—a progress report. *Pestic. Monit. J.* **3**: 124–135.

March, R. B., T. R. Fukuto, R. L. Metcalf, and M. G. Maxon. 1956. Fate of P^{32}-labeled malathion in the laying hen, white mouse, and American cockroach. *J. Econ. Entomol.* **40**: 185–195.

Markin, G. P., H. L. Collins, and J. Davis. 1974. Residues of the insecticide mirex in terrestrial and aquatic invertebrates following a single aerial application of mirex bait, Louisiana—1971–1972. *Pestic. Monit. J.* **8**: 131–134.

Markin, G. P., J. C. Hawthorne, H. L. Collins, and J. H. Ford. 1974a. Levels of mirex and some other organochlorine residues in seafood from Atlantic and Gulf coastal states. *Pestic. Monit. J.* 7: 139–143.

Marking, L. L., and W. L. Mauck. 1975. Toxicity of paired mixtures of candidate forest insecticides to rainbow trout. *Bull. Environ. Contam. Toxicol.* 13: 518–523.

Marshall, K. B. 1975. The spruce budworm and the dollar in New Brunswick. *The Forestry Chronicle.* August: 143–145.

Martin, H. 1972. *Pesticide manual* (3rd ed.). British Crop Protection Council. P. 535.

Martin, H., and C. R. Worthing. 1977. *Pesticide manual* (5th ed.). British Crop Protection Council. P. 593.

Martin, J. E. 1944. *Ribes* eradication effectively controls white pine blister rust. *J. For.* 42: 255–260.

Mathre, D. E. 1971. Mode of action of oxathiin systemic fungicides structure-activity relationships. *J. Agric. Food Chem.* 19: 872–874.

Matsumura, F., and G. M. Boush. 1966. Malathion degradation by *Trichoderma viride* and a *Pseudomonas* species. *Science* 153: 1278.

Matsumura, F., and G. M. Boush. 1967. Dieldrin: Degradation by soil microorganisms. *Science* 156: 959–961.

Matsumura, F., and K. C. Patil. 1969. Adenosine triphosphatases sensitive to DDT in synapses of rat brain. *Science* 166: 121–122.

Mattingly, P. F., R. W. Crosskey, and K. G. V. Smith. 1973. Summary of arthropod vectors. In: *Insects and other arthropods of medical importance*. Smith, K. G. V. (Ed.). British Museum (Natural History) London. Pp. 497–532.

McCallan, S. E. A., A. Hartsell, and F. Wilcoxon. 1936. Hydrogen sulfide injury to plants. *Contrib. Boyce Thompson Inst.* 3: 13–38.

McCaskey, T. A., A. R. Stemp, B. J. Liska, and W. J. Stadelman. 1968. Residues in egg yolks and raw and cooked tissues from laying hens administered selected chlorinated hydrocarbon insecticides. *Poultry Sci.* 47: 564–569.

McCormick, L. L., and A. E. Hiltbold. 1966. Microbiological decomposition of atrazine and diuron in soil. *Weeds* 14: 77–82.

McCraren, J. P., O. B. Cope, and L. Eller. 1969. Some chronic effects of diuron on bluegills. *Weed Sci.* 17: 497–504.

McEwen, F. L. 1977. Pesticide residues and agricultural workers: An overview. In: *Pesticide management and insecticide resistance*. Watson, D. L., and A. W. A. Brown (Eds.). Academic Press, New York. Pp. 37–49.

McLane, M. A. R., and C. Hall. 1972. DDE thins screech owl eggshells. *Bull. Environ. Contam. Toxicol.* 8: 65–68.

McLeese, D. W. 1974. Olfactory response and fenitrothion toxicity in lobsters (*Homarus americanus*). *J. Fish. Res. Bd. Can.* 31: 1127–1131.

McLeese, D. W. 1974a. Toxicity of phosphamidon to american lobsters (*Homarus americanus*) held at 4 and 12°C. *J. Fish. Res. Bd. Can.* 31: 1556–1558.

Meeks, R. L. 1968. The accumulation of ^{36}Cl ring-labeled DDT in a freshwater marsh. *J. Wildl. Mgmt.* 32: 376–398.

Meikle, R. W., E. A. Williams, and C. T. Redeman. 1966. Metabolism of tordon herbicide (4-amino-3,5,6-trichloropicolinic acid) in cotton and decomposition in soil. *J. Agric. Food Chem.* **14**: 384–387.

Menn, J. J., and M. Beroza (Eds.). 1972. Insect juvenile hormones, chemistry and action. Academic Press, New York. P. 341.

Menzie, C. M. 1969. *Metabolism of pesticides.* USDI Bureau of Sport Fisheries & Wildlife Special Science Report—Wildlife No. 127. Washington, D.C. P. 487.

Metcalf, C. L., W. P. Flint, and R. L. Metcalf. 1951. *Destructive and useful insects.* McGraw-Hill Book Co., New York. P. 1071.

Metcalf, R. L. 1971. The chemistry and biology of pesticides. In: *Pesticides in the environment.* Vol. I, Part I. White-Stevens, R. (Ed.). Marcel Dekker, Inc., New York. Pp. 1–144.

Metcalf, R. L. 1973. A century of DDT. *J. Agric. Food Chem.* **21**: 511–519.

Metcalf, R. L. 1977. Model ecosystem approach to insecticide degradation: A critique. *Ann. Rev. Entomol.* **22**: 241–261.

Metcalf, R. L., and J. R. Sanborn. 1975. Pesticides and environmental quality in Illinois. *Ill. Nat. Hist. Surv. Bull.* **31**: 377–436.

Metcalf, R. L., G. K. Sangha, and I. P. Kapoor. 1971. Model ecosystem for the evaluation of pesticide biodegradability and ecological magnification. *Environ. Sci. Technol.* **5**: 709–713.

Mick, D. L., T. D. Gartin, and K. R. Long. 1970. A case report: Occupational exposure to the insecticide naled. *J. Iowa Med. Soc.* June 1970: 395–396.

Milby, T. H., J. E. Mitchell, and T. S. Freeman. 1969. Seasonal neonatal hyperbilirubinemia. *Pediatrics* **43**: 601–605.

Miles, J. R. W. 1968. Arsenic residues in agricultural soils of southwestern Ontario. *J. Agric. Food Chem.* **16**: 620–622.

Miles, J. R. W., and C. R. Harris. 1971. Insecticide residues in a stream and controlled drainage system in agricultural areas of southwestern Ontario. *Pestic. Monit. J.* **5**: 289–294.

Miles, J. R. W., and C. R. Harris. 1973. Organochlorine insecticide residues in streams draining agricultural, urban and resort areas of Ontario, Canada. *Pestic. Monit. J.* **6**: 363–368.

Miles, J. R. W., C. R. Harris, and P. Moy. 1978. Insecticide residues in water, sediment, and fish of the drainage system of the Holland Marsh, Ontario, Canada, 1972–1975. *J. Econ. Entomol.* **71**: 125–131.

Milius, P. 1976. A pesticide that cripples mind and body. *Honolulu Star-Bulletin and Advertiser.* December 6: G-23.

Miller, C. A., and E. G. Kettela. 1975. Aerial control operations against the spruce budworm in New Brunswick, 1952–1973. In: *Aerial control of forest insects in Canada.* Prebble, M. L. (Ed.). Department of the Environment, Ottawa. Pp. 94–112.

Miller, F. M., and E. D. Gomes. 1974. Detection of DCPA residues in environmental samples. *Pestic. Monit. J.* **8**: 53–58.

Miller, L. P., S. E. A. McCallan, and R. M. Weed. 1953. Quantitative studies on the role of hydrogen sulfide formation in the toxic action of sulfur to fungus spores. *Contrib. Boyce Thompson Inst.* **17**: 151–171.

Mills, H. R. 1952. Death in the Florida marshes. *Audobon* **54**: 285–290.

Mitchell, B. J. F., and G. R. Stephenson. 1973. The selective action of picloram in red maple and white ash. *Weed Res.* **13**: 169–173.

Miura, T., and R. M. Takahashi. 1973. Insect developmental inhibitors. 3. Effects on nontarget aquatic organisms. *J. Econ. Entomol.* **66**: 917–922.

Moore, N. W. 1965. Pesticides and birds—a review of the situation in Great Britain in 1965. *Bird Study* **12**: 222–251.

Moreland, D. E., and M. R. Boots. 1971. Effects of optically active 1-(α methylbenzyl)-3-(3,4-dichlorophenyl) urea on reactions of mitochondria and chloroplasts. *Pl. Physiol.* (Baltimore) **47**: 53–58.

Moreland, D. E., and J. L. Hilton. 1976. Actions on photosynthetic systems. In: *Herbicides: Physiology, biochemistry, ecology* (2nd ed.). Vol. 1. Audus, L. J. (Ed.). Academic Press, London. Pp. 493–524.

Moreland, D. E., S. S. Malhotra, R. D. Gruenhagen, and E. H. Shokraii. 1969. Effects of herbicides on RNA and protein synthesis. *Weed Sci.* **17**: 556–562.

Morgan, H. G. 1976. Sublethal effects of diazinon on stream invertebrates. Ph.D. Thesis, University of Guelph, Guelph, Ontario.

Moriarty, F. 1975. Exposure and residues. In: *Organochlorine pesticides: Persistent organic pollutants.* Moriarty, F. (Ed.). Academic Press, London, New York, San Francisco. Pp. 29–72.

Morrod, R. S. 1976. Effects on plant cell membrane structure and function. In: *Herbicides: Physiology, biochemistry, ecolocy* (2nd ed.). Vol. 1. Audus, L. J. (Ed.). Academic Press, London. Pp. 281–304.

Mosser, J. L., T. C. Teng, W. G. Walther, and C. F. Wurster. 1974. Interactions of PCBs, DDT and DDE in a marine diatom. *Bull. Environ. Contam. Toxicol.* **12**: 665–668.

Moubry, R. J., J. M. Helm, and G. R. Myrdal. 1968. Chlorinated pesticide residues in an aquatic environment located adjacent to a commercial orchard. *Pestic. Monit. J.* **1**: 27–29.

Mount, D. I., and G. J. Putnicki. 1966. Summary report of the 1963 Mississippi fish kill. *Trans. North Amer. Wildl. Nat. Resources Conf.* **31**: 177–184.

Muirhead-Thompson, R. C. 1971. *Pesticides and freshwater fauna.* Academic Press, London & New York. P. 248.

Mukula, J. 1970. Weed control in cereal grains of northern Europe. In: *FAO International Conference on Weed Control.* Weed Science Society of America. Pp. 68–78.

Mulhern, B. M., W. L. Reichel, L. N. Locke, T. C. Lamont, A. Belisle, E. Cromartie, G. F. Bagley, and R. M. Prouty. 1970. Organochlorine residues and autopsy data from bald eagles, 1966–1968. *Pestic. Monit. J.* **4**: 141–144.

Muncy, R. J., and A. D. Oliver, Jr. 1963. Toxicity of ten insecticides to the red crawfish, *Procambarus clarki* (Girard). *Trans. Amer. Fish. Soc.* **92**: 428–431.

Murphy, D. A., and L. J. Korschgen. 1970. Reproduction, growth, and tissue residues of deer fed dieldrin. *J. Wildl. Mgmt.* **34**: 887–903.

Murphy, P. G., and J. V. Murphy. 1971. Correlations between respiration and direct uptake of DDT in the mosquito fish *Gambusia affinis*. *Bull. Environ. Contam. Toxicol.* **6**: 581–588.

Musial, C. J., O. Hutzinger, V. Zitko, and J. Crocker. 1974. Presence of PCB, DDE and DDT in human milk in the provinces of New Brunswick and Nova Scotia, Canada, 1974. *Bull. Environ. Contam. Toxicol.* **12**: 258–267.

National Academy of Sciences. 1970. *Vertebrate pests: Problems and control*. Nat. Acad. Sci., Washington, D.C. P. 153.

National Academy of Sciences. 1971. *Chlorinated hydrocarbons in the marine environment*. Nat. Acad. Sci., Washington, D.C.

National Academy of Sciences. 1975. *Contemporary pest control practices and prospects*. Vol. 1. Nat. Acad. Sci., Washington, D.C. P. 506.

National Academy of Sciences. 1975a. *Pest control: An assessment of present and alternative technologies*. Vol. IV. *Forest Pest Control*. Nat. Acad. Sci., Washington, D.C. P. 170.

National Research Council of Canada. 1974. *Picloram: The effects of its use as a herbicide on environmental quality*. Nat. Res. Coun. Can. No. 13684. P. 128.

National Research Council. 1974a. *Chlordane: Its effects on Canadian ecosystems and its chemistry*. Nat. Res. Coun. Can. No. 14094. P. 189.

National Research Council. 1975. *Endosulfan: Its effects on environmental quality*. Nat. Res. Coun. Can. No. 14098. P. 100.

National Research Council. 1975a. *Methoxychlor: Its effects on environmental quality*. Nat. Res. Coun. Can. No. 14102. P. 164.

National Research Council. 1975b. *Fenitrothion: The effects of its use on environmental quality and its chemistry*. Nat. Res. Coun. Can. No. 14104. P. 162.

Netboy, A. 1969. Atlantic salmon. *Sea Frontiers* **15**: 66–77.

Neumeyer, J., D. G. Gibbons, and H. Trask. 1969. Sales of pesticides will top $3 billion by '75. *Chem. Week.* **104**: 38–68.

Newsome, W. H. 1976. Residues of four ethylenebis (dithiocarbamates) and their decomposition products on field-sprayed tomatoes. *J. Agric. Food Chem.* **24**: 999–1001.

Nickerson, P. R., and K. R. Barbehenn. 1975. DDT residues in starlings, 1974. *Pestic. Monit. J.* **9**: 1.

Niemczyk, H. D., and K. O. Lawrence. 1973. Japanese beetle: Evidence of resistance to cyclodiene insecticides in larvae and adults in Ohio. *J. Econ. Entomol.* **66**: 520–521.

Nishizawa, Y., K. Fujii, T. Kadota, J. Miyamoto, and H. Sakamoto. 1961. Studies on the organophosphorus insecticides. Part VII. Chemical and biological properties of new low toxic organophosphorus insecticide *0,0*-dimethyl-*0*-(3-methyl-4-nitrophenyl) phosphorothioate. *Agric. Biol. Chem.* (Tokyo) **25**: 605–610.

Norstrum, R. J., A. E. McKinnon, and A. S. W. deFreitas. 1976. A bioenergetics-based model for pullutant accumulation by fish. Simulation of PCB and methyl-mercury residue levels in Ottawa River yellow perch (*Perca flavescens*). *J. Fish. Res. Bd. Can.* **33**: 248–267.

Norstrum, R. J., A. E. McKinnon, A. S. W. deFreitas, and D. R. Miller. 1975. Pathway definition of pesticide and mercury uptake by fish. *Environ. Quality & Safety* **4**: 811–815.

Nowicki, H. G., J. F. Myrtle, and A. W. Norman. 1972. Effects of organochlorine insecticides on metabolism of cholecalciferol (vitamin D_3) in rachitic cockerel. *J. Agric. Food Chem.* **20**: 380–384.

Nowicki, H. G., R. G. Wong, J. F. Myrtle, and A. W. Norman. 1972a. Inhibition of biological activity of cholecalciferol (vitamin D_3) by *0,p'*-DDT or *p,p'*-DDT in rachitic cockerel. *J. Agric. Food Chem.* **20**: 376–380.

O'Brien, L. P., and G. N. Prendeville. 1972. Shoot zone uptake of soil-applied herbicides in *Pisum sativum* L. *Weed Res.* **12**: 248–253.

O'Brien, R. D. 1967. *Insecticides: Action and metabolism.* Academic Press, New York. P. 332.

Odsjo, T., and J. Sondell. 1976. Reproductive success in ospreys *Pandion haliaetus* in southern and central Sweden, 1971–1973. *Ornis Scandinavica* **7**: 71–84.

Oestreicher, M. I., D. H. Shuman, and C. F. Wurster. 1971. DDE reduces medullary bone formation in birds. *Nature* **229**: 571.

Ogilvie, D. M., and J. M. Anderson. 1965. Effect of DDT on temperature selection by young Atlantic salmon, *Salmo salar. J. Fish Res. Bd. Can.* **22**: 503–512.

Ogilvie, D. M., and D. L. Miller. 1976. Duration of a DDT-induced shift in the selected temperature of Atlantic salmon. *Bull. Environ. Contam. Toxicol.* **16**: 86–89.

Olney, C. E., W. E. Donaldson, and T. W. Kerr. 1962. Methoxychlor in eggs and chicken tissues. *J. Econ. Entomol.* **55**: 477–479.

O'Niell, P. M., and B. E. Langlois. 1976. Effect of heptachlor on the growth, viability and respiration of *Staphylococcus aureus. Bull. Environ. Contam. Toxicol.* **16**: 330–338.

Ontario Pesticides Act. 1973.

Orgill, M. M., G. A. Schmel, and M. R. Petersen. 1976. Some initial measurements of airborne DDT over Pacific Northwest forests. *Atmos. Environ.* **10**: 827–834.

Oser, B. L. 1971. Toxicology of pesticides to establish proof of safety. In: *Pesticides in the environment.* Vol. I, Part II. White-Stevens, R. (Ed.). Marcel Dekker, Inc., New York. Pp. 411–456.

Ottoboni, A. 1969. DDT: The world has been doused with it for 25 years. With what result? *California Health.* August: 1–2; 15.

Oudbier, A. J., A. W. Bloomer, H. A. Price, and R. L. Welch. 1974. Respiratory route of pesticide exposure as a potential health hazard. *Bull. Environ. Contam. Toxicol.* **12**: 1–9.

Owens, L. D. 1969. Toxins in plant disease: Structure and mode of action. *Science* **165**: 18–25.

Owens, R. G. 1969a. Organic sulfur compounds. In: *Fungicides, an advanced treatise.* Vol. II. Torgeson, D.C. (Ed.). Academic Press, New York. Pp. 147–301.

Owens, R. G., and J. H. Rubinstein. 1964. Chemistry of the fungicidal action of tetramethylthiuram disulfide (thiram) and ferbam. *Contrib. Boyce Thompson Inst.* **22**: 241–257.

Ozburn, G. W., and F. O. Morrison. 1962. Development of a DDT-tolerant strain of laboratory mice. *Nature* **196**: 1009–1010.

Pampana, E. J., and P. F. Russell. 1955. Malaria: A world problem. *Chronicle Wld. Hlth. Org.* **9**: 31.

Pape, B. E., and M. J. Zabik. 1970. Photochemistry of selected 2-chloro and 2-methyl thio-4,6-di(alkylamino)-*s*-triazine herbicides. *J. Agric. Food Chem.* **18**: 202–207.

Parka, S. J., and O. F. Soper. 1977. The physiology and mode of action of the dinitroaniline herbicides. *Weed Sci.* **25**: 79–87.

Parka, S. J., and H. M. Worth. 1965. The effects of trifluralin on fish. *Proc. Southern Weed Conf.* **18**: 469–473.

Parker, C. 1963. Factors affecting the selectivity of 2,3-dichloroallyl diisopropylthiocarbamate (Di-allate) against *Avena* spp. in wheat and barley. *Weed Res.* **3**: 259–276.

Parker, J. W. 1976. Pesticides and eggshell thinning in the Mississippi kite. *J. Wildl. Mgmt.* **40**: 243–248.

Paul, B. S., and V. P. Vadlamudi. 1976. Teratogenic studies of fenitrothion on white leghorn chick embryos. *Bull. Environ. Contam. Toxicol.* **15**: 223–229.

Peakall, D. B. 1967. Pesticide-induced enzyme breakdown of steroids in birds. *Nature* (London) **216**: 505–506.

Peakall, D. B. 1969. Effect of DDT on calcium uptake and vitamin D metabolism in birds. *Nature* **224**: 1219–1220.

Peakall, D. B. 1970. *p,p'*-DDT: Effect on calcium metabolism and concentration of estradiol in the blood. *Science* **168**: 592–594.

Peakall, D. B. 1975. Physiological effects of chlorinated hydrocarbons on avian species. In: *Environmental dynamics of pesticides.* Haque, R., and V. H. Freed (Eds.). Plenum Press, New York & London. Pp. 343–360.

Peakall, D. B. 1976. DDT in rainwater in New York following application in the Pacific Northwest. *Atmos. Environ.* **10**: 899–900.

Peakall, D. B., T. J. Cade, C. M. White, and J. R. Haugh. 1975. Organochlorine residues in Alaskan peregrines. *Pestic. Monit. J.* **8**: 255–260.

Pearce, P. A. 1968. *Effects on bird populations of phosphamidon and sumithion used for spruce budworm control in New Brunswick and hemlock looper control in Newfoundland in 1968: A summary statement.* Canadian Wildlife Service Report No. 14. P. 56.

Peterle, T. J. 1969. DDT in Antarctic snow. *Nature* **224**: 620.

Peters, J. W., and R. M. Cook. 1973. Effects of atrazine on reproduction in rats. *Bull. Environ. Contam. Toxicol.* **9**: 301–304.

Peters, R. A. 1963. Biochemical lesions and lethal synthesis. Pergamon Press, Oxford. P. 321.

Peterson, C. A., and L. V. Edgington. 1975. Uptake of the systemic fungicide methyl 2-benzimidazolecarbamate and the fluorescent dye PTS by onion roots. *Phytopathology* **65**: 1254–1259.

Peterson, R. L., G. R. Stephenson, and B. J. F. Mitchell. 1974. Effects of picloram on shoot anatomy of red maple and white ash. *Weed Res.* **14**: 227–229.

Petrosini, G. 1962. Degradazione naturale dei dithiocarbamati. *Notiziario Sulle Malattie Delle Piante* **59**: 59–66.

Phillips, W. M. 1968. Persistence and movement of 2,3,6-TBA in soil. *Weed Sci.* **16:** 144–148.

Pickett, A. D., and N. A. Patterson. 1963. Arsenates: Effect on fecundity in some Diptera. *Science* **140:** 493–494.

Pimentel, D. 1971. *Ecological effects of pesticides on non-target organisms.* U.S. Government Printing Office, Washington, D.C. P. 220.

Pimentel, D. 1973. Extent of pesticide use, food supply, and pollution. *J. N.Y. Entomol. Soc.* **81:** 13–33.

Pimentel, D., E. C. Terhune, W. Dritschilo, D. Gallahan, N. Kinner, D. Nafus, R. Peterson, N. Zareh, J. Misiti, and O. Haben-Schaim. 1977. Pesticides, insects in foods, and cosmetic standards. *Biosci.* **27:** 178–185.

Plapp, F. W., and J. E. Casida. 1958. Bovine metabolism of organophosphorus insecticides: Metabolic fate of 0,0-dimethyl 0-(2,4,5-trichlorophenyl) phosphorothioate in rats and a cow. *J. Agric. Food Chem.* **6:** 662–667.

Platonow, N. S., and L. H. Karstad. 1973. Dietary effects of polychlorinated biphenyls on mink. *Can. J. Comp. Med.* **37:** 391–400.

Pocker, Y., M. W. Beug, and V. R. Ainardi. 1971. Coprecipitation of carbonic anhydrase by 1,1-bis(p-chlorophenyl)-2,2,2-trichloroethane, 1,1-bis(p-chlorophenyl)-2,2-dichloroethylene, and dieldrin. *Biochem.* **10:** 1390–1396.

Pocker, Y., W. M. Beug, and V. R. Ainardi. 1971a. Carbonic anhydrase interaction with DDT, DDE, and dieldrin. *Science* **174:** 1336–1339.

Polishuk, Z. W., M. Wassermann, D. Wassermann, Y. Gronef, S. Lazarovici, and L. Tomatis. 1970. Effects of pregnancy on storage of organochlorine insecticides. *Arch. Environ. Hlth.* **20:** 215–217.

Poonawalla, N. H., and F. Korte. 1964. Metabolism of insecticides. VIII. Excretion, distribution and metabolism of α-chlordane[14]C by rats. *Life Sci.* **3:** 1497–1500.

Porter, R. D., and S. N. Wiemeyer. 1969. Dieldrin and DDT: Effects on sparrow hawk eggshells and reproduction. *Science* **165:** 199–200.

Porter, R. D., and S. N. Wiemeyer. 1972. DDE at low dietary levels kills captive American kestrels. *Bull. Environ. Contam. Toxicol.* **8:** 193–199.

Post, G., and R. A. Leasure. 1974. Sublethal effect of malathion to three salmonid species. *Bull. Environ. Contam. Toxicol.* **12:** 312–319.

Potter, J. L., and R. D. O'Brien. 1964. Parathion activation by livers of aquatic and terrestrial vertebrates. *Science* **144:** 55–60.

Prasad, R., and D. Travnick. 1976. Development of a low pressure trunk-injection apparatus for prevention of Dutch elm disease. Proceedings of the 33rd Annual Meeting of the Canadian Pest Management Society. Pp. 23–35.

Prebble, M. L. (Ed.). 1975. *Aerial control of forest insects in Canada.* Department of the Environment, Ottawa, Canada. P. 329.

Prestt, J., and D. A. Ratcliffe. 1970. Effects of organochlorine insecticides on European birdlife. Proceedings of the XVth International Ornithological Congress, 1970. Pp. 486–513.

Probst, G. W., T. Golab, and W. L. Wright. 1975. Dinitroanilines. In: *Herbicides:*

Chemistry, degradation and mode of action (2nd ed.). Vol. 1. Kearney, P. C., and D. D. Kaufman (Eds.). Marcel Dekker Inc., New York. Pp. 453–500.

Putnam, A. R., and S. K. Ries. 1968. Factors influencing the phytotoxicity and movement of paraquat in quackgrass. *Weed Sci.* **16**: 80–83.

Quinby, G. E., K. C. Walker, and W. F. Dunham. 1958. Public health hazards in the use of organic phosphorus insecticides in cotton culture in the Delta area of Mississippi. *J. Econ. Entomol.* **51**: 831–838.

Radosevich, S. R. 1973. Differential sensitivity of two common groundsel biotypes (*Senecio vulgaris* L.) to several s-triazine herbicides. *Res. Rept. Western Soc. Weed Sci.* **26**: 174.

Radosevich, S. R., and S. G. Conrad. 1977. Competition between s-triazine susceptible and resistant biotypes of three plant species. Weed Science Society of America. Abstract No. 158, p. 77.

Ratcliffe, D. A. 1967. Decrease in eggshell weight in certain birds of prey. *Nature* **215**: 208–210.

Ratcliffe, D. A. 1969. Population trends of the peregrine falcon in Great Britain. In: *Peregrine falcon populations: Their biology and decline*. Hickey, J. J. (Ed.). University of Wisconsin Press, Madison. Pp. 239–269.

Ratcliffe, D. A. 1970. Changes attributable to pesticides in egg breakage frequency and eggshell thickness in some British birds. *J. Appl. Ecol.* **7**: 67–115.

Rathus, E. M. 1973. The effect of pesticide residues on humans. In: *Pesticides and the environment: A continuing controversy*. Deichmann, W. B. (Ed.). Intercontinental Medical Book Corp., New York & London. Pp. 23–31.

Read, D. C. 1960. Effect of soil treatments of heptachlor and parathion on predators and parasites of root maggots attacking rutabagas on Prince Edward Island. *J. Econ. Entomol.* **53**: 932–935.

Read, S., and W. P. McKinley. 1961. DDT and DDE content of human fat. *Arch. Environ. Hlth.* **3**: 209–211.

Ream, C. H. 1976. Loon productivity, human disturbance, and pesticide residues in northern Minnesota. *The Wilson Bull.* **88**: 427–432.

Reidinger, R. F., Jr., and D. G. Crabtree. 1974. Organochlorine residues in golden eagles, United States—March 1964–July 1971. *Pestic. Monit. J.* **8**: 37–43.

Reimold, R. J. 1975. Chlorinated hydrocarbon pesticides and mercury in coastal biota, Puerto Rico and the U.S. Virgin Islands—1972–1974. *Pesric. Monit. J.* **9**: 39–43.

Reinert, R. E. 1970. Pesticide concentrations in Great Lakes fish. *Pestic. Monit. J.* **3**: 233–240.

Reinert, R. E., D. Stewart, and H. L. Seagran. 1972. Effects of dressing and cooking on DDT concentrations in certain fish from Lake Michigan. *J. Fish. Res. Bd. Can.* **29**: 525–529.

Reinke, J., J. F. Uthe, and D. Jamieson. 1972. Organochlorine pesticide residues in commercially caught fish in Canada—1970. *Pestic. Monit. J.* **6**: 43–49.

Reynolds, P. J., I. L. Lindahl, H. C. Cecil, and J. Bitman. 1976. A comparison of DDT and methoxychlor accumulation and depletion in sheep. *Bull. Environ. Contam. Toxicol.* **16**: 240–247.

Rhodes, R. C., I. J. Belasco, and H. L. Pease. 1970. Determination of mobility and adsorption of agrichemicals on soils. *J. Agric. Food Chem.* **18**: 524–528.

Richard, J. J., G. J. Junk, M. J. Avery, N. L. Nehring, J. S. Fritz, and H. J. Svec. 1975. Analysis of various Iowa waters for selected pesticides: atrazine, DDE, and dieldrin—1974. *Pestic. Monit. J.* **9**: 117–123.

Richter, S. B. 1961. (To Vesicol Chemical Corp.). U.S. Patent No. 3,013,054.

Riddiford, L. M. 1972. Juvenile hormone and insect embryonic development: Its potential role as an ovicide. In: *Insect juvenile hormones, chemistry and action.* Menn, J. J., and M. Berosa (Eds.). Academic Press, New York. Pp. 95–111.

Ries, S. K., H. Chmiel, D. R. Dilley, and P. Filner. 1967. The increase in nitrate reductase and protein content of plants treated with simazine. *Proc. Nat. Acad. Sci. U.S.* **58**: 526–532.

Risebrough, R. W., and D. W. Anderson. 1975. Some effects of DDE and PCB on mallards and their eggs. *J. Wildl. Mgmt.* **39**: 508–513.

Risebrough, R. W., J. Davis, and D. W. Anderson. 1970. Effects of various chlorinated hydrocarbons. In: *The biological impact of pesticides in the environment.* Gillett, J. W. (Ed.). Oregon State University Environmental Health Science Service No. 1. Pp. 40–53.

Risebrough, R. W., R. J. Huggett, J. J. Griffin, and E. D. Goldberg. 1968. Pesticides: Transatlantic movements in the Northeast Trades. *Science* **159**: 1233–1235.

Ritchey, S. J., R. W. Young, and E. O. Essary. 1967. The effects of cooking on chlorinated hydrocarbon residues in chicken tissues. *J. Food Sci.* **32**: 238–240.

Ritchey, S. J., R. W. Young, and E. O. Essary. 1969. Cooking methods and heating effects on DDT in chicken tissues. *J. Food Sci.* **34**: 569–571.

Robbins, R. C., and J. Kastelic. 1961. Fate of tetramethyl thiuram disulfide in the digestive tract of the ruminant animal. *J. Agric. Food Chem.* **9**: 256–260.

Robinson, J. 1969. Organochlorine insecticides and bird populations in Britain. In: *Chemical fallout.* Miller, M. W., and G. G. Berg (Eds.). C C Thomas, Springfield, Ill. Pp. 113–169.

Robinson, J. 1970. Pharmacodynamics of dieldrin in birds. In: *The biological impact of pesticides in the environment.* Gillett, J. W. (Ed.). Oregon State University Environmental Health Science Series No. 1. Pp. 54–58.

Robinson, J., A. Richardson, A. N. Crabtree, J. C. Couson, and G. R. Potts. 1967. Organochlorine residues in marine organisms. *Nature* **214**: 1307–1311.

Roelofs, W. L. 1976. Pheromones. In: *The future for insecticides—needs and prospects.* Metcalf, R. L., and J. J. McKelvey, Jr. (Eds.). John Wiley & Sons, New York. Pp. 445–461.

Roelofs, W. L., and R. T. Cardé. 1974. Sex pheromones in the reproductive isolation of lepidopterous species. In: *Pheromones.* Birch, M. C. (Ed.). American Elsevier Publishing Co. Inc., New York. Pp. 96–114.

Roots, B. J., and C. L. Prosser. 1962. Temperature acclimation and the nervous system in fish. *J. Exp. Biol.* **39**: 617–629.

Rosato, P., and D. E. Ferguson. 1968. The toxicity of endrin-resistant mosquitofish to eleven species of vertebrates. *Biosci.* **18**: 783–784.

Rosene, W., Jr., and D. W. Lay. 1963. Disappearance and visibility of quail remains. *J. Wildl. Mgmt.* **27**: 139–142.

Rudd, R. L. 1964. *Pesticides and the living landscape.* Faber and Faber, London. P. 320.

Russell, P. F. 1959. Insects and the epidemiology of malaria. *Ann. Rev. Entomol.* **4**: 415–434.

Rye, R. P., Jr., and E. L. King, Jr. 1976. Acute toxic effects of two lampricides to twenty-one freshwater invertebrates. *Trans. Amer. Fish. Soc.* **105**: 322–326.

Saha, J. G., and A. K. Sumner. 1971. Organochlorine insecticide residues in soil from vegetable farms in Saskatchewan. *Pestic. Monit. J.* **5**: 28–31.

Sameoto, D. D., D. C. Darrow, and S. Guildford. 1975. DDT residues in euphausiids in the upper estuary of the Gulf of St. Lawrence. *J. Fish. Res. Bd. Can.* **32**: 310–314.

Sand, P. F., G. B. Wiersma, H. Tai, and L. J. Stevens. 1971. Preliminary study of mercury residues in soils where mercury seed treatments have been used. *Pestic. Monit. J.* **5**: 32–33.

Sanders, H. O. 1969. Toxicity of pesticides to the crustacean *Gammarus lacustris*. USDI Bureau of Sport Fisheries & Wildlife Technical Paper 25. P. 18.

Sanders, H. O. 1970. Pesticide toxicities to tadpoles of the western chorus frog, *Pesudacris triseriata*, and Fowler's toad, *Bufo woodhousii fowleri*. *Copeia* **2**: 246–251.

Sanders, H. O., and O. B. Cope. 1966. Toxicity of several pesticides to two species of cladocerans. *Trans. Amer. Fish. Soc.* **95**: 165–169.

Sanders, H. O., and O. B. Cope. 1968. The relative toxicities of several pesticides to naiads of three species of stoneflies. *Limnol. Oceanogr.* **13**: 112–117.

Sandifer, S. H., J. E. Keil, J. F. Finklea, and R. H. Gadsden. 1972. Pesticide effects on environmentally exposed workers: A summary of four years of observation of industry and farm workers in South Carolina. *Ind. Med. Surg.* **41**: 9–12.

Savage, E. P., J. D. Tessari, J. W. Malberg, H. W. Wheeler, and J. R. Bagby. 1973. Organochlorine pesticide residues and polychlorinated biphenyls in human milk, Colorado, 1971–1972. *Pestic. Monit. J.* **7**: 1–5.

Schafer, E. W., R. B. Brunton, and N. F. Lockyer. 1975. The effect of subacute and chronic exposure to 4-amino pyridine on reproduction in coturnix quail. *Bull. Environ. Contam. Toxicol.* **13**: 758–764.

Schafer, E. W., Jr., R. B. Brunton, N. F. Lockyer, and D. J. Cunningham. 1975a. The chronic toxicity of methiocarb to grackles, doves, and quail and reproductive effects on quail. *Bull. Environ. Contam. Toxicol.* **14**: 641–647.

Schouwenburg, W. J., and K. J. Jackson. 1966. A field assessment of the effects of spraying a small coastal coho salmon stream with phosphamidon. *Can. Fish. Culturist* **37**: 35–43.

Schultz, D. P., and P. D. Harman. 1974. Residues of 2,4-D in pond waters, mud, and fish, 1971. *Pestic. Monit. J.* **8**: 173–179.

Schultz, D. P., and E. W. Whitney. 1974. Monitoring 2,4-D residues at Loxahatchee National Wildlife Refuge. *Pestic. Monit. J.* **7**: 146–152.

Schulze, J. A., D. B. Manigold, and F. L. Andrews. 1973. Pesticides in selected western streams—1968-1971. *Pestic. Monit. J.* 7: 73–84.

Schweizer, E. E. 1967. Toxicity of DSMA soil residues to cotton and rotational crops. *Weeds* 15: 72–76.

Schwimmer, M., and D. Schwimmer. 1968. Medical aspects of phycology. In: *Algae, man and the environment.* Jackson, D. F. (Ed.). Syracuse University Press, Syracuse, New York. Pp. 279–358.

Scifres, C. J., O. C. Burnside, and M. K. McCarty. 1969. Movement and persistence of picloram in pasture soils. *Weed Sci.* 17: 486–488.

Scott, H. D., and R. E. Phillips. 1973. Absorption of herbicides by soybean seed. *Weed Sci* 21: 71–76.

Scott, J. M., J. A. Wiens, and R. R. Claeys. 1975. Organochlorine levels associated with a common murre die-off in Oregon. *J. Wildl. Mgmt.* 39: 310–320.

Scott, T. G., Y. L. Willis, and J. A. Ellis. 1959. Some effects of a field application of dieldrin on wildlife. *J. Wildl. Mgmt.* 23: 409–427.

Seba, D. B., and E. F. Corcoran. 1969. Surface slicks as concentrations of pesticides in the marine environment. *Pestic. Monit. J.* 3: 190–193.

Seba, D. B., and J. M. Prospero. 1971. Pesticides in the lower atmosphere of the Northern Equatorial Atlantic Ocean. *Atmos. Environ.* 5: 1043–1050.

Sellers, L. G., and P. A. Dahm. 1975. Chlorinated hydrocarbon insecticide residues in ground beetles (*Harpalus pensylvanicus*) and Iowa soil. *Bull. Environ. Contam. Toxicol.* 13: 218–222.

Shafer, M. A. M., and M. E. Zabik. 1975. Dieldrin, fat and moisture loss during the cooking of beef loaves containing texturized soy protein. *J. Food Sci.* 40: 1068–1071.

Shane, M. S. 1948. Effect of DDT spray on reservoir biological balance. *J. Amer. Water Works Assoc.* 40: 333–336.

Sharom, M. S. 1974. The behavior and fate of metribuzin in eight Ontario soils. M.Sc. Thesis, University of Guelph, Guelph, Ontario.

Shaw, S. B. 1971. Chlorinated hydrocarbon pesticides in California sea otters and harbor seals. *Calif. Fish & Game* 57: 290–294.

Shea, K. P. 1974. Nerve damage—the return of "ginger Jake." *Environment* 16: 6–10.

Sherburne, J. A., and J. B. Dimond. 1969. DDT persistence in wild hares and mink. *J. Wildl. Mgmt.* 33: 944–948.

Sherman, R. W. 1973. "Artifacts" and "mimics" of DDT and other organochlorine insecticides. *J. N.Y. Entomol. Soc.* 81: 152–163.

Shimabukuro, R. H., and H. R. Swanson. 1969. Atrazine metabolism, selectivity and mode of action. *J. Agric. Food Chem.* 17: 199–205.

Siegel, S. M., and L. A. Halpern. 1965. Effects of peroxides on permeability and their modification by indoles, vitamin E and other substances. *Plant Physiol.* 40: 792–796.

Sijpesteijn, A. K., H. M. Dekhuijzen, and J. W. Vonk. 1977. Biological conversion of fungicides in plants and microorganisms. In: *Antifungal compounds. Vol. 2. Interactions in biological and ecological systems.* Siegel, M. R., and H. D. Sisler (Eds.). Marcel Dekker, Inc., New York. Pp. 91–147.

Sileo, L., L. K. Karstad, R. Frank, M. V. H. Holdrinet, E. Addison, and H. E. Braun. 1977. Organochlorine poisoning of ring-billed gulls in southern Ontario. *J. Wildlife Dis.* **13**: 313–322.

Simmons, S. W. 1969. Some health-related needs in pesticide investigations. *Indust. Med. Surg.* **38**: 60–63.

Simpson, J. G., and W. G. Gilmartin. 1970. An investigation of elephant seal and sea lion mortality on San Miguel Island. *Biosci.* **20**: 289.

Sims, G. G., C. E. Cosham, J. R. Campbell, and M. C. Murray. 1975. DDT residues in cod livers from the Maritime provinces of Canada. *Bull. Environ. Contam. Toxicol.* **14**: 505–512.

Sims, J. J., H. Mee, and D. C. Erwin. 1969. Methyl 2-benzimidazolecarbamate, a fungitoxic compound isolated from cotton plants treated with methyl 1-(butylcarbamoyl)-2-benzimidazolecarbamate (benomyl). *Phytopathology* **59**: 1775–1776.

Sims, J. L., and F. K. Pfaender. 1975. Distribution and biomagnification of hexachlorophene in urban drainage areas. *Bull. Environ. Contam. Toxicol.* **14**: 214–220.

Sinow, J., and E. Wei. 1973. Ocular toxicity of paraquat. *Bull. Environ. Contam. Toxicol.* **9**: 163–168.

Sladen, W. J. L., C. M. Menzie, and W. L. Reichel. 1966. DDT residues in Adelie penguins and a crabeater seal from Antarctica. *Nature* **210**: 670–673.

Smalley, E. B., C. J. Meyers, R. N. Johnson, B. C. Fluke, and R. Vieau. 1973. Benomyl for practical control of Dutch elm disease. *Phytopathology* **63**: 1239–1252.

Smith, A. J. 1967. The effect of the lamprey larvicide 3-trifluoromethyl-4-nitrophenol on selected aquatic invertebrates. *Trans. Amer. Fish. Soc.* **96**: 410–413.

Smith, D. C. 1971. Pesticide residues in the total diet in Canada. *Pestic. Sci.* **2**: 92–95.

Smith, D. C., R. Leduc, and C. Charbonneau. 1973. Pesticide residues in the total diet in Canada III. *Pestic. Sci.* **4**: 211–214.

Smith, D. C., E. Sandi, and R. Leduc. 1972. Pesticide residues in the total diet in Canada II. *Pestic. Sci.* **3**: 207–210.

Smith, G. R., C. A. Porter, and E. G. Jaworski. 1966. Uptake and metabolism of C^{14} labeled α-chloroacetamides by germinating seeds. Abstracts of 152nd Meeting, American Chemical Society A42, New York.

Smith, M. I., and E. Elvove. 1930. Pharmacological and chemical studies in the cause of so-called ginger paralysis. *Pub. Hlth. Rpts.* **45**: 1703–1716.

Smith, R. B., Jr., J. K. Finnegan, P. S. Larson, P. F. Sahyoun, M. L. Dreyfuss, and H. B. Haag. 1953. Toxicologic studies on zinc and disodium ethylene bisdithiocarbamates. *J. Pharmacol. Exp. Ther.* **109**: 159–166.

Smith, W. E., K. Funk, and M. E. Zabik. 1975. Effects of cooking on concentrations of PCB and DDT compounds in Chinook (*Oncorhynchus tshawyscha*) and Coho (*O. kisutch*) salmon from Lake Michigan. *J. Fish. Res. Bd. Can.* **30**: 702–706.

Snel, M., and L. V. Edgington. 1968. Fungitoxicity, uptake and translocation of two oxathiin systemic fungicides in bean. *Phytopathology* **58**: 1068.

Snel, M., B. vonSchmeling, and L. V. Edgington. 1970. Fungitoxicity and structure-activity relationships of some oxathiin and thiazole derivatives. *Phytopathology* **60**: 1164–1169.

Snow, F. H. 1895. Contagious diseases of the chinch bug. 4th Annual Report of the Director of the Experimental Station, University of Kansas. P. 46.

Sodergren, A. 1975. Monitoring DDT and PCB in airborne fallout. In: *Environmental quality and safety*. Coulston, F., and F. Korte (Eds.). Georg Thieme, Stuttgart. Pp. 803–810.

Somers, E. 1963. The uptake of dodine acetate by *Neurospora crassa*. *Mededel. Landbouwhogeschool Opzoekingssta. Staat Gent.* **28**: 580–589.

Somers, J. D., E. T. Moran, B. S. Reinhart, and G. R. Stephenson. 1974. Effect of external application of pesticides to the fertile egg on hatching success and early chick performance. 1. Pre-incubation spraying with DDT and commercial mixtures of 2,4-D: picloram and 2,4-D : 2,4,5-T. *Bull. Environ. Contam. Toxicol.* **11**: 33–38.

Somers, J. D., E. T. Moran, Jr., and B. S. Reinhart. 1974a. Effect of external application of pesticides to the fertile egg on hatching success and early chick performance. 3. Consequences of combining 2,4-D with picloram and extremes in contamination. *Bull. Environ. Contam. Toxicol.* **11**, 511–516.

Somerville, H. J. 1973. Microbial toxins. *Ann. N.Y. Acad. Sci.* **217**: 93–108.

Souza-Machado, V., J. D. Bandeen, G. R. Stephenson, and K. I. N. Jensen. 1977. Differential atrazine interference with the Hill reaction of isolated chloroplasts from *Chenopodium album* L. biotypes. *Weed Res.* **17**: 407–413.

Spear, R. C., D. L. Jenkins, and T. H. Milby. 1975. Pesticide residues and field workers. *Environ. Sci. Technol.* **9**: 308–313.

Spear, R. C., W. J. Popendorf, J. T. Leffingwell, and D. Jenkins. 1975a. Parathion residues on citrus foliage. Decay and composition as related to worker hazard. *J. Agric. Food Chem.* **23**: 808–810.

Spencer, E. Y. 1973. *Guide to the chemicals used in crop protection* (6th ed.). Research Branch, Agriculture Canada, Ottawa. P. 542.

Spencer, W. F. 1975. Movement of DDT and its derivatives into the atmosphere. *Res. Rev.* **59**: 91–117.

Spencer, W. F., and M. M. Cliath. 1974. Factors affecting vapor loss of trifluralin from soil. *J. Agric. Food Chem.* **22**: 987–991.

Spencer, W. F., M. M. Cliath, K. R. Davis, R. C. Spear, and W. J. Popendorf. 1975. Persistence of parathion and its oxidation to paraoxon on the soil surface as related to worker reentry into treated crops. *Bull. Environ. Contam. Toxicol.* **14**: 265–272.

Spencer, W. F., W. J. Farmer, and M. M. Cliath. 1973. Pesticide volatilization. *Res. Rev.* **49**: 1–47.

Sprankle, P., W. F. Meggitt, and D. Penner. 1975. Absorption, mobility and microbial degradation of glyphosate in soil. *Weed Sci.* **23**: 229–234.

Stahler, L. M., and E. I. Whitehead. 1950. The effect of 2,4-D on potassium nitrate levels in leaves of sugar beets. *Science* **112**: 749–751.

Staiff, D. C., G. K. Irle, and W. C. Felsenstein. 1973. Screening of various absorbents for protection against paraquat poisoning. *Bull. Environ. Contam. Toxicol.* **10**: 193–199.

Stanley, C. W., J. E. Barney II, M. R. Helton, and A. R. Yobs. 1971. Measurement of atmospheric levels of pesticides. *Environ. Sci. Technol.* **5**: 430–435.

Starr, H. G., Jr., F. D. Aldrich, W. D. McDougall III, and L. M. Mounce. 1974. Contribution of household dust to the human exposure of pesticides, 1974. *Pestic. Monit. J.* **8**: 209–212.

Statistics Canada. 1970. *Quarterly Bulletin of Agricultural Statistics.* January–March. P. 14.

Statistics Canada. 1972. Pest control products, sales by Canadian registrants. P. 9.

Statistics Canada. 1975. Sales of pest control products by Canadian registrants. Information Canada, Ottawa.

Stebbings, R. E. 1970. Bats in danger. *Oryx* **10**: 311–312.

Steinhaus, E. A. 1949. *Principles of insect pathology.* McGraw-Hill Book Co. Inc., New York. P. 757.

Steinhaus, E. A. 1951. Possible use of *B. thuringiensis* Berliner as an aid in the biological control of the alfalfa caterpillar. *Hilgardia* **20**: 359–381.

Steinhaus, E. A. 1956. Microbial control—the emergence of an idea. *Hilgardia* **26**: 107–160.

Steinhaus, E. A. 1959. On the improbability of *Bacillus thuringiensis* Berliner mutating to forms pathogenic to vertebrates. *J. Econ. Entomol.* **52**: 506–508.

Stephenson, G. R., and R. Y. Chang. 1978. Comparative activity and selectivity of herbicide antidotes. In: *Chemistry and action of herbicide antidotes.* Pallos, F. M., and J. E. Casida (Eds.). Academic Press, New York. Pp. 35–61.

Stevens, L. J., C. W. Collier, and D. M. Woodham. 1970. Monitoring pesticides in soils from areas of regular, limited, and no pesticide use. *Pestic. Monit. J.* **4**: 145–166.

Stevenson, J. 1976. Reasons for judgement: Bridges Brothers Ltd. and Forest Protection Ltd. Supreme Court of New Brunswick, Queen's Bench Division, June 14.

Stewart, C. A. 1972. Atmospheric circulation of DDT. *Science* **177**: 724–725.

Stewart, D. K. R., and K. G. Cairns. 1974. Endosulfan persistence in soil and uptake by potato tubers. *J. Agric. Food Chem.* **22**: 984–986.

Stickel, L. F. 1946. Field studies of a *Peromyscus* population in an area treated with DDT. *J. Wildl. Mgmt.* **10**: 216–218.

Stickel, L. F. 1951. Wood mouse and box turtle populations in an area treated annually with DDT for 5 years. *J. Wildl. Mgmt.* **15**: 161–164.

Stickel, L. F. 1973. Pesticide residues in birds and mammals. In: *Environmental pollution by pesticides.* Edwards, C. A. (Ed.). Plenum Press, London & New York. Pp. 254–312.

Stickel, L., and W. Stickel. 1969. Distribution of DDT residues in tissues of birds in relation to mortality, body condition, and time. *Indus. Med.* **38**: 44–53.

Stickel, L., W. Stickel, and R. Christensen. 1966. Residues of DDT in brains and bodies of birds that died on dosage and in survivors. *Science* **151**: 1549–1551.

Stickel, W. H., L. F. Stickel, and J. W. Spann. 1969. Tissue residues of dieldrin in relation to mortality in birds and mammals. In: *Chemical fallout*. Miller, M. W., and G. G. Berg (Eds.). C C Thomas, Springfield, Ill. Pp. 174–204.

Stoll, A., and J. Renz. 1942. Uber scillirosid, din gegen nager spezifisch wirksames gift der roten meerzwiebel. (18 über herzglukoside). *J. Helv. Chem. Acta*. **25**: 43– 64.

Strang, R. H., and R. L. Rogers. 1971. A microradiographic study of C^{14}-trifluralin absorption. *Weed Sci*. **19**: 363–369.

Street, J. C. 1969. Organochlorine insecticides and the stimulation of liver microsome enzymes. In: *Biological effects of pesticides in mammalian systems*. Kraybill, H. F. (Ed.). *Ann. N.Y. Acad. Sci*. **160**, 274–290.

Street, J. C. 1969a. Pesticides—the changing scene. *Utah Sci*. September. Pp. 59–62.

Stringer, A., and C. H. Lyons. 1974. The effect of benomyl and thiophanate-methyl on earthworm populations in apple orchards. *Pestic. Sci*. **5**: 189–196.

Stringer, A., and M. A. Wright. 1973. The effect of benomyl and some related compounds on *Lumbricus terrestris* and other earthworms. *Pestic. Sci*. **4**: 165–170.

Stringer, G. E., and R. G. McMynn. 1958. Experiments with toxaphene as fish poison. *Can. Fish. Culturist*. **23**: 39–47.

Stringer, G. E., and R. G. McMynn. 1960. Three years' use of toxaphene as a fish toxicant in British Columbia. *Can. Fish. Culturist* **28**: 37–44.

Stroud, R. H., and R. G. Martin. 1968. *Fish conservation highlights 1963–1967*. Sport Fishing Institute, Washington, D.C. P. 147.

Suffling, R. 1976. Selected ecological factors influencing brush control using tordon 101 herbicide. Ph.D. Thesis, University of Guelph, Guelph, Ontario.

Suffling, R., D. W. Smith, and G. D. Sirons. 1974. Lateral loss of picloram and 2,4-D from a forest podsol during rainstorms. *Weed Res*. **14**: 301–304.

Sundaram, K. M. S. 1973. Degradation dynamics of fenitrothion in aqueous systems. Chemical Control Research Institute (Environment, Canada). Information Reference CC-X-44. P. 19.

Swan, A. A. B. 1969. Exposure of spray operators to paraquat. *Br. J. Indust. Med*. **16**: 322–329.

Swanson, C. R., R. E. Kadunce, R. A. Hodgson, and D. S. Frear. 1966. Amiben metabolism in plants. I. Isolation and identification of an *N*-glucosyl complex. *Weeds* **14**: 319–323.

Sweetser, P. B., and C. W. Todd. 1961. The effect of monuron on oxygen liberation in photosynthesis. *Biochim. et Biophys. Acta*. **51**: 504–508.

Swift, J. E. 1971. *Agricultural chemicals—harmony or discord for food, people, environment*. University of California, Berkeley. (Cited by Bevenue, 1976.)

Tabor, E. C. 1965. Pesticides in urban atmospheres. *J. Air Pollut. Control Assoc*. **15**: 415–418.

Tabor, E. C. 1966. Contamination of urban air through the use of insecticides. *Trans. N.Y. Acad. Sci*. Ser. II. **28**: 569–578.

Tagatz, M. E. 1976. Effect of mirex on predator-prey interaction in an experimental estuarine ecosystem. *Trans. Amer. Fish. Soc*. **105**: 546–549.

Tarrant, K. R., and J. O'G. Tatton. 1968. Organochlorine pesticides in rainwater in the British Isles. *Nature* **219**: 725–727.

Taschenberg, E. F., G. L. Mack, and F. L. Gambrell. 1961. DDT and copper residues in a vineyard soil. *J. Agric. Food Chem.* **9**: 207–209.

Terracini, B., P. J. Cabral, and M. C. Testa. 1973. A multigeneration study on the effects of continuous administration of DDT to BALB/c mice. In: *Pesticides and the environment: A continuing controversy.* Deichmann, W. B. (Ed.). Intercontinental Medical Book Corp., New York & London. Pp. 77–85.

Tette, J. P. 1974. Pheromones in insect population management. In: *Pheromones.* Birch, M. C. (Ed.). American Elsevier Publishing Co. Inc., New York. Pp. 399–410.

Tette, J. P., E. H. Glass, J. L. Brann Jr., and P. A. Arneson. 1975. A brief summary of New York State apple pest management pilot-application project 1973–1975. (Personal communication.)

Tewfik, M. S., and W. C. Evans. 1966. The metabolism of 3,5-dinitro-*o*-cresol (DNOC) by soil microorganisms. *Biochem. J.* **99**: 31–32.

Thomas, J. A., and J. W. Lloyd. 1973. Organochlorine pesticides and sex accessory organs of reproduction. In: *Pesticides and the environment: A continuing controversy.* Deichmann, W. B. (Ed.). Intercontinental Medical Book Corp., New York & London. Pp. 43–51.

Thompson, A. R. 1970. Effects of nine insecticides on the number and biomass of earthworms in pasture. *Bull. Environ. Contam. Toxicol.* **5**: 577–586.

Thompson, R. P. H., G. M. Stathers, C. W. T. Pilcher, A. E. M. McClean, J. Robinson, and R. Williams. 1969. Treatment of unconjugated jaundice with dicophane. *Lancet* **2**: 4–6.

Tierney, G. C. 1947. Death of trees following defoliation by gypsy moths in Connecticut Valley Town of Massachusetts. USDA Bureau of Entomology and Plants. Quarterly Progress Report.

Tomatis, L., N. Day, V. Turosow, and R. T. Charles. 1972. The effect of long-term exposure to DDT on CF₁ mice. *Int. J. Cancer.* **10**: 489–506.

Torstensson, N. T. L., J. Stark, and B. Göransson. 1975. The effect of repeated applications of 2,4-D and MCPA on their breakdown in soil. *Weed Res.* **15**: 159–164.

Triolo, A. J., and J. M. Coon. 1966. Toxicologic interactions of chlorinated hydrocarbon and organophosphate insecticides. *Agric. Food Chem.* **14**: 549–555.

Truffaut, G., and I. Pastac. 1935. British Patent No. 424,295.

Tu, C. M., and J. R. W. Miles. 1976. Interactions between insecticides and soil microbes. *Res. Rev.* **64**: 17–65.

Tu, C. M., J. R. W. Miles, and C. R. Harris. 1968. Soil microbial degradation of aldrin. *Life Sci.* **7**: 311–322.

Tucker, B. V., D. E. Pack, and J. N. Ospenson. 1967. Adsorption of bipyridylium herbicides in soil. *J. Agric. Food Chem.* **15**: 1005–1008.

Tucker, R. K., and D. G. Crabtree. 1970. *Handbook of toxicity of pesticides to wildlife.* U.S. Fish & Wildlife Service, Bureau of Sport Fisheries & Wildlife, Resource Publication No. 84. P. 131.

Turtle, E. E., A. Taylor, E. N. Wright, R. J. P. Thearle, H. Egan, W. H. Evans, and N. M.

Soutar. 1963. The effects on birds of certain chlorinated insecticides used as seed dressings. *J. Sci. Food Agric.* **14**: 456–577.

Turusov, V. S., N. E. Day, L. Tomatis, E. Gati, and R. T. Charles. 1973. Tumors in CF-1 mice exposed for six consecutive generations to DDT. *J. Nat. Cancer Inst.* **51**: 983–995.

Tweedy, B. G., and N. Turner. 1966. The mechanism of sulfur reduction by conidia of *Monilinia fructicola. Contrib. Boyce Thompson Inst.* **23**: 255–265..

Tyler, B. M. J. 1978. Selected arthropod populations in cornfields with different tillage, with emphasis on the northern corn rootworm (*Diabrotica longicornis* Say) and its prey-predator relationships. Ph.D. Thesis, University of Guelph, Guelph, Ontario.

Ukeles, R. 1962. Growth of pure cultures of marine phytoplankton in the presence of toxicants. *Appl. Microbiol.* **10**: 532–537.

Ullstrup, A. J. 1972. The impacts of the southern corn leaf blight epidemics of 1970–1971. *Ann. Rev. Phytopath.* **10**: 37–50.

United States Department of Agriculture. 1965. *Losses in agriculture.* Agriculture Handbook No. 291. U.S. Government Printing Office, Washington, D.C.

United States Department of Agriculture. 1972. *Extent and cost of weed control and an evaluation of important weeds.* 1968, USDA Agriculture Research Services.

United States Department of Interior. Federal Water Pollution Control Admin. 1968. *Water quality criteria.* Report of the National Technical Advisory Committee. Pp. 20, 37, 116.

Vance, B. D., and A. W. Maki. 1976. Bioconcentration of Dibrom by *Stigeoclonium pachydermum. Bull. Environ. Contam. Toxicol.* **15**: 601–607.

Vanderstoep, J., and J. F. Richards. 1969. The changes in eggshell strength during incubation. *Poultry Sci.* **49**: 276–285.

Van Klingeren, B., J. H. Koeman, and J. L. van Haaften. 1966. A study on the hare (Lepus europeus) in relation to the use of pesticides in a polder in the Netherlands. *J. Appl. Ecol.* 3(Suppl.): 125–131.

Van Overbeek, J. 1964. Survey of mechanisms of herbicide action. In: *The physiology and biochemistry of herbicides.* Audus, L. J. (Ed.). Academic Press, London. Pp. 387–400.

Van Raalte, H. G. S. 1973. Of hematomas mice and man. In: *Pesticides and the environment: A continuing controversy.* Deichmann, W. B. (Ed.). Intercontinental Medical Book Corp, New York & London. Pp. 245–251.

Van Velzen, A. C., W. B. Stiles, and L. F. Stickel. 1972. Lethal mobilization of DDT by cowbirds. *J. Wildl. Mgmt.* **36**: 733–739.

Various authors. 1974. Evaluation of commercial preparations of *Bacilus thuringiensis* with and without chitinase against spruce budworm. Information Report CC-X-59.

Chemical Control Research Institute, Ottawa. P. 233.

Varty, I. W. 1975. Forest spraying and environmental integrity. *The Forestry Chronicle* **51**: 146–149.

Veith, G. D. 1975. Baseline concentrations of polychlorinated biphenyls and DDT in Lake Michigan fish, 1971. *Pestic. Monit. J.* **9**: 21–29.

Vinson, S. B., C. E. Boyd, and D. E. Ferguson. 1963. Resistance to DDT in the mosquitofish, *Gambusia afinnis*. *Science* **139**: 217–221.

Von Rumker, R., E. W. Lawless, A. F. Meiners, K. A. Lawrence, G. L. Kelso, and F. Horay. 1975. Production, distribution, use and environmental impact potential of selected pesticides. EPA 540/1-74-001. P. 439.

Wade, R. A. 1969. Ecology of juvenile tarpon and effect of dieldrin on two associated species. USDI Bureau of Sport Fisheries & Wildlife Technical Paper 41. P. 85.

Waibel, P. E., B. S. Pomeray, and E. L. Johnson. 1955. Effect of Arasan-treated corn on laying hens. *Science* **121**: 401–402.

Walker, C. R. 1963. Endothal derivatives as aquatic herbicides in fishery habitats. *Weeds* **11**: 226–232.

Walker, C. R. 1965. Diuron, fenuron, monuron, neburon, and TCA mixture as aquatic herbicides in fish habitats. *Weeds* **13**: 297–301.

Wallace, G. J. 1959. Insecticides and birds. *Audobon* **61**: 10–12.

Wallace, J. B., and U. E. Brady. 1971. Residue levels of dieldrin in aquatic invertebrates and effect of prolonged exposure on populations. *Pestic. Monit. J.* **5**: 295–300.

Wallace, M. E. 1971. An unprecedented number of mutants in a colony of wild mice. *Environ. Pollut.* **1**: 175–184.

Wallace, M. E., P. Knights, and A. O. Dye. 1976. Pilot study of the mutagenicity of DDT in mice. *Environ. Pollut.* **11**: 217–222.

Wallace, R. R., A. S. West, A. E. R. Downe, and H. B. N. Hynes. 1973. The effects of experimental blackfly (Diptera: Simuliidae) larviciding with Abate, Dursban and methoxychlor on stream invertebrates. *Can. Entomol.* **105**: 817–831.

Walsh, C. E., C. W. Miller, and P. T. Heitmuller. 1971. Uptake and effects of dichlobenil in a small pond. *Bull. Environ. Contam. Toxicol.* **6**: 279–288.

Ward, T. M., and R. P. Upchurch. 1965. The role of the amido group and adsorption mechanisms. *J. Agric. Food Chem.* **13**: 334–340.

Ware, G. W., E. J. Apple, W. P. Cahill, P. D. Gerhardt, and K. R. Frost. 1969. Pesticide drift: II. Mist blower vs aerial application of sprays. *J. Econ. Entomol.* **62**: 844–846.

Ware, G. W., B. J. Estesen, and W. P. Cahill. 1971. DDT moratorium in Arizona— agricultural residues after 2 years. *Pestic. Monit J.* **5**: 276–280.

Ware, G. W., B. J. Estesen, and W. P. Cahill. 1974. DDT moratorium in Arizona— agricultural residues after 4 years. *Pestic. Monit. J.* **8**: 98–101.

Ware, G. W., B. J. Estesen, W. P. Cahill, and K. R. Frost. 1972. Pesticide drift: V. Vertical drift from aerial applications. *J. Econ. Entomol.* **65**: 590–592.

Ware, G. W., and E. E. Good. 1967. Effect of insecticides on reproduction in the laboratory mouse. II. Mirex, telodrin and DDT. *Toxicol. Appl. Pharmacol.* **10**: 54–61.

Ware, G. W., and D. P. Morgan. 1976. Worker reentry safety: IX. Techniques of determining safe reentry intervals for organophosphate-treated cotton fields. *Res. Rev.* **62**: 79–100.

Ware, G. W., D. P. Morgan, B. J. Estesen, and W. P. Cahill. 1975. Establishment of reentry intervals for organophosphate-treated cotton fields based on human data: III. 12

to 72 hours post-treatment exposure to monocrotophos, ethyl- and methyl parathion. *Arch. Environ. Contam. Toxicol.* 3: 289–306.

Ware, G. W., D. P. Morgan, B. J. Estesen, W. P. Cahill, and D. M. Whitacre. 1973. Establishment of reentry intervals for organophosphate-treated cotton fields based on human data. I. Ethyl- and methyl parathion. *Arch. Environ. Contam. Toxicol.* 1: 48–59.

Warner, R. E., R. R. Peterson, and L. Borgman. 1966. Behavioral pathology in fish: A quantitative study of sublethal pesticide toxication. *J. Appl. Ecol.* 3(Suppl.): 223–247.

Wassermann, M., L. Tomatis, D. Wassermann, N. E. Day, and M. Djavaherian. 1974. Storage of organochlorine insecticides in adipose tissue of Ugandans. *Bull. Environ. Contam. Toxicol.* 12: 501–508.

Wassermann, M., L. Tomatis, D. Wassermann, N. E. Day, Y. Groner, S. Lazarovici, and D. Rosenfeld. 1974a. Epidemiology of organochlorine insecticides in the adipose tissue of Israelis. *Pestic. Monit. J.* 8: 1–7.

Webb, J. L. 1966. *Enzyme and metabolic inhibitors.* Vol. 3. Academic Press, New York. P. 1028.

Webb, R. E., and F. Horsfall, Jr. 1967. Endrin resistance in the pine mouse. *Science* 156: 1762.

Weber, J. B., T. J. Monaco, and A. D. Worsham. 1973. What happens to herbicides in the environment? *Down to Earth* 29: 12–14.

Weed Science Society of America. 1974. *Herbicide Handbook.* P. 430.

Weil, C. 1976. Evaluation and interpretation of toxicological and carcinogenic studies of pesticides. Presented at the National Meeting of the Entomological Society of America, Honolulu, November 28–December 2.

Weis, J. S., and P. Weis. 1975. Retardation of fin regeneration of *Fundulus* by several insecticides. *Trans. Amer. Fish. Soc.* 104: 135–137.

Wessels, J. S. C., and R. Vander Veen. 1956. The action of some derivatives of phenylurethan and of 3-phenyl-1,1-dimethylurea on the Hill reaction. *Biochim. et Biophys. Acta.* 19: 548–549.

West I. 1964. Pesticides as contaminants. *Arch. Environ. Hlth.* 9: 626–631.

West, I. 1966. Biological effects of pesticides in the environment. In: *Organic pesticides in the environment.* Gould, R. F. (Ed.). Advances in Chemistry Series 60: 38–58.

Westlake, W. E., and F. A. Gunther. 1966. Occurrence and mode of introduction of pesticides in the environment. In: *Organic pesticides in the environment.* Gould, R. F. (Ed.). Advances in Chemistry Series 60: 110–121.

Wheatley, G. A. 1973. Pesticides in the atmosphere. In: *Environmental pollution by pesticides.* Edwards, C. A. (Ed.). Plenum Press, London & New York. Pp. 365–408.

Wheatley, G. A., and J. A. Hardman. 1965. Indications of the presence of organochlorine insecticides in rainwater in central England. *Nature* 207: 486–487.

White, D. H. 1976. Nationwide residues of organochlorines in starlings, 1974. *Pestic. Monit. J.* 10: 10–17.

Wichman, J. R., and W. R. Byrnes. 1970. Tolerance of black walnut and yellow poplar to soil applied herbicides. Abstracts of Weed Science Society of America. P. 33.

Wiersma, G. B., W. G. Mitchell, and C. L. Stanford. 1972. Pesticide residues in onions and soil—1969. *Pestic. Monit. J.* 5: 345–347.

Wiersma, G. B., P. F. Sand, and R. L. Schutzmann. 1971. National soils monitoring program—six states, 1967. *Pestic. Monit. J.* 5: 223–227.

Wiersma, G. B., and H. Tai. 1974. Mercury levels in soils of the Eastern United States. *Pestic. Monit. J.* 7: 214–216.

Wigglesworth, V. B. 1945. DDT and the balance of nature. *Atlantic Monthly* 176: 107–113.

Wilbur, G. W. 1976. River polluters fined $17 million. *Honolulu Star-Bulletin.* October 6.

Willett, K. C. 1963. Trypanosomiasis and the tsetse fly problem in Africa. *Ann. Rev. Entomol.* 8: 197–214.

Williams, C. M. 1967. Third generation pesticides. *Sci. Amer.* 217: 13–17.

Willis, G. H., J. F. Parr, R. I. Papendick, and S. Smith. 1969. A system for monitoring atmospheric concentrations of field-applied pesticides. *Pestic. Monit. J.* 3: 172–176.

Wills, G. R., D. E. Davis, and H. H. Funderburk, Jr. 1963. The effect of atrazine on transpiration in corn, cotton and soybeans. *Weeds* 11: 253–255.

Wilson, D. M., and P. C. Oloffs. 1974. Residues in plants grown in soils treated with experimental insecticide Velsicol HCS-3260. *Can. J. Plant Sci.* 54: 203–209.

Wolfe, J. L., and B. R. Norment. 1973. Accumulation of mirex residues in selected organisms after an aerial treatment, Mississippi—1971–1972. *Pestic. Monit. J.* 7: 112–116.

Wolman, A. A., and A. J. Wilson, Jr. 1970. Occurrence of pesticides in whales. *Pestic. Monit. J.* 4: 8–10.

Woodward, R. B., M. P. Cava, W. D. Ollis, A. Hunger, H. U. Daeniker, and K. Schenker. 1954. The total synthesis of strychnine. *J. Amer. Chem. Soc.* 76: 4749–4751.

Woodwell, G. M., P. P. Craig, and H. A. Johnson. 1971. DDT in the biosphere: Where does it go. *Science* 174: 1101–1107.

Woodwell, G. M., C. F. Wurster, Jr., and P. A. Isaacson. 1967. DDT residues in an east coast estuary: A case of biological concentration of a persistent insecticide. *Science* 156: 821–824.

Woolson, E. A. 1976. Organoarsenical herbicides. In: *Herbicides: Chemistry, degradation and mode of action* (2nd ed.). Vol. 2. Kearney, P. C., and D. D. Kaufman (Eds.). Marcel Dekker, Inc., New York. Pp. 741–776.

Woolson, E. A., J. H. Axley, and P. C. Kearney. 1971. The chemistry and phytotoxicity of arsenic in soils: I. Contaminated field soils. *Soil Sci. Soc. Amer. Proc.* 35: 938.

World Health Organization. 1968. *Evaluation of insecticides for vector control* Part I. WHO/VBC/66.66.

World Health Organization. 1973. *Safe use of pesticides.* WHO Technical Report Series No. 513. Pp. 42–43.

Wurster, C. F., Jr. 1968. DDT reduces photosynthesis by marine phytoplankton. *Science* 159: 1474–1475.

Wurster, C. F., Jr. 1969. Con DDT: It's polluting all the world. *Congressional Record—Senate*. May 5, 1969. **115**: 11350–11351.

Wurster, D. H., C. F. Wurster, Jr., and W. N. Strickland. 1965. Bird mortality following DDT spray for Dutch elm disease. *Ecology* **46**: 488–499.

Yardick, M. K., K. Funk, and M. E. Zabik. 1971. Dieldrin residues in bacon cooked by two methods. *J. Agric. Food Chem.* **19**: 491–494.

Yeo, R. R. 1967. Dissipation of diquat and paraquat, and effects on aquatic weeds and fish. *Weeds* **15**: 42–46.

Yeo, R. R. 1970. Dissipation of endothal and effects on aquatic weeds and fish. *Weed Sci.* **18**: 282–284.

Young, D. R., D. J. McDermott, and T. C. Heesen. 1976. Aerial fallout of DDT in southern California. *Bull. Environ. Contam. Toxicol.* **16**: 604–611.

Youngson, C. R., C. A. I. Goring, R. W. Meikle, H. H. Scott, and J. D. Griffith. 1967. Factors influencing the decomposition of Tordon herbicide in soils. *Down to Earth* **23**: 3–11.

Yu, C-C., and J. R. Sanborn. 1975. The fate of parathion in a model ecosystem. *Bull. Environ. Contam. Toxicol.* **13**: 543–550.

Yule, W. N., and A. D. Tomlin. 1971. DDT in forest streams. *Bull. Environ. Contam. Toxicol.* **5**: 479–488.

Zabik, M. J., B. E. Pape, and J. W. Bedford. 1971. Effect of urban and agricultural pesticide use on residue levels in the Red Cedar River. *Pestic. Monit. J.* **5**: 301–308.

Zavon, M. R., C. H. Hine, and K. D. Parker. 1965. Chlorinated hydrocarbon insecticides in human body fat in the United States. *J. Amer. Med. Assoc.* **193**: 837–839.

Zavon, M. R., R. Tye, and L. Latorre. 1969. Chlorinated hydrocarbon insecticide content of the neonate. In: *Biological effects of pesticides in mammalian systems*. Kraybill, H. F. (Ed.). *Ann. N.Y. Acad. Sci.* **160**: 196–200.

Zeller, H. D., and H. N. Wyatt. 1967. Selective shad removal in southern reservoirs. American Fisheries Society Reservoir Fisheries Symposium. Pp. 405–414.

Zimmerman, P. W., and A. E. Hitchcock. 1942. Substituted phenoxy and benzoic acid growth substances and the relation of structure to physiological activity. *Contr. Boyce Thompson Inst.* **12**: 321–343.

Zoecon Corp. 1974. Product profile: Altosid®, Mosquitoes. Technical Bulletin Z-1-03. Pp. 3.

Page numbers in **boldface** indicate where the index item is discussed in greatest depth.